Science, Race, and Ethnicity

Science, Race, and Ethnicity

Readings from Isis and Osiris

EDITED BY JOHN P. JACKSON, JR.

The University of Chicago Press
Chicago and London

The essays in this volume originally appeared in *Isis* and *Osiris*. Acknowledgment of the original publication date can be found on the first page of each article.

The University of Chicago Press, Chicago, 60637
The University of Chicago Press, Ltd., London

Printed in the United States of America
ISBN: (cl) 0-226-38934-0
ISBN: (pa) 0-226-38935-9

Library of Congress Cataloging-in-Publication Data

Science, race, and ethnicity : readings from Isis and Osiris / edited by John P. Jackson, Jr.
p. cm.
Includes bibliographical references and index.
ISBN 0-226-38934-0 (alk. paper) — ISBN 0-226-38935-9 (pbk. : alk.paper)
1. Race. 2. Science and civilization. 3. Science — Social aspects. 4. Eugenics. 5. Race awareness. 6. Race discrimination. I. Jackson, John P., 1961- II. Isis. III. Osiris (Bruges, Belgium)

GN269 .S394 2002
305.8—dc21 2002067320

The paper used in the publication meets the minimum requirements of American National Standard for Information Sciences—Permanence of Paper for Printed Library Materials, ANSI Z39.48.1984.

CONTENTS

Editor's Foreword

John P. Jackson, Jr.

IN HIS PRIZE-WINNING HISTORY of slavery in North America, Ira Berlin writes, "Of late, it has become fashionable to declare that race is a social construction. In the academy, this precept has gained universal and even tiresome assent."[1] The history of science is no exception to Berlin's claim. While claims about the knowledge produced by physicists and chemists may generate controversy, few dispute that scientific claims about the reality of race and racial differences are, by and large, social constructs. The essays collected in this volume demonstrate the different ways that scientists have made claims about racial reality. While the authors of these essays differ in their geographic and temporal objects of study, they all seem to agree that scientific ideas about race cannot be separated from their political, ideological, and institutional locations.

Racial thinking and scientific thinking, after all, grew together. Recent scholarship in the history of the idea of race has claimed that "race" is a recent development in Western thought. The notion that the human species can be divided into biologically discrete units and that these can be hierarchically arranged is a product of ideas that only came to fruition in the sixteenth century and gained power in the centuries since.[2] Historians of science will note that the idea of race grew concomitantly with the growth of modern science in the Western tradition. The drive to categorize the new discoveries in the age of exploration and cultural notions of "savage" and "civilization" were enmeshed from the start. Hence, few would dispute that "folk" ideas about race were embedded into scientific ideas from the very beginning of the scientific study of race.

There is a danger in falling into a teleological mode of thinking, however, in making claims about how race and science were entangled from the sixteenth century onward. As the essays in this volume make clear, there is no straightforward path between racial ideas and social policies. For example, the scientific claim that a particular race is more prone to disease has been used to justify the assimilation of that race into the dominant one, the creation of a separate homeland for the suspect race, as well as for the willful extermination of that race. There is no logical necessity involved here, only a series of historically contingent circumstances that lead to particular outcomes.

A final implication of the social construction of scientific knowledge about race should be mentioned. Peter Novick contends that most historical literature on race

[1] Ira Berlin, *Many Thousands Gone: The First Two Centuries of Slavery in North America* (Cambridge Mass.: Harvard University Press, 1998), p. 1.

[2] Ivan Hannaford, *Race: The History of an Idea in the West* (Baltimore Md.: Johns Hopkins University Press, 1996); Audrey Smedley, *Race in North America: Origin and Evolution of a Worldview* (Boulder Colo.: Westview, 1999).

outside the history of science written since the 1960s has been "based on the objective truth of scientific antiracism." Historians who write on the "social construction of race" in fields outside the history of science often take as their starting point the "objective truth" of the nonexistence of the races. From this starting point, historians attempt to discover how race has gained meaning through various social mechanisms. For example, David Roedinger, currently one of the most influential historians writing on issues of race that were raised in the last decade, bases his work on the assumption that "race is given meaning through the agency of human beings in concrete historical and social contexts, and is not a biological or natural category." Legal historian Ian F. Haney Lopez argues that "Races are not biologically differentiated groupings but, rather, social constructions." The assumption here is that "biological or natural" categories are not, themselves, creations of "the agency of human beings." But, as the essays in this volume demonstrate, historians of science have excelled in showing how biological categories are just such human constructions. One of my hopes in assembling this volume is to have the history of science join this larger scholarly debate on the social construction of race.[3]

THE ESSAYS

Nancy Leys Stepan argues in her essay that much of the science of human variation in the nineteenth and early twentieth century used an argument by analogy: the lower, more childlike races also were the more feminine. Stepan's essay sets the stage for notions of how social ideas about "savagery" and "civilization" are enrolled to provide support for various scientific racial ideas.

The essays by Kentwood Wells, Malcolm Kottler, and Neal Gillespie explore how race, a pre-evolutionary idea, played a part in nineteenth-century scientific debates about evolution. Wells notes how ideas about climate and disease informed early ideas about how the races might have arisen through natural selection. Kottler and Gillespie explore, in different ways, how notions of progress, civilization, and savagery were used in fierce debates about the status of evolutionary theory. Kottler underscores how racialized savages became key pieces of evidence in Alfred Russel Wallace's proposed solution to the question regarding the evolution of "man's" moral and spiritual sensibilities. Gillespie offers a similar story about the role of the savage (often overtly portrayed in racial terms) in one proposed solution to the ancient dilemma regarding human progress to civilization versus its degeneration into savagery.

The political ideas of progress and civilization that were ineluctably ingrained in the scientific debates of nineteenth-century evolutionary theory continued into twentieth-century political/scientific debates. Yet, there was not just one way that science could be enrolled to support a political position or vice versa. Mitchell Hart's essay explores ways in which Jews could use science for both sides of the political debate about the virtues of "racial" assimilation versus separatism. Jews who

[3] Peter Novick, *That Noble Dream: The "Objectivity Question" and the American Historical Profession* (Cambridge, England: Cambridge University Press, 1988), p. 348; David Roedinger, *Toward the Abolition of Whiteness* (London: Verso Press, 1994), p. 2; Ian F. Haney Lopez, *White By Law: Legal Construction of Race.* (New York : New York University Press, 1996), p. xiii.

argued for assimilation at the end of the nineteenth century could use science to undercut anti-Semitic ideas about the essential racial nature of Jewry. Similarly, Zionists found ample scientific evidence for their call for the creation of a separate Jewish homeland to minimize the medical risks posed by attempts to assimilate into Eastern European culture. Jennifer Michael Hecht explores the hotly contested theories of Georges Vacher de Lapouge's science of "anthroposociology." Lapouge's call for a Darwinian state based on racial purity was an important influence on the *Rassenhygenists,* such as Hans F.K. Guenther, who later borrowed heavily from Lapouge. Hecht explores various reactions to Lapouge in late nineteenth and early twentieth century France—some writers calling for a return to Catholicism as an answer to the materialistic and atheistic Lapouge; others calling for a more refined understanding of Darwinism that did not lead to Lapouge's eugenic state. Edward Beardsley explores the early attempts to use science for antiracist ends in the United States. Exploring the careers of physician Burt G. Wilder and anthropologist Franz Boas, Beardsley traces how these men claimed that a more rigorous scientific understanding of racial differences led to the conclusion that there were no racial differences of any significance. Beardsley further notes that both men were propelled by their scientific conclusions to social action aimed to reduce race prejudice and discrimination in the United States. John P. Jackson, Jr. explores the difficulties of this move from the scientific arena to the social arena. His study of the mid-twentieth century American Jewish Congress demonstrates the difficulty in translating antiracist social science regarding prejudice and discrimination into the courtroom.

The activities described by Jackson are, among other things, a response by American Jews to the horrors of Nazi Germany. No discussion of the interplay of race, science, and politics would be complete without a treatment of the "biological state" created in Germany in the 1930s. The essays by Sheila Faith Weiss make clear that the Nazi racial state's relationship with the science of eugenics is more complicated than often believed. Early German *Rassenhygiene* was more concerned with class than with race, although there was an overlap between those concepts. Many German eugenicists, most notably William Schallmayer, explicitly rejected the Nordic supremacy that Hitler embraced during the Nazi period. The story of the move from *Rassenhygiene* in the Wilhelmine period and the Aryan supremacy of Nazi *Rassenhygiene* is a complex one that Weiss presents with admirable thoroughness in her two articles. Peter Weingart's study of the Kaiser Wilhelm Institute (KWI) for Anthropology, Human Heredity, and Eugenics traces the debates that went on in the German scientific community concerning the relationship between racial ideology and racial science. Again, the message here is that the horrors of the Third Reich were not an inevitable outcome of *Rassenhygiene* but an outcome contingent on, among other things, the political opportunism of KWI's scientific leaders.

Weingart's essay underscores the importance of the institutional bases of racial science, a theme extended in the next section of the volume. Elizabeth Williams explores the failure of nineteenth-century French anthropological institutions to continue racial science into the twentieth-century with the decline of Paul Broca's influence on the field as well as a decline in racialist thinking generally. Garland Allen's study of the Eugenics Records Office in the United States traces the rise and fall of the nerve center for North American eugenics in the first half of the twentieth century. Like Weiss and Weingart, Allen cautions us about conflating eugenic thought and racist thought, and offers a much more complex picture of the relationship between the two ideological outlooks. The essays by Stuart McCook and Joan Mark offer different perspectives on the institutionalization of science. Institutionalizing any discipline results in boundary work wherein only professionals within the discipline are qualified to speak. Both McCook and Mark relate stories of how the institutionalization of science resulted in the exclusion of racially marked individuals. McCook's story of Paul Du Chaillu demonstrates how the standards of evidence within the field sciences excluded a racially marked suspect's views regarding the importance of the gorilla as a key to a scientific understanding of the human race's place in nature. Mark tells a similar story about the status of knowledge claims made by Francis La Flesche, an American Indian who was attempting to contribute to anthropological knowledge of American Indian culture in the late nineteenth-century United States.

The questionable status of Francis La Flesche in American anthropology was an outcome of European dominance of North America. Of course, North America was not the only place that Europeans came to dominate, and racial science reflected the European colonial experience in other parts of the world. Patricia Lorcin traces the complex intertwining of political ideas of colonial administration with medical and scientific ideas about race during the French administration of Algeria in the nineteenth century. Harriet Deacon relates a similar story for South Africa. She carefully unpacks the relationship between racial theories and racial practices in the racist regime of South Africa's Cape Colony. Tessa Morris-Suzuki has a different point in her close study of racial science in Japan. She notes how Japanese racial thinkers freely borrowed ideas from European racial theorists. The borrowing, though, was not a simple transportation of ideas, but was mediated by a number of cultural factors unique to the Japanese experience.

The essays collected here demonstrate that much excellent work has been done in the history of racial science. That a large number of essays are of recent vintage indicates that there is a thriving research community dedicated to the study of the interaction of science and racial ideas.

Race and Gender: The Role of Analogy in Science

By Nancy Leys Stepan

METAPHOR OCCUPIES a central place in literary theory, but the role of metaphors, and of the analogies they mediate, in scientific theory is still debated.[1] One reason for the controversy over metaphor, analogy, and models in science is the intellectually privileged status that science has traditionally enjoyed as the repository of nonmetaphorical, empirical, politically neutral, universal knowledge. During the scientific revolution of the seventeenth century, metaphor became associated with the imagination, poetic fancy, subjective figures, and even untruthfulness and was contrasted with truthful, unadorned, objective knowledge—that is, with science itself.[2]

In the twentieth century logical positivists also distinguished between scientific and metaphoric language.[3] When scientists insisted that analogies or models based on analogies were important to their thinking, philosophers of science tended to dismiss their claims that metaphors had an *essential* place in scientific utterances. The French theoretical physicist Pierre Duhem was well known for his criticism of the contention that metaphor and analogies were important to *explanation* in science. In his view, the aim of science was to reduce all theory to mathematical statements; models could aid the process of scientific discovery, but once they had served their function, analogies could be discarded as extrinsic to science, and the theories made to stand without them.[4]

One result of the dichotomy established between science and metaphor was that obviously metaphoric or analogical science could only be treated as

[1] A metaphor is a figure of speech in which a name or descriptive term is transferred to some object that is different from, but analogous to, that to which it is properly applicable. According to Max Black, "every metaphor may be said to mediate an analogy or structural correspondence": see Black, "More About Metaphor," in *Metaphor and Thought,* ed. Andrew Ortony (Cambridge: Cambridge Univ. Press, 1979), pp. 19–43, on p. 31. In this article, I have used the terms *metaphor* and *analogy* interchangeably.

[2] G. Lakoff and M. Johnson, *Metaphors We Live By* (Chicago/London: Univ. Chicago Press, 1980), p. 191. Scientists' attacks on metaphor as extrinsic and harmful to science predate the Scientific Revolution.

[3] See A. J. Ayer, *Language, Truth and Logic* (New York: Dover, 1952), p. 13.

[4] On Duhem, see Carl H. Hempel, *Aspects of Scientific Explanation and Other Essays in the Philosophy of Science* (New York: Free Press, 1965), pp. 433–477. Hempel agrees with Duhem's view that "all references to analogies or analogical models can be dispensed with in the systematic statement of scientific explanations" (p. 440).

"prescientific" or "pseudoscientific" and therefore dismissable.[5] Because science has been identified with truthfulness and empirical reality, the metaphorical nature of much modern science tended to go unrecognized. And because it went unrecognized, as Colin Turbayne has pointed out, it has been easy to mistake the model in science "for the thing modeled"—to think, to take his example, that nature *was* mechanical, rather than to think it was, metaphorically, seen as mechanical.[6]

More recently, however, as the attention of historians and philosophers of science has moved away from logical reconstructions of science toward more "naturalistic" views of science in culture, the role of metaphor, analogies, and models in science has begun to be acknowledged.[7] In a recent volume on metaphor, Thomas S. Kuhn claims that analogies are fundamental to science; and Richard Boyd argues that they are "irreplaceable parts of the linguistic machinery of a scientific theory," since cases exist in which there are metaphors used by scientists to express theoretical claims "for which no adequate literal paraphrase is known."[8] Some philosophers of science are now prepared to assert that metaphors and analogies are not just psychological aids to scientific discovery, or heuristic devices, but constituent elements of scientific theory.[9] We seem about to move full circle, from considering metaphors mere embellishments or poetic fictions to considering them essential to scientific thought itself.

Although the role of metaphor and analogy in science is now recognized, a critical theory of scientific metaphor is only just being elaborated. The purpose of this article is to contribute to the development of such a theory by using a particular analogy in the history of the life sciences to explore a series of related questions concerning the cultural sources of scientific analogies, their role in scientific reasoning, their normative consequences, and the process by which they change.

RACE AND GENDER: A POWERFUL SCIENTIFIC ANALOGY

The analogy examined is the one linking race to gender, an analogy that occupied a strategic place in scientific theorizing about human variation in the nineteenth and twentieth centuries.

[5] For this point see Jamie Kassler, "Music as a Model in Early Science," *History of Science,* 1982, *20*:103–139.

[6] Colin M. Turbayne, *The Myth of Metaphor* (Columbia: Univ. South Carolina Press, 1970), p. 24.

[7] General works on metaphor and science include Philip Wheelwright, *Metaphor and Reality* (Bloomington: Indiana Univ. Press, 1962); Max Black, *Models and Metaphor* (Ithaca, N.Y.: Cornell Univ. Press, 1962); Mary Hesse, *Models and Analogies in Science* (Notre Dame, Ind.: Univ. Notre Dame Press, 1966); Richard Olson, ed., *Science as Metaphor* (Belmont, Calif.: Wadsworth, 1971); W. M. Leatherdale, *The Role of Analogy, Model and Metaphor in Science* (Amsterdam: North-Holland, 1974); Ortony, ed., *Metaphor and Thought* (cit. n. 1); and Roger S. Jones, *Physics as Metaphor* (Minneapolis: Univ. Minnesota Press, 1982). Warren A. Shibles, *Metaphor: An Annotated Guide and History* (Whitewater, Wisconsin: Language Press, 1971), gives an extensive introduction and guide to the general problem of metaphor, language, and reality.

[8] Thomas S. Kuhn, "Metaphor in Science," in *Metaphor and Thought,* ed. Ortony, pp. 409–419, on p. 414; and Richard Boyd, "Metaphor and Theory Change: What Is 'Metaphor' a Metaphor For?" *ibid.,* pp. 356–408, on p. 360.

[9] For a defense of the centrality of analogy to science see N. R. Campbell, "What Is a Theory?" in *Readings in the Philosophy of Science,* ed. Baruch A. Brody (Englewood Cliffs, N.J.: Prentice-Hall, 1970), pp. 252–267. Shibles, in *Metaphor,* p. 3, also argues that each school of science "is based on a number of basic metaphors which are then expanded into various universes of discourse."

As has been well documented, from the late Enlightenment on students of human variation singled out racial differences as crucial aspects of reality, and an extensive discourse on racial inequality began to be elaborated.[10] In the nineteenth century, as attention turned increasingly to sexual and gender differences as well, gender was found to be remarkably analogous to race, such that the scientist could use racial difference to explain gender difference, and vice versa.[11]

Thus it was claimed that women's low brain weights and deficient brain structures were analogous to those of lower races, and their inferior intellectualities explained on this basis.[12] Woman, it was observed, shared with Negroes a narrow, childlike, and delicate skull, so different from the more robust and rounded heads characteristic of males of "superior" races. Similarly, women of higher races tended to have slightly protruding jaws, analogous to, if not as exaggerated as, the apelike, jutting jaws of lower races.[13] Women and lower races were called innately impulsive, emotional, imitative rather than original, and incapable of the abstract reasoning found in white men.[14] Evolutionary biology provided yet further analogies. Woman was in evolutionary terms the "conservative element" to the man's "progressive," preserving the more "primitive" traits found in lower races, while the males of higher races led the way in new biological and cultural directions.[15]

Thus when Carl Vogt, one of the leading German students of race in the middle of the nineteenth century, claimed that the female skull approached in many respects that of the infant, and in still further respects that of lower races, whereas the mature male of many lower races resembled in his "pendulous" belly a Caucasian woman who had had many children, and in his thin calves and flat thighs the ape, he was merely stating what had become almost a cliché of the science of human difference.[16]

So fundamental was the analogy between race and gender that the major modes of interpretation of racial traits were invariably evoked to explain sexual traits. For instance, just as scientists spoke of races as distinct "species," incapable of crossing to produce viable "hybrids," scientists analyzing male-female differences sometimes spoke of females as forming a distinct "species," individual members of which were in danger of degenerating into psychosexual hybrids

[10] See Nancy Stepan, *The Idea of Race in Science: Great Britain, 1800–1960* (London: Macmillan, 1982), esp. Ch. 1.

[11] No systematic history of the race-gender analogy exists. The analogy has been remarked on, and many examples from the anthropometric, medical, and embryological sciences provided, in Stephen Jay Gould, *The Mismeasure of Man* (New York: W. W. Norton, 1981), and in John S. Haller and Robin S. Haller, *The Physician and Sexuality in Victorian America* (Urbana: Univ. Illinois Press, 1974).

[12] Haller and Haller, *The Physician and Sexuality*, pp. 48–49, 54. Among the several craniometric articles cited by the Hallers, see esp. J. McGrigor Allan, "On the Real Differences in the Minds of Men and Women," *Journal of the Anthropological Society of London*, 1869, *7*:cxcv–ccviii, on p. cciv; and John Cleland, "An Inquiry into the Variations of the Human Skull," *Philosophical Transactions, Royal Society*, 1870, *89*:117–174.

[13] Havelock Ellis, *Man and Woman: A Study of Secondary Sexual Characters* (1894; 6th ed, London: A. & C. Black, 1926), pp. 106–107.

[14] Herbert Spencer, "The Comparative Psychology of Man," *Popular Science Monthly*, 1875–1876, *8*:257–269.

[15] Ellis, *Man and Woman* (cit. n. 13), p. 491.

[16] Carl Vogt, *Lectures on Man: His Place in Creation, and in the History of the Earth* (London: Longman, Green, & Roberts, 1864), p. 81.

when they tried to cross the boundaries proper to their sex.[17] Darwin's theory of sexual selection was applied to both racial and sexual difference, as was the neo-Lamarckian theory of the American Edward D. Cope.[18] A last, confirmatory example of the analogous place of gender and race in scientific theorizing is taken from the history of hormone biology. Early in the twentieth century the anatomist and student of race Sir Arthur Keith interpreted racial differences in the human species as a function of pathological disturbances of the newly discovered "internal secretions" or hormones. At about the same time, the apostle of sexual frankness and well-known student of sexual variation Havelock Ellis used internal secretions to explain the small, but to him vital, differences in the physical and psychosexual makeup of men and women.[19]

In short, lower races represented the "female" type of the human species, and females the "lower race" of gender. As the example from Vogt indicates, however, the analogies concerned more than race and gender. Through an intertwined and overlapping series of analogies, involving often quite complex comparisons, identifications, cross-references, and evoked associations, a variety of "differences"—physical and psychical, class and national—were brought together in a biosocial science of human variation. By analogy with the so-called lower races, women, the sexually deviate, the criminal, the urban poor, and the insane were in one way or another constructed as biological "races apart" whose differences from the white male, and likenesses to each other, "explained" their different and lower position in the social hierarchy.[20]

It is not the aim of this article to provide a systematic history of the biosocial science of racial and sexual difference based on analogy. The aim is rather to use the race-gender analogy to analyze the nature of analogical reasoning in science itself. When and how did the analogy appear in science? From what did it derive its scientific authority? How did the analogy shape research? What did it mean when a scientist claimed that the mature male of many lower races resembled a mature Caucasian female who had had many children? No simple theory of resemblance or substitution explains such an analogy. How did the analogy help construct the very similarities and differences supposedly "discovered" by scientists in nature? What theories of analogy and metaphor can be most effectively applied in the critical study of science?

THE CULTURAL SOURCES OF SCIENTIFIC METAPHOR

How particular metaphors or analogies in science are related to the social production of science, why certain analogies are selected and not others, and why

[17] James Weir, "The Effect of Female Suffrage on Posterity," *American Naturalist,* 1895, *29*:198–215.

[18] Charles Darwin, *The Descent of Man, and Selection in Relation to Sex* (London: John Murray, 1871), Vol. II, Chs. 17–20; Edward C. Cope, "The Developmental Significance of Human Physiognomy, *Amer. Nat.,* 1883, *17*:618–627.

[19] Arthur Keith, "Presidential Address: On Certain Factors in the Evolution of Human Races," *Journal of the Royal Anthropological Institute,* 1916, *64*:10–33; Ellis, *Man and Woman* (cit. n. 13), p. xii.

[20] See Nancy Stepan, "Biological Degeneration: Races and Proper Places," in *Degeneration: The Dark Side of Progress,* ed. J. Edward Chamberlin and Sander L. Gilman (New York: Columbia Univ. Press, 1985), pp. 97–120, esp. pp. 112–113. For an extended exploration of how various stereotypes of difference intertwined with each other, see Sander L. Gilman, *Difference and Pathology: Stereotypes of Sexuality, Race, and Madness* (Ithaca, N.Y.: Cornell Univ. Press, 1985).

certain analogies are accepted by the scientific community are all issues that need investigation.

In literature, according to Warren Shibles, striking metaphors just come, "like rain."[21] In science, however, metaphors and analogies are not arbitrary, nor merely personal. Not just any metaphors will do. In fact, it is their lack of perceived "arbitrariness" that makes particular metaphors or analogies acceptable as science.

As Stephen Toulmin recently pointed out, the constraints on the choice of metaphors and analogies in science are varied. The nature of the objects being studied (e.g., organic versus nonorganic), the social (e.g., class) structure of the scientific community studying them, and the history of the discipline or field concerned all play their part in the emergence of certain analogies rather than others and in their "success" or failure.[22] Sometimes the metaphors are strikingly new, whereas at other times they extend existing metaphors in the culture in new directions.

In the case of the scientific study of human difference, the analogies used by scientists in the late eighteenth century, when human variation began to be studied systematically, were products of long-standing, long-familiar, culturally endorsed metaphors. Human variation and difference were not experienced "as they really are, out there in nature," but by and through a metaphorical system that structured the experience and understanding of difference and that in essence created the objects of difference. The metaphorical system provided the "lenses" through which people experienced and "saw" the differences between classes, races, and sexes, between civilized man and the savage, between rich and poor, between the child and the adult. As Sander Gilman says in his book *Seeing the Insane,* "We do not see the world, rather we are taught by representations of the world about us to conceive of it in a culturally acceptable manner."[23]

The origin of many of the "root metaphors" of human difference are obscure. G. Lakoff and M. Johnson suggest that the basic values of a culture are usually compatible with "the metaphorical structure of the most fundamental concepts in the culture."[24] Not surprisingly, the social groups represented metaphorically as "other" and "inferior" in Western culture were socially "disenfranchised" in a variety of ways, the causes of their disenfranchisement varying from group to group and from period to period. Already in ancient Greece, Aristotle likened women to the slave on the grounds of their "natural" inferiority. Winthrop Jordan has shown that by the early Middle Ages a binary opposition between blackness and whiteness was well established in which blackness was identified with baseness, sin, the devil, and ugliness, and whiteness with virtue, purity, holiness, and beauty.[25] Over time, black people themselves were compared to apes, and their childishness, savageness, bestiality, sexuality, and lack of intellectual capacity stressed. The "Ethiopian," the "African," and especially the "Hottentot"

[21] Shibles, *Metaphor* (cit. n. 7), p. 15.

[22] Stephen Toulmin, "The Construal of Reality: Criticism in Modern and Postmodern Science," *Critical Inquiry,* 1982, *9*:93–111, esp. pp. 100–103.

[23] Sander L. Gilman, *Seeing the Insane* (New York: John Wiley, 1982), p. xi.

[24] Lakoff and Johnson, *Metaphors We Live By* (cit. n. 2), p. 22. The idea of root metaphors is Stephen Pepper's in *World Hypothesis* (Berkeley/Los Angeles: Univ. California Press, 1966), p. 91.

[25] Winthrop D. Jordan, *White over Black: American Attitudes toward the Negro, 1550–1812* (New York: Norton, 1977), p. 7.

were made to stand for all that the white male was not; they provided a rich analogical source for the understanding and representation of other "inferiorities." In his study of the representation of insanity in Western culture, for instance, Gilman shows how the metaphor of blackness could be borrowed to explicate the madman, and vice versa. In similar analogical fashion, the laboring poor were represented as the "savages" of Europe, and the criminal as a "Negro."

When scientists in the nineteenth century, then, proposed an analogy between racial and sexual differences, or between racial and class differences, and began to generate new data on the basis of such analogies, their interpretations of human difference and similarity were widely accepted, partly because of their fundamental congruence with cultural expectations. In this particular science, the metaphors and analogies were not strikingly new but old, if unexamined and diffuse. The scientists' contribution was to elevate hitherto unconsciously held analogies into self-conscious theory, to extend the meanings attached to the analogies, to expand their range via new observations and comparisons, and to give them precision through specialized vocabularies and new technologies. Another result was that the analogies became "naturalized" in the language of science, and their metaphorical nature disguised.

In the scientific elaboration of these familiar analogies, the study of race led the way, in part because the differences between blacks and whites seemed so "obvious," in part because the abolition movement gave political urgency to the issue of racial difference and social inequality. From the study of race came the association between inferiority and the ape. The facial angle, a measure of hierarchy in nature obtained by comparing the protrusion of the jaws in apes and man, was widely used in analogical science once it was shown that by this measure Negroes appeared to be closer to apes than the white race.[26] Established as signs of inferiority, the facial angle and blackness could then be extended analogically to explain other inferior groups and races. For instance, Francis Galton, Darwin's cousin and the founder of eugenics and statistics in Britain, used the Negro and the apish jaw to explicate the Irish: "Visitors to Ireland after the potato famine," he commented, "generally remarked that the Irish type of face seemed to have become more prognathous, that is, more like the negro in the protrusion of the lower jaw."[27]

Especially significant for the analogical science of human difference and similarity were the systematic study and measurement of the human skull. The importance of the skull to students of human difference lay in the fact that it housed the brain, differences in whose shape and size were presumed to correlate with equally presumed differences in intelligence and social behavior. It was measurements of the skull, brain weights, and brain convolutions that gave apparent precision to the analogies between anthropoid apes, lower races, women, criminal types, lower classes, and the child. It was race scientists who provided the new technologies of measurement—the callipers, cephalometers, craniometers, craniophores, craniostats, and parietal goniometers.[28] The low facial angles at-

[26] Stepan, *The Idea of Race in Science*, pp. 6–10.

[27] Francis Galton, "Hereditary Improvement," *Fraser's Magazine*, 1873, 7:116–130.

[28] These instruments and measurements are described in detail in Paul Topinard, *Anthropology* (London: Chapman & Hall, 1878), Pt. II, Chs. 1–4.

tributed by scientists starting in the 1840s and 1850s to women, criminals, idiots, and the degenerate, and the corresponding low brain weights, protruding jaws, and incompletely developed frontal centers where the higher intellectual faculties were presumed to be located, were all taken from racial science. By 1870 Paul Topinard, the leading French anthropologist after the death of Paul Broca, could call on data on sexual and racial variations from literally hundreds of skulls and brains, collected by numerous scientists over decades, in order to draw the conclusion that Caucasian women were indeed more prognathous or apelike in their jaws than white men, and even the largest women's brains, from the "English or Scotch" race, made them like the African male.[29] Once "woman" had been shown to be indeed analogous to lower races by the new science of anthropometry and had become, in essence, a racialized category, the traits and qualities special to woman could in turn be used in an analogical understanding of lower races. The analogies now had the weight of empirical reality and scientific theory. The similarities between a Negro and a white woman, or between a criminal and a Negro, were realities of nature, somehow "in" the individuals studied.

METAPHORIC INTERACTIONS

We have seen that metaphors and analogies played an important part in the science of human difference in the nineteenth century. The question is, what part? I want to suggest that the metaphors functioned as the science itself—that without them the science did not exist. In short, metaphors and analogies can be constituent elements of science.

It is here that I would like to introduce, as some other historians of science have done, Max Black's "interaction" theory of metaphor, because it seems that the metaphors discussed in this essay, and the analogies they mediated, functioned like interaction metaphors, and that thinking about them in these terms clarifies their role in science.[30]

By interaction metaphors, Black means metaphors that join together and bring into cognitive and emotional relation with each other two different things, or systems of things, not normally so joined. Black follows I. A. Richards in opposing the "substitution" theory of metaphor, in which it is supposed that the metaphor is telling us indirectly something factual about the two subjects—that the metaphor is a *literal comparison,* or is capable of a literal translation in prose. Richards proposed instead that "when we use a metaphor, we have two thoughts of different things active together and supported by a single word or phrase, whose meaning is the resultant of their interaction." Applying the interaction theory to the metaphor "The poor are the negroes of Europe," Black paraphrases Richards to claim that "our thoughts about the European poor and American negroes are 'active together' and 'interact' to produce a meaning that is a resultant of that interaction."[31] In such a view, the metaphor cannot be simply

[29] *Ibid.,* p. 311.

[30] Black, *Models and Metaphor* (cit. n. 7), esp. Chs. 3 and 13. See also Mary Hesse, *Models and Analogies in Science* (cit. n. 7); Hesse, "The Explanatory Function of Metaphor," in *Logic, Methodology and Philosophy of Science,* ed. Y. Bar-Hillel (Amsterdam: North-Holland, 1965), pp. 249–259; and Boyd, "Metaphor and Theory Change" (cit. n. 8).

[31] Black, *Models and Metaphor,* p. 38, quoting I. A. Richards, *Philosophy of Rhetoric* (Oxford: Oxford Univ. Press, 1938), p. 93.

reduced to literal comparisons or "like" statements without loss of meaning or cognitive content, because meaning is a product of the interaction between the two parts of a metaphor.

How do these "new meanings" come about? Here Black adds to Richards by suggesting that in an interaction metaphor, a "system of associated commonplaces" that strictly speaking belong only to one side of the metaphor are applied to the other. And he adds that what makes the metaphor effective "is not that the commonplaces shall be true, but that they should be readily and freely evoked."[32] Or as Mary Hesse puts it in *Models and Analogies in Science,* these implications "are not private, but are largely common to a given language community and are presupposed by speakers who intend to be understood."[33] Thus in the example given, the "poor of Europe" are seen in terms strictly applicable only to the "Negro" and vice versa. As a consequence, the poor are seen like a "race apart," savages in the midst of European civilization. Conversely, the "Negro" is seen as shiftless, idle, given to drink, part of the social remnant bound to be left behind in the march toward progress. Both the ideas of "savagery" and of "shiftlessness" belong to familiar systems of implications that the metaphor itself brings into play.

Black's point is that by their interactions and evoked associations both parts of a metaphor are changed. Each part is seen as more like the other in some characteristic way. Black was primarily interested in ordinary metaphors of a culture and in their commonplace associations. But instead of commonplace associations, a metaphor may evoke more specially constructed systems of implications. Scientists are in the business of constructing exactly such systems of implications, through their empirical investigations into nature and through their introduction into discourse of specialized vocabularies and technologies.[34] It may be, indeed, that what makes an analogy suitable for scientific purposes is its ability to be suggestive of new systems of implications, new hypotheses, and therefore new observations.[35]

In the case of the nineteenth-century analogical science of human difference, for instance, the system of implications evoked by the analogy linking lower races and women was not just a generalized one concerning social inferiority, but the more precise and specialized one developed by years of anthropometric, medical, and biological research. When "woman" and "lower races" were analogically and routinely joined in the anthropological, biological, and medical literature of the 1860s and 1870s, the metaphoric interactions involved a complex system of implications about similarity and difference, often involving highly technical language (for example, in one set of measurements of the body in different races cited by Paul Topinard in 1878 the comparisons included measures in each race of their height from the ground to the acromion, the epicondyle, the styloid process of the radius, the great trochanter, and the internal malleolus). The systems of implications evoked by the analogy included questions of com-

[32] Black, *Models and Metaphor,* p. 4.

[33] Hesse, *Models and Analogies in Science* (cit. n. 7), pp. 159–160.

[34] See Turbayne, *Myth of Metaphor* (cit. n. 6), p. 19, on this point.

[35] Black himself believed scientific metaphors belonged to the pretheoretical stage of a discipline. Here I have followed Boyd, who argues in "Metaphor and Theory Change" (cit. n. 8), p. 357, that metaphors can play a role in the development of theories in relatively mature sciences. Some philosophers would reserve the term "model" for extended, systematic metaphors in science.

parative health and disease (blacks and women were believed to show greater degrees of insanity and neurasthenia than white men, especially under conditions of freedom), of sexual behavior (females of "lower races" and lower-class women of "higher races," especially prostitutes, were believed to show similar kinds of bestiality and sexual promiscuity, as well as similar signs of pathology and degeneracy such as deformed skulls and teeth), and of "childish" characteristics, both physical and moral.[36]

As already noted, one of the most important systems of implications about human groups developed by scientists in the nineteenth century on the basis of analogical reasoning concerned head shapes and brain sizes. It was assumed that blacks, women, the lower classes, and criminals shared low brain weights or skull capacities. Paul Broca, the founder of the Société d'Anthropologie de Paris in 1859, asserted: "In general, the brain is larger in mature adults than in the elderly, in men than in women, in eminent men than in men of mediocre talent, in superior races than in inferior races. . . . Other things being equal, there is a remarkable relationship between the development of intelligence and the volume of the brain."[37]

Such a specialized system of implications based on the similarities between brains and skulls appeared for the first time in the phrenological literature of the 1830s. Although analogies between women and blackness had been drawn before, woman's place in nature and her bio-psychological differences from men had been discussed by scientists mainly in terms of reproductive function and sexuality, and the most important analogies concerned black females (the "sign" of sexuality) and lower-class or "degenerate" white women. Since males of all races had no wombs, no systematic, apparently scientifically validated grounds of comparison between males of "lower" races and women of "higher" races existed.

Starting in the 1820s, however, the phrenologists began to focus on differences in the shape of the skull of individuals and groups, in the belief that the skull was a sign faithfully reflecting the various organs of mind housed in the brain, and that differences in brain organs explained differences in human behavior. And it is in the phrenological literature, for almost the first time, that we find women and lower races compared directly on the basis of their skull formations. In their "organology," the phrenologists paid special attention to the organ of "philoprogenitiveness," or the faculty causing "love of offspring," which was believed to be more highly developed in women than men, as was apparent from their more highly developed upper part of the occiput. The same prominence, according to Franz Joseph Gall, was found in monkeys and was particularly well developed, he believed, in male and female Negroes.[38]

By the 1840s and 1850s the science of phrenology was on the wane, since the organs of the brain claimed by the phrenologists did not seem to correspond with the details of brain anatomy as described by neurophysiologists. But although the

[36] For an example of the analogous diseases and sexuality of "lower" races and "lower" women, see Eugene S. Talbot, *Degeneracy: Its Causes, Signs, and Results* (London: Walter Cott, 1898), pp. 18, 319–323.

[37] Paul Broca, "Sur le volume et la forme du cerveau suivant les individus et suivant les races," *Bulletin de la Société d'Anthropologie Paris,* 1861, *2*:304.

[38] Franz Joseph Gall, "The Propensity to Philoprogenitiveness," *Phrenological Journal,* 1824–1825, *2*:20–33.

specific conclusions of the phrenologists concerning the anatomical structure and functions of the brain were rejected, the principle that differences in individual and group function were products of differences in the shape and size of the head was not. This principle underlay the claim that some measure, whether of cranial capacity, the facial angle, the brain volume, or brain weight, would be found that would provide a true indicator of innate capacity, and that by such a measure women and lower races would be shown to occupy analogous places in the scale of nature (the "scale" itself of course being a metaphorical construct).

By the 1850s the measurement of women's skulls was becoming an established part of craniometry and the science of gender joined analogically to race. Vogt's *Lectures on Man* included a long discussion of the various measures available of the skulls of men and women of different races. His data showed that women's smaller brains were analogous to the brains of lower races, the small size explaining both groups' intellectual inferiority. (Vogt also concluded that within Europe the intelligentsia and upper classes had the largest heads, and peasants the smallest.)[39] Broca shared Vogt's interest; he too believed it was the smaller brains of women and "lower" races, compared with men of "higher" races, that caused their lesser intellectual capacity and therefore their social inferiority.[40]

One novel conclusion to result from scientists' investigations into the different skull capacities of males and females of different races was that the gap in head size between men and women had apparently widened over historic time, being largest in the "civilized" races such as the European, and smallest in the most savage races.[41] The growing difference between the sexes from the prehistoric period to the present was attributed to evolutionary, selective pressures, which were believed to be greater in the white races than the dark and greater in men than women. Paradoxically, therefore, the civilized European woman was less like the civilized European man than the savage man was like the savage woman. The "discovery" that the male and female bodies and brains in the lower races were very alike allowed scientists to draw direct comparisons between a black male and a white female. The male could be taken as representative of both sexes of his race and the black female could be virtually ignored in the analogical science of intelligence, if not sexuality.

Because interactive metaphors bring together a *system* of implications, other features previously associated with only one subject in the metaphor are brought to bear on the other. As the analogy between women and race gained ground in science, therefore, women were found to share other points of similarity with lower races. A good example is prognathism. Prognathism was a measure of the protrusion of the jaw and of inferiority. As women and lower races became analogically joined, data on the "prognathism" of females were collected and women of "advanced" races implicated in this sign of inferiority. Havelock Ellis, for instance, in the late nineteenth-century bible of male-female differences *Man and Woman,* mentioned the European woman's slightly protruding jaw as a trait, not of high evolution, but of the lower races, although he added that in white women the trait, unlike in the lower races, was "distinctly charming."[42]

[39] Vogt, *Lectures on Man* (cit. n. 16), p. 88. Vogt was quoting Broca's data.

[40] Gould, *Mismeasure of Man* (cit. n. 11), p. 103.

[41] Broca's work on the cranial capacities of skulls taken from three cemeteries in Paris was the most important source for this conclusion. See his "Sur la capacité des cranes parisiens des divers époques," *Bull. Soc. Anthr. Paris,* 1862, *3*:102–116.

[42] Ellis, *Man and Woman* (cit. n. 13), pp. 106–107.

Another set of implications brought to bear on women by analogy with lower races concerned dolichocephaly and brachycephaly, or longheadedness and roundheadedness. Africans were on the whole more longheaded than Europeans and so dolichocephaly was generally interpreted as signifying inferiority. Ellis not surprisingly found that on the whole women, criminals, the degenerate, the insane, and prehistoric races tended to share with dark races the more narrow, dolichocephalic heads representing an earlier (and by implication, more primitive) stage of brain development.[43]

ANALOGY AND THE CREATION OF NEW KNOWLEDGE

In the metaphors and analogies joining women and the lower races, the scientist was led to "see" points of similarity that before had gone unnoticed. Women became more "like" Negroes, as the statistics on brain weights and body shapes showed. The question is, what kind of "likeness" was involved?

Here again the interaction theory of metaphor is illuminating. As Black says, the notion of similarity is ambiguous. Or as Stanley Fish puts it, "Similarity is not something one finds but something one must establish."[44] Metaphors are not meant to be taken literally but they do imply some structural similarity between the two things joined by the metaphor, a similarity that may be new to the readers of the metaphoric or analogical text, but that they are culturally capable of grasping.

However, there is nothing obviously similar about a white woman of England and an African man, or between a "criminal type" and a "savage." (If it seems to us as though there is, that is because the metaphor has become so woven into our cultural and linguistic system as to have lost its obviously metaphorical quality and to seem a part of "nature.") Rather it is the metaphor that permits us to see similarities that the metaphor itself helps constitute.[45] The metaphor, Black suggests, "selects, emphasizes, suppresses and organizes features" of reality, thereby allowing us to see new connections between the two subjects of the metaphor, to pay attention to details hitherto unnoticed, to emphasize aspects of human experience otherwise treated as unimportant, to make new features into "signs" signifying inferiority.[46] It was the metaphor joining lower races and women, for instance, that gave significance to the supposed differences between the shape of women's jaws and those of men.

Metaphors, then, through their capacity to construct similarities, create new knowledge. The full range of similarities brought into play by a metaphor or analogy is not immediately known or necessarily immediately predictable. The metaphor, therefore, allows for "discovery" and can yield new information through empirical research. Without the metaphor linking women and race, for example, many of the data on women's bodies (length of limbs, width of pelvis, shape of skull, weight or structure of brain) would have lost their significance as signs of inferiority and would not have been gathered, recorded, and interpreted

[43] Alexander Sutherland, "Woman's Brain," *Nineteenth Century,* 1900, *47*:802–810; and Ellis, *Man and Woman,* p. 98. Ellis was on the whole, however, cautious about the conclusions that could be drawn from skull capacities and brain weights.

[44] Stanley Fish, "Working on the Chain Gang: Interpretation in the Law and Literary Criticism," in *The Politics of Interpretation,* ed. W. J. T. Mitchell (Chicago: Univ. Chicago Press, 1983), p. 277.

[45] Max Black, as cited in Ortony, *Metaphor and Thought* (cit. n. 1), p. 5.

[46] Black, *Models and Metaphor* (cit. n. 7), p. 44.

in the way they were. In fact, without the analogies concerning the "differences" and similarities among human groups, much of the vast enterprise of anthropology, criminology, and gender science would not have existed. The analogy guided research, generated new hypotheses, and helped disseminate new, usually technical vocabularies. The analogy helped constitute the objects of inquiry into human variation—races of all kinds (Slavic, Mediterranean, Scottish, Irish, yellow, black, white, and red), as well as other social groups, such as "the child" and "the madman." The analogy defined what was problematic about these social groups, what aspects of them needed further investigation, and which kinds of measurements and what data would be significant for scientific inquiry.

The metaphor, in short, served as a program of research. Here the analogy comes close to the idea of a scientific "paradigm" as elaborated by Kuhn in *The Structure of Scientific Revolutions;* indeed Kuhn himself sometimes writes of paradigms as though they are extended metaphors and has proposed that "the same interactive, similarity-creating process which Black has isolated in the functioning of metaphor is vital also in the function of models in science."[47]

The ability of an analogy in science to create new kinds of knowledge is seen clearly in the way the analogy organizes the scientists' understanding of causality. Hesse suggests that a scientific metaphor, by joining two distinct subjects, implies more than mere structural likeness. In the case of the science of human difference, the analogies implied a similar *cause* of the similarities between races and women and of the differences between both groups and white males. To the phrenologists, the cause of the large organs of philoprogenitiveness in monkeys, Negroes, and women was an innate brain structure. To the evolutionists, sexual and racial differences were the product of slow, adaptive changes involving variation and selection, the results being the smaller brains and lower capacities of the lower races and women, and the higher intelligence and evolutionarily advanced traits in the males of higher races. Barry Barnes suggests we call the kind of "redescription" involved in a metaphor or analogy of the kind being discussed here an "explanation," because it forces the reader to "understand" one aspect of reality in terms of another.[48]

ANALOGY AND THE SUPPRESSION OF KNOWLEDGE

Especially important to the functioning of interactive metaphors in science is their ability to neglect or even suppress information about human experience of the world that does not fit the similarity implied by the metaphor. In their "similarity-creating" capacity, metaphors involve the scientist in a selection of those aspects of reality that are compatible with the metaphor. This selection process is often quite unconscious. Stephen Jay Gould is especially telling about the ways in which anatomists and anthropologists unself-consciously searched for and selected measures that would prove the desired scales of human superiority and inferiority and how the difficulties in achieving the desired results were surmounted.

Gould has subjected Paul Broca's work on human differences to particularly

[47] Thomas S. Kuhn, *The Structure of Scientific Revolutions* (Chicago: Univ. Chicago Press, 1962; 2nd ed., 1973), esp. Ch. 4; and Kuhn, "Metaphor in Science" (cit. n. 8), p. 415.

[48] Barry Barnes, *Scientific Knowledge and Sociological Theory* (London: Routledge & Kegan Paul, 1974), p. 49.

thorough scrutiny because Broca was highly regarded in scientific circles and was exemplary in the accuracy of his measurements. Gould shows that it is not Broca's measurements per se that can be faulted, but rather the ways in which he unconsciously manipulated them to produce the very similarities already "contained" in the analogical science of human variation. To arrive at the conclusion of women's inferiority in brain weights, for example, meant failing to make any correction for women's smaller body weights, even though other scientists of the period were well aware that women's smaller brain weights were at least in part a function of their smaller body sizes. Broca was also able to "save" the scale of ability based on head size by leaving out some awkward cases of large-brained but savage heads from his calculations, and by somehow accounting for the occasional small-brained "geniuses" from higher races in his collection.[49]

Since there are no "given" points of measurement and comparison in nature (as Gould says, literally thousands of different kinds of measurements can theoretically be made of the human body), scientists had to make certain choices in their studies of human difference. We are not surprised to find that scientists selected just those points of comparison that would show lower races and women to be nearer to each other and to other "lower" groups, such as the anthropoid apes or the child, than were white men. The maneuvers this involved were sometimes comical. Broca, for instance, tried the measure of the ratio of the radius to the humerus, reasoning that a high ratio was apish, but when the scale he desired did not come out, he abandoned it. According to Gould, he even almost abandoned the most time-honored measure of human difference and inferiority, namely, brain weights, because yellow people did well on it. He managed to deal with this apparent exception to the "general rule of nature" that lower races had small heads by the same kind of specious argumentation he had used with small-brained geniuses. Broca claimed that the scale of brain weights did not work as well at the upper end as at the lower end, so that although small brain weights invariably indicated inferiority, large brain weights did not necessarily in and of themselves indicate superiority![50]

Since most scientists did recognize that the brain weights of women were in fact heavier in proportion to their body weights than men, giving women an apparent comparative advantage over men, not surprisingly they searched for other measures. The French scientist Léonce Pierre Manouvrier used an index relating brain weight to thigh bone weight, an index that gave the desired results and was in confirmation with the analogies, but that even at the time was considered by one scientist "ingenious and fantastic but divorced from common sense."[51] Even more absurd when viewed from the distance of time was the study mentioned by Ellis by two Italians who used the "prehensile" (i.e., apish) character of the human toe to compare human groups and found it was greater in normal white women than in white men, and also marked in criminals, prostitutes, idiots, and of course lower races.[52]

One test of the social power (if not the scientific fruitfulness) of an analogy in science seems in fact to be the degree to which information can be ignored, or

[49] Gould, *Mismeasure of Man* (cit. n. 11), pp. 73–112. For another example see Stephen Jay Gould, "Morton's Ranking of Race by Cranial Capacity," *Science,* 1978, *200*:503–509.

[50] Gould, *Mismeasure of Man,* pp. 85–96.

[51] Sutherland, "Woman's Brain" (cit. n. 43), p. 805.

[52] Ellis, *Man and Woman* (cit. n. 13), p. 53.

interpretation strained, without the analogy losing the assent of the relevant scientific community. On abstract grounds, one would expect an analogy of the kind being discussed here, which required rather obvious distortions of perception to maintain (at least to our late twentieth-century eyes), to have been abandoned by scientists fairly quickly. Since, however, interactive metaphors and analogies direct the investigators' attention to some aspects of reality and not others, the metaphors and analogies can generate a considerable amount of new information about the world that confirms metaphoric expectations and direct attention away from those aspects of reality that challenge those expectations. Given the widespread assent to the cultural presuppositions underlying the analogy between race and gender, the analogy was able to endure in science for a long time.

For instance, by directing attention to exactly those points of similarity and difference that would bring women and lower races closer to apes, or to each other, the race-gender metaphor generated data, many of them new, which "fit" the metaphor and the associated implications carried by it. Other aspects of reality and human experience that were incompatible with the metaphor tended to be ignored or not "seen." Thus for decades the Negro's similarity to apes on the basis of the shape of his jaw was asserted, while the white man's similarity to apes on the basis of his thin lips was ignored.

When contrary evidence could not be ignored, it was often reinterpreted to express the fundamental valuations implicit in the metaphor. Gould provides us with the example of neoteny, or the retention in the adult of childish features such as a small face and hairlessness. A central feature of the analogical science of inferiority was that adult women and lower races were more childlike in their bodies and minds than white males. But Gould shows that by the early twentieth century it was realized that neoteny was a positive feature of the evolutionary process. "At least one scientist, Havelock Ellis, did bow to the clear implication and admit the superiority of women, even though he wriggled out of a similar confession for blacks." As late as the 1920s the Dutch scientist Louis Bolk, on the other hand, managed to save the basic valuation of white equals superior, blacks and women equal inferior by "rethinking" the data and discovering after all that blacks departed more than whites from the most favorable traits of childhood.[53]

To reiterate, because a metaphor or analogy does not directly present a preexisting nature but instead helps "construct" that nature, the metaphor generates data that conform to it, and accommodates data that are in apparent contradiction to it, so that nature is seen via the metaphor and the metaphor becomes part of the logic of science itself.[54]

CHANGING METAPHORS

Turbayne, in his book *The Myth of Metaphor,* proposes as a major critical task of the philosopher or historian of science the detection of metaphor in science. Detection is necessary because as metaphors in science become familiar or com-

[53] Gould, *Mismeasure of Man,* pp. 120–121.

[54] Terence Hawkes, *Metaphor* (London: Methuen, 1972), p. 88, suggests that metaphors "will retrench or corroborate as much as they expand our vision," thus stressing the normative, consensus-building aspects of metaphor.

monplace, they tend to lose their metaphorical nature and to be taken literally. The analogical science of human difference is a particularly striking example. So familiar and indeed axiomatic had the analogies concerning "lower races," "apes," and "women" become by the end of the nineteenth century that in his major study of male-female differences in the human species, Ellis took almost without comment as the standards against which to measure the "typical female" on the one hand "the child," and on the other "the ape," "the savage," and the "aged human." The tendency for metaphors to become dogmatic and to be seen as literally true and nonmetaphoric is particularly strong in science because of the identification of the language of science with the language of objectivity and reality.

The confusion of metaphor for reality in science would be less important if metaphors did not have social and moral consequences in addition to intellectual ones. This aspect of metaphoric and analogic science is often overlooked in discussions of paradigms, models, and analogies in science, in which the main focus tends to be on the metaphor as an intellectual construct with intellectual consequences for the doing of science. But metaphors do more than this. Metaphors shape our perceptions and in turn our actions, which tend to be in accordance with the metaphor. The analogies concerning racial and gender and class differences in the human species developed in the biosocial sciences in the nineteenth century, for instance, had the social consequences of helping perpetuate the racial and gender status quo. The analogies were used by scientists to justify resistance to efforts at social change on the part of women and "lower races," on the grounds that inequality was a "fact" of nature and not a function of the power relations in a society.

Another reason, then, for uncovering or exposing metaphor in science is to prevent ourselves from being used or victimized or captured by metaphors.[55] The victims of the analogical science of human difference were the women and the human groups conceptualized as "lower" races. Their exclusion from the community of scientists doing the analogizing was, to a large extent, part of the same social division of labor that produced scientific theories of natural inferiority. It was an exclusion that made identifying and challenging metaphors of natural inequality very difficult.

That the analogy between race and gender was eventually discarded (though not until will into the twentieth century) raises the interesting question of how metaphors in science change. For if metaphors are part of the logical structure of science, changes in metaphor will bring about changes in science. Ever since Kuhn published *The Structure of Scientific Revolutions* in 1962, of course, the problem of change has been central to any critical theory of science. Kuhn's contribution was to show that the substitution of one "paradigm" (defined as the beliefs, values, or techniques of a scientific community) by another was a complex historical event that could not be reduced to straightforward questions concerning the increased rationality, comprehensiveness, or logic of one paradigm over another. His work raised important questions about the relationship between scientific theories and empirical reality, about the grounds on which one paradigm is accepted and another rejected, and about the roots of change. He

[55] Turbayne, *Myth of Metaphor* (cit. n. 6), p. 27.

proposed that paradigms were not in fact simple reflections of reality but complex human constructions. Above all, Kuhn stressed the idea that all scientific knowledge is "embedded in theory and rules" which the scientist learns as a member of a scientific community.

Nevertheless, despite the emphasis he gave to the scientific community, Kuhn's own explanation of scientific revolutions tended toward the "intellectualist" rather than the sociological. He concentrated attention on the ways in which the scientific paradigm itself generates, through the normal process of puzzle solving within the paradigm, anomalies that eventually cause the breakdown of the paradigm and its replacement by another. On his own admission, Kuhn paid little attention to the role of social, political, or economic factors in the generation of new metaphors, and therefore new meanings, in science.

Recent work in the history and sociology of science, however, in part under the stimulus of Kuhn's work, has tended to stress the importance of the scientific community itself, as a sociological and political as well as scientific entity, for the generation and rejection of metaphors and analogies. The hope is that close historical and sociological investigation will begin to indicate in what ways particular representations or metaphors of nature are related to the social structure—class organization, professional socialization, interests—of the scientific community. The suggestion is being made that the root metaphors held by a particular scientific community or school of thought can become unsatisfactory, not merely because the data generated by the metaphor do not "fit" the metaphor, but because, for political or social or economic reasons, social formations change and new aspects of reality or human experience become important, are "seen," and new metaphors introduced.

The full implication of my own studies of the changes that eventually occurred in the analogical science of human difference is indeed along these lines—namely, that changes in political and social life were closely tied to the new metaphors of human similarity and equality, as opposed to metaphors of difference and inequality, that were proposed in the human sciences after World War II.[56] The subject of metaphoric change is obviously one requiring much further study.

A BRIEF CONCLUSION

In this essay I have indicated only some of the issues raised by a historical consideration of a specific metaphoric or analogical science. There is no attempt at completeness or theoretical closure. My intention has been to draw attention to the ways in which metaphor and analogy can play a role in science, and to show how a particular set of metaphors and analogies shaped the scientific study of human variation. I have also tried to indicate some of the historical reasons why scientific texts have been "read" nonmetaphorically, and what some of the scientific and social consequences of this have been.

Some may argue I have begged the question of metaphor and analogy in science by treating an analogical science that was "obviously pseudoscientific." I maintain that it was not obviously pseudoscientific to its practitioners, and that

[56] Stepan, *Race in Science* (cit. n. 10), Chs. 6 and 7.

they were far from being at the periphery of the biological and human sciences in the nineteenth and early twentieth centuries. I believe other studies will show that what was true for the analogical science of human difference may well be true also for other metaphors and analogies in science.

My intention has also been to suggest that a theory of metaphor is as critical to science as it is to the humanities. We need a critical theory of metaphor in science in order to expose the metaphors by which we learn to view the world scientifically, not because these metaphors are necessarily "wrong," but because they are so powerful.

Spirit manifestation witnessed at a seance by Alfred Russel Wallace, who testified in court to his belief in it (The Sketch, *May 1, 1907, p.65*).

Alfred Russel Wallace, the Origin of Man, and Spiritualism

By Malcolm Jay Kottler

INTRODUCTION

IT HAS BEEN FORGOTTEN, ignored, or perhaps never known by historians of science that in the second half of the nineteenth century a considerable number of renowned scientists were favorably disposed toward such psychical phenomena as telepathy, clairvoyance, precognition, levitation, slate writing, spirit communication, spirit materialization, and spirit photography. Among the confirmed believers in the reality of these phenomena was Alfred Russel Wallace. Wallace's belief in psychical phenomena and their spiritualist interpretation should be especially interesting to the historian of biology, because it deeply influenced his evolutionary thought.

A study of Wallace's involvement with spiritualism has revealed the true origin of his well-known divergence from Darwin on the origin of man. In published papers concerning the origin of man Wallace never included among the troubling facts his or others' spiritualist experiences, yet his own conviction that natural selection was insufficient to explain the origin of man and that man's origin required the action of higher intelligences guiding the laws of organic development "in definite directions and for special ends" arose from his experiences at seances beginning in July 1865. By November 1866 he was fully convinced of the reality of psychical phenomena and for several years tried, unsuccessfully, to interest fellow scientists in joining him at seances to further investigate the phenomena. Having failed to persuade them of the validity and meaning of this new evidence concerning the nature and, indirectly, the origin of man, Wallace was forced to exclude it from his discussion of man, from the time of his first public announcement of doubts about the sufficiency of natural selection in the origin of man (1869) through the next two decades and the publication of his monumental work *Darwinism* (1889). Despite this restriction on his argument, Wallace presented a formidable case, based on considerations of utility only, against the action of natural selection alone in man's development. Only then did he introduce spiritualism into his published papers, as an explanation for those features of man found inexplicable by natural selection.

Wilma George in her biography of Wallace has made a valuable beginning in the treatment of Wallace's belief in spiritualism and its effect upon his views on the origin of man. But her discussion is incomplete, especially with regard to Wallace's published papers on the origin of man and Wallace's attempts to interest fellow scientists in seance phenomena and to counter their skepticism.[1] Therefore in this paper I wish to discuss Wallace's published contributions on the origin of man through *Darwinism* as well as his concern with spiritualism in the critical period 1865–1869. During this period Wallace formed his lifelong belief that Darwin's, and his own, principle of natural selection, though sufficient in the origin of *other* species,[2] was inadequate for the origin of man. Lastly I wish to consider the high points of Wallace's involvement with spiritualism from 1870 through his death in 1913.

WALLACE AND THE ORIGIN OF MAN (1857–1889)

About one year after he had begun his correspondence with Darwin (1857), Wallace raised the question of the origin of man. In his reply to Wallace, Darwin wrote, "You ask whether I shall discuss Man: I think I shall avoid the whole subject, as so surrounded with prejudices, though I fully admit that it is the highest and most interesting problem for the naturalist."[3] H. L. McKinney has persuasively argued that Wallace was led to his belief in species transformation by means of natural selection through his interest in ethnology and the origin of man.[4] Therefore it seems safe to conclude that in 1858, when papers on natural selection by Darwin and Wallace were jointly presented to the Linnaean Society, Wallace considered man's origin to have been by descent with modification by means of natural selection only.

[1] Wilma George, *Biologist Philosopher: A Study of the Life and Writings of Alfred Russel Wallace* (New York: Abelard-Schuman, 1964), pp. 8, 72–74, 93–94, 120, 157, 235–250, 277–278, 282, 284. James Marchant, *Alfred Russel Wallace: Letters and Reminiscences* (London: Cassel, 1916), Vol. II, pp. 181–186. The most complete discussion of Wallace's views on natural selection and the origin of man, but with little reference to his belief in spiritualism, is Loren Eiseley, *Darwin's Century* (Garden City: Anchor, 1961), pp. 287–324. Brief accounts of Wallace's involvement with spiritualism are S. Smith, "Alfred R. Wallace—Scientific Enthusiast," *Tomorrow*, 1960, *8*:95–104, and N. Fodor, "Dr. Alfred Russel Wallace," in *Encyclopedia of Psychic Science* (London: Arthurs, 1933). Two invaluable treatments of psychical phenomena and research in the nineteenth century, with great relevance to Wallace's spiritualist experiences, are A. Gauld, *The Founders of Psychical Research* (New York: Schocken, 1968) and R. G. Medhurst and K. M. Goldney, "William Crookes and the Physical Phenomena of Mediumship," *Proceedings of the Society for Psychical Research*, 1964, *54*:25–157. Since my paper was completed I have learned of two recent, detailed studies of Wallace's belief in spiritualism and its relationship to his views on the origin of man: F. M. Turner, "Between Science and Religion: The Reaction to Scientific Naturalism in Late Victorian England" (Ph.D. Dissertation, Yale University, 1971), pp. 79–122, and R. Smith, "Alfred Russel Wallace: Philosophy of Nature and Man," *The British Journal for the History of Science*, 1972, *6*:177–199.

[2] Besides the origin of man, Wallace made one other interesting exception to the all-sufficiency of natural selection. He was converted by the highly controversial experimental work of H. C. Bastian to a belief in abiogenesis and heterogenesis which acted, in lieu of natural selection, in the origin of the lower forms of life. The rate of evolution by means of natural selection was considered a function of the complexity of interactions between species, and consequently the development of the early, lower forms of life had been thought exceedingly slow. However, recent calculations of the physicists had vastly reduced the amount of time available for the evolution of *all* forms of life on earth. If natural selection alone had acted, the development of the lower forms of life would have consumed most of the time available. Wallace welcomed abiogenesis and heterogenesis because they accelerated the evolution of these lower forms and thereby freed most of the available time for the evolution of the multitude of higher forms of life. A. R. Wallace, "The Beginnings of Life," *Nature*, 1872, *6*:302–303. Marchant, *Wallace*, Vol. I, pp. 273–278. Eiseley, *Darwin's Century*, pp. 233–244.

[3] Marchant, *Wallace*, Vol. I, p. 133.

[4] H. Lewis McKinney, *Wallace and Natural Selection* (New Haven: Yale University Press, 1972), pp. 80–96.

Natural selection and the mind of man, 1864

Wallace first expressed his views on the origin of man in public and in print six years later (1864). Peter Vorzimmer feels that Lyell's *Antiquity of Man* (1863) was the indirect stimulus to Wallace, while Lyell's presidential address to the summer meeting of the British Association for the Advancement of Science (BAAS) in 1864 was the direct stimulus. In preparation for his address, Lyell wrote to Wallace "as to the division of the Malay Archipelago into two regions, and the relation of this division to the races of man. . . ."[5] Wallace rather quickly gathered together his ideas on the origin of man, for he presented a paper to the Anthropological Society of London on March 1, 1864.[6] Wallace himself cited Herbert Spencer's *Social Statics* as the stimulus for his new ideas about man's development. Confusion surrounds the contents of this paper, because Wallace reprinted it with brief though significant modifications in a collection of essays published in 1870. Thus Vorzimmer has mistakenly dated Wallace's belief in the insufficiency of natural selection, and the necessity for divine intervention, in the origin of man to 1864 and this paper. But Eiseley, De Beer, and George have correctly noted that nowhere in the original 1864 paper did Wallace invoke anything remotely non-natural to explain man's origin. Wallace no doubt contributed to the confusion by stating, incorrectly, in his autobiography that his divergence from Darwin respecting the origin of man was "first intimated" in this 1864 paper.[7]

In the 1864 paper Wallace's leading idea was that man's mind—specifically his intellectual and moral nature—had "shielded" his body from the action of natural selection and thereby put an end to his structural change. This new cause in man's development effectively solved two major problems surrounding the origin of man: (1) did the races of man belong to one species or was each race a species in itself? and (2) why did man's body, with the exception of his skull, so closely resemble the bodies of extant apes, while his skull and mental capacities so widely diverged from those of the same apes?

The strongest evidence put forward by the so-called polygenists in answer to the first question was that remains from ancient Egyptian tombs, about five thousand years old, indicated as much difference *then* between the Negro and Semitic races as *now*. This seemingly contradicted the monogenist hypothesis that the further back in time one went the more alike the different races became until, finally, one reached the point at which man began his existence as a species on this earth and there was only one race. This contradiction was especially marked the more one restricted the age of the earth and the age of man on it. The Egyptian tomb evidence carried little weight with Wallace for two reasons. He believed that man's antiquity far antedated five thousand years and that at some distant time prior to ancient Egypt racial divergence had come

[5] A. R. Wallace, *My Life: A Record of Events and Opinions* (London: Chapman and Hall, 1905), Vol. I, p. 417.

[6] A. R. Wallace, "The Origin of Human Races and the Antiquity of Man Deduced from the Theory of 'Natural Selection,' " *Journal of the Anthropological Society of London*, 1864, *2*: clviii–clxx. Wallace's own summary of the paper appeared in *Natural History Review*, 1864, *II. 4*: 328–336. John C. Greene, *The Death of Adam* (New York: Mentor, 1961), pp. 311–315. Eiseley, *Darwin's Century*, pp. 304–309.

[7] Wallace, *My Life*, Vol. I, p. 418 and Vol. II, p. 17. Marchant, *Wallace*, Vol. I, p. 240 and Vol. II, pp. 183–184. P. J. Vorzimmer, *Charles Darwin: The Years of Controversy* (Philadelphia: Temple University, 1970), p. 190. Gavin De Beer, *Charles Darwin: A Scientific Biography* (Garden City: Anchor, 1965), p. 214. George, *Biologist Philosopher*, pp. 71, 241.

to an end, as had nearly all structural modification in all men, because of the operation of the new cause—the mind of man. Therefore it was fully possible for the differences between human races to have remained unchanged over five thousand years but to have been increasingly smaller in earlier and earlier periods. Similarly, man's body would be only slightly different from an ape's body if this new cause, putting an end to structural change in man, had begun to act at an early stage in man's development.

Wallace illustrated the manner in which this new cause acted to suspend the influence of natural selection on man's body by comparing the survival responses of animals and man to important environmental changes. A change in climate might require a thicker fur or a layer of fat for an animal's survival. Man, in contrast, could survive by means of warmer clothing or shelter without undergoing bodily change. A change in abundance of the food species might require a change in diet and corresponding changes in bodily weaponry (claws, teeth) and internal digestive anatomy for an animal's survival. Man in an early period could survive by means of a better weapon or trap or by hunting in a (larger) group without undergoing bodily change. At the same time his possession of fire enabled him to render many different food species palatable and thereby increase the natural food supply available. Man in a later period would be independent of natural fluctuations in the abundance of a potential food species, because agriculture and domestication of animals provided him with a sure, ready food supply. Thus by means of intellect alone, man, with an unchanged body, could maintain his harmony with a changing universe and survive. Natural selection acted so powerfully on animals because the individual was isolated, on its own. With a slight injury or weakness, an individual might not survive. But man's social and sympathetic feelings checked the action of natural selection in eliminating the weaker among men. Thus by his moral nature alone, man, with an unchanged body, could survive the harsh standard imposed by natural selection on the body of every individual of an animal species.[8]

This stage in man's development having been reached, the focus of natural selection's operation shifted from man's bodily nature to his intellectual and moral nature:

> . . . every slight variation in his mental and moral nature which should enable him better to guard against adverse circumstances, and combine for mutual comfort and protection, would be preserved and accumulated.[9]

The more intellectual and moral races displaced the lower and more degraded. Natural selection acted on man's mental organization and led "to the more perfect adaptation of man's higher faculties to the conditions of surrounding nature and to the exigencies of the social state."[10]

Quite clearly Wallace explained by natural selection the further development of man's intellectual and moral nature once it had reached a "fairly developed" stage. In the paper he did not consider the cause of the development up to the "fairly developed" stage, but in the discussion that followed presentation of the paper, Wallace expressed the belief that animals possessed an intellect, too, and argued that unless they did, one

[8] Wallace, "Origin of Human Races," pp. clviii–clxiv, clxvi, clxix.

[9] *Ibid.*, p. clxiv.

[10] *Ibid.*, p. clxix.

faced "immense difficulty."[11] In other words, Wallace sought the origin of man's intellectual and moral nature in man's nonhuman ancestors and considered its development to the "fairly developed" stage—at which point it became the new cause in man's development—to have been the result of natural selection.

Wallace noted an interesting corollary of this position. If the totality of man were the product of natural selection, then the large increase in brain size in man, as compared to the apes, had to have occurred slowly, just as any other large change effected by natural selection. Since early man—that is, man whose body was still subject to natural selection—had arisen before this increase in brain size had proceeded very far, Wallace felt that traces of early man should be found in the Miocene period "when not a single mammal possessed the same form as any existing species." If man arose in a later period—when mammals were already very similar to extant species—then man would have altered in bodily structure (for example, attained his erect posture) while animal species remained virtually constant. Yet Wallace's entire theory held that man's bodily structure had been constant while the bodies of animal species had changed and not *vice versa*. There was no evidence of man at such an early geological period, but the incompleteness of the fossil record from the appropriate part of the world could explain that.[12]

Wallace sent a copy of his paper to Darwin. The "Anthropologicals" had not appreciated it much, but Wallace hoped Darwin would be able to agree with him. Indeed, Darwin was: in his letter of May 28, 1864, he praised Wallace for his new "great leading idea."[13] However, Darwin was concerned with the prospect of another priority dispute with Wallace. Until receipt of Wallace's paper, Darwin had not realized that once again the two of them were thinking about the same problem—this time the origin of man. Nevertheless, Darwin made an offer to Wallace of his notes on man. Wallace, in fact, had no immediate plans to write more about man and declined Darwin's offer.[14] According to Vorzimmer, Darwin was most afraid Wallace would "scoop" him on sexual selection, which Darwin had only briefly discussed in the *Origin* (1859). In his May 28, 1864, letter Darwin expressed his view that "a sort of sexual selection has been the most powerful means of changing the races of man." And in his *Descent* (1871) he used sexual selection extensively to account for interracial differences. Wallace initially accepted the action of sexual selection in nature but, in time, rejected it. Even in May 1864 he made clear to Darwin his opinion that sexual selection had not been very important in human evolution.[15] Wallace also received

[11] The discussion is printed in the *J. Anthropol. Soc. London*, 1864, *2*: clxx–clxxxvii.

[12] Wallace, "Origin of Human Races," pp. clxvi–clxvii; *My Life*, Vol. I, pp. 419–420. Marchant, *Wallace*, Vol. I, pp. 157–158.

[13] Marchant, *Wallace*, Vol. I, pp. 152–155. Darwin also praised Wallace in a letter to J. D. Hooker. Greene is clearly exaggerating when he claims Darwin was "disturbed" by Wallace's views. Darwin had qualms about minor points only and made some suggestions which Wallace agreed to. F. Darwin and A. C. Seward, eds., *More Letters of Charles Darwin* (London: John Murray, 1903), Vol. II, pp. 31–32. Marchant, *Wallace*, Vol. I, pp. 155–158. Greene, *Death of Adam*, pp. 315–316. Kentwood D. Wells, "William Charles Wells and the Races of Man," *Isis*, 1973, *64*: 223–224. Turner, "Between Science and Religion," p. 89.

[14] Marchant, *Wallace*, Vol. I, pp. 155, 158.

[15] *Ibid.*, pp. 154–155, 157. Vorzimmer, *Darwin*, pp. 191–202. Wallace, *My Life*, Vol. II, pp. 17–20.

praise from Lyell and Spencer for his leading idea, though Lyell had doubts about Wallace's inferred Miocene antiquity for man.[16] So matters stood—so Wallace's friends thought—for five years.

Higher intelligences and the inadequacy of natural selection, 1869–1870

In March 1869 Wallace must have given Darwin a hint that his mind had changed about man. For on March 27, 1869, Darwin wrote to Wallace, "I shall be intensely curious to read the *Quarterly*: I hope you have not murdered too completely your own and my child."[17] The *Quarterly* referred to was the forthcoming April 1869 issue in which Wallace reviewed new editions of two books by Lyell.[18] At the end of his review Wallace took the opportunity to reveal his new thoughts on the origin of man. They were presented as ostensibly the result of a utilitarian analysis of man's unique features. Darwin had emphasized in the *Origin* that natural selection was a principle of utility. Natural selection could not preserve a harmful structure; indeed, the presence of such a structure in a species would be "fatal" to his theory. Furthermore, natural selection was a principle of present utility and relative perfection only; it could not provide for future use. A structure was not preserved by natural selection because it would be valuable in future generations if in the present it was valueless. In addition, there could be no accumulation of favorable variations by natural selection to provide a more efficient (perfect) structure if a less efficient structure was sufficient in the present struggle for existence.[19]

Wallace, in his review, considered some facts about prehistoric and savage races of man which natural selection as a principle of present utility and relative perfection was unable to explain. One such fact, characteristic of all animals (including man), was consciousness. The origin of consciousness was inexplicable to Wallace by evolution, much less natural selection. Wallace delayed treatment of this problem until 1870. The origin of man's intellectual and moral nature appeared to be as unique an event as the origin of consciousness and therefore equally inexplicable by natural selection. The development of this nature had been the direct result of the development of man's brain. Wallace's belief in phrenology left no doubt about this relationship.

Thus the large brain size in prehistoric and savage races could not be explained by natural selection, nor could three other physical features: the hand, the external form, and the organs of speech. Natural selection could not account for these four physical features, because each one was present in prehistoric and savage races in a more highly developed state than was required in the struggle for existence. As a principle of relative perfection only, natural selection could not explain these highly developed states. At least one feature (the hairlessness of man's external form) was actually

[16] Marchant, *Wallace*, Vol. I, p. 158 and Vol. II, pp. 18–19. Wallace, *My Life*, Vol. I, pp. 418–419. Eiseley has noted the very favorable reception accorded Wallace's new idea by Spencer, C. Wright, J. McCosh, E. S. Morse, E. R. Lankester, and J. Fiske (*Darwin's Century*, p. 313).

[17] Marchant, *Wallace*, Vol. I, p. 241.

[18] A. R. Wallace, "Geological Climates and the Origin of Species," *London Quarterly Review* (American ed.), 1869, *126*:187–205. R. Hooykaas, *The Principle of Uniformity in Geology, Biology and Theology* (Leiden: E. J. Brill, 1963), pp. 115–117, 174. W. Irvine, *Apes, Angels, & Victorians* (New York: McGraw-Hill, 1955), pp. 186–187. Eiseley, *Darwin's Century*, pp. 310–312. De Beer, *Darwin: A Scientific Biography*, p. 214. George, *Biologist Philosopher*, pp. 242–246. Greene, *Death of Adam*, p. 316. The two books by Lyell were *Principles of Geology* (10th ed.) and *Elements of Geology* (6th ed.).

[19] Charles Darwin, *On the Origin of Species* (London: John Murray, 1859), pp. 199–201, 204.

harmful to man and therefore could not have resulted from natural selection. Furthermore, all four features were useful to civilized man; in fact they were prerequisites for the civilized state. Therefore their presence in early man, to whom they were useless in such highly developed states, bespoke provision for future use and thus could not have been due to natural selection.

1. Brain size

Wallace believed that the brain was the organ of mind and that brain size was a sound measure of intellectual and moral capacities. Prehistoric and savage races possessed brains almost as large as those of civilized man. But the higher moral faculties, pure intellect, and refined emotions—all made possible by such a large brain—were useless to prehistoric and savage races because they bore no relation to wants, desires, or welfare. Nevertheless, these capabilities were latent because occasionally they were manifested by savage men; given enough time in civilization, these latent capabilities would become patent in living savages. But taking into account the savage's actual needs, Wallace felt natural selection could have provided a brain just "a little superior to that of an ape." There could be no doubt that man's highly developed intellectual and moral nature was useful, indeed necessary, to him in the civilized state, and so the possession by the prehistoric and savage races of the same nature, of no use to them, meant there had been provision for mankind's future. Such provision was impossible by the action of natural selection.

2. Hand

Just as the prehistoric and savage races had the same large brain as civilized man, they had the same perfect hand. But the savage had no need for such perfection and was incapable of fully using his perfect hand, so natural selection could not have provided man with his hand. (Wallace recognized that this argument extended beyond the savage to the Quadrumana. Apes, for example, also possessed a hand more perfect than required.) Since man's arts and sciences ultimately depended on his marvelous hand, and since these were among the chief characteristics of civilization, civilization required man's hand. Again provision for civilized man's future was apparent in the hand of the prehistoric and savage races of man.

3. External form

Wallace envisioned five aspects of external form which could not be explained by natural selection: erect posture; delicate, expressive features; marvelous beauty of form; symmetry of form; and smooth, naked skin. All were useless to prehistoric and savage man—the last positively harmful. Natural selection, consequently, could account for none. But all were essential for civilized existence. Expressiveness of face and beauty of form were essential for civilized man's refined emotions and aesthetic ideas. Naked skin, though harmful to prehistoric and savage man, was most useful to civilized man: it stimulated his inventive and constructive faculties to devise clothing and shelter and helped develop feelings of modesty and thereby contributed to man's moral nature. There was a clear sign of provision for the future in man's external form.

4. Organs of speech

Wallace had no doubt that savages were as vocally able as higher races, but the lowest savage had no use for speech. Needless to say, civilization depended absolutely on man's power of speech. The conclusion was the same as before: natural selection could not account for the presence of the organs of speech.

If natural selection failed so completely to explain these four physical attributes of man, what explanation did exist for them? Wallace's answer was a startling one: there existed some power guiding the action of the "great laws of organic development" in definite directions, for special ends. Man's own guidance of nature for his own ends was a model for this power. Wheat, the seedless banana, breadfruit, the Guernsey milk cow, and the London dray horse—all products of artificial selection—were so like the unaided productions of nature that Wallace felt sure some

> . . . being who had mastered the laws of development of organic forms through past ages [would refuse] to believe that any new power had been concerned in their production and scornfully [reject] the theory that in these few cases a distinct intelligence had directed the action of the laws of variation, multiplication, and survival for his own purposes.

Continuity with respect to effects would be observed despite a discontinuity with respect to causes. Wallace asserted that a Higher Intelligence guiding the laws of organic development for nobler ends in the case of human development was in perfect harmony with science. This was especially so because a strict utilitarian analysis of certain features of man had shown the inadequacy of natural selection to explain them.[20]

Darwin's response was rapid and incredulous:

> If you had not told me I should have thought that they had been added by someone else. As you expected, I differ grievously from you, and I am very sorry for it. I can see no necessity for calling in an additional and proximate cause in regard to Man.

Darwin marked his copy of the review with "a triply underlined 'No' and with a shower of notes of exclamation." He also wrote quickly to Lyell of his disappointment in Wallace.[21]

Wallace understood such a response, for he felt he himself would have reacted similarly to such ideas a few years before. But he was not swayed by Darwin's adverse opinion. In what he later called the most extreme statement of his position, Wallace wrote to Lyell (April 28, 1869):

> It seems to me that if we once admit the necessity of *any* action beyond 'natural selection' in developing man, we have no reason whatever for confining that agency to his brain. On the mere doctrine of chances it seems to me in the highest degree improbable that so many points of structure, all tending to favour his mental development, should concur in man alone of all animals. If the erect posture, the freedom of the anterior limbs from purposes of locomotion, the powerful and opposable thumb, the naked skin, the great symmetry of form, the perfect organs of speech, and, in his mental faculties, calculation of numbers, ideas of symmetry, of justice, of abstract reasoning, of the infinite, of a future state, and many others, cannot be shown to be each and all *useful* to man in the very lowest state of civilization—how are we to explain their co-existence in him alone of the whole series of organized beings? Years ago I saw in London a bushman boy and girl, and the girl played very nicely on the piano. Blind Tom, the half-idiot negro slave, had a 'musical ear' or brain, superior, perhaps, to that of the best living musicians.

---

[20] Wallace, "Geological Climates," pp. 204–205.

[21] Marchant, *Wallace*, Vol. I, pp. 240, 243. F. Darwin, ed., *The Life and Letters of Charles Darwin* (London: John Murray, 1887), Vol. III, p. 117.

Unless Darwin can show me how this latent musical faculty in the lowest races can have been developed through *survival* of the fittest, can have been of *use* to the individual or the race, so as to cause those who possessed it in a fractionally greater degree than others to win in the struggle for life, I must believe that some other power (than natural selection) caused that development. It seems to me that the *onus probandi* will lie with those who maintain that man, body and mind, could have been developed from a quadrumanous animal by 'natural selection.'[22]

Wallace's transformation was discussed at the summer meeting of the BAAS and reported with approval in the religious popular press.[23] In the following year (1870) Wallace elaborated on his position in two essays inserted at the conclusion of a collection of ten essays.[24] One of Wallace's reasons for publishing the collection was his knowledge that Darwin's *Descent* was about to appear (1871) and that on certain matters to be discussed in the *Descent* he differed from Darwin. The first of the two essays ("IX. The Development of Human Races under the Law of Natural Selection") was a reprint of the 1864 paper, for the most part unchanged. The significant change appeared at the end of the essay. In the original version Wallace had concluded with a utopian vision of the future of mankind as natural selection continued to act on man's intellectual and moral nature to produce finer and finer men:

While his external form will probably ever remain unchanged, except in the development of that perfect beauty which results from a healthy and well organized body, refined and ennobled by the highest intellectual faculties and sympathetic emotions, his mental constitution may continue to advance and improve till the world is again inhabited by a single homogeneous race, no individual of which will be inferior to the noblest specimens of existing humanity. Each one will then work out his own happiness in relation to that of his fellows; perfect freedom of action will be maintained, since the well balanced moral faculties will never permit any one to transgress on the equal freedom of others; restrictive laws will not be wanted, for each man will be guided by the best of laws; a thorough appreciation of the rights, and a perfect sympathy with the feelings, of all about him; compulsory government will have died away as unnecessary (for every man will know how to govern himself), and will be replaced by voluntary associations for all beneficial public purposes; the passions and animal propensities will be restrained within those limits which most conduce to happiness; and mankind will have at length discovered that it was only required of them to develope the capacities of their higher nature, in order to convert this earth, which had so long been the theatre of their unbridled passions, and the scene of unimaginable misery, into as bright a paradise as ever haunted the dreams of seer or poet.[25]

---

[22] Marchant, *Wallace*, Vol. I, pp. 243–244. Wallace, *My Life*, Vol. I, pp. 427–428. Lyell was pleased with Wallace's new position, even though he was not particularly impressed with Wallace's argument against natural selection from bodily structures. He wrote to Darwin: "I rather hail Wallace's suggestion that there may be a Supreme Will and Power which may not abdicate its function of interference but may guide the forces and laws of Nature." K. Lyell, ed., *Life, Letters and Journal of Sir Charles Lyell* (London: John Murray, 1881), Vol. II, p. 442. Hooykaas, *Principles of Uniformity*, p. 117.

[23] A. Ellegard, "Darwin and the General Reader," *Acta Universitatis Gothoburgensis*, 1958, 7:84.

[24] A. R. Wallace, *Contributions to the Theory of Natural Selection* (London:Macmillan, 1870). Reprinted in A. R. Wallace, *Natural Selection and Tropical Nature: Essays on Descriptive and Theoretical Biology* (London:Macmillan, 1891).

[25] Wallace, "Origin of Human Races," pp. clxix–clxx. See "Mr. Wallace on Natural Selection Applied to Anthropology," *The Anthropological Review*, 1867, *5*:103–105.

By 1870 Wallace was doubtful about natural selection's ability to produce such a future. The mediocre were, after all, the ones who reproduced most prolifically in civilized nations despite the fact that there was an indubitable advance, "on the whole a steady and a permanent one—both in the influence on public opinion of a high morality, and in the general desire for intellectual elevation." Wallace was led to invoke an

> . . . inherent progressive power of those glorious qualities which raise us so immeasurably above our fellow animals, and at the same time afford us the surest proof that there are other and higher existences than ourselves, from whom these qualities may have been derived, and towards whom we may be ever tending.[26]

The only other relevant change in the essay was Wallace's inclusion of the words "from some unknown cause" to explain the development of man's mind from its near-animal condition to the point at which it began to shield man's body from natural selection. Therefore this essay in its new form was contradictory. It still included passages describing natural selection's accumulation of slight variations in man's intellectual and moral nature leading to ever-higher human types. But in its final paragraph it referred to an inherent progressive power of development in man's intellectual and moral nature handed down from on high. With such an inherent power, man's intellectual and moral nature was independent of external conditions and the "chance" appearance of favorable variations. Therefore it was independent of and inexplicable by natural selection.

The final essay ("X. The Limits of Natural Selection as Applied to Man") was newly written, and so it lacked the contradictions of the previous essay. In the review of Lyell's books Wallace had mentioned two fundamentally new aspects of life which natural selection could not explain. He had discussed only the origin of man in the 1869 review. In this 1870 essay he discussed both the origin of man and the origin of consciousness. Whereas Wallace had laid greatest stress in the review on physical structures of man which could not be accounted for by natural selection, he chose in the essay to emphasize man's higher intellectual and moral nature and the problems it posed for natural selection. But Wallace still believed there were physical structures, aside from the large brain which was the necessary substratum of man's intellectual and moral nature, which were not due to natural selection. Man's hairlessness was the physical feature most strongly indicative of the action of a power other than natural selection, because hairlessness was actually harmful to prehistoric and savage man. Hair in mammals protected against severe climates and was most plentiful on the back, yet the most hairless part of man's body was his back. The savage coped with his climate by covering his back, though he might not cover any other part of his body. The unmistakable conclusion Wallace drew was that the savage "missed" his hairy back and that his naked skin was detrimental to him.

Wallace could imagine some objections to this argument, but he dismissed them all. Perhaps man did not really need a hairy back because of his erect posture. Aside from leaving unexplained why man then covered his back, the objection failed to consider man's stooped posture, which exposed the back to the elements. Perhaps hair had been useful but had nevertheless been eliminated because it had been correlated

[26] Wallace, *Contributions*, pp. 330–331.

with a very harmful structure which natural selection had eliminated. "Correlation of growth" was one of Darwin's ways to explain the persistence of useless structures. They were physiologically linked to useful ones and thereby preserved by natural selection. Wallace in imagining this objection considered the correlation of useful and very harmful structures with the resultant elimination of both. But Wallace found it hard to believe that in man alone a correlation between hair and a harmful structure had arisen. Furthermore, even if hair had been eliminated because of correlation of growth, why did not one observe reversion to a hairy condition in colder climates once the harmful structure had been eliminated?[27]

Certain mental and moral features of man were even more difficult to account for by natural selection. Wallace selected five: mathematical ability, ability to form abstract ideas, ability to perform complex trains of reasoning, aesthetic qualities, and moral qualities. In all cases savages possessed latent capabilities, but in no case did their needs require these capabilities. In only rare instances were these capabilities ever used. The savage did not need to form or use abstract ideas, because his language contained no words for them. The savage did not reason on any subject not appealing immediately to the senses; nor did he need to foresee beyond the simplest necessities. Six years earlier Wallace had emphasized the shielding effects of man's mind. He did not have to develop larger claws or teeth to cope with a larger prey animal; he could adapt by means of his intellect—by constructing a better spear. There was no denying that savages made and used weapons, but, in 1870, Wallace wondered if they "exhibited more mind in using them than do many lower animals." The savage seemed to function generally on the level of animal intellect. The jaguar was as ingenious and thoughtful in the capture of fish as any savage. Various behavioral traits in the wolf, jackal, fox, antelope, monkey, field mouse, beaver, and orang-utan were also indicative of as much "care and forethought bestowed by many savages in similar circumstances."[28]

There seemed to be simply no question about the uselessness of many latent mental faculties in savages. There was no use for

> . . . the capacity to form ideal conceptions of space and time, of eternity and infinity—the capacity for intense artistic feelings of pleasure, in form, colour, and composition—and for those abstract notions of form and number which render geometry and arithmetic possible.

Wallace was even doubtful that civilized man had fully employed these capabilities, so they appeared more appropriate to civilized man's future, not only the future of prehistoric man. Lastly, man's moral sense—his conscience—was utterly inexplicable on grounds of utility. The *practice* of honesty might be understandable on such grounds. But man's moral sense included the feeling of sanctity for such things as honesty. How could this feeling arise from considerations of utility? In the case of honesty it was even hard to argue that its practice resulted from its utility.

> The utilitarian sanction for truthfulness is by no means very powerful or universal. Few laws enforce it. No very severe reprobation follows untruthfulness. In all ages and countries, falsehood has been thought allowable in love, and laudable in war; while, at the

[27] *Ibid.*, pp. 344–349. Vorzimmer, *Darwin*, p. 214. Eiseley, *Darwin's Century*, p. 314.

[28] Wallace, *Contributions*, pp. 340–343.

> present day it is held to be venial by the majority of mankind, in trade, commerce, and speculation. A certain amount of untruthfulness is a necessary part of politeness in the east and west alike, while even severe moralists have held a lie justifiable, to elude an enemy or prevent a crime.

Wallace's conclusion was that a feeling of right or wrong in man was antecedent to any experience of utility. This feeling was then attached to certain "acts of universal utility or self-sacrifice." Therefore such a feeling, preceding experience of utility, could not be accounted for by natural selection.[29]

A consideration of all these facts led inescapably to the conclusion of the 1869 review that some higher intelligences had been necessary for man's development. The most objectionable aspect to Wallace of this inescapable conclusion was that the laws governing the rest of the universe were then insufficient to produce the "ultimate aim and outcome of all organized existence—intellectual, ever-advancing, spiritual man." The solution, Wallace surmised, lay in the probability that these laws were also under "the controlling action of such higher intelligences."[30]

The remainder of the essay dealt with the origin of consciousness. Wallace took issue with Huxley's belief that thoughts were the result of molecular change in protoplasm. If molecules themselves lacked consciousness, then complex arrangements of them could not produce consciousness. It is interesting in this regard to note Wallace's belief that the origin of life could be explained by complex arrangements of lifeless molecules. The difference between the two cases lay in our ability to conceive a transition from inert to vital by means of "a specific combination and co-ordination of the matter and the forces that compose the universe . . ." as contrasted to our inability to conceive such a transition from unconscious to conscious. (Wallace did not insist, though, that life had arisen in a purely physical manner; see the discussion below.) There had to be conscious beings "outside of, and independent of" matter. The addition of such beings to unconscious matter resulted in consciousness. Wallace philosophized further. Matter was an impossibility; it was really just a manifestation of force. Force, in turn, could be of two sorts: natural force and will force. The latter was apparent in the power of the will to direct natural forces in the body. A power to direct implied the exertion of force by the will. Wallace speculated that all force was really will force. If so, the universe was the will of higher intelligences or one Supreme Intelligence.[31]

Several months before the essays were published, Wallace informed Darwin of the nature of their contents. Darwin began his lament before he had a chance to read. "But I groan over Man—you write like a metamorphosed (in retrograde direction) naturalist, and you the author of the best paper that ever appeared in the *Anthropological Review*! Eheu! Eheu! Eheu!—Your miserable friend," and several weeks later, "I must add that I have just re-read your article in the *Anthropological Review* and *I defy* you to upset your own doctrine.[32]

[29] *Ibid.*, pp. 351–354.

[30] *Ibid.*, pp. 359–360.

[31] *Ibid.*, pp. 362, 365–368, 372 A–C. Pearson considered Wallace's ideas a "singularly feeble contribution." K. Pearson, *The Grammar of Science* (London: J. M. Dent, 1937), p. 342.

[32] Marchant, *Wallace*, Vol. I, pp. 250–251.

Wallace's critics

Instead of attempting to be complete in my survey of the critical reaction of evolutionists to Wallace's new ideas on the origin of man, I will offer a representative sampling of the nature and tone of the responses. The reviewers of Wallace's *Contributions* were very hostile to the concluding essay. Several did not bother to dispute Wallace's facts; they simply rejected outright his repugnant conclusions. The Saturday Reviewer, in fact, almost admitted that natural selection was not "a universal solvent for all the mysteries of organized being" and included among the mysteries "the stages of man's intellectual and moral progress." Nevertheless, the introduction of "some occult or spiritual agency or force in nature and man, prior and superior to all law, and exterior to the unity of cosmical order" was "parting company with science."[33]

But other reviewers did criticize Wallace's facts as well as his conclusions. They denied his claim that nakedness was harmful. Both E. Claparède and A. Dohrn suggested the possibility that man had begun clothing himself before he lost his hair. Thereafter hair had become useless and was lost. Darwin in his *Descent* even tried to find utility in hairlessness: hair was harmful in the tropics where man originated, or possibly hair was harmful because it hosted ticks and other infestations. But eventually Darwin did acknowledge the strength of Wallace's case for the harmfulness of man's naked skin and finally settled on sexual selection as the cause of hairlessness, especially in females. C. Wright, contrary to Claparède and Dohrn, agreed with Wallace that nakedness preceded clothing and admitted the uselessness of man's hairless skin, but he felt this was compatible with natural selection. Wallace had overlooked the possible correlation of hairlessness with brain size and the resultant preservation of both because of the great value of increased brain size.[34]

Huxley and Darwin differed from Wallace as to the value of a large brain to prehistoric and savage races. Huxley even quoted from another essay by Wallace in the *Contributions* to show the great mental challenges in a savage's daily life. He concluded that "in complexity and difficulty . . . the intellectual labor of a 'good hunter or warrior' considerably exceeds that of an ordinary Englishman." Wright did not deny that the savage possessed many unneeded and unused latent mental capabilities, but he wondered if the mere possession of language did not require the large brain of the savage. Dohrn's critique was rather ironic. To demonstrate the utility of a large brain to a savage, and thereby make possible its origin by natural selection, Dohrn invoked one of the supplementary hypotheses to natural selection—the inherited effect of the use and disuse of parts. He argued that if the brain were not used, it would degenerate. Since it had not degenerated, it must have been used.[35]

Wright disputed Wallace's assertion that the practice of honesty could not have

[33] *Saturday Review*, 1870, *29*:710. *Westminster Review*, 1870, *94*:195. *Nature*, 1870, *2*:472–473.

[34] E. Claparède, "Remarques à propos de l'ouvrage de M. Alfred Russel Wallace sur la Théorie de la Sélection Naturelle," *Bibliothèque Universelle de Genève*, 1870, *38*:186. A. Dohrn, *The Academy*, 1871, *2*:160. Charles Darwin, *The Descent of Man and Selection in Relation to Sex* (New York: Appleton, 1871), Vol. I, pp. 143–144 and Vol. II, pp. 359–362. Darwin, *The Descent* (2nd ed., New York: A. L. Burt, 1874), pp. 64–65. C. Wright, *North American Review*, 1870, *111*: 292. Wallace was not impressed by Darwin's counterarguments. Marchant, *Wallace*, Vol. II, p. 31.

[35] T. H. Huxley, "Mr. Darwin's Critics," *The Contemporary Review*, 1871, *18*:470–471. Darwin, *The Descent*, Vol. I, p. 132. L. Huxley, ed., *Life and Letters of Sir Joseph Dalton Hooker* (New York: Appleton, 1918), Vol. I, p. 130. Wright, *loc. cit.*, pp. 295–297. Dohrn, *loc. cit.*, p. 159.

arisen from experiences of utility. He sought the "uncalculating, uncompromising moral imperative" in what was good for the race, rather than in an individual's own experience of utility. The bee sting was analogous: it was good for the species but disadvantageous (even fatal) to the individual. Nevertheless the bee sting was clearly, to Wright, within the province of natural selection. Darwin's explanation of man's moral nature followed the same lines. Considerations of utility to the group accounted for its origin. Man's moral nature ultimately derived from his social instinct. Acts in the best interest of the tribe were approved and held moral, whereas acts contrary to that interest were disapproved and held immoral. Man's conscience arose from the mental struggle which ensued when a self-beneficial act was perceived as injurious to the tribe.[36]

Having discussed Wallace's facts and in their own opinions refuted them, these reviewers proceeded to reject his conclusions. But a closer examination of their critiques reveals that they were not entirely satisfied with their own arguments. The next resort was *reductio ad absurdum* and ridicule. Claparède marvelled at Wallace's inability to explain a hairless man by natural selection, since he could derive by natural selection a *hairy* mammal and a *feathery* bird from a *scaly* reptile. Furthermore, did Wallace intend to explain the hairless mammals (elephant, rhinoceros, hippopotamus, and whale) by some superior force rather than natural selection? Claparède also wondered if Wallace assumed that a higher intelligence was required to produce the singing voice of male birds, since he considered one necessary to produce the musical voices of men and especially women? Was it not more probable that sexual selection had operated in both instances: by female choice in birds and *vice versa* in man? Or as an analogy to the presumably unused but highly developed brain of the savage, Claparède pointed to the well-developed larynx of nonsinging birds and asked if Wallace intended to explain this larynx by the action of a superior force providing for future, singing birds. Was it not more reasonable to conclude that the nonsinging birds had once been singers, had lost the singing habit, but retained the larynx? Similarly, could not the savage have degenerated from a higher form of man who had used his large brain? Wright also suggested the possibility of degeneration (see below).[37]

Dohrn and Claparède proposed that a large brain did not reflect mental power anyway. A savage could then have a large brain and yet not possess unneeded and unused capabilities. Did Wallace really think an elephant or whale was more intelligent than man? After all, they possessed larger brains than any man. Huxley must not have been entirely sure the savage was so much brighter than the ordinary Englishman, for he admitted the possibility that the savage possessed a bigger brain than his needs required. But he wondered what difference that made, for surely a porpoise possessed a bigger brain than it needed, and more surely a wolf possessed a bigger one than it needed. Would Wallace then conclude that the larger-than-necessary brain of the wolf was provision for the wolf's future as the more intelligent dog?[38]

Critics of natural selection like G. Buckle, A. Bennett, and the Dublin Reviewer wrote, not surprisingly, more favorably of Wallace's new views on the origin of man. But even they had criticisms. Wallace had not gone far enough in questioning the

[36] Wright, *loc. cit.*, pp. 299–300. Darwin, *The Descent*, Vol. I, pp. 68–70.

[37] Claparède, *loc. cit.*, pp. 184–188; Wright, *loc. cit.*, p. 294.

[38] Dohrn, *loc. cit.*, p. 159. Claparède, *loc. cit.*, pp. 187–188. Huxley, *loc. cit.*, pp. 471–472.

efficacy of natural selection in nature. The clergyman evolutionist C. Kingsley felt the same way and wrote to Wallace to tell him so. Only the Edinburgh Reviewer had unqualified praise for Wallace's argument.[39]

Wallace did not succumb to his many critics; he enjoyed controversy. He respected the contributions of Darwin and Wright, but he was repelled by Claparède's technique of ridicule, though Darwin in his *Descent* considered Claparède's critique most effective and able. In a letter to *Nature* Wallace rebutted Claparède convincingly, saying that he had failed to point out that the reptile's scales, the bird's feathers, and the mammal's hair were all adaptive. Similarly, the hairless mammals (excluding man) were not harmed in the least by the absence of hair and were protected in other ways—with a thick skin, for example. The problem for the Darwinian with respect to man was the harmfulness of hairlessness. After all, the mammoth and woolly rhino were proof that in cold climates there was reversion to a hairy condition. But this had not happened in man. In his essay Wallace had already discounted the operation of sexual selection in the origin of man's voice. Savage males did not select mates with any consideration for their voices, and savage females did not choose a mate. In the case of both sexes, sexual selection could not have acted. As for the analogy between the complex larynx capable of song in nonsinging birds and the large, unused brain of savage man, Wallace had a ready answer. There was no evidence that this larynx had ever been developed before birds began to sing. So Claparède could be right that whenever non-singers possessed this larynx it was a sign of degeneracy. But the burden was on Claparède to prove that man's brain had become highly developed only when it had been needed—in other words to prove that prehistoric races had needed and used their large brains even though extant savages did not. Wallace had little difficulty putting his finger squarely on Claparède's main thesis: "the theory of Natural Selection *must* apply equally to man and the rest of Nature, or to neither." Against this all-or-nothing argument, Wallace remarked that Darwin had only claimed that, on analogy, plants and animals had a common origin. Wallace now turned Claparède's logic against him: "Mr. Claparède . . . would, I presume, say that, either all animals or plants must be descended from one common ancestor or, that no two species are thus descended." Despite the cogency of Wallace's refutation, Claparède's aphorism in criticism of Wallace—"man was God's domestic animal"—caught on. In an appendix to the *Contributions* Wallace tried to make clear that intelligences other than God could have been active agents in man's development.[40]

Wallace's only other response to his critics appeared in a review of Darwin's *Descent*. Darwin had argued that man's physical defenselessness had been possible partly because of his large brain (and partly because of his freedom from dangerous enemies—something of a contradiction!). Therefore his brain had gradually increased in size by means of natural selection. Wallace simply could not accept this reasoning.

[39] G. Buckle, "Natural Selection Insufficient to the Development of Man," *Popular Science Review*, 1871, *10*: 14–24. A. W. Bennett, "The Theory of Natural Selection from a Mathematical Point of View," *Nature*, 1870, *3*:32–33. "Evolution and Faith," *Dublin Review*, 1871, N.S. *17*:6. His Wife, ed., *Charles Kingsley: His Letters and Memories of his Life* (London:Macmillan, 1902), Vol. IV, p. 77. "Darwin *on the Descent of Man*," *Edinburgh Review*, 1871, *134*:195–235. Wallace, *My Life*, Vol. II, pp. 62–63.

[40] Marchant, *Wallace*, Vol. I, pp. 253–255, 259 and Vol. II, pp. 31–32. Darwin, *The Descent*, Vol. I, p. 132. A. R. Wallace, "Man and Natural Selection," *Nature*, 1870, *3*:9. Claparède, *loc. cit.*, p. 182. Wallace, *Contributions*, pp. 372–372A. *Nature*, 1870, *2*:472. Pearson, *Grammar of Science*, p. 342.

Many animals had been exposed to the same external dangers as man, yet not one had developed an especially large brain. "Man could have acquired very little of his superiority by a struggle with animals." Darwin had also adopted the leading idea of Wallace's 1864 paper. Brain size increased through intergroup natural selection. But for such selection to occur, two things were required: a large population and a large area. All agreed, according to Wallace, that man's development had been restricted to a small area; otherwise he would have diverged into several species. The real problem was the vast extent of the existing differences between man and animals. In the same tone as he had written to Lyell in 1869 Wallace now concluded,

> His absolute erectness of posture, the completeness of his nudity, the harmonious perfection of his hands, the almost infinite capacities of his brain, constitute a series of correlated advances too great to be accounted for by the struggle for existence of an isolated group of apes in a limited area.[41]

Wallace's views were well known to all by now. In only one public statement between 1871 and *Darwinism* (1889) did he refine his argument; this statement was his 1876 address to the BAAS as President of the Biology Section.[42] In the address, Wallace made one interesting concession to his critics, but in doing so he actually strengthened his case against natural selection. Claparède and Wright had suggested that the large brain of the savage could have reflected a prior stage in man's development during which the brain had been used. The savage no longer used it but still possessed it. This deterioration or degeneration theory of the savage was common at the time, especially since it was repugnant to many that European man could have ever passed through a savage stage.[43] Wallace suddenly took notice of this theory when he read an address by Albert Mott delivered in 1873. Wallace was completely convinced by Mott's evidence, and a part of the BAAS address was devoted to a presentation of the degeneration theory. Mott's evidence included the works of the Easter Islanders and the American Indians. Wallace added the works of the Egyptian pyramid builders on the basis of Piazzi Smyth's numerology. All represented far greater achievements than anything in the civilizations that succeeded them. Though such a theory removed the problem of the savage's failure to use the latent capabilities of his brain, it in no way damaged Wallace's brain-size argument against the adequacy of natural selection in the origin of man.

It was now clear that man had reached a high point in his intellectual and moral development in the "very remote" past. How much time had man had to reach that point? In his 1864 paper Wallace had noted the necessity of a very great antiquity for man if natural selection alone had acted in his development. St. G. Mivart's work had shown that no one ape was any closer to man than any other ape, therefore the line leading to man must have originated before any divergence had occurred among the apes. Since there was evidence of such divergence in the Miocene, thus, in support of

[41] Darwin, *The Descent*, Vol. I, p. 151. A. R. Wallace, *The Academy*, 1871, *2*:182–183. Marchant, *Wallace*, Vol. I, pp. 256–260. Eiseley, *Darwin's Century*, pp. 293–295.

[42] *Report of the Forty-Sixth Meeting of the British Association for the Advancement of Science held at Glasgow in September 1876, Transactions*, 1876:100–119. Reprinted in full in A. R. Wallace, *Tropical Nature and Other Essays* (London: Macmillan, 1878), pp. 249–303. Reprinted in part in Wallace, *Natural Selection and Tropical Nature*, pp. 416–432 and as "Difficulties of Development as Applied to Man," *Popular Science Monthly*, 1876, *10*:60–72. Eiseley, *Darwin's Century*, pp. 310, 312.

[43] Eiseley, *Darwin's Century*, pp. 297–302.

Wallace's 1864 conclusion, man must have begun his development very early, assuming natural selection alone had acted. In 1864 there was no evidence of this antiquity. Twelve years later, despite extensive explorations, the oldest known crania were still the ones discovered about thirty years before and were from a relatively recent geological period. Since they were nearly as large as modern crania, they could not have belonged to early man. These explorations had failed to unearth older remains or missing links. Wallace's ultimate conclusion was that if such evidence was not discovered, "it will be at least a presumption that [man] came into existence at a much later date, and by a much more rapid process of development." If, indeed, man's origin was more recent than the Miocene, a much shorter period existed than most envisaged for man's evolution. In a relatively short period of time, the brain of prehistoric races had to have attained near-modern size. Natural selection was totally incapable of such a feat.[44]

Wallace's last statement of his belief that natural selection was insufficient to explain the origin of man, in an influential publication, appeared in *Darwinism* (1889).[45] Vorzimmer believes the timing of the publication of this work—after Darwin's death in 1882—was not an accident. With one exception Wallace had come to be more a Darwinian selectionist than Darwin. Darwin had gradually bowed before many of the criticisms of his adversaries and accordingly supplemented the action of natural selection with other processes such as sexual selection and the inheritance of acquired characters. Wallace, in contrast, had no use for these supplementary hypotheses and refused to restrict the action of natural selection.[46] The one exception was, of course, man. *Darwinism* presented the curious picture of fourteen chapters of neo-Darwinism followed by a last chapter of anti-Darwinism. In the fifteenth chapter there was one noteworthy change in Wallace's earlier position. Wallace no longer doubted that man's body, except his brain, had developed by natural selection; his disagreement with Darwin was restricted to man's intellectual and moral nature. Wallace was willing to accept Darwin's demonstration of continuity from animal to man with respect to this nature. There were rudiments of man's intellectual and moral nature throughout the animal world, and savages occupied an intermediate position between animals and civilized man. But Wallace denied that continuity proved the operation of natural selection in the origin of man's intellectual and moral nature. In his 1869 review Wallace had used the analogy of artificial selection to demonstrate that continuity of effects did not require a continuity of cause(s). In *Darwinism* he employed a geological analogy. For a long time geologists had considered only two factors modeling the surface of the earth: volcanoes and the elements. Then the action of glaciers became appreciated, an action which was perfectly continuous with that of volcanoes and the elements, but obviously involving a new agency.[47]

In the case of man's intellectual faculties Wallace offered a new and independent proof of natural selection's inability to produce them. Characters developed by

[44] A. Mott, "On the Origin of Savage Life: Opening Address Read before the Literary and Philosophical Society of Liverpool, October 6th, 1873," *The Academy*, 1874, *5*:66. *Report*, pp. 113–118.

[45] A. R. Wallace, *Darwinism: An Exposition of the Theory of Natural Selection with Some of its Applications* (London:Macmillan, 1889). Wallace's last major work, *The World of Life* (1910), in which this belief was fully developed and extended to all of nature, cannot be considered an influential work (see discussion below).

[46] Vorzimmer, *Darwin*, pp. 210–212.

[47] Wallace, *Darwinism*, pp. 461–463.

natural selection were present in all individuals of a species and were relatively invariable. All savages had about the same running speed, bodily strength, acuteness of vision. But there were great inequalities concerning intellectual qualities among individuals. Such inequalities were incompatible with the action of natural selection. Wallace was able to draw support from another great neo-Darwinian of the day, August Weismann. Weismann could not envisage any life-or-death value attached to any of man's talents. He rejected their origin in the inherited effects of use and disuse of parts. He could only explain them as "bye-products" of the human mind. In 1870 Wright had made a similar suggestion. Wallace found this vague conception to be no explanation at all, partly because of his phrenological belief that human talents represented distinct mental faculties which corresponded to definable parts of the brain.[48]

The alternative Wallace proposed in 1889 was a more developed expression of his 1869/1870 conclusions. A spiritual essence in man, capable of progressive development, was the most acceptable explanation for man's intellectual and moral nature. The fact that a new cause had acted twice before in the organic world—in the origin of life and the origin of consciousness—made it very likely that a new cause had acted a third time in the origin of man's intellectual and moral nature. There existed an unseen universe, a world of spirit. The degree of spiritual influx determined the state of living matter—unconscious, conscious, or intellectual. Wallace included in this spiritual world gravitation, cohesion, chemical force, radiant force, and electricity in order to achieve the unity he had earlier missed in nature (1870). The purpose of the world was the "development of the human spirit in association with the human body." In fact the whole universe was a "grand, consistent whole adapted in all its parts to the development of spiritual beings capable of indefinite life and perfectibility."[49]

WALLACE AND SPIRITUALISM

In all his publications concerning the origin of man from 1869 through 1889 Wallace conveyed the impression that the facts adduced from a utilitarian analysis of certain unique features of man were the sole grounds for his conclusion that natural selection was inadequate to explain man's development. Wallace's belief in spiritualism had entered these publications only at the end of each as the explanation for those features of man which the utilitarian analysis had demonstrated natural selection could not account for. Wallace never suggested that his belief in spiritualism had been in any way the cause of his doubts about the efficacy of natural selection in the origin of man. But as early as the 1870s, Anton Dohrn, in his short paper "Englische Kritiker und Anti-kritiker über den Darwinismus," felt that the intense religiosity dominant among the English had ultimately been behind Wallace's divergence from Darwin. In Wallace's case this national religious conviction had been expressed through a belief in spiritualism.[50] I, too, believe that Wallace's spiritualist beliefs were the origin of his

[48] *Ibid.*, pp. 469–472. A. Weismann, *Essays upon Heredity and Kindred Biological Problems* (Oxford: Clarendon, 1891), Vol. I, pp. 96–99. Wright, *loc. cit.*, pp. 297–298. Pearson, *Grammar of Science*, p. 165.

[49] Wallace, *Darwinism*, pp. 473–477.

[50] A. Dohrn, "Englische Kritiker und Anti-kritiker über den Darwinismus," *Das Ausland*, 1871, (Nr. 49):1153–1155. A. R. Wallace, *On Miracles and Modern Spiritualism* (London: James Burns, 1875), p. vi. Wallace, *My Life*, Vol. II, p. 295.

doubts about the ability of natural selection to account for all of man. In the remainder of this paper I will discuss Wallace's involvement with psychical phenomena and spiritualism and examine how it influenced his views on the origin of man in the critical period 1865–1869.

In 1864 Wallace was firmly committed to the action of natural selection alone in the development of man. In 1869 he first expressed publicly his new point of view that natural selection was unable to explain the origin of man and that higher intelligences guiding man's development were required. Something happened between 1864 and 1869 to change his mind: the crucial event was Wallace's conversion to spiritualism. Wallace believed that spiritualism was incompatible with his earlier view (1864) that natural selection alone had acted in the development of man. Indeed, Wallace wrote to Darwin in 1869 that his new view was solely the result of his new belief in spiritualism.

Wallace's first acquaintance with the phenomena of spiritualism occurred in July 1865. Shortly thereafter Wallace began to read extensively in the spiritualist literature. Within a little more than a year (November 1866) he had become convinced of the reality of the phenomena and, not long thereafter, of their spiritualist interpretation. It is important to distinguish Wallace's belief in the reality of psychical phenomena from the spiritualist interpretation of those phenomena, because it was the incompatibility Wallace perceived between spiritualism and the origin of man by means of natural selection alone which led him to his new views on man. The alternative psychic-force interpretation would not necessarily have forced Wallace to reject his 1864 belief in the adequacy of natural selection in the origin of man, for this interpretation held that psychical phenomena resulted from the action of a previously unknown *natural* force. The interpretation was noncommittal about the existence of spirits, or even opposed their existence.[51] On the other hand, according to the spiritualist interpretation, incorporeal intelligences—spirits independent of matter—were the active agents responsible for seance phenomena. Wallace rejected the psychic-force interpretation of psychical phenomena in favor of the spiritualist interpretation once he had become convinced of the reality of spirit communication and spirit manifestation. These phenomena demanded "survival" after bodily death and thereby established the existence of incorporeal intelligences and a duality of "organised spiritual form" and physical body within man. The essence of man was his spirit: "if you leave out the spiritual nature of man you are not studying man at all." Natural selection could not explain this spirit, which possessed an inherent tendency of progressive development and the ability to interact powerfully with mind and ordinary matter "as must revolutionise [materialist] philosophy."

Perhaps the strongest clue to the actual origin of Wallace's new views on man is a letter written to Darwin after Darwin had read Wallace's 1869 review.

[51] W. Crookes, *Researches in the Phenomena of Spiritualism*, reprinted in R. G. Medhurst, coll., *Crookes and the Spirit World* (New York: Taplinger, 1972), pp. 128–129. A second valuable source on Crookes' involvement with spiritualism is E. E. Fournier d'Albe, *The Life of Sir William Crookes* (London: T. Fisher Unwin, 1923), pp. 174–239. It appears that Crookes began his formal investigation of seance phenomena as a spiritualist but emerged from the investigation unconvinced of "survival." Then, late in life, he returned to a belief in spiritualism. Medhurst and Goldney, "Crookes," pp. 127–133. Medhurst, *Crookes*, pp. 227–248.

distinct points as to demonstrate a large amount of truth—both in the principle and in the details—of the method by which they were produced." The phrenologists had correctly diagnosed Wallace's close attention to facts and his readiness to theorize upon them; his fondness for argument and slowness to be convinced; his lack of wit and of mathematical ability; his love of music, but lack of ear and sense of time; his lack of self-confidence, despite some vanity and more ambition; and his possession of concentrative powers but lack of verbal memory. The first two "combined with large *Ideality* and *Wonder* (as indicated by both phrenologists) giving a strong love of the beauties and the mysteries of nature, [furnished] the explanation of [his] whole scientific work and writings."[57] Thus at an early age, prior to his life in science, Wallace had become acquainted with and committed to two heresies, solely on the basis of personal experience.

A year later (1848) Wallace embarked on his first voyage (South America) and was gone four years. In 1854 he was off again (to the Far East) and was gone eight more years. During this second voyage Wallace discovered the principle of natural selection, independently of Darwin. Wallace had continued to practice mesmerism on the Indians of South America during his first voyage, while during the second voyage he began to hear of the phenomena of spiritualism.[58]

The beginnings of modern spiritualism are traced back to upstate New York in 1848 when mysterious rapping noises occurred in the presence of some of the children of the Fox family of Hydesville. By means of code, these raps were able to communicate intelligently and apparently about matters unknown to anyone present. The spiritualist movement spread rapidly in America and within a few years had begun to grow in England. From 1852 to 1854, when Wallace was back in England between voyages, the first American mediums were making visits to England. But spiritualism in England suffered a temporary setback in 1853 and only fully revived in 1859 when Wallace was already in the East. This setback was the work of scientists, especially Faraday, who investigated table turning, one of the most frequently observed physical phenomena of seances. In at least one instance Faraday successfully demonstrated that the motive force for the turning had come from the sitters: unconscious muscular movements from hands placed on the table had produced the turning the sitters had expected and desired. In the same year W. B. Carpenter explained the mental phenomena of seances (for example, intelligent raps) by the medium's ability to detect unconscious muscular movements by the sitters (so-called muscle reading) when the medium was on the right track in an inquiry. Another "exposure" from this year was the clever work of G. H. Lewes, who obtained the rapped reply "Y-E-S" from the medium Mrs. Hayden to the written, but unspoken, question: "Is Mrs. Hayden an impostor?"[59]

Wallace's reaction while in the East to news of spiritualist phenomena was highly skeptical:

[57] *Ibid.*, pp. 174–177. Wallace, *My Life*, Vol. I, pp. 257–262. H. Spencer, *An Autobiography* (New York: Appleton, 1904), Vol. I, pp. 227–231.

[58] Wallace, *My Life*, Vol. II, pp. 275–276; *Miracles*, p. 124.

[59] Gauld, *Founders*, pp. 3–31, 66–87. Michael Faraday, "Experimental Investigations of Table-Turning," *The Athenaeum*, 1853, pp. 801–803. B. Jones, ed., *The Life and Letters of Michael Faraday* (Philadelphia: Lippincott, 1870), Vol. II, pp. 307–308. W. B. Carpenter, "Electrobiology and Mesmerism," *London Q. Rev.*, 1853, *93:* 501–557. G. H. Lewes, "The Rappites Exposed," *The Leader*, 1853, *4*:261–263.

> During my eight years' travels in the East I heard occasionally, through the newspapers, of the strange doings of the spiritualists in America and England, some of which seemed to me too wild and *outré* to be anything but the ravings of madmen. Others, however, appeared to be so well authenticated that I could not at all understand them, but concluded, as most people do at first, that such things *must* be either imposture or delusion.[60]

Shortly after his return to England in 1862 Wallace paid a visit, with his friend Bates, to Herbert Spencer. Both had read Spencer and were hopeful he could give some clue to the origin of life—a problem left unsolved by Darwin. They were quite disappointed by Spencer, who believed the problem

> was too fundamental . . . to even think of solving at present. We did not yet know enough of matter in its essential constitution nor of the various forces of nature; and all [we] could say was that everything pointed to its having been a development out of matter—a phase of that continuous process of evolution by which the whole universe had been brought to its present condition.[61]

One recent biographer of Wallace has concluded that Wallace then turned to mediums and spiritualism for answers to this and other fundamental questions.[62] It is true that eventually he found his answers in spiritualism, but from 1862 to 1865 there is no evidence of any interest by Wallace in spiritualism.

Conversion to spiritualism, 1865–1869

Nevertheless, the things he had heard intrigued him. "Being aware, from my own knowledge of Mesmerism, that there were mysteries connected with the human mind which modern science ignored because it could not explain, I determined to seize the first opportunity on my return home to examine into these matters." The opportunity came in July 1865 when Wallace attended his first seances at the home of a skeptical lawyer friend. Only Wallace, the friend, and the friend's family were present. There was no medium. Wallace's notes from the time described the phenomena:

> Sat with my friend, his wife, and two daughters, at a large loo table, by daylight. In about half an hour some faint motions were perceived, and some faint taps heard. They gradually increased; the taps became very distinct, and the table moved considerably, obliging us all to shift our chairs. Then a curious vibratory motion of the table commenced, almost like the shivering of a living animal. I could feel it up to my elbows. These phenomena were variously repeated for two hours. On trying afterwards, we found the table could not be voluntarily moved in the same manner without a great exertion of force, and we could discover no possible way of producing the taps while our hands were upon the table.

Wallace continued to attend seances with these friends, making "experiments" all the time to elucidate the phenomena. Wallace required one after another to leave the table. The raps continued, until finally Wallace was alone at the table and there were still "two dull taps or blows, as with a fist on the pillar or foot of the table, the vibration of which I could feel as well as hear." Wallace was convinced there had been no deception. His notes ended, "These experiments have satisfied me that there is an unknown power developed from the bodies of a number of persons placed in connection by sitting round a table with all their hands upon it."[63]

[60] Wallace, *My Life*, Vol. II, p. 276; *Miracles*, p. 124.

[61] Wallace, *My Life*, Vol. II, pp. 23–24.

[62] A. Williams-Ellis, *Darwin's Moon* (London: Blackie, 1966), pp. 183–184.

[63] Wallace, *Miracles*, pp. 124–127.

Wallace's first seances with a medium followed just two months later. The medium was Mrs. Marshall, the most renowned British medium from 1858 to 1868. Wallace went with skeptical friends. He now encountered mental as well as physical phenomena: the physical phenomena included table levitation and the movement of other objects in defiance of gravity; the mental phenomena included the communication of names, ages, and other particulars of the relatives of those present at the seance, presumably totally unknown to the medium. The communication was accomplished just as it had been in 1848. The letters of the alphabet were indicated and raps sounded at the correct letters to spell out the name and place of death of one of Wallace's brothers, as well as the name of the last mutual friend to see the brother. Wallace also witnessed a combined physical/mental phenomenon, a prelude to slate writing (see below). A piece of paper and a pencil were placed under the center foot of a table, and after some raps sounded, the paper was removed and a name was found written on it.

Wallace was aware of the possibility of deception and accordingly performed tests to rule it out. In the case of the table movements he investigated the furniture in advance of the seance and assured himself that they were ordinary pieces. Then he placed them wherever he pleased prior to the seance. In the case of the communications by raps, Wallace knew the capability of certain persons to detect slight, even unconscious, movements made by interested parties during the indication of letters of the alphabet. Thus he made every effort to avoid such movements. Since in one instance the medium, through the raps, spelled out the sought-for name backwards, Wallace was convinced she could not have been muscle reading. Finally, in the case of the paper and pencil, Wallace secretly marked the paper, so he knew there had been no substitution of a paper with a name written on it for the blank paper put under the table foot.[64]

About this time Wallace began reading the spiritualist literature extensively and discovered to his surprise that the most reputable persons had become convinced of the reality of seance phenomena. He decided to bring together their testimony in a magazine article entitled "The Scientific Aspect of the Supernatural," which was also privately printed as a pamphlet (1866). Wallace omitted his own experiences from this piece, because he had yet to obtain evidence for the phenomena in his own home. But such evidence was not long in coming. At first the phenomena were not very impressive, but Wallace was still inclined to believe they were not produced by the efforts of those present at the seance.[65] Only in November 1866 did he and friends succeed in finding, among their numbers and in Wallace's own home, someone in whose presence conclusive phenomena occurred. The discovery of this new medium, Miss Nichol, was actually made by Wallace's sister.[66] Raps, table tilting and levitation, the production of musical sounds, and most remarkable of all the production of flowers and fruit (so-called apports) were witnessed. The most elaborate test Wallace ever instigated to rule out deception occurred with Miss Nichol at this time. To ensure that table levita-

[64] Gauld, *Founders*, pp. 71–72. Fodor, *op. cit.*, "Mrs. Mary Marshall." Wallace, *Miracles*, pp. 128–131; *My Life*, Vol. II, p. 277. At a later seance with Mrs. Marshall (1867) Wallace actually spoke to the medium's "spirit controls," the ubiquitous John and Katie King, even while the medium was out of the room. B. Coleman, "Passing Events—The Spread of Spiritualism," *The Spiritual Magazine*, 1867, *II. 2*:349–350. Medhurst and Goldney, "Crookes," pp. 33–34.

[65] Wallace, *Miracles*, pp. 119, 131–132.

[66] F. Sims, "A New Medium," *Spiritual Mag.*, 1867, *II. 2*:49–51.

tion was not the simple result of the medium's foot lifting up the table, Wallace

> . . . prepared the table before [the] second trial without telling any one, by stretching some thin tissue paper between the feet an inch or two from the bottom of the pillar, in such a manner that any attempt to insert the foot must crush and tear the paper. The table rose as before, resisted pressure downwards, as if it was resting on the back of some animal, sunk to the floor, and in a short time rose again, and then dropped suddenly down. [Wallace] now with some anxiety turned up the table, and, to the surprise of all present, showed them the delicate tissue stretched across altogether uninjured! Finding that this test was troublesome as the paper or threads had to be renewed every time, and were liable to be broken accidentally before the experiment began, [Wallace] constructed a cylinder of hoops and laths, covered with canvas. The table was placed within this as in a well, and, as it was about eighteen inches high, it effectually kept feet and ladies' dresses from the table.[67]

Again the table levitated.

The apport of flowers was so marvelous that Wallace preserved the flowers and "attached to them the attestation of all present that they had no share, as far as they knew, in bringing the flowers into the room." Wallace wrote a short description of this December 1866 seance, which was published in the *Spiritual Magazine*: in all there were "15 chrysanthemums, 6 variegated anemones, 4 tulips, 5 orange berried solanums, 6 ferns, of two sorts, 1 *Auricula sinensis*, with 9 flowers—37 stalks in all." The freshness, coldness, and dewiness of the flowers precluded the possibility that they had been brought into the room by any member of the party, for over an hour had passed in a warm room before the production of the flowers. There was only one entrance to the seance room, and there was no sound of an outsider bringing in the flowers at the time of their appearance. There was some "very diffused light" which made the table visible, so any outsider should have been seen as well as heard. None was.[68]

After the experiences with Miss Nichol in late 1866 and the first half of 1867, Wallace was sure of the reality of the phenomena and inclined toward their explanation on spiritualist grounds. Facts had "beaten" him.[69] He felt the same facts would affect fellow scientists similarly. The zeal with which Wallace tried in the next two years to interest scientist friends in observing the phenomena is clear evidence of the completeness of his conversion to spiritualism by 1867. I feel that Wallace's omission of the facts of spiritualism from his discussions of the origin of man was the direct result of his total failure to interest such friends. Thus it is worth examining Wallace's "missionary" efforts.

Wallace sent his 1866 pamphlet to Huxley and invited him to the weekly Friday seance Wallace held with Miss Nichol and his friends (November 1866). He felt sure Huxley would be shocked by Wallace's interest in this "new branch of Anthropology," but he wanted Huxley to see the phenomena for himself "before finally deciding that we are all mad." Huxley replied that he was not shocked; nor was he "disposed to issue a Commission of Lunacy" against Wallace. He even thought Wallace might be right. But Huxley was just not interested.

[67] Wallace, *Miracles*, pp. 133–134, 162–163.

[68] A. R. Wallace, "A Postscript to 'A New Medium'," *Spiritual Mag*., 1867, *II. 2*:52. For other seances Wallace attended with Miss Nichol, see Coleman, *op. cit*., pp. 254–255, 349; Wallace, *My Life*, Vol. II, p. 292; *Miracles*, pp. 134–136, 163–164. Fodor, *op. cit*., "Apports" and "Mrs. Samuel Guppy II."

[69] Wallace, *Miracles*, pp. vii, 125.

> I never cared for gossip in my life, and disembodied gossip, such as these worthy ghosts supply their friends with, is not more interesting to me than any other. As for investigating the matter, I have half-a-dozen investigations of infinitely greater interest to me to which any spare time I may have will be devoted. I give it up for the same reason I abstain from chess—it's too amusing to be fair work, and too hard work to be amusing.

Needless to say, Wallace objected to Huxley's characterization of all seance phenomena as "gossip."

> As for the "gossip" you speak of, I care for it as little as you can do, but what I do feel an intense interest in is the exhibition of *force* where force has been declared *impossible*, and of *intelligence* from a source the very mention of which has been deemed an *absurdity*. Faraday has declared (apropos of this subject) that he who can prove the existence or exertion of force, if but the lifting of a single ounce, by a power not yet recognised by science, will deserve and assuredly receive applause and gratitude. . . . I believe I can now show such a force, and I trust some of the physicists may be found to admit its importance and examine into it.[70]

But Huxley never attended a seance with Wallace.

However, Huxley had previously attended seances with others and after Wallace's (rejected) invitation he attended some seances as well. At some time before 1863 he had investigated the medium Mrs. Hayden whom Lewes had "exposed" in 1853. Despite the fact that Huxley and Lewes were convinced Mrs. Hayden was a fraud, Augustus De Morgan was converted to spiritualism by her.[71] In 1870 William Crookes publicly announced his intention to investigate spiritualism and a year later (1871) began to publish the results of his experiments on the new psychic force. Over twenty years before, Darwin had expressed disgust for such matters as mesmerism and clairvoyance, but after reading one of Crookes' articles he was very perplexed. He wrote: "Nothing is so difficult to decide as where to draw a just line between scepticism and credulity." Now he hoped G. G. Stokes would accept Crookes' offer to jointly investigate the phenomena.[72]

The next year (1872) Darwin's cousin Francis Galton participated in several seances with the mediums D. D. Home and Kate Fox (one of the Fox children in 1848), which were part of Crookes' investigation of spiritualism. Galton approached spiritualism as "rubbish"; however, he was confounded and staggered by the phenomena he observed and was "very disinclined to discredit them." Galton was "convinced the affair [was] no matter of vulgar legerdemain and [believed] it well worth going into. . . ." In addition to the usual seance phenomena Galton had the opportunity to witness something totally new and "confidential." In a letter to Darwin,

[70] Marchant, *Wallace*, Vol. II, pp. 187–188. Wallace, *Miracles*, pp. 214–217. M. Faraday, "Observations on Mental Education," 1854, in E. R. Lankester, ed., *Science and Education* (London: Heinemann, 1917), pp. 39–74.

[71] L. Huxley, ed., *The Life and Letters of Thomas Henry Huxley* (New York: Appleton, 1901), Vol. I, p. 451. S. E. De Morgan, ed., *Memoirs of Augustus De Morgan* (London: Longmans, Green, 1882), pp. 221–222. Wallace, *Miracles*, pp. 83–84. Fodor, *op. cit.*, "Dr. Robert Chambers" and "Mrs. W. R. Hayden." Gauld, *Founders*, pp. 67–68.

[72] Francis Darwin, ed., *The Life and Letters of Charles Darwin*, Vol. I, pp. 373–374. Darwin and Seward, eds., *More Letters of Charles Darwin*, Vol. II, p. 443. Medhurst, *Crookes*, pp. 41–42. The date of Darwin's letter to Lady Derby is probably Oct. 1871 rather than 1874 as suggested by Darwin and Seward. In Jan. 1872 Darwin wrote to Galton, "Have you seen Mr. Crookes? I hope to Heaven you have, as I for one should feel entire confidence in your conclusion." K. Pearson, ed., *The Life, Letters and Labours of Francis Galton* (Cambridge: Cambridge University Press, 1924), Vol. II, p. 147.

Galton wrote

> What will interest you very much, is that Crookes has needles (of some material not yet divulged) which he hangs *in vacuo* in little bulbs of glass. When the finger is *approached* the needle moves, sometimes (?) by attraction, sometimes by repulsion. It is not affected at all when the operator is jaded but it moves most rapidly when he is bright and warm and comfortable after dinner. Now different people have different power over the needle and Miss F. [medium] has extraordinary power. I moved it myself and saw Crookes move it, but I did not see Miss F. (*even* the warmth of the hand cannot radiate through glass). Crookes believes he has hold of quite a grand discovery....

This discovery was Crookes' radiometer. Judging from Galton's comment that the medium had the most power over the needle, it would seem as if Crookes and Galton initially considered the motion of the radiometer needle as evidence of a psychic force. Galton hoped that Darwin would join him in an investigation of psychical phenomena in which just the two of them and the medium Home would be present. The investigation never took place.[73]

Darwin finally did attend a seance in January 1874 with the medium C. Williams, at the home of one of his sons. Lewes and his wife (George Eliot) were present along with Darwin's cousins Hensleigh Wedgwood (a spiritualist) and Galton. After a while Darwin became hot and tired, and so he left to rest before the astounding phenomena took place. All sorts of objects jumped about the room (Galton called it a good seance). Afterwards, Darwin "came downstairs, and saw all the chairs, &c., on the table, which had been lifted over the heads of those sitting round it." Darwin was puzzled; nevertheless he declared, "The Lord have mercy on us all, if we have to believe in such rubbish." About a week later a smaller, more carefully organized seance was held with Williams. Huxley attended incognito, and though he did not publicly expose the medium to the other sitters, he and George Darwin were completely satisfied Williams was a cheat and imposter. Darwin was pleased and relieved by Huxley's account. "Now to my mind an enormous weight of evidence would be requisite to make one believe in anything beyond mere trickery...."[74]

After Huxley's refusal Wallace invited W. B. Carpenter to a seance. Wallace could not guarantee anything on the first visit, and so he hoped Carpenter would come at least six times. Carpenter did come once to a seance with Miss Nichol at which there were some weak raps but nothing else. Carpenter was passive throughout the seance and never returned.[75]

Wallace turned next to Tyndall. In 1864 Tyndall had attended his first seance after Faraday, beseiged by spiritualists since 1853, had refused an invitation but then transferred it to Tyndall. Tyndall's account of the seance is very amusing and, if reliable (Wallace had doubts), illustrates the extreme credulity of some nineteenth-

[73] Pearson, ed., *Life of Galton*, Vol. II, pp. 63–65; also pp. 53, 66. Medhurst and Goldney, "Crookes," p. 95.

[74] F. Darwin, ed., *Life of Darwin*, Vol. III, pp. 186–188. L. Huxley, ed., *Life of Huxley*, Vol. I, pp. 452–456. H. Litchfield, ed., *Emma Darwin, A Century of Family Letters 1792–1896* (London: John Murray, 1915), Vol. II, pp. 216–217. Marchant, *Wallace*, Vol. II, p. 198.

[75] Wallace, *My Life*, Vol. II, p. 278; *Miracles*, p. 225. Carpenter had attended one seance about a year before (Dec. 1865) with Wallace at the home of Mr. Marshman. Samuel Butler, who was also present, thought it was "transparent humbug." He noted Wallace swallowed everything, while Carpenter was properly contemptuous. H. F. Jones, *Samuel Butler: Author of Erewhon* (London: Macmillan, 1919), Vol. I, pp. 126–127, 316–318. Wallace, *My Life*, Vol. II, pp. 296–298.

century seance sitters.[76] Tyndall's response to Wallace's pamphlet was not promising: he read it with "deep disappointment" and continued,

> I see the usual keen powers of your mind displayed in the treatment of this question. But mental power may show itself, whether its material be facts or fictions. It is not lack of logic that I see in your book, but a willingness that I deplore to accept data which are unworthy of your attention. This is frank—is it not?

Tyndall replied to Wallace's seance invitation by asking permission to investigate the phenomena just as he would "in other departments of nature." Wallace gave his permission but asked Tyndall to sit passively for two or three seances with Miss Nichol and *then* apply whatever tests he chose. Tyndall came but ignored Wallace's requests. He insisted on sitting at a distance from the table and joked with the fun-loving Miss Nichol. There were raps, but Tyndall wanted something more remarkable. Nothing more happened and Tyndall, like Carpenter, never returned.[77]

Wallace's last try was Lewes, to whom he also sent his pamphlet. Lewes was very busy and, because of his previous negative experience, incredulous. But he reluctantly agreed to come if he could fully investigate and bring someone like Spencer. As with Tyndall, Wallace agreed to these conditions as long as Lewes sat passively at the first seance. Lewes never came at all.[78]

Wilma George has commented that Wallace's insistence on passivity at the first few seances played into the hands of (fraudulent) mediums who tried to determine the attitudes and critical powers of the seance sitters before engaging in any trickery. Once a medium discovered an investigative nonbeliever at the seance, the phenomena presumably stopped to avoid exposure. It is true that such practices existed among mediums. But Wallace's request was not completely unreasonable. Galton concurred, "I really believe the truth of what they [mediums] allege, that people who come as men of science are usually so disagreeable, opinionated and obstructive and have so little patience, that the seances rarely succeed with them." The refusal of Wallace's scientist friends to abide by his request reflected their a priori skepticism or disbelief more than an objective desire to freely investigate.[79]

In May 1868 Tyndall wrote to the *Pall Mall Gazette* about the medium Home, the most famous medium of modern times. Home had never been exposed. Tyndall claimed that Home had only escaped exposure by avoiding investigations by scientists. In 1861 Faraday had agreed to attend seances with Home. The seances were never held, because, according to Tyndall, Home had refused to participate. Tyndall's charge precipitated a month-long interchange between spiritualists, including Home, and skeptics. Home, in fact, had never been aware of the proposition. Faraday had agreed to investigate, but only if Home would agree to certain conditions such as

> 7. If the effects are miracles, or the work of spirits, does he [Home] admit the utterly contemptible character, both of them and their results, up to the present time, in respect either of yielding information or instruction, or supplying any force or action of the least value to mankind?

---

[76] J. Tyndall, "Science and the 'Spirits'," *Fragments of Science for Unscientific People* (New York: Appleton, 1871), pp. 402–409. Wallace, *Miracles*, pp. 144–145. F. Podmore, *Modern Spiritualism* (London: Methuen, 1902), Vol. II, p. 147. Fodor, *op. cit.*, "Mrs. Newton Crosland."

[77] Wallace, *My Life*, Vol. II, pp. 278–281; *Miracles*, p. 225.

[78] Wallace, *My Life*, Vol. II, p. 281.

[79] George, *Biologist Philosopher*, p. 248. Pearson, ed., *Life of Galton*, Vol. II, pp. 64–65.

Robert Bell, who, through Emerson Tennent, had made the proposal to Faraday, then broke off negotiations since Faraday was not approaching the matter with the proper attitude. Tyndall proceeded to make an offer to investigate Home under the same conditions Faraday had. Lewes eventually joined the correspondence, claiming that mediums evaded investigations by scientists because scientists had successfully demonstrated how tables were turned, raps were sounded, and ropes were untied, in perfectly normal ways. Lewes suggested that Tyndall sit with any medium one time and propose three questions to decide the whole matter.[80]

Wallace was quite upset by the letters from Tyndall and Lewes, because he had invited them to investigate Miss Nichol the preceding year. They, not the medium, had been guilty of evasion. Wallace wrote to the *Pall Mall Gazette* stating that one reputable scientist, Cromwell Varley, had already investigated Home and satisfied himself of the absence of fraud. Seeking to put an end to the correspondence, the editor of the *Pall Mall Gazette* refused to publish Wallace's letter. Wallace promptly castigated Lewes for publishing in a journal which refused critical replies. In the end Tyndall did learn through Wallace of Varley's investigations. During the *Pall Mall Gazette* dispute Tyndall had written to Wallace about the possibility of Varley's performing some tests at a seance. Wallace forwarded Tyndall's letter to Varley but questioned rightly the decisiveness of a single test, positive or negative, as proposed by Tyndall and Lewes. Wallace recognized that no single case was conclusive, but he insisted that many cases had survived scrutiny.

> During the last two years I have witnessed a great variety of phenomena under such varied conditions that each objection as it arose was answered by other phenomena. The further I inquire, and the more I see, the more impossible becomes the theory of imposture or delusion. I *know* that the facts are real natural phenomena, just as certainly as I know any other curious facts in nature.[81]

The reaction of scientists to Crookes' investigations in the early 1870s confirmed Wallace's skepticism about Tyndall's sincerity.

> When Mr. Crookes ... first announced that he was going to investigate so-called spiritual phenomena, many public writers were all approval; for the complaint had long been that men of science were not permitted by mediums to inquire too scrupulously into the facts. One expressed "profound satisfaction that the subject was about to be investigated by a man so well qualified;"—another was "gratified to learn that the matter is now receiving the attention of cool and clear-headed men of recognised position in science;"—while a third declared that "no one could doubt Mr. Crookes' ability to conduct the investigation with rigid philosophical impartiality." But these expressions were evidently insincere—were only meant to apply, in case the result was in accordance with the writers' notions of what it ought to be.... But when the judge, after a patient

[80] The correspondence in the *Pall Mall Gazette* extended from May 5 through May 25, 1868. Letters from Tyndall appeared on May 5, 7, 9, 18, and 25; from Home on May 6 and 11; from others on May 12, 16, 19 (Lewes), 20, 21, and 22 (editorial). Most of the correspondence was reprinted in *Spiritual Mag.*, 1868, *II. 3*:254–228, 325–332, 380–382.

[81] Wallace, *My Life*, Vol. II, pp. 282–283, 291–293. Having received Tyndall's letter via Wallace, Varley wrote to Tyndall describing in detail his 1860 and 1864 seances with Home. Later Varley visited Tyndall and told him that he [Tyndall] threw seance phenomena into confusion as if he were a great magnet. Crookes agreed. *Spiritual Mag.*, 1868, *II. 3*: 273–278. *Report on Spiritualism of the Committee of the London Dialectical Society* (London: Longmans, Green, Reader, and Dyer, 1871), p. 265. Fournier d'Albe, *op. cit.*, pp. 209–210.

> trial lasting several years, decided against them, and their accepted prophet blessed the hated thing as an undoubted truth, their tone changed; and they began to suspect the judge's ability, and to pick holes in the evidence on which he founded his judgment.[82]

Wallace's inability to interest Huxley, Carpenter, Tyndall, and Lewes in seance phenomena put a damper on his missionary zeal, but these failures in no way diminished his own interest. For his abilities as a keen observer and theoretician, plus his scientific reputation, Wallace was invited to join, in January 1869, the Committee of the London Dialectical Society to investigate the phenomena alleged to be spiritual manifestations. The committee also tried to interest Huxley, Carpenter, Tyndall, and Lewes in the investigation but fared no better than Wallace had. Huxley's letter rejecting the invitation of the committee to participate is a classic example of his sarcastic wit, style, and attitude toward spiritualism:

> I regret that I am unable to accept the invitation of the Council of the Dialectical Society to cooperate with a Committee for the investigation of 'Spiritualism'; and for two reasons. . . . In the second place, I take no interest in the subject. The only case of 'Spiritualism' I have had the opportunity of examining into for myself [the mediumship of Mrs. Hayden], was as gross an imposture as ever came under my notice. But supposing the phenomena to be genuine—they do not interest me. If any body would endow me with the faculty of listening to the chatter of old women and curates in the nearest cathedral town, I should decline the privilege, having better things to do. And if the folk in the spiritual world do not talk more wisely and sensibly than their friends report them to do, I put them in the same category. The only good I can see in a demonstration of the truth of 'Spiritualism' is to furnish an additional argument against suicide. Better live a crossing-sweeper than die and be made to talk twaddle by a 'medium' hired at a guinea a seance.

In the report of the committee (1871) part of a paper delivered by Wallace to the entire society on various arguments against the occurrence of miracles was included. Wallace was one of the editors of the report and asked questions at the examination of some witnesses. He also witnessed many seance phenomena under test conditions when no paid mediums were present.[83]

The critical period 1865–1869 ended with Wallace's participation on the committee. Wallace continued to demonstrate a great interest in spiritualism in the years after 1870, and I will conclude my discussion with some of the forms that interest took during the rest of Wallace's life. The years 1865–1869 prove that Wallace's belief in spiritualism was certainly not a quirk of his last years;[84] the decade of the 1870s is further testimony to that fact.

The confirmed spiritualist, 1870–1913

In 1871 Wallace attended his first seance with Home, unquestionably the most celebrated psychical medium of modern spiritualism. The year before, Crookes had begun his investigation of Home and seance phenomena, and Wallace was invited to one of these seances which occurred at the home of Miss Douglas, to whom Wallace had been introduced by R. Chambers in 1869. Since Wallace was the only one present

[82] Wallace, *Miracles*, pp. 174–175. *Cf.* Medhurst, *Crookes*, pp. 35–36.

[83] Wallace, *My Life*, Vol. II, p. 276. *Report on Spiritualism*, pp. vi, 82–90, 183, 210, 225, 227; 229–230, 278–279 (Huxley), 266 (Carpenter), 265 (Tyndall), 230, 263–265 (Lewes). Wallace, *Miracles*, p. 214. Fodor, *op. cit.*, "Dialectical Society."

[84] Eiseley, *Darwin's Century*, p. 296.

who had never sat with Home before, he was given every opportunity to examine the phenomena. While the table levitated, Wallace personally looked underneath with a candle to be sure Home's feet were distant from any part of the table. Similarly, while Home held the top end of an accordion with one hand (placing the other hand in full view on top of the table), Wallace requested the playing of "Home Sweet Home" and observed the accordion playing a few bars of this air by itself, "a shadowy yet defined hand on the keys" at the bottom end of the accordion.[85]

In 1874 Wallace was invited to contribute an article on spiritualism to the *Fortnightly Review*, a periodical founded in the 1860s by Lewes to acquaint the educated public with scientific rationalism.[86] After the publication of "A Defence of Modern Spiritualism" Wallace received invitations to seances with the well-known mediums K. Cook, W. Eglinton, and F. Monck. At these he finally began to experience the more advanced phenomena of spiritualism, including spirit manifestations.[87]

The full-form materialization or manifestation of a spirit was a new seance phenomenon in the 1870s. It so happened that Mrs. Guppy (formerly Miss Nichol) whom Wallace had discovered was the first to introduce such materializations. In order to accumulate enough "energy" for the materialization of a full figure, the medium was confined to a small enclosure, the cabinet. Furthermore, since light presumably interfered with the materialization, the enclosure was kept totally dark. Eventually full figures became visible (materialized) outside the cabinet and often persisted for long periods of time, touching and conversing with the seance sitters.[88] Critics of spiritualism charged that the manifested spirits were the mediums themselves who, under the cover of total darkness, had left the enclosure to roam among the gullible sitters outside. By changing their clothes, among other things, these mediums succeeded in tricking the sitters into believing spirits had materialized from thin air. However, Wallace convinced himself that the spirits he had observed could not possibly have been the mediums in disguise. Cook had pierced ears, but the spirit materializing during her seances did not. Haxby had a shorter foot and a shorter body than his spirit, the Indian Abdullah. In Eglinton's case, a thorough and impromptu search of the medium and the room after the seance failed to produce the clothing of the spirit which Eglinton would have had to wear had the medium been the spirit.

The most amazing materializations were those few which actually occurred in light with both the medium and the materialized figure in full view together. At a seance with Monck in 1877 Wallace did indeed witness a complete materialization in daylight during which Monck and the materialized figure gradually separated to a distance of six feet before the figure was slowly reabsorbed by the medium. Though such a phenomenon no doubt appeared to others to be "midsummer madness," Wallace was convinced that under such circumstances the medium could not have been guilty of fraud. In fact, thirty years later (1907) Wallace testified in court on Monck's behalf.

Archdeacon Colley had participated in 1876 in the exposure of the medium Eglinton who, like Monck, specialized in materializations. In the same year others claimed to have exposed Monck as well, but Colley defended Monck's mediumship. Many years

[85] Marchant, *Wallace*, Vol. II, pp. 189–190. Wallace, *My Life*, Vol. II, pp. 286–287. Medhurst, *Crookes*, pp. 169–171.

[86] A. R. Wallace, "A Defence of Modern Spiritualism," *The Fortnightly Review*, 1874, *N.S.*, *15*:630–657, 785–807. Wallace, *My Life*, Vol. II, p. 295. Gauld, *Founders*, p. 63.

[87] Wallace, *My Life*, Vol. II, pp. 327–331. Marchant, *Wallace*, Vol. II, pp. 193–195.

[88] Gauld, *Founders*, pp. 79–83.

later, in 1906, Colley challenged the magician J. N. Maskelyne to duplicate Monck's materialization in good light. Maskelyne was well known for his claim that he could imitate, by trickery, all seance phenomena. Maskelyne proceeded to make good his claim on stage, but Colley was not satisfied with the performance and refused to pay Maskelyne the one thousand pounds promised in the challenge. In the lawsuit that followed, Wallace testified that he had experienced the same phenomena described by Colley. Besides, he had seen Maskelyne's imitation and characterized it as "perfectly ludicrous" and an "absurd travesty." The court decided in Colley's favor, largely because of Wallace's support.[89] (See Fig. 1.)

Photo. Elliott and Fry.

A PRINCE OF SCIENCE FOR THE PLAINTIFF: DR. ALFRED RUSSEL WALLACE,

Who declares that he had witnessed similar phenomena to those which Archdeacon Colley claims to have witnessed.

Figure 1.* Illustrated London News, *May 4, 1907, p. 673.

Spirit photography was also a new phenomenon of mediumship in the 1870s and a development clearly related to materializations. Thus it is not surprising that the first spirit photograph taken in England, like the first materialization, occurred in the presence of Mrs. Guppy. At this time such photographs were not pictures taken of spirits visible to sitters at seances; rather, when a picture was taken in the presence of a medium, faces (even full forms) that had not been visible appeared on the photographic plates. In 1874 Wallace accompanied Mrs. Guppy to the photographer after raps at a seance alerted Wallace to the possibility of obtaining a spirit photograph of his deceased mother. The raps were prophetic, since Wallace did obtain a spirit photograph which a brother, hostile to spiritualism, agreed resembled their mother. Wallace's involvement with mediums extended beyond seances. Both Wallace's health and his son's health were poor, so Wallace visited a medium for the purpose of healing. He

[89] Trial testimony is reprinted in *The Times*, Apr. 27, 1907. "Archdeacon Colley's Challenge to the Conjurer Maskelyne" and "Archdeacon *versus* Conjurer—a Challenge and a Lawsuit," in *The Annals of Psychical Science*, 1906, *4*:333–335 and 1907, *5*:397–398. During his testimony Wallace also "explained" the (supposed) 1876 exposure of Monck. "Monck was not caught in the act of trickery. Monck was a guest on the occasion, and a demand was made that he should be searched, and he departed through the window" [laughter]. For photographs and sketches of the trial, see *The Illustrated London News*, 1907, *130*:673 and *Proceedings of the National Laboratory of Psychical Research*, 1929, *1*, pt. 2, Plt. 18. For the exposures of Eglinton and Monck and

took the medium's advice, and the conditions of father and son were, in fact, improved.[90]

Wallace continued to lament the refusal of scientists to investigate the phenomena. He did advise the few who approached him. And he vigorously rebutted the published critiques of spiritualism by those friends, especially Carpenter, who had snubbed his overtures in the previous decade (1860s). St. George Mivart's interest in spiritualism apparently arose from Wallace's 1866 pamphlet. When in 1870 Mivart was going to Naples, where Mrs. Guppy was staying, Wallace happily provided a letter of introduction on Mivart's request. Mivart attended three seances at which Mrs. Guppy obliged with her specialty, the production of flowers in a closed room. Mivart was not entirely convinced, but he was favorably inclined. Wallace felt Mivart was to blame for not obtaining more conclusive results, since he had been impatient and had tried to dictate the type of phenomena. Mivart visited Lourdes four years later and wrote to Wallace of his belief in the reality of the miracles that had supposedly occurred there. But Mivart never made public his beliefs. A letter appeared in *Nature* in 1880 signed "M." in which the author wondered if brain vibrations transmitted through the ether might explain the established facts of thought transference, clairvoyance, and mesmerism. Wallace always believed "M." was Mivart.[91]

Wallace's most vigorous defense of spiritualism began in 1874 when his article in the *Fortnightly Review* appeared. In that article Wallace had asserted,

> My position, therefore, is that the phenomena of Spiritualism in their entirety do *not* require further confirmation. They are proved, quite as well as any facts are proved in other sciences; and it is not denial or quibbling that can disprove any of them, but only fresh facts and accurate deductions from those facts. When the opponents of Spiritualism can give a record of their researches approaching in duration and completeness to those of its advocates; and when they can discover and show in detail, either how the phenomena are produced or how the many sane able men here referred to have been deluded into a coincident belief that they have witnessed them; and when they can prove the correctness of their theory by producing a like belief in a body of equally sane and able unbelievers,—then, and not till then, will it be necessary for spiritualists to produce fresh confirmations of facts which are, and always have been, sufficiently real and indisputable to satisfy any honest and persevering inquirer.[92]

Carpenter in 1871 had vehemently attacked Crookes' psychical research. In 1875 in a new edition of his *Mental Physiology* Carpenter repeated his twenty-year-old arguments against the reality of psychical phenomena and specifically criticized Wallace. Wallace responded to the criticisms in the appendix to his book *On Miracles and Modern Spiritualism* (1875), which brought together Wallace's paper on miracles delivered to the Dialectical Society, his "The Scientific Aspect of the Supernatural," and "A Defence of Modern Spiritualism."[93]

---

other seances of Monck attended by Wallace, see Fodor, *op. cit.*, "William Eglinton" and "Rev. Francis Ward Monck"; *Journal of the Society for Psychical Research*, 1889–1890, *4*:143–145; *The Spectator*, Oct. 6, 1877, pp. 1239–1240.

[90] Wallace, *Miracles*, pp. 190–192. Fodor, *op. cit.*, "Frederick A. Hudson." Wallace, *My Life*, Vol. II, p. 397. Marchant, *Wallace*, Vol. II, p. 241. George, *Biologist Philosopher*, p. 157.

[91] Wallace, *My Life*, Vol. II, pp. 300–305, 309–310. M., "A Speculation Regarding the Senses," *Nature*, 1880, *21*:323–324.

[92] Wallace, *Miracles*, pp. 204–205.

[93] W. B. Carpenter, "Spiritualism and its Recent Converts," *London Q. Rev.* (American ed.), 1871, *131*:161–189. W. B. Carpenter, *Principles of Mental Physiology* (New York: Appleton, 1875), pp. 626–627. Wallace, *Miracles*, pp. 31–32, 225–227.

In the summer of 1876 Wallace and Carpenter "clashed" again during the BAAS meeting, on the occasion of the delivery of a paper by William Barrett on thought transference in the mesmeric trance. In view of the great controversy surrounding this paper, it is worth remarking that the young physicist Barrett, once an assistant of Tyndall, was very skeptical of the reality of the more fantastic seance phenomena such as the elongation and levitation of the medium's body. He preferred to attribute reports of such phenomena to the power of suggestion exerted by the medium upon the seance sitters.[94] Barrett had submitted his paper to the Biology Section, of which Wallace was president (see above for his presidential address). The committee of the section chose not to report on the paper but by a small majority referred it to the section's Anthropology Department, of which Wallace was chairman. The committee of the department was also divided, but by a majority of one—Wallace's deciding vote—the reading of the paper was approved. Carpenter arrived during the paper, and the discussion that followed was heated. Lord Rayleigh was especially active in the discussion. He had attended his first seances in 1874 after learning of Crookes' investigation, and though he forever remained undecided as to the reality of psychical phenomena, he was convinced that ridicule of investigations of those phenomena was wrong. Since Wallace had personally witnessed some of the same phenomena as Barrett (such as phreno-mesmerism), he described his experiences with mesmerism many years before and recommended that a committee be appointed to study the phenomena. The BAAS later chose to print only the title of Barrett's paper in its report.[95]

This lively session of the otherwise rather uneventful meeting of the BAAS was vividly reported by the press and was followed by a lengthy correspondence in *The Times*. In addition to Barrett's paper, the mediumship of H. Slade was at the heart of the debate. In the discussion of the paper Rayleigh had cited his own recent experiences with the American medium Slade, whose specialty was slate writing. In one of the more common forms, two clean slates were placed face to face with a bit of pencil in between. Eventually scratching noises were heard, and when the two slates were separated, writing or drawing was found where previously there had been nothing. Rayleigh had brought a professional conjuror to help detect any fraud in the manipulation of the slates. No fraud was detected, and the conjuror's only contribution was the suggestion that the phenomena "might have something to do with electricity."[96]

[94] This "hallucination" explanation of observations at seances had been put forward before. In 1872 Wallace had specifically disputed it in *Nature*, arguing that it was the medium who showed all the signs of being in a trance (thus subject to suggestion), not the seance sitters, who were fully alert. A. R. Wallace, "Ethnology and Spiritualism," *Nature*, 1872, *5*:363–364. Wallace, *Miracles*, pp. 123–124.

[95] *The Times*, Sept. 13, 19 (Wallace), 20, 22, 1876. Marchant, *Wallace*, Vol. II, pp. 195–196. Only seven years later, after the formation of the Society for Psychical Research, did Barrett's paper, "On Some Phenomena Associated with Abnormal Conditions of Mind," appear in print: *Proceedings of the Society for Psychical Research*, 1882–1883, *1*:238–244. Two years before the paper was published Barrett asked Wallace's opinion of the card-guessing experiments done on thought transference by the Sidgwicks. The experiments had yielded mostly negative results. Wallace offered some interesting comments on the difficulties involved in the card-guessing kind of experiment. Marchant, *Wallace*, Vol. II, pp. 200–201.

[96] F. Podmore, *op. cit.*, Vol. II, p. 89. R. J. Strutt, Fourth Baron Rayleigh, *Life of John William Strutt, Third Baron Rayleigh* (augmented ed., Madison:University of Wisconsin Press, 1968), pp. 65–68, 409. Medhurst and Goldney, "Crookes," pp. 90–94.

Just the day before the presentation of Barrett's paper, the biologist E. R. Lankester had attended a seance with Slade. Although Lankester later claimed during his lawsuit that he had gone to the seance unprejudiced, it seems as if he had for some time been intent on exposing the mediums Herne and Williams. Then in the summer of 1876 Slade came to England and immediately created a sensation. At the first seance Lankester was passive and led Slade to believe he was a believer in the phenomena. But a few days later he returned with a colleague for a second seance, at which he was sure he had detected fraud. The next day (Sept. 16) Lankester's letter to *The Times* appeared in which he exposed Slade. He also called Wallace's action on behalf of Barrett's paper "more than questionable" and said the meeting of the BAAS had been thereby "degraded." Wallace replied (Sept. 19), defending his behavior. He then described his own visit to Slade (in August) which had been "completely unlike" what Lankester presumably observed. Lankester eventually decided to bring suit against Slade. The trial, which extended through October, was of great public interest. On the last day Wallace testified for the defense, describing his three seances with Slade (two since the trial had begun!) at which he had found no evidence of imposture. But the court ruled in favor of Lankester against Slade.[97]

Late in 1876 the London Institution invited Carpenter to lecture on spiritualism. The lectures when published set off a very vehement exchange of views between Carpenter and Wallace. During his dispute with Wallace, Carpenter also took up the hatchet again with Crookes. Before the battle finally ran down, over a year after it had begun, the controversy was spread across the pages of five different journals.[98]

Carpenter's argument consisted of three major criticisms. (1) There was a strong a priori improbability about psychical phenomena. In such matters the judgment of sense outweighed the evidence of the senses. (2) There were many known cases of fraud, revealed by detection or confession. Similar to such cases were the imitations by

[97] For the debate over Slade's mediumship see *The Times*, Sept. 16 (Lankester), 18, 19 (Wallace), 20, 21 (Lankester and Slade), and 23 (Slade), 1876. The prosecution of Slade was reported with verbatim transcripts in *The Times* on Oct. 3, 11, 21, 23, 28, 30 (Wallace), and Nov. 1, 1876. C. C. Massey, "Translator's Preface" to J. C. F. Zöllner, *Transcendental Physics* (London: W. H. Harrison, 1880), pp. xxviii–xxxix. Houdini, *A Magician Among the Spirits* (New York: Harper, 1924), pp. 80–82. Gauld, *Founders*, pp. 124–127. A contemporary letter from George Romanes to Darwin about the Slade episode is of interest:

> Lankester seems to have doubled up Slade in fine style. I suppose the latter has always trusted to his customers not liking to resort to violent methods. His defence in the "Times" about the locked slates was unusually weak. 'Once a thief always a thief' applies, I suppose, to his case; but it is hard to understand how Wallace could not have seen him inverting the table on his head. In this we have another of those perplexing contradictions with which the whole subject appears to be teeming. I do hope next winter to settle for myself the simple issue between Ghost *versus* Goose.

E. Romanes, ed., *The Life and Letters of George John Romanes* (London: Longmans, Green, 1896), p. 46.

[98] W. B. Carpenter, "Mesmerism, Odylism, Table-Turning, and Spiritualism Considered Historically and Scientifically," *Fraser's Magazine*, 1877, *N.S. 15*:382–405. W. B. Carpenter, *Mesmerism, Spiritualism, &c. Historically and Scientifically Considered* (New York: Appleton, 1877). A. R. Wallace, *Quarterly Journal of Science*, 1877, *N.S. 7*:391–416. W. B. Carpenter, "Psychological Curiosities of Spiritualism," *Fraser's Mag.*, 1877, *N.S. 16*:541–564, 806. A. R. Wallace, "Psychological Curiosities of Scepticism," *Fraser's Mag.*, 1877, *N.S. 16*:694–706. W. B. Carpenter, *Nature*, 1877, *16*:546–547 and *Nature*, 1877, *17*:8–9, 26–27, 81, 122–123. A. R. Wallace, *Nature*, 1877, *17*:8, 44, 101. W. B. Carpenter, "The Curiosities of Credulity," *The Athenaeum*, Dec. 22, 1877, pp. 814–815 and "The Psychological Curiosities of Credulity," *Athenaeum*, Jan. 26, 1878, p. 122. A. R. Wallace, "The Curiosities of Credulity," *Athenaeum*, Jan. 12, 1878, pp. 54–55 and "The Psychological Curiosities of Credulity," *Athenaeum*, Feb. 2, 1878, p. 157.

two 1876 letters from Romanes to Darwin, which he had been shown during his North American trip (1887), much to Romanes' amazement and chagrin. What, then, Wallace retorted, about the Romanes of "incapacity and absurdity"?[105]

The Society for Psychical Research was founded in 1882, and Wallace was one of its early members. His participation, however, was always very limited. He repeatedly rejected suggestions from William Barrett that he make himself available for the society's presidency, because he feared that his reputation as a "crank" and "faddist" would injure the reputation of the new society. Wallace's reluctance to accept the presidency did not mean he was ashamed of his belief in spiritualism. When once asked "whether he believed that light and proof would come from occultism," he responded, with a smile, "Why are you afraid of the term spiritualism? . . . I am a spiritualist, and I am not in the least frightened of the name!"[106] Wallace became very much concerned about the skepticism of many of the society's members. He was also dissatisfied with the manner in which they conducted their investigations. They treated mediums as if they were on trial and applied their own conditions from the beginning of the investigations; instead they should have treated the mediums with consideration and patiently followed "the advice of the intelligence" working through them. Crookes had proceeded in this latter way and successfully obtained "striking results, under the most stringent conditions and subject to the most varied tests." But members of the Society for Psychical Research had obtained few results using their own methods and had frequently become convinced the mediums were impostors. Wallace was always skeptical of many claims of exposure and, in fact, all of his communications to the society were defenses of the legitimacy of suspected mediums.[107]

Wallace's involvement with spiritualism did momentarily increase during his 1886/1887 trip to the birthplace of spiritualism—America. He was engaged in many discussions of the subject with Oliver Wendell Holmes, William James, Elliott Coues, and others. He attended seances in Boston, Washington D.C., and San Francisco, where he delivered a popular lecture entitled "If a Man die, shall he live again?" The trip led directly to Wallace's last publications on the phenomena of spiritualism—two articles for the Boston *Arena* on recent evidence for apparitions. These were incorporated, four years later, into the third edition of *Miracles and Modern Spiritualism.*[108]

With the exception of the American trip, Wallace attended very few seances after 1880. But in 1896 he was visited by a medium while ill and given a seance. The medium's "controls" made several predictions all of which seemed highly improbable to Wallace because of his poor health. But all three were fulfilled once Wallace's health improved. Needless to say, Wallace could not imagine the possibility of chance fulfillment.

[105] Marchant, *Wallace*, Vol. II, 215. Wallace, *My Life*, Vol. II, pp. 317–326.

[106] Marchant, *Wallace*, Vol. II, pp. 208, 210–211. H. Begbie, "Master Workers XVII. Dr. Alfred Russel Wallace," *The Pall Mall Magazine*, 1904, *34*:76.

[107] Marchant, *Wallace*, Vol. II, p. 204. Wallace, *My Life*, Vol. II, p. 294. *J. Soc. Psychical Res.*, 1887–1888, *3*:273–288, 313–317; 1889–1890, *4*:143–144; 1891–1892, *5*:43; 1893–1894, *6*:33–36; 1899–1900, *9*:22–30, 56–57. *Proc. Soc. Psychical Res.*, 1899, *14*:373.

[108] Wallace, *My Life*, Vol. II, pp. 115, 117, 160, 210, 213, 337–349. George, *Biologist Philosopher*, pp. 235–236. A. R. Wallace, "Psychography in the Presence of Mr. Keeler," *Psychical Review*, 1891, *1*:16–18. *Arena*, 1890–1891, *3*:129–146, 257–274. At this same time (1892), Wallace contributed the article "Spiritualism" to the 10th ed. of *Chambers' Encyclopedia* (Philadelphia: Lippincott, 1892), Vol. IX, pp. 645–649.

> "There's a divinity that shapes our ends, rough-hew them as we will;" and those who have reason to know that spiritual beings can and do influence our thoughts and actions, will see in such directive incidents as these examples of such influences.[109]

Spiritualism was, to Wallace, the science of the spiritual nature of man, but during the second half of his life spiritualism also became his religion. Wallace found in spiritualism much more than an explanation for various human features that natural selection could not account for. But I do not believe Wallace's initial involvement with matters of the human mind, including spiritualism, was motivated by religious sentiment. Unquestionably his readiness to investigate seance phenomena resulted from his experiences with phrenology and mesmerism in the 1840s; he refused to dismiss seance phenomena as a priori impossible, as incredible as they might seem, because of these early experiences. Wallace was emphatic that his acceptance of the reality of psychical phenomena as undisputed facts was the product of his investigation of those phenomena. Yet at first Wallace hesitated to accept the spiritualist interpretation of those facts: there was "no place in [his] fabric of thought" for spirits. However, more facts "beat" him. The reality of spirit communication and spirit manifestation convinced Wallace of the existence of spirits and revealed to him the nature of spirit life. Wallace insisted that a fear of death had not pressured him to accept the spiritualist interpretation which contended that man's spirit survived his bodily death.[110]

Nevertheless, having accepted this spiritualist interpretation because of undeniable facts, Wallace received great solace from it. His belief in spiritualism relieved him

> . . . from the crushing mental burthen imposed upon those who—maintaining that we, in common with the rest of nature, are but products of the blind eternal forces of the universe, and believing also that the time must come when the sun will lose his heat and all life on the earth necessarily cease—have to contemplate a not very distant future in which all this glorious earth—which for untold millions of years has been slowly developing forms of life and beauty to culminate at last in man—shall be as if it had never existed; who are compelled to suppose that all the slow growths of our race struggling towards a higher life, all the agony of martyrs, all the groans of victims, all the struggles for freedom, all the efforts towards justice, all the aspirations for virtue and the well-being of humanity, shall absolutely vanish, and, "like the baseless fabric of a vision, leave not a wrack behind."[111]

Furthermore, the moral teachings of spiritualism were eagerly welcomed by Wallace, since they complemented the ethical code he already adhered to. These teachings were primarily derived from trance-speaking mediums who were presumably in communication with spirits of the dead.[112] Wallace was an extremely high-minded, ethical individual. Clear signs of his deep concern for the plight of his fellow human beings are discernible from his early life. Eventually this concern was expressed in Wallace's ardent advocacy of such social heresies as land nationalization, socialism, and anti-militarism.[113] The Christian concept of the afterlife was unsatisfying to Wallace. He

[109] Wallace, *My Life*, Vol. II, pp. 234, 397–400. Marchant, *Wallace*, Vol. II, pp. 223–224.

[110] Wallace, *Miracles*, pp. vii, 108, 125, 221; *My Life*, Vol. II, pp. 349–350. Crookes' involvement with spiritualism began soon after the death of his brother in 1867. Fournier d'Albe, *op. cit.*, pp. 133–134. It has been suggested to me that the great personal tragedy of his brother Herbert's death in 1851 while both were in South America may have "driven" Wallace to spiritualism.

[111] Wallace, *Darwinism*, pp. 476–477.

[112] Wallace, *Miracles*, pp. 108–118, 213–223.

[113] Wallace, *My Life*, Vol. I, pp. 79–105 and Vol. II, pp. 235–274. George, *Biologist Philosopher*, pp. 219–225.

rejected the meting out of rewards and punishments by an external power. This was an "arbitrary system . . . dependent on stated acts and beliefs only, as set forth by all dogmatic religions." But Wallace found the spiritualist conception in full accord with his humanitarian ethical beliefs as well as his beliefs in "continuity" throughout nature. The happiness or misery of the spirit after death depended entirely on the extent to which an individual had exercised and cultivated his mental and moral faculties while he was alive. The spirit started, after death, from the level of intellectual and moral development attained on earth.[114] There was a ready place for spiritualism in Wallace's personal nature, in addition to Nature as a whole.

Wallace's last three large works all reflected the place of spiritualism in nature. *The Wonderful Century* described the successes and failures of science in the nineteenth century. The failures included science's lack of appreciation for phrenology, mesmerism, and psychical phenomena.[115] Any brief discussion of *Man's Place in the Universe* and *The World of Life* would totally fail to convey the permeating influence of Wallace's belief in spiritualism. In these two scientific works Wallace fully incorporated that belief into the discussion of scientific issues. Consequently these works are interesting, but coming so late in Wallace's life—past his prime—they were not influential.

Both *Man's Place in the Universe* and *The World of Life* sought to demonstrate the existence and constant action, possibly through thought transference, of Mind throughout the universe. On earth this action prepared and provided for man.[116] Wallace was very much impressed with the diversity in each realm of nature. He believed this diversity was only intelligible in terms of the action of a purposeful Mind guiding organic development in preparation for man's existence on earth.[117] In the vegetable realm Wallace considered the many kinds of wood which happened to be "so exactly suited to the needs of civilised man that it [was] almost doubtful if he could have reached civilisation without them." But as far as Wallace could determine, the many qualities of these different kinds of wood had been unnecessary and useless to the plants from which they were derived and to the lower animals that had coexisted with them before the origin of man. In the animal realm Wallace perceived the entire course of vertebrate evolution to be preparation by a guiding Mind for man. He was especially struck by the circumstance that animals which man had eventually domesticated "should have been slowly evolved so as to reach their full development at the very time when [man] became able to profit by them. . . ." A foreseeing Mind had "so

[114] Wallace, *Miracles*, pp. 101, 108–118, 213–223; *My Life*, Vol. I, p. 88. "A Visit to Dr. Alfred Russel Wallace, F. R. S.," *The Bookman*, 1898, *13*:122–123.

[115] Wallace, *The Wonderful Century*, pp. 159–212. But most of the space allotted to failures of science dealt with a new heresy of Wallace, his belief that smallpox vaccination was not only useless but actually the cause of smallpox deaths rather than their cure. *Nature* refused to notice Wallace's book, despite his innumerable contributions in previous years. The editor, Norman Lockyer, was very hostile to several of Wallace's heresies. Lockyer had refused an invitation from Wallace in Oct. 1865 to attend a seance, but he confessed to the physicist and spiritualist Oliver Lodge in 1907 to having personally had a psychic experience. *Ibid.*, pp. 213–324. Marchant, *Wallace*, Vol. II, p. 206. A. J. Meadows, *Science and Controversy: A Biography of Sir Norman Lockyer* (Cambridge, Mass.: M.I.T. Press, 1972), p. 10.

[116] A. R. Wallace, *Man's Place in the Universe* (New York: McClure, Phillips, 1903). A. R. Wallace, *The World of Life* (London: Chapman and Hall, 1910). George, *Biologist Philosopher*, pp. 274–278, 281–284. Marchant, *Wallace*, Vol. II, pp. 89–90, 93–98, 101–102, 121–122, 178–179.

[117] Wallace, *The World of Life*, pp. 278–284, 390–400.

directed and organised . . . life, in all its myriad forms, as, in the far-off future, to provide all that was most essential for the growth and development of man's spiritual nature." Even in the inorganic realm Wallace could discern the action of Mind. Wallace noted that most of the elements were exceedingly rare, and only fourteen were essential for the existence of the earth and life upon it. The remaining elements, then numbering over sixty, appeared to be useless until Wallace considered their relation to man. The ancient metals, for example, had proven to be necessary for civilization. Wallace could only conclude that these elements had been produced specifically for the purpose of promoting man's development.[118]

Wallace and psychical research

Was Wallace a competent investigator of seance phenomena? This final question is worth asking in view of Wallace's insistence that his belief in spiritualism resulted from his own investigation of the phenomena and not, contrary to Dohrn's charges, from "preconceived or theoretical opinions." Indeed Wallace stated ". . . the cardinal maxim of Spiritualism is, that every one must find out the truth for himself." Wallace's advice to a Dr. Edwin Smith was

> If you want to *know* anything about Spiritualism you should experiment yourself with a select party of earnest enquirers—personal friends. When you have thus satisfied yourself of the existence of a considerable range of the physical phenomena and of many of the obscurities and difficulties of the inquiry, you may use the services of public mediums, without the certainty of imputing every little apparent suspicious circumstance to trickery, since you will have seen similar suspicious facts in your private circle where you *knew* there was no trickery.[119]

Wallace's biological work had proven he was a keen observer of nature, and for this reason his acceptance of psychical phenomena baffled fellow evolutionists, who were disbelievers. At the same time, however, Wallace was not, on his own admission, an experienced experimentalist.

Wilma George has stressed Wallace's naïveté and gullibility and, by implication, impugned his observations and investigations:

> No one could convince him that seances were a fraud. If a man said he had seen his dead son it was a fact. If a medium said he could raise spirits it was a fact. Transferring to others his own attribute of innocent generosity, he was deceived by them and by himself. All men conformed to his own standards of integrity.[120]

Though this statement is an exaggeration, there is definitely some truth to it. A long-time friend and neighbor of Wallace remarked that Wallace had an

> . . . apparently unfailing confidence in the goodness of human nature. No man nor woman but he took to be in the main honest and truthful, and no amount of disappointment—not even losses of money and property incurred through this faith in others' virtues—had the effect of altering this mental habit of his.[121]

Contemporary psychical researchers felt that because of his trusting nature Wallace was too credulous in his investigations of spiritualism. Thus after a seance he had attended with Wallace, William James told Josiah Royce, who much later told E. B.

[118] *Ibid.*, pp. 325–326, 280–284, 357–361.
[119] Marchant, *Wallace*, Vol. II, p. 210. Wallace, *Miracles*, p.223; *My Life*, Vol. II, p. 350.
[120] George, *Biologist Philosopher*, p. 244.
[121] Marchant, *Wallace*, Vol. II, p. 109.

Poulton, "It is a curious thing to see Wallace plunging head foremost into a flood which we Americans only allow just to wet our feet."[122]

Wallace believed that he had never met a medium who was a scoundrel. He acknowledged the existence of fraudulent mediums, but he vigorously defended the honesty of *all* the mediums of his own experience. Even though all of these mediums, with the possible exception of the illustrious Home, were exposed as frauds by others, Wallace retained his initial conviction as to their integrity and continued to attend their seances. However, it is important to understand that exposures of mediums were not always clearcut.[123] Thus Wallace was never alone in the defense of a medium. To conclude that all of his observations and investigations were unreliable just because the mediums were, at some time, "exposed" by someone else is unjustified. But the fact that Wallace *never* witnessed fraud at a seance does strongly suggest that at some seances he was the victim of deception.

Once Wallace had become convinced of the reality of the phenomena from his seances with Miss Nichol he became extremely reluctant to reject any reports of phenomena which resembled what he had personally observed and authenticated. "Once demonstrate that genuine mediumship exists in any case, and the whole argument of assuming imposture in every case falls to the ground." Wallace was able to carry this principle to extremes. In order to prove that slate writing was a fraud the admitted conjuror S. J. Davey posed as a slate-writing medium in the mid-1880s. Seance sitters could not detect fraud and wrote convincingly of the legitimacy of the phenomena; however, the psychical researcher R. Hodgson to whom Davey had divulged the means of trickery was present at the same seances and was able to follow the deception. He could demonstrate that the reports of seance sitters were, in fact, very incomplete and inaccurate despite their words of absolute assurance as to what had transpired. Wallace never attended a seance with Davey, but by this time he had observed slate writing with several mediums—his seance with F. Evans during his American tour being one of the most astonishing cases of slate writing on record. Since the phenomena at Davey's seances were so similar to those of his own experience,

[122] Poulton, *loc. cit.*, p. xxix. F. Turner has quoted, from an unpublished letter, F. Myers' low opinion of Wallace's critical judgment at seances:

> His worst credulity as to the good faith of cheating mediums belongs to a separate compartment of his mind—or rather forms a part of his innocent generosity of nature, an unwillingness to believe that anyone will do anything wrong.

However, Myers' colleague R. Hodgson shuddered at Myers' credulity!

> *Myers* (bless his dear soul!) *can* be as sceptical as anyone about some individual person or thing, but if he once gets his sympathies enlisted,—his evidence isn't worth 2 straws. This is part and parcel of his big, poetic divine genuine soul, & he can't help it!

Turner, "Between Science and Religion," p. 107. Gauld, *Founders*, p. 233.

[123] The case of the medium Charles Williams sharply illustrates this point. In the 1870s a number of prominent psychical researchers—Crookes, the young Myers, the Russians A. Aksakov and A. Butlerov—were convinced from personal experience of the genuineness of Williams' mediumship. Some of Romanes' 1876 psychical experiences occurred at a seance with Williams. But at the very same time Huxley was completely satisfied from his observations that Williams was a fraud, and Romanes soon changed his mind when he thought he saw Williams cheating. The medium F. Herne with whom Williams had held joint seances early in his career was openly exposed more than once, thus casting doubt on Williams, too. And Williams himself was even exposed. Yet Williams continued his mediumship undaunted into the twentieth century, and, for what it is worth, Home, who was extremely critical of rival mediums, vouched for Williams. Fodor, *op. cit.*, "Charles Williams" and "Frank Herne." Medhurst and Goldney, "Crookes," pp. 44–47, 49–50, 103.

Wallace refused to accept Davey's assertion that he was just a conjuror and not a medium. Wallace's logic was quite a shock to psychical researchers who concluded from the similarity that the mediums were actually conjurors, rather than that Davey was actually a medium. By this time it was (almost) logically impossible to prove to Wallace that any medium was a fraud.[124]

Eusapia Palladino, unlike Davey, was a professed medium—next to Home perhaps the most renowned medium of the nineteenth century (see Fig. 2.). She had been investigated in 1894 on a small island owned by the physiologist (Nobel prize winner) and spiritualist Charles Richet. The investigators, including Richet, Myers, and the

Figure 2. Table levitation at a seance by the medium Eusapia Palladino (photograph courtesy of Mr. A. H. Wesencraft, Harry Price Library, University of London).

physicist Oliver Lodge, reported positively. But in the subsequent year in England, Palladino failed to convince many of her sitters, which this time included the physicists Rayleigh and J. J. Thomson. In some instances she was detected using fraud. The magician Maskelyne had been present at the English seances and in October 1895 wrote to the *Daily Chronicle* of his own negative conclusions.

[124] *J. Soc. Psychical Res.* 1891–1892, *5*:43 and 1893–1894, *6*:33–47. R. Hodgson, "Mr. Davey's Imitations by Conjuring of Phenomena sometimes attributed to Spirit Agency," *Proc. Soc. Psychical Res.*, 1892, *8*:253–310. Gauld, *Founders*, pp. 204–207. Fodor, *op. cit.*, "William Eglinton" and "Magicians." It should be kept in mind that Davey thought he might possess legitimate telepathic powers, though he forever insisted that the slate writing itself was pure trickery.

Though Wallace had never attended a seance with Palladino, he came to her defense. He had already reported the phenomena observed on Richet's island in *Miracles and Modern Spiritualism*. As for the Cambridge experiments, Palladino could and would cheat, perhaps unconsciously, unless she was properly controlled; thus the burden, according to Wallace, was on her investigators to prevent her from cheating. If given the opportunity, she might cheat, especially if her psychical powers were running low. During the seances Hodgson had purposely relaxed the controls hoping to catch her using fraud. Under these circumstances Wallace was hardly surprised that Palladino, or her "control," had resorted to fraud. Nor did the occurrence of (unconscious) fraud in this situation call into question the validated phenomena when trickery had been excluded.

In essence Wallace was arguing that once satisfied of the legitimacy of a mediumship, no subsequent evidence of fraud could affect the original judgment.[125] These attitudes toward the Davey and Palladino phenomena were not conducive to conclusive investigation. Thus it is not difficult to see why Wallace's personal experiences had little influence on the unconvinced. However, unless one totally rejects the possibility that at least some psychical phenomena are real, it is impossible to dismiss *all* of Wallace's experiences as "mental hallucination" or "insanity."

CONCLUSION

Wallace's interest in matters spiritual, mystical, or religious has generally been vaguely associated with his divergence from Darwin on the origin of man. I have tried to document a causal relationship. In his autobiography Wallace noted four major areas in which he differed from Darwin. The first and, to Wallace, foremost of these areas was the origin of man's intellectual and moral nature. Wallace found natural selection inadequate to explain it and introduced the action of higher intelligences to purposefully guide man's development. To a certain extent Darwin also found natural selection inadequate, but he rejected the addition of any non-natural cause for man. Instead he supplemented the action of natural selection with sexual selection and the inherited effects of the direct action of the environment and of the use and disuse of parts.

I believe this difference of opinion arose as the direct result of Wallace's conversion to spiritualism, which occurred at the same time the divergence over the origin of man did. In a letter to Darwin from this period (1869) Wallace specifically attributed his new view of man to his new belief in the reality of psychical phenomena and their spiritualist interpretation. The letter reflected Wallace's firm belief that there was an incompatibility between the spiritualist interpretation of psychical phenomena and the development of man by means of natural selection alone, which forced him to modify his earlier views. The very existence of an *immaterial* spirit, responsible for "the enormous influence of ideas, principles, and beliefs" in man's life, could not be explained by the struggle for *material* existence. The inherent progressive power of development that characterized man's mind was incompatible with natural selection. In addition, the

[125] Gauld, *Founders*, pp. 221–245. Wallace, *Miracles* (3rd ed., London: Nichols, 1901), pp. 103–104. E. Clodd, *Pioneers of Evolution* (New York: Appleton, 1897), p. 149. Medhurst and Goldney, "Crookes," pp. 31–32, 144–148.

ability of an immaterial spirit to interact with matter in a purposeful manner was incompatible with a materialism linked to "purposeless" natural selection.

But from the first half of this paper it is clear that Wallace was able to present a case against natural selection, which his contemporaries considered formidable, without reference to psychical phenomena or spiritualism. It is tempting therefore to conclude that perhaps, contrary to my thesis, Wallace had two independent grounds for his divergence—scientific and spiritual. According to this alternative viewpoint, Wallace *originally* concluded that natural selection was inadequate in the origin of man on the basis of his utilitarian analysis of various human features. Thus Wallace's simultaneous discovery of spiritualism was not the origin of his doubts about natural selection's sufficiency in man's development, but spiritualist phenomena provided further evidence, and in spiritualism Wallace found an explanation for those human features inexplicable by natural selection on purely utilitarian grounds.

I find this reconstruction of Wallace's thought on the origin of man in the years 1864–1869 unlikely. It fails to explain what prompted Wallace's new analysis of man. In 1864 Wallace had unequivocally supported natural selection's sufficiency in the development of *all* of man. It is true that in his 1864 paper Wallace considered man "in some degree a new and distinct order of beings" and "a being . . . in some degree superior to nature, inasmuch as he knew how to control and regulate her action, and could keep himself in harmony with her. . . ." But this uniqueness of man was solely due to the nature of his mind and, as Wallace emphasized throughout his 1864 paper, man's mind was the product of natural selection alone. I have discussed Wallace's 1864 position that natural selection was responsible for the development of man's intellectual and moral nature both before and after that nature had effectively shielded man's body from the further action of natural selection. Therefore it would be a mistake to interpret Wallace's description of man as "a being apart" to be the initial evidence for his 1869 conclusion that natural selection was inadequate in the origin of man. His 1864 position could not have logically led him to perform the utilitarian analysis of man's unique physical and mental features. Wallace's statement in his autobiography that his divergence from Darwin over the origin of man was "first intimated" in 1864—that is, prior to his first seance (1865)—must be rejected as incorrect. Certainly Darwin and other evolutionists, who lavished praise on Wallace for the great new ideas in the 1864 paper, recognized no signs of doubt in Wallace's mind as to the sufficiency of natural selection in man's development.[126]

The alternative viewpoint also fails to explain a curious aspect of Wallace's new analysis of man. The kinds of scientific arguments employed by Wallace against natural selection were hardly new with him; some had already been applied to man, while most had already been applied to animals. Wallace rejected virtually every claim that various *animal* features were inexplicable by natural selection because they were more highly developed than required or because they were useful to future generations but useless in the present. His refusal to apply the same reasoning to animals that he applied to man speaks forcefully for my contention that the root of his belief that

[126] Both F. Turner and R. Smith (see n. 1) contend that Wallace's 1864 paper does foreshadow his later (1869) doubts about the sufficiency of natural selection in the origin of man. However Smith's argument is vitiated by his exclusive use of the 1870 modified version of Wallace's 1864 paper.

natural selection could not account for the whole of man was in spiritualism, not science. In 1869 Wallace did recognize that the hand of apes was more highly developed than required, and in 1870 he did attempt to subsume all natural forces under will force. But only late in life, in *The World of Life*, did Wallace extend the action of higher intelligences to all organic development.

It has been suggested to me that Wallace's conception of natural selection differed from Darwin's conception; unlike Darwin's, it could not in fact explain man's intellectual and moral nature. According to this viewpoint, the source of Wallace's recognition of natural selection's inadequacy in the origin of man was his own conception of the nature of natural selection rather than his belief in spiritualism. Wallace was being perfectly logical, therefore, in his rejection of natural selection's sufficiency in the origin of man.

I find this reconstruction unlikely, too. Wallace did disagree with Darwin over several matters in the years 1864–1869. Each disagreement did revolve around the adequacy of natural selection, and Vorzimmer has already noted the differing conceptions. The three major disagreements concerned sexual selection, the origin of interspecific hybrid sterility, and the origin of man. Darwin perceived sexual selection as a supplement to natural selection; Wallace rejected the need for sexual selection. Thus Wallace defended the efficacy of natural selection in regard to a class of phenomena while Darwin denied it. The same pattern emerged with respect to the origin of hybrid sterility. According to Darwin natural selection could not explain it; such sterility was the incidental byproduct of speciation. But Wallace believed natural selection could account for it and again credited natural selection with more power than Darwin did.[127] Only in the case of the origin of man did Wallace find natural selection less adequate than Darwin did. But Darwin, too, supplemented the action of natural selection in the evolution of man.

The mere fact that Darwin and Wallace disagreed over the adequacy of natural selection in the cases of two evolutionary problems aside from the origin of man in no way suggests that they were destined to disagree also with respect to the origin of man. Besides, in the two cases of sexual selection and the origin of hybrid sterility Wallace supported natural selection's adequacy against Darwin's doubts, while in the case of the origin of man, Wallace had the graver doubts about natural selection's sufficiency.

The differing conceptions of natural selection emerge from the disagreement over the role of natural selection in the origin of hybrid sterility. Darwin conceived of natural selection as applied to individuals only, at least in this problem. Darwin "could not see how the inability to breed properly, to breed in lesser numbers, or to yield abnormal offspring could be selected as advantageous to any organism." Wallace, in contrast, conceived of natural selection as applied to both individuals and groups of individuals (for example, the entire species). There could be selection for a trait—sterility—advantageous to a group of individuals—incipient species—but disadvantageous to an individual within the group.[128] But I fail to see how Wallace's conception of natural selection could not account for the origin of man while Darwin's could. Indeed, with respect to the origin of man's moral nature, Darwin himself resorted to the "good of the species" argument. Unless it can be shown how Wallace's

[127] Vorzimmer, *Darwin*, pp. 188–209. George, *Biologist Philosopher*, pp. 200–204, 80–83. Wallace, *My Life*, Vol. II, pp. 17–20.

[128] Vorzimmer, *Darwin*, p. 207.

conception of natural selection logically forced him to the conclusion that natural selection was inadequate in the origin of man, this reconstruction also fails to explain what prompted Wallace's new analysis of man.

If, then, Wallace *originally* came to the conclusion that natural selection was inadequate to explain certain physical, mental, and moral features of man on the basis of his belief in spiritualism, how should his scientific arguments offered in support of his position be interpreted? Possibly he really believed in them. Having come to a conclusion from the direction of spiritualism, Wallace discovered another route—science. Since his colleagues rejected spiritualism, Wallace attempted to convince them of the validity of his new view with a strict utilitarian analysis of man. According to this "compromise" Wallace's belief in spiritualism did give rise to his first doubts about natural selection in the origin of man. Wallace hoped fellow scientists would perceive the significance of psychical phenomena for a complete understanding of human development. When he realized their outright rejection of the reality of those phenomena he was compelled to find reasons more acceptable to them for questioning natural selection's adequacy. Thus spiritualism stimulated Wallace to reconsider the utility of various human features, and the results of this new analysis (a foregone conclusion?) reinforced his earlier doubts which had been created by spiritualism alone.

On the other hand, Wallace may never have been truly convinced himself by his utilitarian analysis. This is a real possibility. Pearson was puzzled by Wallace's arguments and commented that Wallace's "pages read as if he had invented his difficulties in order to justify his beliefs."[129] In *Darwinism* Wallace rejected many of the arguments he had earlier employed against the efficacy of natural selection in the origin of man. The physical features of man, once considered inexplicable by natural selection, were now conceded to be explicable by it. Wallace "retreated" to man's intellectual and moral nature—the original stumbling block, with respect to man, for other evolutionists. Yet in this same work and throughout the last two decades of the nineteenth century Wallace was the most ardent champion of neo-Darwinism—that is, of the sufficiency of natural selection alone as the mechanism of evolution. Of course his advocacy of natural selection's hegemony did not extend to man. This striking duality certainly suggests an extra-scientific basis for Wallace's belief in the inadequacy of natural selection in the origin of man. It also highlights Wallace's insecurity about his scientific grounds for challenging natural selection's sufficiency in man's development.

At the same time it must be remembered that the vast mental gulf existing between man and animals was not easily bridged by selectionists. Darwin's own demonstration that these mental differences were of degree only and not of kind was weak and incomplete.[130] In fact, the natural selectionist's position with respect to man was very much a statement of faith in a continuity of causes in nature. Wallace, on the other hand, repeatedly remarked that continuity of effects did not necessarily imply a continuity of cause(s). In this context Wallace's problems with the origin of man, side by side with his satisfaction with natural selection in the origin of animal species, should be understandable. Indeed, Wallace was the first to perceive the difficulties in

[129] Pearson, *Grammar of Science*, p. 343.

[130] H. W. Conn, *Evolution of To-day* (New York: G. P. Putnam, 1894), pp. 327–338.

explaining a huge, rapid increase in human brain size by natural selection's accumulation of slight favorable variations.[131]

Therefore I tend to believe Wallace was persuaded by his scientific as well as spiritual arguments against natural selection. Yet I remain convinced that Wallace's belief in the reality of psychical phenomena and their spiritualist interpretation created the initial doubts about natural selection and stimulated his rethinking, on grounds of utility, man's unique features.

Lastly, I do not know whether the work of any other nineteenth-century scientist who firmly believed in the reality of psychical phenomena was so directly influenced by that belief. But I hope this paper, demonstrating the direct influence of Wallace's belief in spiritualism on his evolutionary thought, will encourage the growth of interest among historians of science in the history of psychical research.

[131] Eiseley, *Darwin's Century*, pp. 309–314. Loren Eiseley, *The Immense Journey* (New York: Vintage, 1958), pp. 79–94.

William Charles Wells and the Races of Man

By Kentwood D. Wells

IN 1866, while preparing the fourth edition of the *Origin of Species* for publication, Charles Darwin added to his "Historical Sketch" a brief notice of a paper written by William Charles Wells (1757–1817) in 1813. "In this paper," Darwin wrote, "he distinctly recognizes the principle of natural selection, and this is the first recognition which has been indicated." Darwin tempered his praise for Wells with the remark that "he applies it [natural selection] only to the races of man, and to certain characters alone."[1]

Despite Darwin's statement of the importance of Wells' ideas, relatively little attention has been given to Wells' place in the history of science. Most authors who have discussed him at all have treated him as a forerunner of Darwin who stumbled upon a great truth but failed to fully recognize the significance of his discovery.[2] He is generally considered to have been a man whose ideas came before their time and consequently fell on deaf ears. Great emphasis has been placed on the similarity of his ideas to those of Charles Darwin. Yet few authors have examined the relationship between Wells' ideas and those of his contemporaries. This paper is an attempt to re-evaluate Wells' views on natural selection and race formation from this point of view.

William Wells was born in Charleston, South Carolina, on May 24, 1757. In 1770 he travelled to Edinburgh to study at the university there, but he returned to Charleston the next year to begin an apprenticeship with Dr. Alexander Garden. Wells' family was sympathetic to the British government during the American Revolution, and he moved to Edinburgh in 1775 to begin medical studies. He later settled in London and became a student at St. Bartholomew's Hospital. After receiving an M.D. degree in 1780, Wells returned to the United States for four years, but in 1785 he settled permanently in London.[3]

I would like to express my appreciation to Dr. William B. Provine, Department of History, Cornell University, for reading the manuscript and making numerous helpful suggestions. Dr. K.A.R. Kennedy, of the Department of Anthropology, also read parts of the manuscript. (The author informs us that he is, regretfully, unrelated to his subject. ED.)

[1] Charles Darwin, *The Origin of Species: A Variorum Text*, ed. Morse Peckham (Philadelphia: University of Pennsylvania Press, 1959), p. 61.

[2] Richard Harrison Shryock, "The Strange Case of Wells's Theory of Natural Selection (1813): Some Comments on the Dissemination of Scientific Ideas," in *Studies and Essays in the History of Science and Learning Offered in Homage to George Sarton*, ed. M. F. Ashley Montagu (New York: Henry Schuman, 1944), p. 205.

[3] Norman Moore, "William Charles Wells," *Dictionary of National Biography*, Vol. XX, pp. 1146–1147.

The scant biographical information available on Wells fails to reveal anything particularly notable in his life.[4] He published a number of minor papers in the *Philosophical Transactions of the Royal Society* which have long since been forgotten. In 1792 he published his *Essay on Single Vision With Two Eyes*, which was widely read and admired at the time. In 1814 he published his *Essay on Dew*, for which he received the coveted Rumford Medal of the Royal Society.[5] Yet it is a single paper, a mere twelve pages long, for which Wells is remembered today.

Wells' paper was first read to a meeting of the Royal Society in 1813. It dealt with observations he had made on a piebald woman named Hannah West. He used these observations as a base for speculations on the origin of human racial differences. The paper was subsequently published as an appendix to a new edition of his *Two Essays: One on Dew and the Other on Single Vision With Two Eyes* (1818), with the improbable title, "An Account of a Female of the White Race of Mankind, Part of Whose Skin Resembles that of a Negro; with Some Observations on the Causes of the Differences in Colour and Form Between the White and Negro Races of Men."[6]

Wells noted that different races of men were susceptible to different diseases. Thus Europeans in tropical Africa often died from local diseases, while the native population was less affected. The same was true of Africans transferred to northern latitudes. Wells therefore concluded that colonies of Europeans in Africa would tend to die out or become weak and feeble. He believed that the differences in color among races, while not the cause of their varying susceptibility to diseases, was "a sign of some difference in them."[7] He then went on to show how this adaptation to the diseases and climate of a certain region could account for the formation of human races. Noting that "among men, as well as among other animals, varieties of a greater or less magnitude are constantly occurring," Wells stated that man was able to select these variations to improve his domesticated animals. He continued,

> But what is here done by art, seems to be done, with equal efficacy, though more slowly, by nature, in the formation of varieties of man, fitted for the country which they inhabit. Of the accidental varieties of man, which would occur among the first few and scattered inhabitants of the middle regions of Africa, some one would be better than the others to bear the diseases of the country. This race would consequently multiply, while the others would decrease, not only from their inability to sustain the attacks of disease, but from their incapacity of contending with their more vigorous neighbors. The colour of this vigorous race I take for granted, from what has been already said, would be dark. But the same disposition to form varieties still existing, a darker and darker race would in the course of time occur, and as the darkest would be best fitted for the climate, this would at length become the most prevalent, if not the only race, in the particular country in which it had originated.[8]

Wells believed that the influence of this natural selection on the color of the human race "would necessarily operate chiefly during its infancy," when men were "a few

[4] See F. L. Pleadwell, "That Remarkable Philosopher and Physician, Wells of Charleston," *Annals of Medical History*, 1934, *4*:128–149, for some anecdotes of Wells' life.

[5] Moore, "William Wells," p. 1147.

[6] William Charles Wells, *Two Essays, One on Dew and the Other on Single Vision With Two Eyes* (Edinburgh: Archibald and Constable, 1818). A reprint of Wells' paper is available in H. Lewis McKinney, *Lamarck to Darwin: Contributions to Evolutionary Biology 1809–1859* (Lawrence, Kansas: Coronado Press, 1971), pp. 21–28.

[7] Wells, *Two Essays*, p. 434.

[8] *Ibid.*, pp. 434–436.

wandering savages." Once men had adopted a "more refined mode of life," they would tend to retain certain customs and practices that would preserve the original race.[9]

Wells' paper attracted very little attention when it first appeared, although I have located two references to it which are worth noting. Alexander Tilloch, who was at that time the editor of *The Philosophical Magazine*, made a point of attending meetings of various scientific societies and reporting on the papers which were presented. In his report of the meeting of the Royal Society on April 1, 1813, Tilloch noted that "Dr. C. Wells communicated an account of Harriet Trest, a woman who has her left shoulder, arm, and hand as black as the blackest African, while all the rest of her skin is very white." He went on to state Wells' conclusion that "blackness of skin is no proof of differences of species and that the sun does not blacken but rather whitens the skin."[10]

The theoretical portion of Wells' paper was presented on April 8, and Tilloch reported it as follows: "The doctor indulged a variety of speculations; supposed with Volney, that the Egyptians were negroes; conceived that the want of civilization contributes to make people black; and referred to various South-sea islanders and others, to sanction this singular fancy." Tilloch added, with obvious annoyance, that "the length of these conjectures prevented the reading of a valuable paper by Professor Berzelius and Dr. Marcet."[11]

Wells' paper received a more sympathetic and detailed treatment in Thomas Thomson's *Annals of Philosophy*. While Tilloch had managed to miss the whole natural selection argument, Thomson gave the essence of Wells' theory in his summary:

> On Thursday the 8th of March [April] Dr. Wells's paper was concluded. He gave his opinion about what occasioned the difference between negroes and whites. It is well known that whites are not so well able to bear a warm climate as negroes; and that they are liable to many diseases in such a situation, from which negroes are free. On the other hand, whites are much better fitted to bear a cold climate than negroes. Suppose a colony of whites transported to the torrid zone, and obliged to subsist by their labour, it is obvious that a great proportion of them would speedily be destroyed by the climate and the colony, in no long period of time, annihilated. The same thing would happen to a colony of negroes transported to a cold climate. Dr. Wells conceives that the black colour of negroes is not the cause of their being better able to bear a warm climate, but merely the sign of some difference in constitution which makes them able to bear such a climate. Suppose a colony of white men carried to the torrid zone; some would be better able to resist the climate than others. Such families would thrive, while the others decayed. These families would exhibit the sign of such a constitution; that is, they would be dark: and as the darker they were, the better they would be able to resist the climate; it is obvious, that the darker varieties would be the most thriving, and that the colony, on that account, would become gradually darker and darker coloured till they degenerated into negroes. The contrary would happen to negroes transported to cold climates.[12]

Here, in a journal published in 1813, is a theory of race formation by natural selection. Yet I have found no firm evidence that this account attracted any attention, just as the paper attracted no attention when published in Wells' widely read *Two Essays* in 1818. In fact, until Darwin referred to the paper in his "Historical Sketch" in 1866, it remained lost in obscurity.

[9] *Ibid.*, p. 436.

[10] [Alexander Tilloch], "Proceedings of Learned Societies," *The Philosophical Magazine*, 1813, *41*:302.

[11] *Ibid.*, pp. 302–303.

[12] [Thomas Thomson], "Proceedings of Philosophical Societies," *Annals of Philosophy*, May 1813, *1*:383.

There has been some discussion of the reasons for Wells' failure to attract any attention to his theory. There has also been considerable controversy over the question of how completely Wells anticipated the Darwin-Wallace view of natural selection. As we shall see, these two problems are closely related. Shryock has argued that Wells "put together two hitherto isolated ideas which had long been familiar to biologists—'natural selection' and the 'origin of species'—in an essentially new combination."[13] Although Wells emphasized the application of his theory to human races, Shryock believed that "Wells plainly envisaged the theory as applicable to animal forms."[14] Similarly Conway Zirkle, in his study of the history of natural selection, states that Wells "obviously understood its wider application."[15] Loren Eiseley has been more cautious in his interpretation. While acknowledging that Wells mentioned animals briefly and clearly anticipated natural selection, Eiseley makes the important observation that there is no evidence that Wells grasped the concept of unlimited organic change in time.[16] Thus it is not at all clear that Wells attributed the origin of new species to natural selection, as Shryock maintained.

In my own view, I believe that Wells not only failed to grasp the idea of species changes through time, but he probably did not intend to apply the process of natural selection to wild animals. He merely stated that "among men, as well as among other animals, varieties of a greater or less magnitude are constantly occurring."[17] He further stated that man could select these varieties in domesticated animals. The only mention of selection "by nature" referred to the formation of human races or varieties.

Thus it is inappropriate to group Wells with men such as Erasmus Darwin, Lamarck, and Robert Chambers, who were concerned with the evolution of species through time. Instead, he belongs with the group of physician-anthropologists whose works were appearing in the first two decades of the nineteenth century. During this period such British physicians as James Cowles Prichard (1786–1848) and William Lawrence (1783–1867) were engaged in speculation on the origin of human races. In the United States men like Samuel Stanhope Smith and a host of later writers also produced treatises explaining the origin of racial differences.[18]

The early years of the nineteenth century marked the emergence of anthropology as a science in its own right. Earlier discussions of human races had been largely speculative, with very little support from actual observations. The great eighteenth-century naturalists, Linnaeus and Buffon, had attempted to classify races according to physical characteristics. Anatomists like Petrus Camper (1722–1789) and John Hunter (1728–1793) had assembled and studied collections of human skulls and quantified physical differences between races. However, it was only in the work of the great German

[13] Shryock, "Strange Case," p. 203.

[14] *Ibid.*

[15] Conway Zirkle, "Natural Selection Before the 'Origin of Species,'" *Proceedings of the American Philosophical Society*, 1941, *84*:106.

[16] Loren Eiseley, *Darwin's Century* (New York: Doubleday, 1959), p. 122. John C. Greene, *The Death of Adam* (Ames, Iowa: Iowa State University Press, 1959), pp. 244–247, also discusses Wells' theory, but he does not speculate on whether Wells understood the wider applications.

[17] Wells, *Two Essays*, p. 434.

[18] On Prichard and Lawrence, see Kentwood D. Wells, "Sir William Lawrence (1783–1867): A Study of Pre-Darwinian Ideas on Heredity and Variation," *Journal of the History of Biology*, 1971, *4*:319–361; on Smith, see Samuel Stanhope Smith, *An Essay on the Causes of the Variety of Complexion and Figure in the Human Species*, ed. Winthrop D. Jordan (Cambridge, Mass.: Harvard University Press, 1965; reprint of 2nd ed., New Brunswick, N.J., 1810).

anatomist Johann Friedrich Blumenbach (1752–1840) that a cohesive science of anthropology began to appear.[19]

In England in 1813 the great debate over the unity versus plurality of the human species, which dominated anthropology at mid-century, had scarcely begun. A few writers, notably Lord Kames (1696–1782) and Charles White (1728–1813), had championed the polygenist doctrine which held that each human race constituted a separate species of man.[20] For the most part, however, the monogenists had the field to themselves. This was due mainly to the influence of Blumenbach, who was the intellectual godfather of the two most important anthropologist-ethnologists of the day, James Cowles Prichard and Sir William Lawrence. These men, following Blumenbach's lead, stated as their basic premise the fact that all human races constituted a single species. Yet Blumenbach, Prichard, and Lawrence could scarcely deny that there were enormous physical differences among the races of man, for these differences were readily observable in nature. It is not surprising, therefore, that much of their attention became focused on the problem of explaining the origin of human racial variations.

In the late eighteenth century Blumenbach made a very significant contribution to the study of variation when he suggested that variations in the human species might be explained by examining analogous variations in domesticated animals. His aim was to show that the differences between human races were no greater than the differences between varieties of a single species of domesticated animal. To prove his point Blumenbach cited numerous examples from both the human and animal worlds, thereby assembling a valuable compendium of information which would serve as a base for future workers.[21]

Prichard's *Researches Into the Physical History of Man* first appeared in 1813, although a shorter version had been published in Latin as a dissertation in 1808.[22] In 1819 Lawrence's *Lectures on Physiology, Zoology, and the Natural History of Man* appeared in print.[23] Prichard's and Lawrence's ideas were very similar, and Lawrence's work owed much to the 1813 edition of Prichard's book.[24] Both men agreed with Blumenbach that variations within the human species were no greater than those within domesticated animals, and they cited many of his examples. However, when they came to consider the causes of these variations, Prichard and Lawrence diverged sharply from their German colleague.

Blumenbach was very traditional in his explanation of the causes of variations.

[19] This early period of anthropology has not been well studied, but a few good sources are available. For an excellent discussion of the pre-scientific period, see Margaret T. Hodgen, *Early Anthropology in the Sixteenth and Seventeenth Centuries* (Philadelphia: University of Pennsylvania Press, 1964); see also D. J. Cunningham, "Anthropology in the Eighteenth Century," *Journal of the Royal Anthropological Society*, 1908, *38*: 10–35; William Stanton, *The Leopard's Spots: Scientific Attitudes Toward Race in America 1815–59* (Chicago: University of Chicago Press, 1960); on Blumenbach, see Thomas Bendyshe, ed., *The Anthropological Treatises of Johann Friedrich Blumenbach* (London: The Anthropological Society, 1865).

[20] Stanton, *Leopard's Spots*, pp. 15–18.

[21] Bendyshe, *Anthropological Treatises of Blumenbach, passim.*

[22] James Cowles Prichard, *Disputatio inauguralis de generis humani varietate* (Edinburgh: Abernethy and Walker, 1808); *Researches Into the Physical History of Man* (London: J. and A. Arch, 1813).

[23] I have used the following edition: William Lawrence, *Lectures on Physiology, Zoology, and the Natural History of Man* (Salem, Mass.: Foote and Brown, 1828).

[24] I have discussed the relationship between Lawrence and Prichard more fully in "Sir William Lawrence," pp. 351–361.

Environmental factors such as climate, food, and way of life were assumed to be capable of producing inheritable changes in the physical constitution of men and animals. Even artificial constriction of the head and other such man-made changes were thought to produce inheritable effects. In Blumenbach's theory these external factors acted gradually on a whole race to produce modifications in skin color or hair texture. Thus, a light-colored race would become gradually darkened after residing in a hot tropical environment.[25] This was simply a reiteration of the time-honored theory of the inheritance of acquired characteristics, which, as Zirkle has shown, had been generally accepted for centuries.[26]

Prichard and Lawrence, on the other hand, emphatically rejected this ancient theory. The reasons for this are unclear, but they were unequivocal in stating their views.[27] "There is no foundation," Prichard wrote, "for the common opinion which supposes the black races of men to have acquired their colour by exposure to the heat of a tropical climate during many ages. . . . The offspring . . . inherit only their connate peculiarities, and not any of the acquired qualities."[28] Similarly, Lawrence declared, "In all the changes which are produced in the bodies of animals by the action of external causes, the effect terminates in the individual; the offspring is not in the slightest degree modified by them."[29]

Prichard and Lawrence suggested that the differences among human races originated from hereditary variations which appeared spontaneously in certain individuals and which were perpetuated through breeding. Thus, while Blumenbach had emphasized environmental factors which affected whole races, Prichard and Lawrence shifted the emphasis to individual variations. As Prichard expressed it,

> The children of the same parents, though often bearing a general resemblance yet exhibit always some difference and frequently a considerable diversity in these respects. . . . By observing that such a tendency to deviation exists even among the individuals of the same family, and that whatever examples of variations may arise, have a general disposition to become hereditary, we appear to make some progress towards an explanation of the diversities of figure, which characterize different races of the human kind.[30]

In his thinking on the origin of variations Wells was in essential agreement with Prichard and Lawrence, although it seems unlikely that any direct influence was involved.[31] Wells stated that black skin color was not a characteristic of a distinct species of man. Furthermore, he rejected environmentalist explanations of the origin of variations and denied that the sun could produce a blackening of the skin.[32] Like Prichard and Lawrence, Wells utilized the analogy between man and domesticated

[25] Bendyshe, *Anthropological Treatises of Blumenbach*, pp. 129, 196–200.

[26] Conway Zirkle, "The Early History of the Idea of Acquired Characteristics and of Pangenesis," *Transactions of the American Philosophical Society*, 1946, N.S. *35*:91–151.

[27] Lawrence accepted the inheritance of acquired characteristics in writings which appeared prior to his 1819 lectures. He seems to have changed his views as a result of the influence of Prichard. See K. Wells, "Sir William Lawrence," pp. 351–361.

[28] Prichard, *Researches* (1813), p. 230.

[29] Lawrence, *Lectures*, p. 436.

[30] Prichard, *Researches* (1813), p. 34.

[31] Wells could not have read Lawrence's 1819 *Lectures* before delivering his own paper in 1813. Wells' paper was given in April 1813, while the preface to Prichard's *Researches* is dated Nov. 1813. Wells could have read Prichard's 1808 dissertation (see n. 22 above), but Wells makes no references to any of his sources. There seems not to have been any influence of Wells on Lawrence, although Prichard may later have read his paper (see n. 37 below).

[32] Wells, *Two Essays*, p. 431.

animals, noting that spontaneous variations "of a greater or less magnitude are constantly occurring" among them.[33]

When he came to a consideration of the mechanism by which such spontaneous variations were preserved in a race, Wells diverged from Prichard and Lawrence. These men noted that man was capable of preserving spontaneous variations in domesticated animals through selection, and they suggested that such principles could be applied to humans as well. They also noted that different ideas of beauty among races might act as a sort of sexual selection and exert an influence on the physical appearance of races. In addition, they were aware of the fact that individuals with similar characteristics might become isolated and form a new race through interbreeding.[34]

Wells also referred to man's ability to perpetuate variations in domesticated animals through selective breeding. However, previous quotations from Wells' work have shown that he clearly understood that a natural process of selection could occur in the formation of human races. In formulating this theory Wells demonstrated a grasp of the concept of adaptation which was notably lacking in the works of Prichard and Lawrence. The survival of a race depended on its ability to adapt to the diseases of the region which it inhabited. Wells was thus able to present a plausible explanation for the distribution of different racial characters as they exist in the modern world. Prichard and Lawrence, while suggesting ways in which variations could be preserved in humans, admitted that they could not satisfactorily explain why particular racial characters had been perpetuated. In this respect Wells' ideas were more "advanced" than those of his contemporaries.

While many of the ideas on variation expressed by these men, particularly Wells, seem remarkably modern to us today, we must remember that all of these anthropologists were concerned with explaining variations within a single species, *Homo sapiens*, and not with the evolution of new species.[35] Current speculations on evolution, as expressed in the writings of Lamarck, Erasmus Darwin, and others, were never mentioned by Prichard, Lawrence, and Wells. It is therefore not surprising that someone like Wells failed to formulate a full-fledged anticipation of Darwinism, for he was functioning in an intellectual world completely different from that of the evolutionists.

It is apparent that the question of why Wells' theory failed to attract much attention is essentially an irrelevant one. This question assumes that his theory would have been recognized by his contemporaries as something bold and innovative. Yet, as I have attempted to demonstrate, Wells' ideas were basically in the mainstream of current anthropological thinking on the nature of variation. True, he had formulated a theory of natural selection which was in advance of the ideas of his contemporaries, but even in this he was not unique, since he did not apply it to the origin of new species. Various concepts of natural selection had been advanced before.[36] Even Prichard, in the

[33] *Ibid.*, p. 434.

[34] Prichard, *Researches* (1813), pp. 38–44; Lawrence, *Lectures*, pp. 387–395.

[35] Various authors have cited Prichard and Lawrence, in addition to Wells, as forerunners of Darwin. See esp. C. D. Darlington, *Darwin's Place in History* (New York: Macmillan, 1955), pp. 16–24; Philip G. Fothergill, *Historical Aspects of Organic Evolution* (London: Hollis and Carter, 1952), pp. 83–86; E. B. Poulton, "A Remarkable Anticipation of Modern Views on Evolution," in *Essays on Evolution* (Oxford: Clarendon Press, 1908), pp. 173–192.

[36] Zirkle, "Natural Selection Before the 'Origin,'" pp. 71–123.

second edition of his great anthropological treatise, expressed ideas remarkably similar to those of Wells.[37] Yet Prichard continued to be concerned only with variations within the human species.

Much of the inclination of historians to read anticipations of Darwinian evolution into the thoughts of earlier writers stems from a failure to consider carefully the ways in which such scientific theories come into being. In addition, there is the constant problem of hindsight colored by an intellectual world in which evolution is taken for granted. While the reality of the *process* of evolution by natural selection has been accepted by modern science, we must bear in mind that the *concept* of natural selection did not exist independently as a sort of Platonic ideal, waiting to be discovered, recognized, or stumbled upon by a Wells or a Darwin or a Wallace. Both Darwin and Wallace had fully accepted the reality of evolution when they formulated the concept of natural selection to provide a mechanism for the origin of new species. Wells was not even thinking in terms of evolution when he hit upon natural selection as a mechanism for preserving human racial variations. To expect him to have anticipated Darwinism, by accident or design, would therefore be unreasonable. It is not that he failed to find the correct answers; he simply had not asked the same questions.

The precise way in which Darwin became aware of Wells' paper has never been fully explained. The story is worth telling, for it provides an interesting sidelight on this episode in the history of ideas. In the fourth edition of the *Origin* Darwin stated, "I am indebted to the Rev. Mr. Brace, of the United States, for having called my attention to the above passage in Dr. Wells' work."[38] In the fifth edition, published in 1869, Darwin modified his acknowledgement to read: "I am indebted to Mr. Rowley, of the United States, for having called my attention, through Mr. Brace, to the above passage in Dr. Wells' work."[39]

Shryock has suggested that "Mr. Brace" may have been Charles Loring Brace, the American clergyman. He noted that Mr. Rowley's identity is unknown, but that "if something were known of the latter's familiarity with Wells's theory, it might throw light on its possible—even if obscure—influence on the biological thought of the mid-nineteenth century."[40]

Darwin apparently first learned of Wells' paper sometime in 1865, when he mentioned in a letter to Joseph Dalton Hooker that "a Yankee" had referred him to it.[41] This "Yankee" was presumably either Brace or Rowley. The "Rev. Mr. Brace" Darwin mentioned in his "Historical Sketch" was undoubtedly Charles Loring Brace,

[37] Prichard, in the 2nd ed. of his *Researches into the Physical History of Mankind*, 2 vols. (London: J. and A. Arch, 1826), recognized a sort of natural selection very similar to the theory advanced by Wells. He postulated that races not adapted to the diseases of a particular region would die out. Prichard may have known of Wells' paper, but I have found no definite evidence to prove that he read it. However, he did read another article in the same volume of the *Annals of Philosophy* which contained the summary of Wells' theory, and he cited it in his book. See Prichard, *Researches* (1826), Vol. II, pp. 550, 573–574, 581.

[38] Darwin, *Origin* (Variorum Text), p. 62.

[39] *Ibid.*

[40] Shryock, "Strange Case," p. 199.

[41] Francis Darwin, ed., *The Life and Letters of Charles Darwin* (London: John Murray, 1887), Vol. III, p. 41. Both Shryock ("Strange Case," p. 199) and Eiseley (*Darwin's Century*, p. 125) incorrectly assume that Darwin learned of Wells' paper in 1860, when the "Historical Sketch" was first written. Morse Peckham's variorum edition of the *Origin* has shown that changes were made in the "Historical Sketch" as well as in the text and that the reference to Wells did not appear until 1866.

as Shryock has suggested, but who was "Mr. Rowley" and how did he come to know of Wells' work?

The most likely answer comes from an article which appeared in the *American Journal of Science and Arts* for September 1868. Written by one Samuel Rowley, it was entitled "A New Theory of Vision."[42] In it are several references to Wells' *Essay Upon Single Vision with Two Eyes*, in the London 1818 edition, the one which contained the appendix on race formation. Unfortunately nothing more is known about Rowley, and I have found nothing else written by him. It seems likely, however, that this is Darwin's "Mr. Rowley, of the United States." Apparently Rowley happened upon the natural selection paper while reading the essay on vision, in preparation for his article, and recognized that it anticipated Darwin's views.

However, Darwin mentioned that he learned of the paper from Mr. Rowley "through Mr. Brace." Did Rowley first send word of the Wells paper to Charles Loring Brace, a clergyman, and if so, why? It seems that Brace was something of an amateur anthropologist, and in 1863 he published a book entitled *The Races of the Old World: A Manual of Ethnology.*[43] What is more, he made a point of praising Darwinian natural selection, and he applied natural selection to the formation of human races.

Brace theorized that in "some very remote age of the past" a tribe might have moved from Asia or elsewhere to eastern Africa. The climate and diseases would be different from those in the tribe's native region. In some offspring slight varieties might appear which fitted "the possessors to resist the destructive influences of the new climate," and these would be more likely to survive. Therefore Brace saw "no difficulty on this supposition—on the Darwinian theory of an imperceptible accumulation of profitable change through long periods of time . . . of accounting for the origin of the negro from the white man, or from the brown, or from some other race."[44]

We can see that Brace was quite clear in his application of natural selection to human races. It therefore seems possible that Rowley called Brace's attention to the Wells paper for his *own* information rather than for Darwin's. Yet one question remains: How did Rowley happen to know Brace? It turns out that they were neighbors! The preface to Brace's book was written at his home in "Hastings-on-the-Hudson" in April 1863.[45] Fortunately for later historians, Rowley included his own address with his article: "Hastings-upon-Hudson, N.Y."[46]

The story of Wells' theory does not end with Darwin's brief mention of it in his "Historical Sketch." Darwin himself had independently developed a mechanism similar to Wells' whereby "negroes and other dark races might have acquired their dark traits by the darker individuals escaping from the deadly influence of the miasma of their native countries during a long series of generations."[47] As early as 1862 he even hit upon the rather novel (although for Darwin, typical) idea of sending questionnaires to Army doctors in all parts of the British Empire inquiring about the different susceptibilities of the British and native soldiers to diseases.[48] Darwin told Wallace of

[42] Samuel Rowley, "A New Theory of Vision," *American Journal of Science and Arts*, 1868, 2nd ser., *46*:153–167.

[43] Charles Loring Brace, *The Races of the Old World: A Manual of Ethnology* (New York: Scribner's, 1863).

[44] *Ibid.*, pp. 496–499.

[45] *Ibid.*, p. iv.

[46] Rowley, "New Theory of Vision," p. 167.

[47] Charles Darwin, *The Descent of Man* (2nd ed.; New York: Hurst and Co., 1874), p. 209.

[48] *Ibid.*, p. 210n.

his theory and questionnaire in a letter written in May 1864.[49] In his reply Wallace stated that while that was all very interesting, he did not expect the results of the inquiry "to be favourable to your view."[50] This is not surprising, since in his 1864 paper on human races and natural selection, which he and Darwin were discussing in these letters, Wallace declared that man's physical constitution was no longer influenced by natural selection.[51] As it happened, Darwin got no replies from his inquiry and later concluded that there was no evidence for his theory.[52]

When Wallace read about Wells in Darwin's "Historical Sketch" he wrote to Darwin, "How curious it is that Dr. Wells should so clearly have seen the principle of Natural Selection fifty years ago, and that it should have struck no one that it was a great principle of universal application in Nature!"[53] A few years later Wallace brought out his *Contributions to the Theory of Natural Selection* (1870), and he mentioned Wells in the preface. The 1864 paper was republished under the title "The Development of the Human Races Under the Law of Natural Selection." Curiously enough, Wallace added one paragraph to the original paper:

> There is one point, however, in which nature will still act upon him as it does on animals, and to some extent, modify his external characters. Mr. Darwin has shown that the colour of the skin is correlated with constitutional peculiarities both in vegetables and animals, so that liability to certain diseases or freedom from them is often accompanied by marked external characters. Now there is every reason to believe that this has acted, and, to some extent, may still continue to act, on man. In localities where certain diseases are prevalent, those individuals of savage races which were subject to them would rapidly die off; while those who were constitutionally free from the disease would survive, and form the progenitors of a new race. These favoured individuals would probably be distinguished by peculiarities of *colour*, with which again peculiarities in the texture or the abundance of *hair* seem to be correlated, and thus may have been brought about those racial differences of *colour*, which seem to have no relation to mere temperature or other obvious peculiarities of climate.[54]

This paragraph was a direct contradiction of the main point of Wallace's paper as originally written. Although the evidence is circumstantial, it seems very likely that Wallace was prompted to add this paragraph after reading Darwin's account of Wells' ideas. Apparently he found Wells' arguments of 1813 so convincing that he decided to make an exception to his belief that natural selection could no longer act on man's body.[55]

[49] James Marchant, ed., *Alfred Russel Wallace: Letters and Reminiscences* (New York: Harpers, 1916), p. 128.

[50] *Ibid.*, p. 129.

[51] Alfred Russel Wallace, "The Origin of Human Races and the Antiquity of Man Deduced From the Theory of Natural Selection," *Journal of the Anthropological Society of London*, 1864, *2*: clxiii–clxv; clxviii–clxix.

[52] Darwin, *Descent*, p. 211.

[53] Marchant, *Wallace*, p. 145.

[54] Alfred Russel Wallace, "The Development of the Human Races Under the Law of Natural Selection," in *Contributions to the Theory of Natural Selection* (London: Macmillan, 1870), p. 316.

[55] Wallace had been interested in the origin of human races since at least 1845, when he read the works of Prichard and Lawrence and was impressed by their ideas on variation. For additional information on Wallace's anthropological interests, see H. Lewis McKinney, "Alfred Russel Wallace and the Discovery of Natural Selection," *Journal of the History of Medicine and Allied Sciences*, 1961, *21*: 333–357; H. Lewis McKinney, *Wallace and Natural Selection* (New Haven, Conn.: Yale University Press, 1972); K. Wells, "Sir William Lawrence," pp. 339–344.

With Wallace's statement we have come full circle to return to our starting point. Nearly sixty years earlier William Wells had formulated a theory of natural selection without giving any consideration to the evolutionary implications of his ideas. Wallace, for whom evolution and natural selection had long since become established facts, thought he detected the ghost of a fellow evolutionist emerging from the shadows of the past. He sought to revive these long-lost ideas by applying them to one aspect of his own evolutionary philosophy. Ironically, in doing so, Wallace was preparing a solution to the only problem Wells had considered—the origin of human racial variations.

The Duke of Argyll, Evolutionary Anthropology, and the Art of Scientific Controversy

By Neal C. Gillespie

THE ENCOUNTER BETWEEN EVOLUTIONARY SCIENCE and traditional religion in the nineteenth-century English-speaking world is conventionally symbolized by the famous meeting of Bishop Samuel Wilberforce and the biologist Thomas Henry Huxley at Oxford in 1860. Another meeting, less familiar but no less filled with portent, occurred at Dundee in 1867. There one of the founders of prehistoric archaeology, Sir John Lubbock, took on another bishop and opened another controversy which, in its course, was actually more revealing of the attitudes of an increasingly secularized science toward biblical religion at that time than its more celebrated predecessor. While Huxley's victory over Wilberforce was a triumph of integrity over obfuscation, the response of Lubbock and his fellow evolutionists to the theory of cultural degeneration, and to its champion, the Duke of Argyll, shows that the use of bias and evasion to defend a favored belief was not limited to the theological opponents of evolution.

The story of evolutionism and degeneration in the nineteenth century is important but complex. Consequently, it cannot be dealt with here. Instead, this essay is concerned with analyzing an example of the practical means by which religion and science were separated, of the ways sometimes employed to screen out ideas hostile to the new evolutionary synthesis. A close study of this almost forgotten confrontation suggests that the need of many Anglo-American evolutionists to make a philosophically necessary break with the "biblical" science of their predecessors sometimes led them (whatever their ultimate wishes might have been) to take polemical advantage of the antagonism between science and religion to protect their new doctrine and to blacken dissent with theological tar. Evolutionary science's role as a destroyer of older beliefs has often been noted. Its own defensiveness and sensitivity to criticism in its early years is less well known and seldom so well illustrated as in this instance.

* * *

Lubbock had not intended to use a meeting of the British Association for the Advancement of Science to start a debate: his purpose was to end one. Thirty years before, and again in 1855, the late Archbishop of Dublin, Richard

Whately, had raised the old question of whether man had risen from primeval savagery or had his ancestors been civilized from the beginning? Lubbock spoke to show from the steadily accumulating data of prehistoric archaeology and anthropology that the Archbishop's belief in the civilization of early men was nonsense. He was soon answered by the Duke of Argyll, who is better remembered as a politician than a scientist or philosopher, although he was a spirited writer in both areas. Other defenders of Whately appeared, and the issue was joined. The question became intertwined with the Darwinian controversy and was not finally laid to rest until near the century's end, when *Homo erectus* from Java provided science with both a "missing link" and a "compleat" savage as man's ancestor.

* * *

Since classical antiquity Europeans who could not believe that man had declined from a perfect state had speculated on his gradual rise to civilization from savagery. From Lucretius to Adam Smith a progressionist or evolutionist model running from savagery through barbarism to civilization was assumed by many to have been the upward path of social development. Archbishop Whately, casting about for a way to show by an appeal to reason the truth of Christian revelation, seized the opportunity offered by the popularity of this progressionist model of human self-creation to urge the startling proposition that the very existence of civilized man proved the truth of the contrary version of the early history of man revealed in Genesis.

Like most, Whately saw a providential scheme unfolding in society's advance. He did not deny social progress. What he did deny was that this progress had developed spontaneously out of savagery. The beginning of the process, he wrote, "is not (as several writers on Political Economy have appeared to suppose) what is properly called the *savage* state." On the contrary, "we have no reason to believe that any community ever did, or ever can, emerge, unassisted by external helps, from a state of utter barbarism into anything that can be called civilization. . . ."[1] History shows no record, he claimed, of any "tribe of savages" civilizing themselves. But it does show numerous cases of tribes who had remained in a stagnant condition for centuries, suggesting thereby an endless immobility; or even more revealing were the savages who possessed tools and institutions far beyond their ability to invent.[2] Here was clear evidence that low savages represented a degeneration from a previous state in some degree civilized. Since Whately thought savages too improvident, immoral, and stupid to improve themselves, and since all known cases of social advance on their part had come about as the result of borrowing from more civilized neighbors, how—and here he pressed home—can we account for the origin of civilization at a time when there were no advanced nations? The answer was obvious: by divine revelation, just as we are told in Genesis. Hence, reason and history corroborate revelation quite independently of any assumption of

[1] Richard Whately, *Introductory Lectures on Political Economy* (2nd ed., London: B. Fellows, 1832), pp. 106–107. See also Richard Whately, "On the Origin of Civilization," *Young Men's Christian Association Lectures, 1854–1855* (London: James Nisbet & Co., 1855), pp. 11–14, 21.

[2] Whately, *Introductory Lectures*, pp. 110–111, 117–118, 129.

the Bible's inspiration.[3] "If [civilization] was *not* the work of a divine instructor," he taunted skeptics, "*produce an instance,* if you can, of a nation of savages *who have civilized themselves.*"[4]

This challenge to believers in the emergence of civilization out of savagery (to say nothing of the later Darwinists) could not be ignored.[5] Other degenerationists supported it, and progressionists rose up to defeat it long before Sir John Lubbock undertook the final demolition at Dundee.[6]

The apparent absurdity of Whately's degenerationist argument should not obscure the fact that in the mid-nineteenth century, in one form or another, it had considerable force and enjoyed wide acceptance in both lay and scientific circles. It is true that most social scientists seem to have subscribed to the progressionist model of savagery–barbarism–civilization, but not all scientific students of the human past did so. It is true, also, that its plausibility rested on a lack of knowledge about ancient man and contemporary non-Western cultures. But this ignorance was shared by progressionist and degenerationist alike.

Anyone interested in the history of human society in those days would have worked with one of these two basic models. Although varied in its individual expressions, the evolutionist, or progressionist, version received classic formulation in Edward Burnett Tylor's *Primitive Culture* (1871). The father of British anthropology postulated a single social entity called civilization or culture which was possessed by all peoples but in various stages of development. Those in the "same grade of civilization" might be compared regardless of distinctions of time, place, or race. This was the famous comparative method. The "phenomena of Culture," wrote Tylor, "may be classified and arranged, stage by stage, in a probable order of evolution."[7] Because the evolutionary process was believed to have begun in savagery, the observation of modern savages (who were close, if not identical, to ancient man) could solve many puzzles about human origins. The rival degenerationist theory denied that the lowest modern savages were to be taken as a model of primeval man, and so denied both the savage theory of human origins and a basic part of the developmental chain assumed by the comparative method. As long as degenerationism remained viable, it presented evolutionary anthropology with a fundamental theoretical challenge. Why Lubbock was moved to attack Whately, then, is no mystery. Despite his attempt to present it as such, this was not a sideshow encounter of science and reaction; it involved basic and important scientific issues.

* * *

[3] *Ibid.*, pp. 111–112, 114–115, 118–120, 121–122, 138, 149. See also Whately, "Origin of Civilization," pp. 19–20, 29.

[4] Whately, *Introductory Lectures,* p. 123.

[5] Whately saw his teaching as clearly fatal to Lamarck's idea of evolution; see Whately, "Origin of Civilization," p. 26.

[6] A popular degenerationist work was W. Cooke Taylor, *The Natural History of Society,* 2 vols. (London: Longmans, 1840). Traditionally prevalent in both religious and scientific circles, the view was still widespread in the periodical literature in both Great Britain and the United States. A vigorous evolutionist attack on Whately and degenerationism prior to Lubbock's was [Robert Chambers], *Vestiges of the Natural History of Creation* (8th ed., London: John Churchill, 1850), esp. pp. 238–240.

[7] Edward Burnett Tylor, *Primitive Culture,* 2 vols. (London: John Murray, 1871), Vol. I, pp. 6–7.

Degeneration, as Lubbock understood it, was the doctrine that "man was, from the commencement, pretty much what he is at present; if possible, even more ignorant of the arts and sciences than now, but with mental qualities not inferior to our own. Savages," it held, are "the degenerate descendants of far superior ancestors."[8] Responding to Whately's taunt to produce a race of self-civilizing savages, Lubbock complained that the Archbishop had framed the question so as to exclude the possibility of proof: an alien eyewitness who could leave a record of such progress would necessarily be civilized and so destroy the required isolation of the savages; while a people sufficiently literate to leave a contemporary record of their own elevation would not be savages.[9] Lubbock evaded this logical cul-de-sac by appealing to the data of anthropology. We have, he claimed, considerable evidence of progress "even among savages," while modern civilization bears witness to its barbaric past in the numerous survivals of savage belief and practice. Among illustrations of the former he mentioned the domestication of the potato and the llama by savage migrants to the New World, iron-working by savage Africans, and that favorite of progressionists, the Australian boomerang, which must have been invented by that savage people, as it is not known to any other culture in the world. And who would deny that the Australians are the most abject savages imaginable? Throughout the civilized world, in the second case, traces of a primitive Stone Age remain, while the development of customs governing the relation between the sexes and social relations generally tells the same story of progress.

In answer to the claim that the lowest modern savages are degenerate survivors of a higher state, Lubbock asked: where are the survivals of this lost culture? Could such a culture disappear without leaving a tradition? If such tribes as the Australians once had domesticated animals and plants, where are the feral descendants? Where are the architectural ruins—or even the tools, weapons, and the virtually indestructible pottery that would indicate more advanced ancestors? Would any tribe once in possession of something as useful as religion ever lose it? Could any race once able to think in abstract terms, to grasp the generalized idea of "color" or "tree," lose that ability? Yet, the lowest modern races lacked both.[10] No one could supply the proofs Whately had demanded, but Sir John found the circumstantial evidence decisively against degenerationism. Nor was his audience less convinced: his peroration invoking continuing and universal progress was received with "loud cheers."[11]

* * *

Few things in the history of scientific controversy have been more misunderstood than the Duke of Argyll's reply to Lubbock's Dundee paper. Contrary to general opinion, then and since, it was not his purpose to defend the views

[8] John Lubbock, "On the Primitive Condition of Man" in *The Origin of Civilisation and the Primitive Condition of Man: Mental and Social Condition of Savages* (New York: D. Appleton, 1870), p. 325. The address is also in *Anthropological Review*, 1968, *6*:1–14, discussion, 15–21; and in *Transactions of the Ethnological Society*, 1868, *6*:328–341.

[9] Lubbock, *Origin of Civilisation*, pp. 326–327.

[10] *Ibid.*, pp. 329–335. On the boomerang question see also Edward Burnet Tylor, *Researches into the Early History of Mankind and the Development of Civilisation* (London: John Murray, 1865), p. 186.

[11] *The* (London) *Times*, Sept. 11, 1867, p. 10.

of Richard Whatley. Nor did he limit himself to a criticism of the points raised in Lubbock's address. His intention was to challenge the theoretical underpinning of evolutionary anthropology: to show that the facts were incompatible with the idea of primeval man as a mindless, brutish savage, and to suggest that the comparative method, which insisted on the analogy of modern primitives and ancient man, was frequently arbitrary in application and unsupported by sufficient evidence.

Primeval Man, which first appeared in 1868 as a series of essays in the periodical *Good Words,* contained many points not addressed directly to the Whately-Lubbock disagreement but not unimportant to the question. Taking up man's origin and antiquity, Argyll denied human evolution (and, at this time, any other kind) and so deprived the progressionists of a biological basis for a belief in man's early savagery, while warning of the Darwinism implicit in Lubbock's views. He next admitted the heretofore unimagined duration of man's existence on the earth, but he saw no conflict between this and the biblical record. Argyll then passed to the matter at hand: the moral and material condition of the first men. He agreed with Lubbock that Whately's demands for proof of savage self-civilization were unreasonable, and he doubted that savages were incapable of self-improvement.[12]

Nonetheless, Argyll faulted Sir John on two vital points. The first was his assumption that the demonstration of a crude level of technological knowledge in primeval times was adequate grounds for inferring a comparably low degree of moral and intellectual life such as that which Lubbock found among modern savages using a similar technology. Sir John, in other words, was guilty of what Glyn Daniel has called the comparative ethnographic fallacy. Lubbock, it should be noted, did seem at odds with himself about how literally this rule should be applied, as were most progressionists.[13] The second assumption was that the more coarse and vicious a custom, the older it was.[14] On these two assumptions, thought Argyll, Lubbock had erected his "savage-theory" of human origins.

Against Sir John's use of the level of technology as an index to mental and moral ability Argyll appealed first to the (to him) undeniable fact that the specific unity of man implied an innate equality. Be it ever so approximate—and the Duke was no egalitarian[15]—this equality was sufficient to prevent the relegation of any race, ancient or modern, to the category of near-brutes. Evolutionists, who, he charged, were constantly using comparative anatomy to narrow the gap between the gorilla and *Homo sapiens,* were too prone to ignore the factor of mind in their comparisons. Argyll, who saw a clear affinity connecting physical form, ecological place, and intellect in all animal species, thought this wrong. Whatever merely physical similarities there might be, the mental chasm between man and brute was infinite. A Hindu skull might be only eleven cubic inches larger than a gorilla's, but the one was the skull

[12] Duke of Argyll, *Primeval Man: An Examination of Some Recent Speculations* (New York: John W. Lovell Co., n.d. [first published in 1868]), pp. 46–47.

[13] John Lubbock, *Pre-Historic Times, As Illustrated by Ancient Remains and the Manners and Customs of Modern Savages* (2nd ed., London: Williams and Norgate, 1869), pp. 416–417, 574–575; but cf. pp. 539 f. Also see Lubbock, *Origin of Civilisation* (5th ed., 1889), pp. 2–3. Glyn E. Daniel, *A Hundred Years of Archaeology* (London: Duckworth, 1950), p. 186.

[14] Argyll, *Primeval Man,* pp. 44–45.

[15] *Ibid.,* p. 22.

of a rational being, the other that of a brute. Given the unfailing correlations of mind and physical form in nature, one must infer that all men having the same body must have essentially the same mind, the same nature, regardless of their technology.

Argyll supported this conclusion by an appeal to prehistoric archaeology: the oldest known human remains (which the Duke failed to name but which were probably the Neanderthal [1857] and Engis [1833] skulls) revealed that as far back as science could trace, man has always been man, generically unique, his ancient races fully within the range of modern variation. There was no paleontological evidence that primeval man had come up from the ape and so no warrant for thinking him mentally less than modern man. Indeed, given his defenselessness, his weak and weaponless body, man with little more mind than a gorilla could not have survived.[16]

The savage theory erred, then, in failing to distinguish between an absence of technological knowledge and a lack of mental faculties. Lubbock spoke of "'utter barbarism,'" but no creature, in his words, "'worthy to be called a Man,'" could have lacked a rational mind: "a mind capable of reason, disposed to reason, and able to acquire, to accumulate, and to transmit knowledge." Such a being, the Duke urged, could not be said "on the ground merely of his ignorance of mechanical arts" to be "in a condition of 'utter barbarism.'" How much less so if he possessed a moral sense. If to reason from a lack of technical knowledge to a lack of intellect was fallacious, how much more so to discover thereby a lack of morality?[17]

Believing that primeval man's first ventures into technology were guided by instinct (to throw a rock or to use a stick as a club Argyll considered as instinctive in man as biting was in a dog), he rejected Whately's thesis of divine instruction as improbable and, at any rate, as beyond proof.[18] But to assert against Whately, as the savage theorists did, that man's earliest material progress was of his own making, involved them in the admission that man's noblest inventions were his first. But far from being the work of mindless savages, making the first tools, taming fire, discovering agriculture, were mental feats of the highest order. "The very earliest inventions of our race," wrote the Duke, "must have been the most wonderful of all, and the richest in the fruits they bore."[19] Interestingly, Charles Darwin, in his disagreement with Alfred Russel Wallace over the evolution of the human brain, concurred. Speaking of early tools, the use of fire, and so forth, he wrote,

> These several inventions, by which man in the rudest state has become so preeminent, are the direct result of the development of his powers of observation, memory, curiosity, imagination, and reason. I cannot, therefore, understand how it is that Mr. Wallace maintains that "natural selection could only have endowed the savage with a brain a little superior to that of an ape."

Among these primeval discoveries Darwin called that of fire and language the greatest "ever made by man."[20]

[16] *Ibid.*, pp. 18–24.

[17] *Ibid.*, pp. 43–44.

[18] *Ibid.*, pp. 47–49, 51. Whately, of course, denied the reality of such instincts: "Origin of Civilization," pp. 18–21.

[19] Argyll, *Primeval Man*, pp. 49–50.

[20] Charles R. Darwin, *The Descent of Man and Selection in Relation to Sex* (London: John Murray, 1871), Vol. I, p. 132.

But this was not the extent of the paradox in the savage theory. Even if we accept modern savages as models of primeval man, Argyll thought, we must still entertain a high opinion of the latter by that analogy, for the implements of the former are frequently "highly ingenious, and sometimes eminently beautiful." In fact, a savage, having few material possessions to start with, might require more intellect and power of observation than a civilized man, certainly not significantly less.[21]

The second point raised against Lubbock was his assumption, shared by other progressionists, that traces of rude, simple, and frequently cruel customs surviving in more highly developed societies showed that those societies had begun in savagery. The most such evidence could prove, objected Argyll, was that the society had passed through one or more periods of barbarism: "it affords no presumption whatever that barbarism was the Primeval Condition of Man, any more than the traces of Feudalism in the laws of modern Europe prove that feudal principles were born with the Human Race." Such customs may well have been, as some claimed, "not Primeval but Medieval," that is, the fruit of time and corruption. Even Lubbock acknowledged a tendency in man to develop base and unnatural customs, for example cannibalism, which practice even he hesitated to call primeval.[22] Similarly, the absence of religion among the most abject modern savages (which Argyll doubted) did not prove that atheism was man's first state; it merely indicated the degree to which a people could decline—just as the elaborate superstitions of modern Hinduism showed how the passage of time and men's imaginations could work the corruption of a pure religion like that of the Vedas.[23]

Nor would the Duke accept the use of stone implements by ancient civilizations as evidence that they had emerged from a savage Stone Age. Neither were the drift implements of glacial Europe evidence of a contemporaneous and universal Stone Age: "it would be about as safe to argue from these implements as to the condition of Man at that time in the countries of his Primeval Home, as it would be in our own day to argue from the habits and arts of the Eskimo as to the state of civilization in London or in Paris." Argyll did not deny that the use of stone had preceded the use of metal as a general fact of history. What he did question was the universal application of a rigidly evolutionist Stone Age–Bronze Age–Iron Age sequence and the further assumption that Stone Age technology was to be equated with a brutish condition of mind and body.[24]

If Argyll rejected the progressionist idea that the lowest contemporary savages were fossils of human evolution, how did he account for them? They were, first, the result of the operation of a principle of degradation in human nature, a principle which all history witnessed and which worked in opposition to a principle of progress. Secondly, they were the victims of severe natural environments. The first men, he speculated, had lived in a warm country where a fruitful nature complemented their undeveloped knowledge of the arts and sciences. In time, population pressure drove the weaker into the remote and less hospitable areas of the world, into the barren wastes, the

[21] Argyll, *Primeval Man*, pp. 45–50.
[22] *Ibid.*, pp. 44–45.
[23] *Ibid.*, pp. 60–63.
[24] *Ibid.*, pp. 58–60.

"woods and rocks" at the extremities of the several continents. These were the savages of today: the Eskimo, the wretched inhabitants of Tierra del Fuego, the Australians, the Tasmanians. Reduced to brutelike condition by their natural surroundings, they had degenerated, losing whatever arts and skills they might have once possessed. Argyll thought it important to note, in support of his view, that these men, degraded as they were, readily showed they had "all the perfect attributes of humanity" when placed in more favorable circumstances. Their degradation, then, was the result of their poor environment and the momentary dominance of man's tendency to evil, not of an innate inability to progress, as Whately and some of his critics mistakenly believed. But, against Lubbock, there was nothing to connect these unfortunates with the original condition of man.[25]

* * *

The Duke prepared the way for the subsequent misrepresentation of *Primeval Man* by stating that Whately's argument, though weak and one which he had never accepted, was stronger than Lubbock's theory of social evolution from savagery.[26] Replying at the Exeter meeting of the British Association in 1869, Sir John set the tone of progressionist response by charging Argyll with merely reproducing the Archbishop's position "with but slight alteration and somewhat protected by obscurity," however aware the Duke might be of certain of its failings, and by implying that Argyll was an isolated and eccentric critic whose main concern was with the protection of orthodox religion.[27]

The substance of Lubbock's reply may be dealt with briefly. To the Duke's question—how does one justify inferring a condition of savagery, which commonly includes a debased mind and moral sense, from a low level of technology—Lubbock was evasive. First he replied by restating his original case: a society without knowledge must be a very barbarous one. Secondly, he suggested the impropriety of measuring the morality of savages:

> Morality implies responsibility, and consequently intelligence. The lower animals are neither moral nor immoral. The lower races of men may be, and are, vicious; but allowances must be made for them. On the contrary . . . the higher the mental power, the more splendid the intellectual endowment, the deeper is the moral degradation of him who wastes the one and abuses the other. . . . savages are more innocent, and yet more criminal, than civilized races; they are by no means in the lowest possible moral condition, nor are they capable of the higher virtues.[28]

This stands in surprising contrast to the position taken earlier in *Pre-Historic Times,* where Lubbock clearly had equated low moral and mental qualities and located both in contemporary savages; it was apparently an *ad hoc* redefinition of morality.[29] Neither Argyll nor Lubbock had been concerned previously with savages as willful wrongdoers but rather with the level of their moral knowledge and practice on the conventional scale of civilization. Now Lubbock seemed

[25] *Ibid.,* pp. 51–58, 61–62.
[26] *Ibid.,* pp. 2, 45, 50–51, 64.
[27] Lubbock, *Origin of Civilisation,* pp. 337–339, 361 n.2.
[28] *Ibid.,* p. 340.
[29] Lubbock, *Pre-Historic Times,* 2nd ed., pp. 560–563. This passage continued to appear in all subsequent editions through the last of 1913.

to be saying that moral considerations were not even a proper category for the evaluation of savage life. Indeed, in The *Origin of Civilisation,* published the following year, Lubbock concluded that some contemporary primitives had no moral feelings at all and that therefore these sentiments could not have belonged to primeval man.[30] Progress, in this view, would be from amorality to morality rather than from immorality to morality. But the tension between amoral and immoral savages is not resolved in Lubbock's work. Still, the Duke's question remained unanswered.

Turning next to the issue of survivals as evidences of primeval savagery, Sir John, after turning aside Argyll's challenge that the most vicious customs are not necessarily the oldest by taxing him with self-contradiction, simply restated his own position:

> My argument, however, was that there is a definite sequence of habits and ideas; that certain customs (some brutal, others not so), which we find lingering on in civilised communities, are a page of past history, and tell a tale of former barbarism, rather on account of their simplicity than their brutality. . . .[31]

But this left Argyll asking, as he had earlier, "a page of past history?" "former barbarism?" when and in what degree? Primeval savagery was still assumed and the assumption still unjustified.

Lubbock then moved rapidly over the remainder of Argyll's criticisms. Alleging that the Scottish peer believed man's primeval state to have been civilization, he ridiculed the notion that the savages of today are degenerate survivors of more advanced ancestors who suffer from environmental deprivation: " 'rocks and woods' " may provide a salutary environment, as Scotland itself shows. Where, he asked, is the evidence of degeneration among the Australians and Eskimos? And if the former declined because of the physical environment, a "gloomy" outlook is thereby suggested for British settlers down under.

Reaffirming other views put forth in his paper at Dundee and quite properly accusing Argyll of misrepresenting his use of the Stone–Bronze–Iron Age schema, Lubbock ended by adding two new arguments for the savage theory. First, using Darwin's observation that "crossed races of men are singularly savage and degraded," he suggested that "perhaps . . . the degraded state of so many half-castes is in part due to reversion to a primitive and savage condition. . . ." This inference rested on the belief that "offspring produced by crossing different varieties tends to revert to the type from which these varieties are descended." The second drew on the often remarked similarity of the minds of children and savages. Naturalists, he said, increasingly agreed that "the development of the individual is an epitome of that of the species." The mental resemblance of savages and children, then, "assumes a singular importance and becomes almost conclusive as regards the question now at issue."[32] Savages were frozen in time at that primeval point which children recreated.

* * *

A number of points should be made in summing up this encounter. First, it is important to notice that both Argyll and Lubbock dismissed Whately's

[30] Lubbock, *Origin of Civilisation,* pp. 269–271.
[31] *Ibid.,* pp. 340–341.
[32] *Ibid.,* pp. 341–361.

contention that savages are incapable of self-improvement and that divine instruction is the only hypothesis with which to explain the beginning of civilization. It also should be acknowledged that Lubbock, who was generally a careful and accurate reasoner, justly reproached Argyll with vagueness, contradiction, and technical ignorance. However, this should not obscure the fact that Lubbock faced neither of the Duke's major points. He offered no systematic justification of the equation of modern savages with primeval man and of the inference of mental and moral capacity from material culture, nor did he justify the assumptions underlying the doctrine of survivals. In fact, Lubbock's reply was little more than a restatement of his Dundee paper, which strategy forced him into a number of revealing acts.

First of all, Lubbock was forced to equate Argyll's position with Whately's, to treat the Duke as a mere reviver of the Archbishop's thesis. This involved at least two important misrepresentations: first, that the Duke believed primeval man to have been civilized; and, secondly, and most important, that Argyll wrote as a champion of religious orthodoxy. A third point, related to the second, was to ignore Argyll's vigorous conviction that what he was doing was science and not theology.

Contemporaneous with the Argyll-Lubbock debate was a sporadic discussion of the problem of defining civilization. Two groups tended to form. One thought of civilization in material terms of progress in science and technology. The absence of these things necessarily meant savagery. Members of this group, more often than not, seem to have been progressionists. The others, who mainly favored degenerationism, insisted that a civilized man should possess reason and moral sensibilities somewhat well developed regardless of his material accomplishments, and that early man was so possessed.[33] Argyll fits neatly into neither group. Like the first, he believed that an extensive knowledge of arts and sciences was a necessary element of civilization. Like the second, he had a high opinion of primeval man. Primeval man, for Argyll, was fully man in his faculties but he was not "civilized," as he possessed only "a knowledge of the simplest arts which were necessary for the sustenance of life . . . a condition of mere childhood."[34] And these, at first, as we have seen, were instinctive. Contrary to his critics, the Duke's point was not that primeval man was civilized but that not being so did not necessarily make him a savage.

As to Argyll's being a defender of orthodoxy, he not only did not represent himself as such, but, if by orthodoxy—a vague term at best—is meant the conventional devotion to the integral unity of Scripture and a narrow literalism in its interpretation, it is clear that he was not orthodox in that sense. In fact, reviews of his books in the orthodox press frequently censured his "advanced" views on theological questions.[35] Repeatedly in *Primeval Man* he

[33] Not exhaustive, but illustrative, are Sebastian Evans, "Sir John Lubbock on the Origin of Civilisation," *Nature*, Mar. 9, 1871, p. 363; J. Gould Avery, "Racial Characteristics as Related to Civilisation" (abstract), *Journal of the Royal Anthropological Institute of Great Britain and Ireland*, 1873, *2*:63–65; James Reddie, "Man, Savage and Civilized—An Appeal to Facts," *Ethnological Journal*, 1865, *1*:158–159; A. B. [reply to Reddie], *Ethnol. J.*, 1866, *2*:224; S. J. Whitmee, "The Degeneracy of Man," *Nature*, May 20, 1875, p. 47; Anon., "Tylor on Primitive Culture," *Edinburgh Review*, 1872, *135*:51.

[34] Argyll, *Primeval Man*, pp. 10–11, 52–57. Duke of Argyll, *The Unity of Nature* (New York: G. P. Putnam's Sons, 1885), pp. 328–384.

[35] See, e.g., *Catholic World*, 1868, *6*:598–600; *ibid.*, 1869, *9*:746; *Month*, 1867, *7*:555–556, 560; *Princeton Review*, 1870, *42*:65–68; from a later period, *Presbyterian and Reformed Review*, 1897,

took a liberal view of the interpretation of biblical passages. Also, he was very careful to disassociate his idea of human corruptibility, on which he based his principle of degeneration, from any theological base.[36] However, his frequent use of biblical phrases and illustrations and his willingness to express openly his Christianity and to treat it as a datum of which science should take cognizance doubtlessly made him appear orthodox to the more secular minded. One suspects that Lubbock, who, it should be mentioned, was not himself hostile to religion,[37] was antagonized more by the biblical flavor of Argyll's expression than by his ideas. Indeed, the latter seem to have slipped by him. In this, Sir John established an archetypical pattern for exchanges of this kind.

* * *

Historians writing about the incident have tended to follow Lubbock's interpretation. Andrew D. White, in his still influential *History of the Warfare of Science and Theology in Christendom* (1896), depicted both Whately and Argyll as motivated solely by religious considerations. He attributed Whately's attack on progressionist anthropology to his "holding a brief for the Church" and suggested that had the Irish primate "been a layman" he "would probably have studied the matter with more care and less prejudice."[38] The Duke he described as the "head and front of the orthodox party," as a "determined supporter of the theological side," as a defender of "the old theological view" of human origins, as "the foremost leader of the theological rear-guard," and attributed his principle of degradation to the influence of Scottish Calvinism. Oddly enough, White did acknowledge the Duke's "liberal views" on Darwinism, which is to say that Argyll was willing to treat the role of natural selection in the origin of new species as a scientific question subject to unprejudiced tests of evidence and logic just as he (and Whately—though with less candor) hoped to see the degeneration hypothesis treated. White, however, followed Lubbock in the not too difficult refutation of Argyll where he was most vulnerable and ignored his questions regarding the equivalence of Stone Age technology with intellectual and moral savagery and the evidence of survivals.[39]

Recent writers have continued to see Whately and Argyll as spokesmen for an orthodox theology and as saying essentially the same thing. George Stocking, for instance, sees the degenerationists using the Bible "as a kind of Kuhnian paradigm for research on the cultural, linguistic, and physical diversity of mankind."[40] This view is not wrong, but it tends to obscure the attempts of degenerationists to argue their case as *science* just as it does the theoretical embarrassments for progressionism that their critique involved. Marvin Harris, to cite another example, writes, "degenerationism—the belief that all contempo-

---

8:303–304. See also Charles Hodge, *What is Darwinism?* (New York: Scribner, Armstrong & Co., 1874), pp. 24–26. On the other hand, apparent religious liberals praised Argyll for his attempts to conciliate science and religion: *Congregational Review*, 1870, *10*:553–565; *Christian Remembrancer*, 1867, *53*:128–137; *Outlook*, June 27, 1896, p. 1209.

[36] Argyll, *Primeval Man*, pp. 10–12, 26–31, 41–42, 61.

[37] Lubbock, *Pre-Historic Times*, 2nd ed., pp. viii–ix, 580–581.

[38] Andrew Dickson White, *A History of the Warfare of Science and Theology in Christendom*, 2 vols. (New York: D. Appleton, 1898), Vol. I, p. 304.

[39] *Ibid.*, pp. 82, 86n–87n, 281n, 282, 305–307.

[40] George W. Stocking, Jr., *Race, Culture, and Evolution* (New York: Free Press, 1968), p. 71.

rary primitives are descended from peoples who enjoyed civilization prior to the construction of the Tower of Babel—was, in its multifarious expressions, little more than an attempt to preserve the authority of Biblical history." For Harris, all degenerationists were merely theological reactionaries.[41]

It would be absurd to deny the prominent place occupied by a concern for religious truth in the minds of degenerationists. But an awareness of this should not lead one to assume that a clear-cut "science versus theology" delineation of the question is proper. Some progressionists so presented it; their example has been followed. But a reading of Argyll's work alone makes the smell of red herring too pronounced to be ignored. In accepting this progressionist version, historians have simply perpetuated the interested interpretation of the more partisan members of the evolutionary party who used the division between science and theology to dismiss Argyll and others like him as theologians intruding into the area of science.

A survey of expressed opinion suggests that contemporary critics of degenerationism may be gathered into three rough groups (with some overlapping, of course): a minority who took the theory seriously enough as science to argue against it and who rejected it as science; those who also refuted the theory as an error in science, but who suggested an underlying theological motive on the part of its advocates; and last, those who simply dismissed the theory as religious reaction.

Representative of the first group are Sir Charles Lyell and Charles Darwin. Lyell, in the *Principles of Geology,* raised the then popular question "Whether man has been degraded from a higher or has risen from a lower stage of civilization." Drawing on Lubbock's Dundee paper, he defended the second alternative but nowhere even hinted that this was other than a scientific question.[42] In *The Descent of Man,* Darwin argued against degenerationism but treated Whately and Argyll with respect, praised some of their points (Whately on the use of language by the lower animals, Argyll's discussion of man's moral nature), admitted that the "problem . . . of the first advance of savages toward civilization is at present much too difficult to be solved," and left the question without imputing to his opponents ulterior motives. He did, however, reject what he took to be their pessimism:

> To believe that man was aboriginally civilized and then suffered utter degradation in so many regions, is to take a pitiably low view of human nature. It is apparently a truer and more cheerful view that progress has been much more general than retrogression; that man has risen, though by slow and interrupted steps, from a lowly condition to the highest standard as yet attained by him in knowledge, morals, and religion.[43]

[41] Marvin Harris, *The Rise of Anthropological Theory* (New York: Crowell, 1968), p. 54. Among others taking the same general view of degenerationism, see Milton Millhauser, *Just Before Darwin, Robert Chambers and Vestiges* (Middleton, Conn.: Wesleyan University Press, 1959), pp. 106–107; Robert H. Lowie, *The History of Ethnological Theory* (New York: Rinehart, 1937), p. 20; Abram Kardiner and Edward Preble, *They Studied Man* (New York: World, 1961), pp. 53–54; in the same vein, but more perceptive, is E. S. Carpenter, "The Role of Archaeology in the 19th Century Controversy between Developmentalism and Degenerationism," *Pennsylvania Archaeologist,* 1950, *20*:13–15.

[42] Charles Lyell, *Principles of Geology* (11th ed., New York: D. Appleton, 1872), Vol. II, pp. 485–486.

[43] Darwin, *Descent of Man,* Vol. I, pp. 50–53, 59, 100n, 149–151, 160–161, 174–177.

It is curious to note that Darwin provided primeval man with a "fall" not too unlike Argyll's notion of corruption. In the earliest times, when men had just evolved into the human state, their instincts were still stronger than their newly emerging powers of reason. Consequently there was no infanticide, no "artificial scarcity of women," and hence no female slavery and polyandry, and no "early betrothals"; there was a free choice of sexual partners, all adults paired, and all children were raised, insofar as possible. With the development of reason, all this changed: "man advanced in his intellectual powers, but . . . retrograded in his instincts."[44]

In addition to Lubbock, the second group may be represented by Edward Burnet Tylor and by Ludwig Büchner, several of whose translated works enjoyed popularity in Great Britain and America. While more than willing to argue the question on the basis of empirical data and convinced that what evidence of degeneration there was applied only to a small number of individual primitive groups and was not a general rule for all savages, Tylor felt that the degeneration hypothesis was essentially a theological one. Dismissing Whately as a "speculative philosopher, whose course is but little obstructed by facts," Tylor confessed, "I do not think I have ever met with a single fact which seems to me to justify the theory . . . that the ordinary condition of the savage is the result of degeneration from a far higher state."[45] In *Primitive Culture,* he warned the reader that

> It must be borne in mind . . . that the grounds on which this theory has been held have generally been rather theological than ethnological. . . . At the present time it is not unusual for the origin of civilization to be treated as a matter of dogmatic theology. It has happened to me more than once to be assured from the pulpit that the theories of ethnologists who consider man to have risen from a low original condition are delusive fancies, it being revealed truth that man was originally in a high condition.[46]

Büchner took much the same line. Degeneration occurs in limited cases, but such weakened tribes soon became extinct. This was a universal law of nature. But the robustness of true savages and their adjustment to their environment show that their condition has been a permanent one; other peoples had simply outstripped them. Progressionism was the only theory in accord with the facts. Degenerationism was merely a restatement of theological conclusions and contrary "to every known fact."[47]

In the third category there appears sharply a tendency which is noticeable also in the second, although matched there by the dignity of counterargument. Degenerationists insisted that the controversy over man's primeval condition was one within science. The more partisan progressionists dismissed it as one of science versus an outmoded theology. Clearly, Argyll saw Whately and Lubbock as scientific opponents and stressed the Archbishop's exclusion of the testimony of Genesis *as revelation* from his argument and insisted no less

[44] *Ibid.*, Vol. II, pp. 351–352, 366. Cf. Argyll, *Unity of Nature,* pp. 386–390.

[45] Tylor, *Researches,* pp. 160, 364–365.

[46] Tylor, *Primitive Culture,* Vol. I, p. 36. For arguments against Whately, see *ibid.*, pp. 41–42 and *Researches,* pp. 161–162; against Argyll, *Primitive Culture,* Vol. I, p. 60; on the true significance of degeneration, *Researches,* pp. 182–184, 190, and *Primitive Culture,* Vol. I, pp. 46–48.

[47] Ludwig Büchner, *Man in the Past, Present and Future,* trans. W. S. Dallas (London: Asher and Co., 1872), pp. 67–71, 275–277, 324–325.

on his own scientific approach.[48] But Whately had a weakness for appealing to the literal letter of Genesis, and even one of Argyll's fairest critics noted his willingness to receive "aid . . . from a due respect to the Mosaic narrative."[49] Such lapses in scientific rigor were not lost on the advocates in category three. Among them the American anthropologist Lewis Henry Morgan denounced degenerationism as "a corollary from the Mosaic cosmology" and as neither necessary or tenable. Lecturing at the Lowell Institute in Boston in 1865, J. P. Lesley, Secretary of the American Philosophical Society, branded degenerationism as a Jewish creation legend and declared "Science cannot resign to a theological conjecture." J. Stahl Patterson, writing in *The Radical,* condemned it as an old "Jewish" view and its advocates Whately and Argyll as social as well as religious reactionaries. John Fiske, Herbert Spencer's leading American disciple, dismissed Whately's "wonderful theory" as "not worth while for any busy man to stop and refute." And the French scientist N. Joly passed over the more biblical forms of the idea as undeserving of "serious reflection" to say nothing of refutation.[50]

* * *

Such impatience with degenerationism was not unrelated to the prevalent belief that the Duke was merely restating Whately's argument, that he was merely a conservative defender of orthodox religion who had been put in his place by Sir John Lubbock. To some extent Argyll had set himself up for this misinterpretation by his seeming reluctance to say that Whately's notion was wrong, although he obviously rejected the idea of miraculous instruction of early man.[51]

Closer attention to the Duke's case might have led to some clarifying discussion of the assumptions of evolutionary anthropology long before this did occur by pointing out the degree to which he was introducing new questions. For, in truth, Whately's position, no less than Lubbock's, depended on the identification of primeval man with the modern savage. If primeval man had not possessed the supposed disability of modern savages, why would he have needed a divine revelation?[52] Along with his attack on the doctrine of savage primeval survivals, Argyll was attempting to re-examine this equation, which Whately accepted even while he denounced it and which progressionists seldom seriously questioned. But one is left with the feeling that more progressionists got their idea of Argyll's views from reading Lubbock's Exeter paper than from reading

[48] Argyll, *Primeval Man,* pp. 2–3, 7–9. See also Whately, *Introductory Lectures,* pp. 28–31.

[49] Whately, "Origin of Civilization," p. 20; anon., "Argyll on Archaic Anthropological Speculations," *Anthropol. Rev.,* 1869, 7:376.

[50] Lewis Henry Morgan, *Ancient Society, or Researches in the Lines of Human Progress from Savagery through Barbarism to Civilization* (New York: Henry Holt, 1878), pp. 7–8. J. P. Lesley, *Man's Origin and Destiny* (Philadelphia: Lippincott, 1869), pp. 133–134; J. Stahl Patterson, "The Metal Ages," *Radical,* 1871, *9*:47; John Fiske, *Outlines of Cosmic Philosophy,* 4 vols. (Cambridge, Mass.: Riverside Press, 1902), Vol. III, pp. 257–258; N. Joly, *Man before Metals* (New York: D. Appleton, 1883), pp. 259–260.

[51] E.g., Argyll, *Primeval Man,* pp. 50–51.

[52] For a discussion of the contradiction in Whately's argument between denying that primeval men were savages and requiring them to have been such by implication, see John Hannah, "Primeval Man," *Contemporary Review,* 1869, *11*:165–166.

Primeval Man—a conclusion reinforced by the fact that of all the degenerationists only the two whom Lubbock attacked have survived in memory. As it was, Charles Darwin spoke for most of his persuasion when he wrote in the *Descent of Man:*

> The arguments recently advanced by the Duke of Argyll and formerly by Archbishop Whately, in favor of the belief that man came into the world as a civilized being and that all savages have since undergone degradation, seem to me weak in comparison with those advanced on the other side.[53]

The ambiguity surrounding the term "civilization" only promoted a continuing confusion of issues; and, with distressing regularity throughout the remainder of the century, the Duke was written off as a naïve religionist who was in over his head. Yet Argyll, for all his flaws as a scientist, had called into question the same supports of nineteenth-century social evolutionism—the comparative method and doctrine of survivals—as Boas, Malinowski, and other anthropological rebels of the twentieth century were to question, but he was not answered. Science, no less than religion, had its ways of dealing with heretics who arose before their time. When Argyll wrote, the new science of man was concerned to win its independence from the biblical tradition. Only later, when a secular outlook was well established, could it turn to a critical evaluation of its evolutionary presuppositions.

[53] Darwin, *Descent of Man,* Vol. I, p. 174.

Racial Science, Social Science, and the Politics of Jewish Assimilation

By Mitchell B. Hart

ABSTRACT

This essay examines the work of Jewish social scientists who in the first decades of this century analyzed modern Jewish life from the perspective of anthropology and medicine. While the historiography of the social and racial sciences has focused almost exclusively on the Jews as objects of these sciences, scholars have begun to explore the degree to which Jews themselves were involved in social and racial scientific research about their own people. Trained in the natural and social sciences, Jewish researchers shared the same conceptual and methodological framework as their non-Jewish counterparts. Yet they had their own social and political agendas, and they used their research to achieve these. This essay demonstrates that Jewish social scientists, while united in their desire to counter scientific anti-Semitism through the use of social science, nonetheless were divided in the political or ideological conclusions they drew from their findings. More specifically, the essay shows how anthropological and medical analyses of Jewry were impelled by the debate between Jewish "assimilationists" and nationalists (Zionists) over Jewish integration in modern society and culture.

WELL BEFORE THE FIRST DECADE OF THE TWENTIETH CENTURY, the "sickness of Jewry" had become more than a metaphor. The "Jewish body," both individual and collective, was something that, for many Jews as well as non-Jews, was diseased and degenerate, an entity that literally required healing. The importance of this image of the diseased Jewish body and of "the Jew" as an object of scientific research and popular representation for the anti-Semitic imagination is well known and fairly well documented. Indeed, when we think about Jews and racial theories of health and disease, we are accustomed to thinking about the Jews first as objects of a largely anti-Semitic racial science and then as victims—primarily of a Nazi regime fueled to a significant degree by that science.[1]

I would like to acknowledge the generous support of the Florida International University Foundation and the university's College of Arts and Sciences for the preparation of this article. I thank Margaret Rossiter and the anonymous referees for their very helpful criticisms and suggestions.

[1] This is so for very good and clearly understood reasons. As historians of German medical and racial thought have shown, it is impossible to understand the Holocaust without comprehending the degree to which racial science and a medicalized racial ideology occupied central positions in Nazi thought and policy. See Paul Weindling, *Health, Race, and German Politics between National Unification and Nazism, 1870–1945* (Cambridge: Cambridge Univ. Press, 1989); Michael Burleigh and Wolfgang Wippermann, *The Racial State: Germany, 1933–1945* (Cambridge: Cambridge Univ. Press, 1991); and Robert Proctor, *Racial Hygiene: Medicine under the Nazis* (Cambridge, Mass.: Harvard Univ. Press, 1988). On the image of the Jewish body see, among many others, Sander Gilman, *The Jew's Body* (London: Routledge, 1991); Proctor, *Racial Hygiene*; and George Mosse, *Toward the Final Solution* (New York: Fertig, 1978).

Without downplaying the significance of racial anti-Semitism, this essay seeks to show that the history of racial science vis-à-vis the Jews is more complex and convoluted. Historians have demonstrated that racial science and eugenics were ideologically heterogeneous into the 1920s, even the 1930s. A racialized interpretation of history and contemporary social problems was by no means the provenance of right-wing conservatives and anti-Semites alone. Eugenics, for instance, was at various times touted by liberals, socialists, and feminists as "progressive." There was no necessary or essential connection between a strong belief in the reality and import of race and any one particular ideological position.[2]

Nor was a belief in the reality and importance of racial identities and differences, and an involvement in research into these matters, confined to one particular ethnic or religious group. Although some Jewish writers were addressing themselves to issues such as Jewish health and disease already in the late eighteenth century, it was in the half century or so between 1880 and 1930 that Jewish scholars and popular writers became especially engaged in the debates over racial identity and difference, heredity and environment. Thus by 1906 Arthur Ruppin, a prominent Jewish social scientist and the head of the Bureau for Jewish Statistics in Berlin, could assert that "almost all inquiries into the social, intellectual, and physical differences between Jews and Christians address the question of whether these differences have their root in the particular racial makeup or in the unique economic and political conditions of the Jews over the past two thousand years. One might designate this question, in fact, as the fundamental problem or issue of social scientific research about the Jews." Ruppin was speaking here not only about the literature produced by non-Jewish authorities, but also about the growing body of work emanating from Jewish demographers, anthropologists, and physicians. Contemporary Jewry was thus an object of scholarly research undertaken by Jews; the inquiry into the racial identity and health of the Jews became a central component of a developing Jewish social science that sought to understand and explain modern Jewry by using social scientific knowledge and methods. It broadened the definition of a Jewish *Wissenschaft* (science or scholarship), shifting the focus of research away from Judaism and its past and onto Jewry in the present. Social scientific analyses of the condition of modern Jewry became, in turn, part of the contemporary debate about fundamental political and ideological matters: civic emancipation, integration and assimilation, anti-Semitism, and Jewish nationalism.[3]

Science, it is important to note, did not speak about the Jews univocally. The narrative was not stable or essentialized; that is, the discourse of Western science, including medi-

[2] Proctor, *Racial Hygiene*, Ch. 1; Weindling, *Health, Race, and German Politics*; and Sheila Faith Weiss, "The Race Hygiene Movement in Germany," *Osiris*, 2nd Ser., 1987, *3*:193–236. The point has been made most recently in Benoit Massin, "From Virchow to Fischer: Physical Anthropology and 'Modern Race Theories' in Wilhelmine Germany," in Volksgeist *as Method and Ethic: Essays on Boasian Ethnography and the German Anthropological Tradition*, ed. George W. Stocking, Jr. (Madison: Univ. Wisconsin Press, 1996).

[3] Arthur Ruppin, "Begabungsunterschiede christlicher und jüdischer Kinder," *Zeitschrift für Demographie und Statistik der Juden*, 1906, *2*(8–9):129–136, on p. 129. Here and elsewhere, all translations are mine unless otherwise indicated. A few recent works have traced the development of a racial scientific discourse among Jewish scholars and considered the political uses to which this was put. See John Efron, *Defenders of the Race: Jewish Doctors and Race Science in Fin-de-Siècle Europe* (New Haven, Conn.: Yale Univ. Press, 1994); Sander Gilman, *The Case of Sigmund Freud: Medicine and Identity at the Fin de Siècle* (Baltimore: Johns Hopkins Univ. Press, 1993); Gilman and Nancy Leys Stepan, "Appropriating the Idioms of Science: The Rejection of Scientific Racism," in *The Racial Economy of Science*, ed. Sandra Harding (Bloomington: Indiana Univ. Press, 1993), pp. 170–193; Mitchell Hart, "Picturing Jews: Iconography and Racial Science," *Studies in Contemporary Jewry*, 1995, *11*:159–175; and Eric L. Goldstein, " 'Different Blood Flows in Our Veins': Race and Jewish Self-Definition in Late Nineteenth-Century America," *American Jewish History*, 1997, *85*(1).

Figure 1. Figures of devils. From Hans F. K. Günther, Rassenkunde des jüdisches Volkes *(1922), 2nd ed. (Munich: Lehmann, 1930), illustrations 37 and 38. The caption reads, "With near Eastern racial facial features." Illustration 37, on the left, is described as "the figure of a devil; from a manuscript in Cambridge." Illustration 38, on the right, is described as "the spirit of evil; a gargoyle from Notre Dame in Paris."*

cine, vis-à-vis the Jews as objects of research and analysis was heterogeneous. For many, of course, there did exist an essential Jewish nature, more often than not cast in negative, even dangerous, terms. According to the German anthropologist Richard Andree, the Jews were important for science because "one can trace them with certainty over thousands of years, and no other racial type has maintained such constant form and remained as immune to time and space. . . . The ancient Jewish physical and mental structures have remained constant." This notion of an essential, immutable Jewish body and soul, unchanged over the millennia, was a commonplace in the anthropological and medical literature.[4] The political and social message of this immutable Jewish nature was clear: the "Jewish body" was racially different and pathological, and opponents of Jewish emancipation and assimilation were correct in insisting that Jews were unfit to be part of a healthy modern nation-state.[5] (See Figures 1 and 2.) To be sure, it would be in Central Europe that this struggle against Jewish emancipation and integration would eventually succeed, with the most horrific consequences. However, as we shall see, this struggle was by no means limited to Central Europe in the decades before the rise of Nazism. Jews in France, England, and the United States, for instance, also confronted an increasingly vocal and organized anti-Semitism, oftentimes framed in racial scientific terms.

Yet the scientific literature produced about the Jews included not only negative but also

[4] Richard Andree, *Zur Volkskunde der Juden* (Bielefeld/Leipzig, 1881), p. 24. For other examples and analysis see Gilman, *Jew's Body* (cit. n. 1); and Hart, "Picturing Jews."

[5] From the 1870s onward social medicine and a racialized anthropology were crucial elements in the struggles over national identity, immigration, and, in the case of countries such as Germany and Italy, national unification. The notions of biological health and vitality and of an organic unity between the human and animal worlds were linked up in academic and popular discourses with the process of national integration. For the German case see Paul Weindling, *Health, Race, and German Politics* (cit. n. 1), p. 6 and *passim*. For Italy see Daniel Pick, *Faces of Degeneration* (Cambridge/New York: Cambridge Univ. Press, 1989), Ch. 5. On this connection between images of the Jewish body and the political arguments against Jewish integration see Gilman, *Jew's Body*.

***Figure 2.** Figure from Hans F. K. Günther,* Rassenkunde des jüdisches Volkes *(1922), 2nd ed. (Munich: Lehmann, 1930), illustration 249. The caption reads, "Jewish group at a spa. In their bearing and countenance they are certainly 'Jewish,' despite the attempt to appear Occidental."*

positive images and analyses, assertions of Jewish health and normalcy as well as disease and pathology. There was no fixed or uniform image of the "Jewish race" and its status or value vis-à-vis other "races."[6] More specifically, as the passage from Ruppin illustrates, there was an ongoing debate, which lasted well into the 1920s, over what the medico-anthropological data revealed about two fundamental issues: identity and causality. Were the Jews a religious community? A nation? A race? Just as significantly: If a race, could Jews be defined as a "pure race"? And what accounts for particular or "peculiar" Jewish traits, if one agreed that such things did exist? Were the Jews a product of nature or nurture, biology or environment? Was their susceptibility to or immunity from particular diseases genetically determined? Or could these be explained by a host of historical and contemporary social factors, including persecution, the adverse living conditions of the ghetto, and the "unhealthy" effects of orthodox religious laws and customs?

The heterogeneous nature of the scientific discourse about Jews allowed Jewish scientists to fashion responses out of elements drawn from this same scientific discourse, to construct their own narratives of "Jewish race," health, and disease. Especially important in this regard was the "environmentalist" explanation for anatomical and physiological traits, which emphasized historical and social forces as determinative and embraced the concept of acquired traits or characteristics. By the late nineteenth century the theory of acquired traits had come to be associated almost exclusively with the early eighteenth-century French naturalist Jean-Baptiste Lamarck—though, as commentators point out, the notion of acquired traits was already universally accepted when Lamarck utilized it. La-

[6] Massin, "From Virchow to Fischer" (cit. n. 2), refers to a "liberal racialism" that held sway in Germany and whose great champion was the renowned physician and physical anthropologist Rudolf Virchow. Although they have received far less attention in the scholarly literature than racial determinists and anti-Semites, there were a number of non-Jewish authorities who argued for a historicosocial interpretation of Jewish identity and the Jewish condition and linked this with a liberal, progressive political agenda. See the discussions in Mosse, *Toward the Final Solution* (cit. n. 1), pp. 91–92; and Erwin H. Ackerknecht, *Rudolf Virchow: Doctor, Statesman, Anthropologist* (Madison: Univ. Wisconsin Press, 1953).

marckism posited a tight connection between an organism and its surroundings and explained evolutionary transformation in an organism as a direct response to environmental conditions or challenges. In order to maintain (or regain) equilibrium or harmony with its particular environment, a species will adapt, either by using individual organs or limbs more frequently and thereby strengthening and enlarging them or by developing new parts to meet the new conditions. The anatomical and physiological changes that occurred as a result of this process of adaptation were, according to this theory, heritable. Significant structural changes, necessitated by environmental challenges, in turn produce transformations at the genetic level. New or altered structures become part of what is inherited by subsequent generations. Lamarckism, as it came to be formulated in the late nineteenth century, was associated almost exclusively with this notion of acquired characteristics, or "soft inheritance." This theory brought Lamarckians into conflict with Mendelians and supporters of August Weismann's germ plasm theory.[7] While Mendelian genetics, of course, eventually proved the more satisfactory theory, the dispute was by no means settled by the first decades of this century. Neither the ascendancy of Darwin's theory of natural selection nor the rediscovery of Mendel's laws of inheritance meant the demise of the notion of acquired traits. As Ludmilla Jordanova has written, for decades after 1900, indeed as late as the 1950s, "prominent biologists continued to defend Lamarck's biological philosophy of the inheritance of acquired characteristics."[8] Certainly, during the period in which Jewish social scientists were relying so heavily upon this sort of environmentalist argument the theory was a legitimate, if contested, one within the scientific community.

Jewish social scientists, however, were by no means entirely consistent in their adoption of this sort of environmentalism. They could, at times, put forth ideas and images of Jewry that relied on essentialist notions, making use of evidence found in the work of writers like Andree. Nonetheless, Jewish social scientists did rely heavily on a Lamarckian environmentalism, so much so that in some circles adherence to Lamarckian theory itself came to be identified as "Jewish."[9] And the theory of acquired traits did serve Jewish social scientists' purposes nicely.[10] It allowed them, in the first place, to explain the particular physical or mental traits often identified as racially Jewish as historically or socially determined. This was an attempt to undermine an explanatory framework that relied on an overly rigid biological determinism of the sort that undergirded a racialized anti-Semitism

[7] See, e.g., Ernst Mayr, *The Growth of Biological Thought* (Cambridge, Mass.: Harvard Univ. Press, 1982), p. 356. On Lamarck and his subsequent interpreters see *ibid.*, pp. 343–360; and Ludmilla J. Jordanova, *Lamarck* (Oxford: Oxford Univ. Press, 1984). The literature on this controversy regarding inheritance is enormous. For general overviews see Mayr, *Growth of Biological Thought*, p. 681 ff; and Proctor, *Racial Hygiene* (cit. n. 1), pp. 30–38.

[8] Jordanova, *Lamarck*, p. 111. To a significant degree the debate was infused with, and shaped by, powerful nationalist sentiments and struggles. On the politics of the debate see Proctor, *Racial Hygiene*, Chs. 1–2.

[9] E.g., the prominent German racial scientist Fritz Lenz argued, in Robert Proctor's words, that "the tendency toward 'Lamarckism' is a genetically selected racial characteristic" of the Jew: Proctor, *Racial Hygiene*, pp. 30–55. I have dealt with the Jewish, and especially Zionist, appropriation of essentialist arguments in Hart, "Picturing Jews" (cit. n. 3).

[10] This is not to say that there were no Jewish scientists who embraced a Mendelian view and applied it to the issues under discussion; they were, however, in the minority. The British Jewish biologist Redcliffe Salaman, for instance, who is probably best known for his epic social history of the potato, published articles on Jews, race, and heredity from a Mendelian perspective. He concluded that "the Jewish facial type . . . is a character which is subject to the Mendelian law of Heredity": Redcliffe Salaman, "Heredity and the Jew," *Journal of Genetics*, 1911, *3*(1):273–292, on p. 285. Salaman was also a Zionist, and he offered up his views on Jewish racial uniformity as evidence of a "genetic" as well as historical link between the Jews and the land of Israel. See Salaman, "Racial Origins of Jewish Types," *Transactions of the Jewish Historical Society of England*, 1922, *9*:163–184; and Salaman, "What Has Become of the Philistines: A Biologist's Point of View," *Palestine Exploration Fund Quarterly*, 1925, pp. 1–17.

aimed at limiting or reversing the civil and social rights attained by Jews over the course of the nineteenth century and restricting the number of Jewish immigrants coming from the Russian Empire. The politics of anti-Semitism assumed different shapes in different countries, each with its own political tradition vis-à-vis Jews and other minority groups. Nonetheless, in fundamental respects the scientific discourse on race and disease was international, and one finds the same or similar images and ideas circulating in academic texts produced in every language in which scientists wrote.[11] Since the "international" anti-Jewish discourse that emerged in the last half of the nineteenth century employed the methods and categories of the social sciences, Jews trained in these disciplines were particularly well equipped to respond.

Like other minority groups, such as blacks and women, Jews sought to insert themselves into the dominant narrative of science as a means of both identification with and differentiation from the dominant narrative and culture. So Jewish social scientists, it is important to note, were not necessarily averse to the idea that one could locate Jewish difference or particularity at the physiological level. They did not altogether disavow the notion of "race" as a category applicable to the Jews. Moreover, they accepted, to different degrees, the conjunction of "Jewish" and "diseased." As we shall see, they agreed with many non-Jewish authorities that, statistically, Jews suffered to a greater extent from a host of identifiable physical and mental disorders.

The ability and need to analyze contemporary Jewry in social scientific terms, during a period in which the social sciences were profoundly influenced by a racialized biology and medicine and in which anti-Semites employed this same scientific discourse, meant that the Jewish and anti-Semitic analyses overlapped in significant ways. Thus even as Jewish social scientists sought to dispel the dominant image of the Jews as diseased, and therefore eugenically dangerous, they put forth many of these same notions. As Sander Gilman and Nancy Leys Stepan have pointed out, in resisting or rejecting the racial and social scientific discourse about Jews, Jewish scientists perforce had to accept this discourse as well:

> As Jews (and other groups stereotyped in the biological and social sciences of the day) were drawn more deeply into the sciences of racial difference, whether in measuring themselves as a race by craniometry and other methods, or by comparing one fraction of the Jewish "race" with another, or by commenting on or contesting the thesis of Jewish pathology and illness, they were tempted simultaneously to embrace and reject the field: to embrace science's methods, concepts, and the promise it held out for discovering knowledge, and to reject, in a variety of ways, the conclusions of science as they appeared to apply negatively to themselves.[12]

Thus for Jewish social scientists the discourse on race presented both challenge and opportunity. As scientists, educated in central and western European universities, they internalized a body of literature that took physical and mental pathologies as signs of a collective identity and difference, reading normalcy or abnormality at the social, national, and racial levels. The social and medical sciences had been used to identify the Jewish body and mind as diseased and degenerate. As Jews, identified with their community or *Volk,* these scientists felt compelled to respond in some way to the increasingly widespread image of the diseased and dangerous Jew.

[11] See Gilman, *Jew's Body* (cit. n. 1); Proctor, *Racial Hygiene* (cit. n. 1); and Mosse, *Toward the Final Solution* (cit. n. 1).

[12] Gilman and Stepan, "Appropriating the Idioms of Science" (cit. n. 3), pp. 178–179.

Yet it would be a mistake to view their interest and involvement in the social sciences solely in reactive terms. Physical anthropology, social and racial hygiene, and demography also offered Jewish social scientists a new language with which to redefine Jewish collective identity at a time when, among Jews and non-Jews alike, this identity was highly contested. Moreover, the environmentalist tradition within the social sciences opened up the possibility of progressive, meliorative change at the collective level. It allowed Jewish social scientists to fill the roles of apologist and reformer, to defend their own people from attacks by anti-Semites and offer suggestions for self-improvement based on the knowledge and insights of science.

The possibilities of reform and regeneration were conceived along a number of different and competing political and ideological lines, impelled by conflicts and debates both between representatives of Jewry and the non-Jewish world and among factions within Jewry itself. The positions taken by Jewish researchers were driven in large measure by ideological commitments and political goals. More specifically, the debate between Jewish social scientists in Europe and the United States over the nature and condition of modern Jewry was carried on between Zionists and integrationists. Zionists were Jewish nationalists who rejected, at the ideological level at least, the integration or assimilation of the Jews into the broader culture and society. Integrationists were those who continued to champion such goals.[13]

The images, ideas, and statistics generated about Jewish bodies and minds were embedded in narratives that sought to make a persuasive ideological or political point. The politics of a social scientific representation of Jewish racial identity and health can be explained in part as emerging out of particular national or local conditions; but they also derived from transnational or international concerns, ideological and intellectual debates and conflicts not limited to any one country. Particular national or more local contexts are of course important in determining the shape and direction that ideas take; just as significant, however, is the intertextuality of developing discourses, the way in which writers draw upon and respond to previous and contemporary scholarship. Of course, the form and content of scholarship oftentimes transcend geographic boundaries. This essay therefore takes a comparative approach, focusing on the multiple, mutually related contexts within which the social scientific debate over the nature of modern Jewish identity emerged and on the political impetus to this scholarship.[14]

[13] By "at the ideological level" I mean to note that despite their opposition to assimilation, Zionists living in Europe or the United States did not feel compelled, as individuals, to resettle in Palestine. More often than not the champions of assimilation referred to themselves as "assimilationists." However, since that term today carries a pejorative sense, I use the more neutral "integrationist." It is worth noting, however, that although their opponents used the term disparagingly, assimilationists did not believe it implied anything negative; quite the opposite. On the polemics of the term "assimilationist" see Ezra Mendelsohn, *On Modern Jewish Politics* (Oxford: Oxford Univ. Press, 1993), Ch. 1.

[14] I am concerned in this essay with the representation, rather than the reality, of Jewish health and disease and with the politics of such constructions. This is not to say that the reality of Jewish poverty and disease, or even of differential rates of particular pathologies, did not exist or cannot be reconstructed. Such work has been done by historians who rely on the statistics and findings of much of the social scientific literature discussed in this essay. I am not denying the value of this research for social historians, although I would argue that utilizing the statistics presents considerable methodological difficulties. Here, however, I am asking different questions of the same texts. For examples of the way social historians use this primary material see Jacob Jay Lindenthal, "*Abi Gezunt*: Health and the Eastern European Jewish Immigrant," *Amer. Jewish Hist.*, 1981, *70*:420–441; and Deborah Dwork, "Health Conditions of Immigrant Jews in the Lower East Side of New York, 1880–1914," *Medical History*, 1981, *25*:1–4.

Zionist social scientists, while by no means monolithic in opinion, agreed that the statistical and anecdotal evidence about contemporary Jewry demonstrated that, in the first place, the Jews were indeed a nation (united in the past and present by common racial traits, among other things) and that, second, acculturation and assimilation had had disastrous effects on the "Jewish body." Zionists embraced the definition of Jewry as a *Volk* (oftentimes using the term *Rasse*, "race," in a descriptive way) and biomedical analyses of the Jewish condition as empirical proof of just such an identity.[15] Physiological traits or conditions provided Zionists with what they believed to be objective markers or signs of peoplehood. Anthropology offered anatomical evidence of difference; medicine established identity and difference through the discourse on pathology. Moreover, the discourse of disease and degeneration provided the framework within which to construct a Jewry in need of nationalist regeneration.

This interest in questions of Jewish racial identity and, more especially, the anxious concern over the health of the "Jewish body" mirrored more general anxieties expressed by social scientists, reformers, and social critics during this period. The notion of degeneration—at the individual, social, national, and racial levels—emerged in the latter half of the nineteenth century as one of the dominant models of explanation for the host of "social pathologies" believed to be plaguing societies. The medical model of degeneracy provided a conceptual language by which social problems could be explained. "Degeneration," in the sense the term came to have after the work of the French physician Bénédict Augustin Morel, indicated not just variation but deterioration. The conflation of the individual body with ideas of the social and national body meant that indications of illness within the individual were taken as signs of social and national decline. Mental and nervous diseases, syphilis, consumption, and alcoholism were believed to be both causes and symptoms of national decline, attacks not just upon the health of particular individuals or even classes but upon the body of the nation as a whole. Decline and degeneracy were also evident—so the argument went—in phenomena such as lower birth rates and rising rates of crime and suicide.[16] Degeneration was the obverse of progress, a rejection of the faith in the inevitability of improvement. Yet even as they bemoaned the present, degeneration theorists also, of course, held open the promise of a brighter future. Regeneration was possible, but it demanded that expert knowledge—produced by these self-same scientists—be translated into social policy.

Like their non-Jewish counterparts, Jewish social scientists committed to the Zionist cause could insert themselves into the nationalist drama as "physicians" to their people, diagnosing ills and offering a "cure" rooted in expert scientific knowledge. For some

[15] This is not say that each and every Zionist was enthusiastic about grounding Jewish identity and solidarity in an anthropological-medical foundation. There were objections to an overreliance on natural scientific arguments, proof itself that such arguments had become commonplace enough to require refutation. For an example see the article by the German Zionist leader Robert Weltsch, "Gelegentlich einer Rassentheorie," *Welt*, 1913, *17*(12):365–367.

[16] On the idea of degeneration, disease, and images of national decline see Pick, *Faces of Degeneration* (cit. n. 5); J. Edward Chamberlain and Sander Gilman, eds., *Degeneration: The Dark Side of Progress* (New York: Columbia Univ. Press, 1985); and Robert Nye, *Crime, Madness, and Politics in Modern France: The Medical Concept of National Decline* (Princeton, N.J.: Princeton Univ. Press, 1984). On England and the United States see Daniel Kevles, *In the Name of Eugenics: Genetics and the Uses of Human Heredity* (New York: Knopf, 1985); on Latin America see Nancy Leys Stepan, *"The Hour of Eugenics": Race, Gender, and Nation in Latin America* (Ithaca, N.Y./London: Cornell Univ. Press, 1991).

Zionists, Jewish degeneration was to be found exclusively in eastern Europe, among the masses of *Ghettojuden.* In 1900, for instance, Max Mandelstamm, a prominent Russian Zionist, delivered a speech before the Fourth Zionist Congress in London on "the physical amelioration of Jewry." Mandelstamm had studied medicine at Khartov University and ophthalmology at the University of Berlin. When he returned to Russia from Germany he ran an eye clinic in Kiev, and it was there in the 1880s that he became active in Zionism. In his speech Mandelstamm made a point of limiting himself to an analysis of the *Ostjuden*, or eastern European Jews. For it was in the East, he insisted, that one could locate "Jewish degeneration." Mandelstamm set out to demonstrate "that the decrepit, miserable, weak bodily constitution of the Jews of the ghetto is the exclusive result of their wretched social and economic situation."[17] He acknowledged that the Jewish masses ought to be characterized as "degenerate," but he embraced an environmental rather than racial explanation for the condition.

In almost every way, he claimed, the Jews compared unfavorably with "normally developed people." Jewish height on average (162 cm) was lower than that of non-Jews (165–170 cm); their limbs were shorter, their chests and muscles less well developed. They were susceptible to all sorts of diseases, suffering disproportionately from tuberculosis, numerous skin ailments, and eye diseases such as trachoma and nearsightedness. Jewish women suffered from particular women's disorders in greater numbers than their non-Jewish counterparts. Nor were Jews especially immune to certain diseases, as a popular notion had it. "As far as so-called immunities of Jews against known epidemic diseases, such as pestilence or cholera in the Middle Ages, this is foolish drivel from an earlier time. No one today would seriously try to justify such unfounded claims."[18]

However, Mandelstamm modified this point moments later, all but accepting the "foolish" notion he had just ridiculed. If Jews did display a marked ability to resist disease (*Widerstandsfähigkeit*), this was due not to any racial peculiarity but, rather, to their moderation in food and drink, their adherence to hygienic laws, and their greater willingness to consult physicians. In particular, their general abstinence from alcohol allowed Jews to resist infection to a greater degree than non-Jews. Yet sadly, he informed his listeners, this ability to resist disease had been decreasing in recent times, as the "better-off Jews assimilate into the non-Jewish population" and adopt their excessive behavior.[19] The end of Jewish isolation, the integration of Jews into the larger society, would therefore lead not to health but to the loss of whatever positive attributes Jews already possessed.

In contrast to their poorly developed bodies, Mandelstamm argued, Jews possessed larger than average brains. "Weak muscles, badly developed respiratory organs, weak bone structure, slight physical strength, little capacity for physical labor. Only the skull is highly developed among them; the cranial capacity of the Jew is on average greater than that of the non-Jewish population." In line with nineteenth-century scientific materialism, Man-

[17] *Stenographisches Protokoll der Verhandlungen des IV Zionisten-Congresses*, 15 Aug. 1900, Vol. 4 (Vienna, 1900), pp. 117–131; Mandelstamm's articles are rpt. as "Rede Dr. Max Mandelstamms," *Welt*, 1900, *4*(35):1–7, on p. 1. The speech was translated into English and published as *How Jews Live* (London, 1900). Mandelstamm had called for the "physical regeneration of the Jews" already in his speech to the Second Zionist Congress in 1898. This talk, together with Max Nordau's on *Muskeljudentum*, fueled enthusiasm in Zionist circles for gymnastics, which many believed essential to building up not only strong Jewish bodies but a vital Jewish national consciousness. On this view and the formation of gymnastics societies in early Zionist circles see George Eisen, "Zionism, Nationalism, and the Emergence of the Jüdische Turnerschaft," *Leo Baeck Institute Yearbook*, 1983, *28*:247–262. On Mandelstamm see *Encyclopedia Judaica* (Jerusalem: Keter, 1971), Vol. 11, pp. 867–869.

[18] "Rede Dr. Max Mandelstamms," p. 4.

[19] *Ibid.*, p. 5.

delstamm assumed a direct relation between the size and shape of the cranium and mental prowess. Their larger cranial capacity, established by the numerous craniometric studies of anthropologists, indicated the greater average intelligence of the Jews.[20]

This was, however, no great cause for celebration. For Mandelstamm, as for many others, the greater mental capacity of the Jews marked a pathological condition. It indicated that Jews suffered disproportionately from nervous and mental disorders that could be traced back to an "overuse of the brain."[21] Hyperintelligence had produced the "nervous" or "insane" Jew. "Abnormality" was not a sign of health—of equality with or even superiority to the non-Jew; rather, it showed the diseased and inferior condition of eastern European Jewry.

In Mandelstamm's view, the physical and moral impoverishment of Jewish life in the East had produced degeneration (*Entartung*) at the physical level. Regeneration could occur only through a complete transformation of eastern Jewry's social and economic conditions. He allowed that this could, theoretically, occur in Russia if the five million Jews there were given the freedoms of movement, occupation, and residence that Jews in the West enjoyed. Since such improvements could not be expected, all that could be hoped and worked for were half-measures: reforms of the cheder (the elementary religious school), instruction in hygiene, better lighted and ventilated rooms, and the establishment of occupational training centers and gymnastic leagues. Complete regeneration demanded a solution possible only beyond the borders of Europe. In concluding his talk, Mandelstamm spoke as both physician and Zionist:

> Ghetto Jewry cannot become either physically or morally healthy in the ghetto. This can only occur outside of the ghetto, on their own land and soil [applause], and indeed in the only single land where it will be possible to secure this goal patiently and calmly; in a land where new powers and energies can lead to physical and intellectual improvement. Only with a mass emigration to Palestine, carefully planned with great foresight . . . can a new and invigorated generation of Jews arise.[22]

Mandelstamm's analysis focused on eastern European Jewry as degenerate and assumed the relative health and normalcy of western European Jews. However, by the first decade of the century Zionists were turning their attention more and more toward western Jewry and identifying degeneration with assimilation. Rather than considering the effects of oppression and isolation on the Jews, social scientific analyses increasingly explored the impact of a freedom and integration facilitated by civil emancipation, urbanization, and *embourgeoisement*.

By the turn of the century the question being asked about Jewish communities in the West, by both Jews and non-Jews, was no longer "Should the Jews be granted freedom and equality?"[23] Rather, social scientists, among others, were inquiring into the effects of such freedom on the state, society, and Jews themselves. What was the impact of emancipation and, even more, assimilation? Zionists were less interested in the impact of Jewish

[20] *Ibid.*, p. 4. On the idea of "Jewish intelligence" as a theme in modern European thought see Sander Gilman, *Smart Jews: The Construction of the Image of Superior Jewish Intelligence* (Lincoln: Univ. Nebraska Press, 1996).

[21] "Rede Dr. Max Mandelstamms," p. 4.

[22] *Ibid.*, p. 6.

[23] There was no one pattern of Jewish emancipation. The process occurred in different ways, with varying degrees of success, in different countries. For a comparative approach to the question see the essays collected in Pierre Birnbaum and Ira Katznelson, eds., *Paths of Emancipation: Jews, States, and Citizenship* (Princeton, N.J.: Princeton Univ. Press, 1995).

assimilation on the general society than in its effects on the character of the Jews themselves. The conclusions they reached were, on the whole, less than sanguine. The Viennese physician Martin Engländer, for instance, revealed a link between assimilation and degeneration in his 1902 study on "the startling frequency of illness among the Jewish race." Engländer found proof of this connection in statistics showing comparatively higher rates of mental and nervous disorders, alcoholism, and suicide among middle-class western European Jews. Hugo Hoppe, another physician, and at the time a noted authority on alcoholism, argued much the same thing in his writings on Jews, alcoholism, and disease. Hoppe cited statistics that showed a marked increase in physical and mental disorders, as well as criminality, among assimilated Jews. This he attributed to the assimilation of modern, gentile mores, chief among them the increasing consumption of alcohol (a pathology, he insisted, to which Jewish academics were especially prone).[24]

Thus emancipation and assimilation threatened to efface the identity and unity of Jews assumed to have existed before the modernization process took hold. In breaking down the social, economic, and physical barriers between Jews and non-Jews (labeled at times "Christians," at times "Aryans") that had been in place for centuries, the forces of modernity posed the greatest danger to Jewish survival. Arthur Ruppin argued that the significance of increasing rates of Jewish conversion and intermarriage lay in its negative "anthropological" or "racial" effect. Ruppin was not only a leading social scientific authority on Jews but also an important Zionist administrator. After resigning as head of the Bureau for Jewish Statistics in 1907 he moved to Palestine to become head of the World Zionist Organization's Palestine office. In books and articles published between 1904 and 1940 he developed a comprehensive and sophisticated social scientific approach to modern Jewish life, focusing on the physical as well as social effects of assimilation. While he was firmly wedded to a sociological, or social determinist, position, Ruppin was nonetheless strongly influenced by current bioracial thinking. Assimilation, he insisted, adversely affected modern Jewry at the physical as well as the cultural and spiritual level. "The number of Jews," he wrote in *The Jews of Today* (1904), "whose facial physiognomy displays none of the traits of the so-called Jewish type, whose morphological type cannot be identified as of Jewish descent is substantial." The high level of "racial mixing" (*Rassenvermischung*) that occurred during the nineteenth century in the West meant that physically, or anthropologically, eastern and western European Jews were more and more different types. For Ruppin, this represented the possibility of an extreme, even absolute, division or chasm between the two Jewries. "Therewith the last bridge, racial unity, connecting eastern and western Jewry, divided as they are already culturally, will be destroyed. And since their bond with eastern European Jewry has vanished and they stand alone, the absorption of western European Jewry will happen quite easily."[25]

Assimilation, for Ruppin as for most Zionist social scientists, was directly responsible for the pathological, abnormal condition of modern Jewry. More precisely, they saw a causal connection between material and social progress and physical and mental pathologies. Reflecting a broader antibourgeois animus that pervaded much of early Zionist ideology, they rejected the notion that political and social freedom, together with higher

[24] Martin Engländer, *Die auffälend häufigen Krankheitserscheinungen der jüdischen Rasse* (Vienna, 1902); Hugo Hoppe, *Krankheit und Sterblichkeit bei den Juden und Nichtjuden* (Berlin, 1903); and Hoppe, "Die Kriminalität der Juden und der Alkohol," *Z. Demog. Statis. Juden*, 1907, *3*(3):38–41.

[25] Arthur Ruppin, *Die Juden der Gegenwart* (Cologne: Jüdischer Verlag, 1904), p. 94. On the significance of social scientific thinking for Ruppin's Zionist work see Derek Penslar, *Zionism and Technocracy: The Engineering of Jewish Settlement in Palestine, 1870–1918* (Bloomington: Indiana Univ. Press, 1991), pp. 80–102.

(i.e., middle class) living standards, would produce a "healthy Jew."[26] Precisely the opposite. Ruppin, for instance, argued that through the Jews' increasingly intensive participation in the modern economy, in science, art, and higher education, the areas of contact between Jews and Christians were becoming ever greater. As a result, the sense of belonging to the Jewish *Volk* was disappearing. This, in turn, must lead to an increase in intermarriage and conversion, further lowering the number of Jews. Ruppin's analysis was echoed by a number of other Zionist social scientists at the time. The Austrian physician and Zionist Ignaz Zollschan concluded that "the only protection against intermarriage, as with conversion, in the Diaspora is the ghetto. In countries where Jews take part in public and economic life as equals, intermarriage is continuously on the rise. This is most frequently the case in the large cities." Zollschan viewed this as the most serious threat facing modern Jewry, speaking of assimilation in terms of "racial suicide."[27]

Just as important, the higher rates of intermarriage negatively affected the physical, racial makeup of Jewry. While Zionist social scientists disagreed among themselves about whether the Jews constituted a "pure race," they nonetheless shared the belief that intermarriage posed a threat to the racial quality as well as the numerical quantity of contemporary Jewry. Intermarriage, and to a lesser extent proselytism (the conversion of non-Jews to Judaism), introduced "foreign blood" into the collective Jewish body, with deleterious effects. In a 1905 article entitled "Mixed Marriages," the editors of the American Zionist journal the *Maccabbean* asserted that "the mixed marriage is more tolerable from a religious than from a racial standpoint." Clearly working with a bioracial, rather than a religious, definition of Jewish identity or "Jewishness," the writers denied that genuine conversion was even possible. True, both Christianity and Judaism have rituals whereby, for instance, a Christian man becomes a Jew and participates to whatever degree in the life of the synagogue and the Jewish home; or a Jewish woman marries a Christian man and takes on the sacraments of the church. To the nationalist, however, "both these conditions are mere theories: he cannot really become a Jew, she cannot really cease to be a Jewess." Echoing the findings of Jewish social scientists, the writers argued that intermarriage produced both a quantitative and a qualitative loss for the Jewish people. Jewry, like other nations, is dependent on numerical strength. "We have survived because our birth rate has always been so well ahead of our death rate."[28] Yet intermarriage yields children who, in most cases, will be raised as non-Jews.

[26] Not every prominent Zionist writer who utilized social scientific categories shared this antibourgeois bias. The well-known cultural critic and Zionist leader Max Nordau, for example, did a great deal to disseminate the notion that abnormality was the defining characteristic of modern Jewish existence. For Nordau, who trained as a physician, physical degeneration, manifested in the deformed and weak Jewish body, was one of the clearest signs of this abnormal condition. However, Nordau's conception of "healthy" and "normal" stemmed in large measure from bourgeois norms. Following his teacher, the criminal anthropologist Cesare Lombroso, he defined normalcy as the middle: the moral, cultural, social, and aesthetic area between the depths of the criminal and insane and the heights of genius. See the analysis by George Mosse in his introduction to Max Nordau, *Degeneration* (Lincoln: Univ. Nebraska Press, 1993); and Gilman, *Jew's Body* (cit. n. 1), pp. 53–54. On Nordau's views on degeneration, Zionism, and bourgeois values see Amos Funkenstein, "Zionism, Science, and History," in *Perceptions of Jewish History* (Berkeley/Los Angeles: Univ. California Press, 1993), pp. 338–350, esp. pp. 340–342. Lombroso had attempted to account for the high percentage of Jews among the insane by referring to the disproportionate number of geniuses among them: Cesare Lombroso, *Der Antisemitismus und die Juden im Lichte der modernen Wissenschaft*, trans. H. Kurella (Leipzig, 1894), pp. 43–50. The Zionist racial anthropologist Leo Sofer repeated this claim in his summary of scientific opinion on Jewish pathologies: Leo Sofer, "Zur Biologie und Pathologie der jüdischen Rasse," *Z. Demog. Statis. Juden*, 1906, *2*(6):85–92, on p. 90.

[27] Ruppin, *Juden der Gegenwart* (cit. n. 25), pp. 266–267; and Ignaz Zollschan, *Das Rassenproblem unter besonderer Berücksichtigung der theoretischen Grundlagen der jüdischen Rassenfrage*, 5th ed. (Vienna, 1925), pp. 478–479 and *passim*.

[28] "Mixed Marriages," *Maccabbean*, May 1905, *7*(5):216.

Qualitatively, intermarriage threatens to dilute the "racial particularity of the Jews." It transforms the physical quality of a race, its anatomical structure; that, in turn, changes the intellectual level or quality, since the one is linked with the other: "The genius of race seeks life and maintenance; we ourselves wish to live, we love our ideas, and our understanding of them for that matter, not someone else's interpretation. For that we need our own bodies, our own blood, our own cast of cranium, and since ideas are eternal we wish our ancestors [*sic*] to be of our stock so that they may be actuated by the same sentiments."[29]

Writing in the Anglo-Jewish journal the *Jewish Review* four years later, Zollschan sounded a more ominous note. He warned that the "high ethnical qualities" of the Jews were in danger of being diminished by "intermarriage with less carefully bred races." In 1911, in a speech to the Tenth Zionist Congress, he cautioned that the "assimilation of blood," which resulted from the high rate of intermarriage, represented "a more difficult, because irreparable, harm to the *Volk* than cultural or emotional assimilation."[30] This "assimilation of blood" was, according to Zollschan's own analysis, the direct result of cultural and social emancipation and assimilation. The integration of Jews into modern bourgeois society—the economic, intellectual, and social freedom found in the West—estranged the Jew from tradition and faith and eliminated the social and economic isolation that kept the Jew from intermixing with the Christian. Freedom produced crisis and decline, a point borne out by statistics showing numerical losses and a qualitative weakening of the race.[31]

Zionist social scientists had discovered, so they believed, a dynamic of pathological development from which European Jews could not escape. "Modernization" carried western Jewry out of the ghetto and up and along the road to political and social freedom and material well-being. But these were precisely the conditions that generated the myriad physical and mental diseases associated with modern life. And although eastern European Jewry had thus far succeeded in escaping the same fate—because of the czarist regime's resistance to economic and political modernization or Westernization—this was only a

[29] *Ibid.*

[30] Ignaz Zollschan, "The Jewish Race Problem," *Jewish Review*, 1912, *2*:391–408, on p. 405; and Zollschan, "Rassenproblem und Judenfrage," *Welt*, 1911, *15*(34):897. The publication of Zollschan's speech in *Die Welt* elicited an angry response from Alfred Waldenburg ("Jüdsiche Rassenhochzucht," *ibid.*, 1911, *15*[37]:976), a Zionist and racial scientist who believed he deserved the credit for first alerting Zionists to the dangers intermarriage posed to the Jewish race. Waldenburg claimed that as early as 1902, in his comparative study of Jewish and non-Jewish races, he had put forth the thesis that Jewish health was directly related to Jewish racial inbreeding. "The Jewish race has achieved a relatively high immunity from cholera, and a number of acute as well as chronic infectious diseases through inbreeding, and this is an hereditary trait which is weakened or diminished, proportionately, by the introduction of non-Jewish blood." Waldenburg argued that intermarriage produced degeneration and led to "racial suicide"; he called for "a superior racial inbreeding" in Palestine. Waldenburg's ideas do appear in Zionist works on Jewish anthropology in which the idea of racial purity plays a paramount role; see, e.g., Aron Sandler, *Anthropologie und Zionismus: Ein populär wissenschaftlicher Vortrag* (Breslau, 1904), pp. 26–27.

[31] See also the analysis in Ignaz Zollschan, "The Significance of the Mixed Marriage," in *Jewish Questions: Three Lectures* (New York, 1914). Zollschan offered statistics on rising intermarriage rates from those countries whose governments produced "confessional statistics"—i.e., permitted counting and measuring along religious lines. There are enormous problems with the way Zollschan and others utilized statistics as evidence. Thus, for example, Zollschan tells us at one point that the number of intermarriages in Prussia rose from 2,100 in 1885 to 5,100 in 1905. However, he fails to provide the other figures, such as absolute growth in the overall population, required to gauge the significance of these numbers. Elsewhere, as in his discussion of Jewish communities in Denmark, Belgium, and Holland, he provides no absolute numbers, only percentages. For instance, he observes that intermarriage increased in Copenhagen (where three-fourths of the Danish Jews lived) from 55.8 percent in the period 1880–1889 to 93.1 percent in 1900–1905. However, since we don't know the actual number of Jews (200? 200,000?), we have no way of knowing what sort of overall quantitative shift—and, in Zollschan's terms, "loss" to Jewry—he is talking about.

temporary reprieve. Russia would soon modernize, eastern Jews would assimilate, and the same sort of collective degenerative process would threaten.

By casting European Jewry as either actually or potentially diseased and identifying the causes with the very condition of modernity, Zionists were able to demonstrate scientifically that Jews literally had no future in the Diaspora. Social laws and forces beyond their control were working to eliminate their unique anatomical traits and their resistance to infectious and contagious illnesses, to increase their rates of mental and nervous disorders, and to lower their fertility rates. Both quantitatively and qualitatively, Zionist social scientists argued, modern Jewry was under attack.

Jewish nationalism insisted it would produce a "new Jew," a "healthy Jew" who would stand as the antithesis of the degenerate Diaspora Jew. This could occur only in Palestine, in a Jewish homeland or state with a Jewish culture and economy. Capitalism, the metropolis, higher secular education, modern culture—these essential components of modernity were not negative in and of themselves. Rather, they were dangerous insofar as they were pillars of a majority gentile culture that lured Jews into assimilating.

Zionists, in fact, sought to recreate European society, in one form or another, in Palestine. The assertion that Jewish nationalism would regenerate the Jew physically and spiritually went hand in hand with the Zionist vision of a physically and morally regenerated Palestine.[32] Since so much of late nineteenth- and early twentieth-century social scientific discourse had been infused with the language of the natural sciences, it is not surprising that we find an interesting conjunction of images related to the Jewish body and the land of Palestine. In their present state, both were unhealthy and unproductive. Again, the sickness of the modern Jew was due in large measure to the sickness of the modern condition in general. The Jews were a product of industrialization, urbanization, and the high degree of social, cultural, and intellectual integration or assimilation into the dominant society that capitalism and the metropolis made possible. The Munich social scientist Felix Theilhaber, for instance, appropriated the language of *völkish* romanticism to describe the negative impact of modern urban capitalist life on German Jewry. Pointing to the "simplicity," "naturalness," and "health" of rural agricultural life, he contrasted the Jew who had not been "refined by overcultivation [*von keinerlei 'Überkultur' beleckt]*, to the Jew who "with his nervous character seeks out the excitements of the city." The modern city, characterized by intense intellectual and economic struggle (the "infection of modern capitalism"), became the space within which the process of Jewish degeneration unfolded. "Jewry is the most urbanized, capitalistic, rationalistic sector of the population"; hence the higher rates within that sector of every sort of physical and social pathology directly reflected the very condition of modernity.[33]

Zionism was by no means an antiurban movement. It was, after all, responsible for building the most modern and European of cities in what would become the State of Israel: Tel-Aviv.[34] Its antiurbanist and antimodernist sentiments were not essential but, rather, emerged out of a social and cultural critique of modern European Jewish life. Nonetheless, there was a strong rural and agricultural bent to much early Zionist writing and iconog-

[32] Michael Berkowitz, *Zionist Culture and West European Jewry before the First World War* (Cambridge: Cambridge Univ. Press, 1993), Ch. 6.

[33] Felix Theilhaber, *Der Untergang der deutschen Juden* (Munich, 1911), pp. 69, 148–152. See also Theilhaber, *Die Schädigung der Rasse durch soziales und wirtschaftliches Aufsteigen beweisen an der Berliner Juden* (Berlin, 1914).

[34] On the beginnings of Tel-Aviv and the urbanist movement within Zionism see Ilan Troen, "Establishing a Zionist Metropolis: Alternative Approaches to Building Tel-Aviv," *Journal of Urban History*, 1991, *18*:10–36.

raphy. And the positive images generated about the land, physical labor, and "rootedness" allowed Zionists to make their arguments about the degeneration of the Jewish body even more forcefully. The physical, as well as economic, unproductiveness of modern Jewry, in this view, was mirrored in the infertility of Palestinian agriculture and industry. The illnesses of the Jew in the Diaspora—tuberculosis, diabetes, neurasthenia—were doubled in the unhygienic and diseased condition of Palestine. These two images, of the diseased and abnormal Jewish body and the unhealthy, unproductive land of Palestine, were intimately related, for the solution to both depended on Zionist mediation. Ultimately, Jewry and Palestine would enjoy a revitalization through Zionist efforts at nation-building. Through their physical labor on the soil, the Jews would regenerate their own bodies and the land of their ancestors.

THE DIASPORA AS CURE: NON-ZIONIST NARRATIVES OF JEWISH HEALTH AND DISEASE

Zionists were not alone among Jews in engaging in social scientific research on the modern Jewish condition and in marshaling the evidence for particular political or ideological purposes. One of the first Jewish scholars to engage in statistical and anthropological studies of Jews, the Anglo-Jewish social scientist Joseph Jacobs, set about demonstrating the racial purity and identity of Jewry and drawing sociopolitical conclusions from this evidence. Jacobs sought to explain the high numbers of brachycephalic, or broad-headed, Jews counted by anthropologists by invoking the Jews' purported heightened mental abilities.[35] He referred to the research of the famed English statistician and eugenicist Karl Pearson, who had demonstrated that brachycephalism implied superior intelligence. The enlarged Jewish head was a product of historical adaptation in a Darwinian struggle for survival. In order to compete successfully, Jews were forced to develop their intellectual capacities, which included involvement in trade and finance, beyond average or "normal" levels; their skulls grew bigger in order to accommodate their oversized brains.

Whereas Max Mandelstamm had understood this Jewish trait as a manifestation of pathology, Jacobs found this morphological adaptationism significant because it demonstrated the racial purity as well as the extraordinary mental powers of Jewry. Jacobs was not a Jewish nationalist; rather, he was a proponent of integration or "assimilationism." He viewed Jewish racial purity and intelligence as proof that, as an elite and highly cultured group, the Jews were perfectly capable of integrating into and contributing to the more general—in this case, English—society. In his racial hierarchy the Jews were equal or superior to the English, Scottish, and "Teutons."[36]

Jacobs, whose work tended much more toward apologia than toward critique and reform, posited, for the most part, a healthy, even superior, Jewish body and mind. However, as I argued earlier, Jewish social scientists more often focused on the negative attributes and qualities of contemporary Jewry; this was as true for integrationists as for Zionists. Unlike the Zionists, however, integrationists made the case for assimilation as an agent of Jewish health and normalcy. Integration facilitated positive transformation at the physical and biological levels, and this transformation in turn was the sign or indication of assimilation's success.

The Zionist and integrationist interpretive frameworks were similar in significant ways. Both conceded that segments of Jewry were "inferior"—that is, physically (and by exten-

[35] Joseph Jacobs, "Are Jews Jews?" *Popular Science Monthly*, 1899, *4*:502–511.
[36] *Ibid.*, pp. 506–507, 508.

sion culturally) enfeebled, suffering in disproportionate numbers from certain diseases. And both relied on a strong environmental determinism in which the physical and moral character of the Jews would be transformed through an encounter with new, healthy surroundings. They differed in their ideas about where such a transformation could and should occur and in their evaluations of the effects of isolation from or assimilation into the larger non-Jewish community. For instance, the German-Jewish health official (*Sanitätsrat*) and head of the communal organization Bnei Brith, Louis Maretzki, believed that the Jews suffered from certain degenerative illnesses, particularly nervous and mental disorders. He, too, stressed the environmental over the racial origins of this degeneration. Yet for Maretzki the "cure" lay in reform at the local, communal level. He urged the Jews to renew themselves as a community (*Gemeinschaft*) within Germany through greater attention to health and hygiene.

The key to such a renewal of Jewish collective health lay not in the ideal of the nation but in the family. "The family," Maretzki wrote, "is the foundation of humanity. It is our entire communal life; it is essential not only to economic unity but also to the foundation of the total social body." Over the past few decades the Jewish family had suffered through a process of degeneration and disintegration. Modernity itself was to blame. ("Our age is merciless in destroying what our fathers held dear.") But for Maretzki "modernity" meant the abandonment and destruction of the middle-class European values of hard work, obedience, and *Bildung*. Above all, modernity had destroyed the healthy patriarchal family structure, the "sacred hearth" from which "morality, the charm of veneration [*Zauber der Ehrwürdigkeit*], and a deeply ingrained *Gemütsfülle* emanate." A return to the traditional organization of the family would serve as a catalyst for Jewish health: "The patriarchal family structure brings tranquility, contentment, and sedateness into the family circle. It is an important and expedient means of battling against agitation of the nerves." In full agreement with those who viewed the city or metropolis as a source of nervous illness, Maretzki argued that "it would be an improvement of the greatest kind to settle more Jews in small towns and villages." The "quiet life" there, the lack of stimulation, was a proven means of combating neurasthenia. Jewry was so tightly bound up with metropolitan urban life, however, that meliorative efforts had to be made there. Enumerating the many positive effects upon health of working in the civil service, Maretzki argued for an increase in the number of Jews serving as lower- and middle-level government employees.[37]

Maretzki's analysis of Jewish health and disease served to legitimate Jewish life at the local, communal level. Social problems, he urged, were to be solved through personal *Bildung* and communal reform. Although he agreed with Zionist social scientists on many points, he nonetheless maintained a faith in the liberal, emancipationist ideal of Jewish integration and progress. Jewish "health," in his view, was attainable within German (and European) society.

The self-critique to be found in this sort of analysis, the concern over a Jewish community too deeply engaged in distinctly modern forms of economic, social, and cultural life, also contained an implicit critique of aspects of modernity, if not modernity as a whole. Thus one finds in the Jewish social scientific discourse elements of the "politics of cultural despair" that Fritz Stern and others have analyzed in the German right.[38] Yet as

[37] Louis Maretzki, "Die Gesundheitsverhältnisse der Juden," in *Statistik der Juden: Eine Sammelschrift* (Berlin: Bureau für Statistik der Juden, 1917), pp. 123–151, on pp. 148, 149, 150.

[38] Fritz Stern, *The Politics of Cultural Despair* (Berkeley: Univ. California Press, 1974). For a modification of Stern's thesis see Jeffrey Herf, *Reactionary Modernism: Technology, Culture, and Politics in Weimar and the Third Reich* (Cambridge/New York: Cambridge Univ. Press, 1984).

significant as this "antimodernist" mode of thought is—since the images of the Jews and "modernity" and of the reactionary right and "antimodernity" seem to be so closely and commonly identified—there were also Jewish scientists who made the argument that Jewish disease and degeneration stemmed from an incomplete modernization and Westernization. In this explanatory framework, Orthodox Jewish beliefs and rituals, in addition to the "traditional" or premodern world in which Orthodox Jews lived, were the great impediments to physical and moral improvement. So for liberal Jewish social scientists such as Hermann Oppenheim and Moritz Benedikt, the representation of the West as healthy was linked closely with the persistent image of the *Ostjude*, or eastern European Jew, as the "diseased Jew" whose "environment" produced a wide variety of psychopathologies. Benedikt attributed the high rates of Jewish mental disease mainly to the continued dominance of Orthodox Judaism. Steeped in traditional religious modes of thought and behavior, the eastern European Jew was literally made ill by his adherence to orthodoxy. "From childhood on," Benedikt wrote in 1918, "especially in religious circles, the dialectics of the Talmud are forced on capable and incapable brains, whereby paralysis is actually bred."[39] For Oppenheim, the combination of poverty and oppression and the overemphasis on traditional religious study explained the high incidence of mental and nervous disorders.

Both analyses represented a continuity with Jewish Enlightenment critiques of traditional Jewish life in eastern Europe, and both sought to reinforce the notion of western European Jewry as the embodiment of health and normalcy. Assimilation into modern Western society and culture was seen as the best hope for a healthy, regenerated Jewry in the future. Difference, judged as a negative condition, was defined as a nonessential state of being, to be overcome through the positive work of integration.

ASSIMILATION AND THE VIRTUES OF AMERICA

This concern with the physical and moral transformation of the eastern European Jew was intimately bound up with the phenomenon of large-scale immigration and the political and cultural struggles it produced. By the last quarter of the nineteenth century the concern over eastern and southern European immigration was being voiced not only in the major cities of Europe but also in the United States. North American social scientists played a crucial role in the developing debate over the nature of these immigrants and their impact on the "body" of America. Eastern and southern European immigrants were of course not the only objects of social scientific and policy concern. Debates over the "racial" identity and health of the nation continued to focus on the place and the impact of blacks; but the impact of Jews and Italians was also of interest and concern.

The focal point of anthropological and medical investigations into the Jewish body was the eastern European immigrant population, which had begun arriving in massive numbers in the 1880s. Anti-Jewish and nativist forces railed at the impoverishment, uncleanliness, and "strange" social and cultural customs of the immigrants. By the end of the nineteenth century, when the rates of immigration soared, a more biological and racial determinist language had also gained greater prominence, so that Jews were increasingly described as "parasites" or "germs" and in similar terms drawn from biology and medicine. These metaphors were given enormous power by the very real dangers posed by contagious

[39] Benedikt is quoted in John Efron, "The " 'Kaftanjude' and the 'Kaffeehausjude': Two Models of Jewish Identity: A Discussion of Causes and Cures among German-Jewish Psychiatrists," *Leo Baeck Inst. Yearbook*, 1992, *37*:169–188, on p. 176.

diseases such as typhus and cholera, diseases associated with immigrants from southern and eastern Europe. Yet as the historian of medicine Howard Markel suggests, the episodic nature of these diseases, and the fact that the panic they engendered passed rather quickly, meant that they were less than fully effective as metaphors of long-term racial and medical danger. Far more effective was the danger associated with the genetic inheritance of negative traits. This provided "a stronger and more permanent metaphor of disease" with which to characterize the danger posed by immigration and, by extension, all "alien" groups.[40] Heritable diseases and defective genes were passed on from generation to generation; hence they were a threat not only to the present but also to the future of the American social body.

Thus American anti-Semitism, like its European counterpart, had a "clearly identifiable genetic component inextricably linked with the restriction of the 'new' immigration of white ethnics from southern and eastern Europe prior to 1924."[41] Clearly influenced by European racialist thinkers such as Joseph-Arthur de Comte Gobineau, Georges Vacher de Lapouges, and Houston Stewart Chamberlain, American racialists identified the Jews as an "Asiatic" or "oriental" race. The Jews were, therefore, deemed incompatible with and unassimilable to the Anglo-Saxon race, which, it was argued, had created and continued to define the United States.

Evidence of this unassimilability was found in Jewish anatomy and physiology, in deformed and disease-prone bodies. This physical deformation—attributed variously to Jewish "inbreeding" (i.e., endogamy) born of racial pride or to historical intermixture with inferior eastern or oriental breeds (e.g., the medieval Khazars)—was genetic or biological and hence presented a grave threat to the racial purity and health of America. Restrictionists in the decades prior to 1924, the year in which the U.S. Congress succeeded in passing highly restrictionist legislation, referred repeatedly to the physical and moral degeneration of Jewry and the threat these posed. For instance, at the turn of the century the racial anthropologist William Z. Ripley wrote, in a work identified at the time as definitive in racial science, that the eastern European Jews suffered from a "physical degeneracy" that was evident in their short stature, deficient lung capacity, and other negative traits. Ripley warned his readers that this "great Polish swamp of miserable human beings . . . threatens to drain itself off into our country as well, unless we restrict its ingress." Manly Simons, medical director in the U.S. Navy, asserted that "as a type Jews are beginning to show mental and physical degradation, as evidenced by the great variability of development, great brilliancy, idiocy, moral perversity, epilepsy, physical deformity, anarchistic and lawless tendencies." And E. A. Ross, writing in 1914, insisted that "on the physical side the Hebrews are the polar opposite of our pioneer breed. Not only are they undersized and weak muscled, but they shun bodily activity and are exceedingly sensitive to pain." In sum, American nativists and anti-Semites, in the words of Alan Kraut, "sought to sketch the Jew as a public health menace, one who might end up on the public relief roles in droves, deficient in the physical vitality to stand the test of the rugged American environ-

[40] For an account of these diseases and the interconnection between epidemics and the politics of immigration see Howard Markel, *Quarantine! East European Jewish Immigrants and the New York City Epidemics of 1892* (Baltimore: Johns Hopkins Univ. Press, 1997).

[41] Robert Singerman, "The Jew as Racial Alien: The Genetic Component of American Anti-Semitism," in *Anti-Semitism in American History*, ed. David A. Gerber (Urbana: Univ. Illinois Press, 1986), pp. 103–128, on p. 103.

ment, as did the pioneer forbearers of the native-born and the sturdier stock that had emigrated to the United States from Northern and Western Europe."[42]

It was the task of social scientists sympathetic to the goals of immigration and integration to provide the empirical, scientific evidence for the viability of assimilation and thereby counter the nativist and anti-Semitic position. American Jewish physicians, psychologists, and sociologists, for instance, acknowledged that Jews suffered at disproportionately high rates from insanity, neurasthenia, and other disorders. "Step into any clinic for nervous diseases in any large city in Europe and America," the neurologist Abraham Myerson wrote in 1920, "and the Jew is unduly represented amongst the patients."[43] Responding both to American nativists, who after World War I relied increasingly on psychological data to prove the inferiority of non-Anglo-Saxon "races," and to the Zionist critique of Jewish assimilation, Myerson and others championed assimilation as the cure for Jewish disorders. The best means of prevention or cure, according to the psychiatrist and Freud translator A. A. Brill, in an article titled "The Adjustment of the Jew to the American Environment," is "a gradual process of Americanization, or assimilation."[44] The sociologist Charles Bernheimer, writing on Jewish health conditions in Philadelphia, conceded that nearly all medical authorities—including Mandelstamm, Engländer, Richard Krafft-Ebing, and Jean-Martin Charcot—recognized that the Jews are prone to nervous and mental disorders; it is a "medical axiom." This predisposition is hereditary, a "racial predisposition transmitted through generations." Nonetheless, a combination of the will to adapt and assimilate and a healthy environment will cure the nervous Jew. "With all his proverbial tenacity of character, the Jew, and especially the Eastern Jew, is physically and psychically extremely plastic, and only needs a reasonably favorable environment to develop into a noble specimen of man. His energy, intelligence, and integrity will solve many of the perplexing economic problems, and in that way the sanitary and hygienic questions will, partly at least, be answered."[45]

Myerson, Brill, and Bernheimer spoke of a Jewish race and Jewish racial traits. Yet they insisted that these traits were mutable, a product of history and environment. In Myerson's

[42] Ripley is quoted in Singerman, "Jew as Racial Alien," p. 106. Allan Chase identifies Ripley as a professor of economics at MIT and Columbia University, but he is remembered as a "raciologist." See Allan Chase, *The Legacy of Malthus* (New York: Knopf, 1976), p. 96 and *passim*. Simons and Ross are quoted in Alan Kraut, *Silent Travelers: Germs, Genes, and the Immigrant Menace* (New York: Basic, 1994), p. 155; see also p. 145. Interestingly, this image of the Jew as small and physically weak, averse to labor and "sensitive to pain," was the opposite of the racial image of the American Negro, who was represented as oversized, frighteningly strong, given over only to physical activities, and immune to pain. Just what the "ideal" qualities were, then, was not fixed but depended on the object of the discourse.

[43] Abraham Myerson, "The 'Nervousness' of the Jew," *Mental Hygiene*, 1920, *4*:65–72, on p. 65. Myerson at the time held a number of professional positions, including assistant professor of neurology at Tufts College Medical School and consulting physician in the Psychopathic Department of Boston State Hospital. The historian of American Jewry Andrew Heinze is currently at work on a study of Jews and the development of American psychology. I have relied in part on his essay "Judaism Confronts Psychology: Abraham Myerson, Joseph Jastrow, and the Psychological Reorientation of American Jewish Thought," presented at the conference on "Jews and the Biological and Social Sciences" held at the Oxford Centre for Hebrew and Jewish Studies in August 1998.

[44] A. A. Brill, "The Adjustment of the Jew to the American Environment," *Mental Hyg.*, 1918, *2*:219–231, on pp. 229–230. In addition see Brill and M. J. Karpas, "Insanity among Jews," *Medical Record*, Oct. 1914, *86*. Brill was born in Austria and immigrated to the United States in 1889, holding positions in psychiatry at Columbia and New York University. He is best known as the first American translator and interpreter of Freud's writings. See "A. A. Brill," in *Dictionary of American Medical Biography*, ed. Martin Kaufman (Westport, Conn.: Greenwood, 1984), Vol. 1, p. 94.

[45] Charles Bernheimer, "Philadelphia" (section on health), in *The Russian Jew in the United States*, ed. Bernheimer (Philadelphia, 1905), pp. 304–317, on pp. 314, 316.

formulation, Jewish characteristics were the product of social, not biological, heredity; and these change as the environment changes. "How quickly racial characters can be changed under a fostering environment may be exemplified by the development of the last generation in Jewish life in the free countries of Europe and America. . . . What persecution could not do throughout the centuries, toleration does in a generation."[46]

Much the same point was made by the eminent German-Jewish anthropologist Franz Boas, who made the Jews an integral part of his studies on the anatomies of immigrants. His studies on head forms are probably the best-known example of a liberal or assimilationist physical anthropology, one that sought a direct influence on the shape of political debate and social policy. Boas's earliest forays into the study of head forms included the Jews. In 1903 he published "Heredity in Headform" in the *American Anthropologist*, examining the relationship between inheritance and head form in light of Mendel's recently rediscovered work. He included data on forty-nine Jewish families compiled by the New York physician and anthropologist Maurice Fishberg. Fishberg's research also found its way into Boas's best-known study of race, immigration, and assimilation, *Changes in Bodily Form of Descendants of Immigrants*. This work on the relation between immigration and anatomical formation constituted part of a much larger study undertaken by numerous social scientists on behalf of the U.S. Congress Immigration Commission. Published in 1911, in forty volumes, the study "symbolized the high point of political propaganda for immigration restriction before the immigration laws were enacted in the twenties."[47]

The political impulse behind Boas's contribution, however, was antirestrictionist. He granted, as he put it in a 1908 letter, that the new immigrants coming into the country from eastern, central, and southern Europe were different from the "tall blonde Northwestern type of European." "And the question has justly been raised, whether this change of physical type will influence the marvelous power of amalgamation that our nation has exhibited for so long a time." Boas suggested that the investigation "be directed towards an inquiry into 1) the assimilation or stability of type, and 2) changes in the characteristics of the development of the individual." Intensive anthropometric investigations would reveal the extent to which forces such as "selection" in the immigrant process, intermarriage, and environmental changes affect bodily form over generations. If, as Boas had come to believe, environmental factors significantly influenced the typology of immigrants, then "all fear of an unfavorable influence of South European immigrants upon the body of our people should be dismissed."[48]

Boas focused his research into "changes in bodily form" on the shape of the head, since this was taken by almost every anthropologist at the time to be the most "stable" of anatomical traits. In 1908 he studied young Russian-Jewish men at City College and two public high schools in New York. As George Stocking writes, this anthropometric inves-

[46] Myerson, " 'Nervousness' of the Jew" (cit. n. 43), p. 71.

[47] Elazar Barkan, *The Retreat of Scientific Racism* (Cambridge: Cambridge Univ. Press, 1992), p. 83. On the impact of Mendelian genetics on Boas's anthropology see J. M. Tanner, "Boas' Contributions to Knowledge of Human Growth and Form," in *The Anthropology of Franz Boas: Essays on the Centennial of His Birth*, ed. Walter Goldschmidt (Menasha, Wis.: American Anthropological Association, 1959). Boas acknowledged Fishberg's early assistance and the fact that it was Fishberg's investigations "among the New York Jews [that] indicated the practicability of the present investigation": Franz Boas, *Changes in Bodily Form of Descendents of Immigrants* (New York: Columbia Univ. Press, 1912), p. 2. Fishberg was also instrumental in Boas's decision to request funding from the Immigration Commission for the 1908–1910 study. See the discussion in George W. Stocking, Jr., *Race, Culture, and Evolution* (Chicago: Univ. Chicago Press, 1982), pp. 174–175.

[48] Franz Boas to J. W. Jenks, 23 Mar. 1908, rpt. in *The Shaping of American Anthropology, 1883–1911: A Franz Boas Reader*, ed. George W. Stocking, Jr. (New York: Basic, 1974), pp. 202, 203; and Boas to Jenks, 31 Dec. 1909, *ibid.*, p. 213. See also Stocking, *Race, Culture, and Evolution*, pp. 161–194.

tigation yielded unexpected results. Rather than the stability of head form that anthropologists took as a given, Boas found significant modifications among immigrant children. This divergence between generations indicated that the physical and social environment exerted a marked impact on the anatomy of these new Americans. Further anthropometric studies of Italian and Jewish immigrant skulls revealed a plasticity of head shape that, Boas argued, demonstrated the powerful impact of environment and culture on anatomy.

> The investigation has shown much more than was anticipated. There are not only decided changes in the rate of development of immigrants, but there is also a far-reaching change in the type—a change which cannot be ascribed to selection or mixture, but which can only be explained as due directly to the influence of environment. . . . It has been stated before that, according to all our experiences, the bodily traits which have been observed to undergo a change under American environment belong to those characteristics of the human body which are considered the most stable. We are therefore compelled to draw the conclusion that if these traits change under the influence of environment, presumably none of the characteristics of the human types that come to America remain stable. The adaptability of the immigrant seems to be very much greater than we had a right to suppose before our investigations were instituted.

As historians of anthropology have made clear, Boas well understood the political implications of his results and was not at all reticent about using them to fight against immigration restrictions.[49]

Boas presented these same findings, albeit in a greatly abbreviated form, in a paper delivered to the First Universal Race Congress, held in London in 1911. The European Jewish immigrant figured prominently here as well: "Thus, among the East European Jews the head of the European-born is shorter than the head of the American-born. It is wider among the European-born than it is among the American-born. At the same time the American-born is taller."[50] Significantly, Boas did not sever the purported link between bodily form, and particularly head form, and mental or moral characteristics or abilities. This link was central to racial anthropology and to racial science in general. A good deal of the power of racial thought derived, of course, from this notion that nonphysical traits, or an individual's internal life, could be deduced from external or physical traits. But Boas did argue that the instability or plasticity of human type meant that one could expect a plasticity of mind and character as well. If environmental forces worked to transform the physical, they would do the same for the mental and spiritual.

THE JEWISH POLITICS OF JEWISH ASSIMILATION

This promotion of the ideas of a plasticity of (Jewish) mind and body and of integration as melioration was intended primarily to shape debate on immigration and to counter nativist arguments made in the name of science. That is, it can be understood first and foremost within the more limited context of American debates over national identity and immigration. In Boas's case, for instance, the Jews were only one component in a broader,

[49] Stocking, *Race, Culture, and Evolution*, p. 176; Boas, *Changes in Bodily Form* (cit. n. 47), p. 2; and Barkan, *Retreat of Scientific Racism* (cit. n. 47), p. 84. On the general intellectual and political context of Boas's work see also Adam Kuper, *The Invention of Primitive Society* (London/New York: Routledge, 1988), Ch. 7.

[50] Franz Boas, "Instability of Human Types," in *Papers on Interracial Problems Communicated to the First Universal Races Congress Held at the University of London, July 26–29, 1911*, ed. Gustav Spiller (Boston, 1912), rpt. in *Shaping of American Anthropology*, ed. Stocking (cit. n. 48), pp. 214–218, on p. 216.

Figure 3. *Portrait of Maurice Fishberg. From Solomon Kagan,* American Jewish Physicians: Biographical Sketches *(Boston, 1942), page 49.*

more inclusive analysis of immigrant groups.[51] Nonetheless, these studies also contributed, however obliquely, to the debate taking place between Zionists and integrationists. Any exhortation of the benefits of assimilation was a repudiation of the nationalist understanding of the collective fate of Jewry in the Diaspora.

These two impulses—the politics of American identity and the politics of Jewish identity—came together explicitly in the work of the physician and anthropologist Maurice Fishberg. (See Figure 3.) Fishberg was the integrationist counterpart of the Zionist social scientist Arthur Ruppin. His books and articles published during the first two decades of

[51] Boas's studies on head forms and physical anthropology contained research on Italians, Native Americans, Eskimos, and others, and this research occupied but a small portion of his intellectual energies. To be sure, a case can, and has, been made for an intimate connection between Boas's Jewish origins, his immigrant background, and his anthropology. Opponents at the time, and some later historians, have seen in the "cultural turn" advanced by Boas and his disciples an identifiable "Jewish element." However, Boas's intellectual and professional focus was never the Jews. On this see, most recently, Gelya Frank, "Jews, Multiculturalism, and Boasian Anthropology," *American Anthropologist*, 1997, *99*:731–745.

this century contain the most comprehensive and systematic social scientific justification for Jewish assimilation. Fishberg had emigrated to the United States from Russia in 1889. By the early 1900s he had established himself, through his writings and professional work, as an authority on contemporary Jewry. In particular, his position as chief medical examiner for the United Hebrew Charities of New York offered him direct access, as both physician and physical anthropologist, to eastern European Jewish immigrants. He conducted anthropometric measurements and examinations for physical and mental disorders.[52] These empirical studies were supplemented with the extensive physical anthropological and medical literature on the Jews produced by European and American authorities.

Fishberg's analysis was impelled, on the one hand, by specific developments within the United States: the mass immigration of eastern European Jews into New York and elsewhere, beginning in the last quarter of the nineteenth century; the extreme poverty of these immigrants, as well as their high rates of disease; and the racial anti-Semitism that took hold in important quarters and prompted the push for far tighter immigration restrictions. In 1905 and again in 1907, at the behest of the congressional committee investigating the immigration question, Fishberg traveled to eastern Europe as anthropological consultant to the United States Bureau of Immigration. He was a friend and colleague of Franz Boas; as already noted, his findings were incorporated into Boas's studies of immigrant head forms. Fishberg, in turn, incorporated Boas's research on head forms into his own writings on Jews.[53]

At the same time, Fishberg's work owed a great deal to intellectual and political developments in Europe. As a member of an international scientific community, he drew heavily on research conducted about Jews throughout Europe and published in Russian, German, and other European journals. Writing in German as well as English, Fishberg addressed his work to an audience both within and beyond the United States.[54] His analysis of modern Jewish life was driven in part by particular national (American) and local (New York) concerns. Yet it was also impelled by his antagonism to European Zionism, to what he believed to be the dangerously mistaken racial definition of Jewry advanced by some Jewish nationalists and the "separatist" conclusions they drew from it. Fishberg's analysis, then, must be understood not only as a product of the particular American political and social context, but also as a contribution to intellectual and political debates occurring mainly in Europe among Jewish elites, including social scientific authorities.

[52] In a footnote at the beginning of his 1905 study on the physical anthropology of eastern European Jews, Fishberg acknowledged the difficulty he had encountered in the past in obtaining "anthropometric measurements of living people." He then thanked Lee K. Frankel, the manager of the United Hebrew Charities of New York, who "afforded me the opportunity to obtain anthropometric measurements in connection with my work as medical examiner of the charities": Maurice Fishberg, *Materials for the Physical Anthropology of the Eastern European Jews* (Memoirs of the American Anthropological and Ethnological Societies) (Lancaster, Pa., 1905), p. 5. On Fishberg see the entry in *Encyclopedia Judaica* (cit. n. 17), Vol. 6, pp. 1328–1329. It is an indication of how quickly Fishberg established himself in New York that in 1904 he earned an entry in the *American Jewish Yearbook*'s "Biographical Sketches": *American Jewish Yearbook* (1904), Vol. 6, p. 93.

[53] See, e.g., Fishberg, *Materials for the Physical Anthropology of the Eastern European Jews*, pp. 139, 141. For more general discussions of the mass Jewish immigration into America see Arthur Hertzberg, *The Jews in America* (New York: Simon & Schuster, 1989), Ch. 10; Irving Howe, *World of Our Fathers* (New York: Simon & Schuster, 1976); and John Higham, *Send These to Me: Jews and Other Immigrants in Urban America*, rev. ed. (New York: Atheneum, 1975).

[54] Previous discussions of Fishberg have focused exclusively on the North American context of his ideas and activities: the conflict over the immigration of eastern European Jews into the United States in the first decades of this century, the debate over the health and disease of Jewry, and the relation between this biophysical status and environmental conditions. The most recent, and fullest, discussion of Fishberg in the American context is Kraut, *Silent Travelers* (cit. n. 42), Ch. 6. For an important work in German see Maurice Fishberg, *Die Rassenmerkmale der Juden: Eine Einfuhrung in ihre Anthropologie* (Munich, 1913).

As we have seen, it was widely asserted that there existed a Jewish "racial pathology" that manifested itself in physical and moral disease and degeneration. Such claims were made by American nativists and anti-Semites—but also by Zionists. Anthropological and medical statistics were marshaled by the former to demonstrate that eastern European Jews were racially alien and physically inferior to "native" Americans; hence their increasing presence posed a threat to the biological and racial health of the American nation.[55] The latter, also armed with statistical data, sometimes agreed with the assessment that modern Jewry was "diseased"; at other times they insisted on the relative health, even superiority, of Jewry, though they believed that advantage was rapidly disappearing. In either case, Zionists asserted that social scientific evidence on the Jewish condition demonstrated that Jewry had no future in the Diaspora, including the United States.[56] The complex forces of emancipation and capitalism had led inevitably to integration and assimilation. These, in turn, produced dangerous levels of decline and degeneration.

In Fishberg's view the conclusion to be drawn from the analyses of both sides, Zionists and anti-Semites, was the same. As he put it in the preface to his best-known work, *The Jews: A Study of Race and Environment* (1911):

> It appears that the prevailing opinion is that the Jews, alleged to have maintained themselves in absolute racial purity for three or four thousand years, may prove hard to assimilate. On the one hand we have those Jews who take great pride in the purity of their breed, and, on the other, the people among whom they live who see a peculiar peril in the prospect of indefinitely harboring an alien race which is not likely to mix with the general population. This apprehension is confirmed by the Jewish nationalists, who look for repatriation in Palestine, or some other territory, thus corroborating the opinion that they are aliens in Europe, encamped for the time being, and waiting for an opportunity to retreat to their natural home in Asia.[57]

"Those Jews" who believed in the purity of the Jewish race were indeed the same Jewish nationalists whose efforts to leave Europe offered clear-cut proof of Jewish racial antagonism to Christian European culture. Fishberg repudiated such avowedly political or ideological interpretations of contemporary Jewish life, even as he offered his own ideologically charged analysis. Only science, he insisted, and not politics, could competently address and answer pressing questions about Jewish racial identity and assimilation. Utilizing European and American anthropological and medical statistical studies, he set out to prove that the Jews were not a "pure race" but, rather, the product of centuries-long racial and ethnic mixing and that the etiology of "Jewish diseases" was social and economic rather than racial. Hence for Fishberg the racial heterogeneity of the Jews allowed for the possibility and probability of assimilation into non-Jewish environments. (See Figure 4.)

In important respects, these assertions were not all that different from those made by Zionist social scientists. Like the Zionists, Fishberg sought to undermine the image of the

[55] On the impact of European racial thinking about the Jews in the United States and its role in debates about immigration see John Higham, *Strangers in the Land*, 2nd ed. (New Brunswick, N.J.: Rutgers Univ. Press, 1988), pp. 138–154; Alan M. Kraut, "Silent Strangers: Germs, Genes, and Nativism in John Higham's *Strangers in the Land*," *Amer. Jewish Hist.*, 1986, *76*:142–158; and Singerman, "Jew as Racial Alien" (cit. n. 41), pp. 103–128.

[56] This did not mean that American Zionists denied the viability of American Jewish life or that they planned on packing up and moving to Palestine themselves. It did mean that there was a widespread belief that the United States could not be a viable solution to the crisis facing the millions of Jews left in eastern Europe. See the discussions in Naomi Cohen, *American Jews and the Zionist Idea* (New York: Ktav, 1975); and Ben Halpern, "The Americanization of Zionism, 1880–1930," in *Essential Papers on Zionism*, ed. Jehuda Reinharz and Anita Shapira (New York/London: New York Univ. Press, 1996).

[57] Maurice Fishberg, *The Jews: A Study of Race and Environment* (New York, 1911), pp. v–vi.

Figure 4. *The Jewish Lads' Brigade, London. From Maurice Fishberg,* The Jews: A Study of Race and Environment *(New York, 1911), figures 127 and 128. The caption reads, "Showing the effects of the environment on the type of the Jew." Figure 127, at the top, is captioned "The raw material." Figure 128, at the bottom, is captioned "What became of them."*

Jew as essentially deformed and diseased, and hence a threat to the health of the nation, by adopting a strategy that granted the debilities and disabilities of Jews while at the same time insisting that these deformities were a product of environment and, to a lesser extent, history. In a 1903 work entitled *Health Problems of the Jewish Poor* Fishberg called attention to the "small stature, the weak chest, the fragile skeleton" of the eastern European Jew. But these were acquired traits, he insisted, and so not necessarily passed on hereditarily. Like other Jewish social scientists who appropriated an environmental model to explain physiological and anatomical traits, he explained ill-health in eastern European Jews by reference to the external oppression imposed by the czarist regime and the internal oppression maintained by a Jewish religious tradition at odds with modernity (although he also gave credence to the vitalizing effects of religious laws and practices). In Russia the Jewish child was placed immediately into the cheder: "The unsanitary and unhygienic surroundings of those schools are well known to everybody, and it is rather surprising that any child is able to visit them daily for several years and come out alive."[58] There was no outdoor activity, he continued, only Bible and Talmud study. The healthy body was invariably "retarded" by such an experience.

"Now contrast this with the condition obtained in the United States," Fishberg wrote, where Jewish children exercised out of doors and schools were "models of hygiene." Even first-generation Jews were in far better physical condition than their eastern European counterparts:

> They are taller, their chests and muscular systems are in better condition than those of their less happy brothers and sisters in eastern Europe. They prove by this that the shortcomings of the Jews in European ghettos are more the result of cruel and malignant abuse, persecution and unjust oppression than a racial characteristic. Wherever he is given a chance to recuperate the Jew at once manifests a more remarkable regeneration of his physical self; he is becoming an entirely new man.

This reference to the Jew becoming a "new man" through assimilation into American life paralleled, and in all likelihood consciously responded to, the well-known Zionist image of the "new Jew" or "new man."[59] The Zionist "new man" would come about, as we have seen, through the regenerative efforts of Jewish nationalism, including a return to the land and agricultural labor and the creation of a majority Jewish polity and culture. Fishberg's vision of regeneration, in contrast, hinged on integration into the already healthy American environment and was structured around just those values inimical to the Russian czarist regime: freedom and tolerance. The social sciences, particularly physical anthropology and social medicine or hygiene, provided him with the tools to demonstrate scientifically that Jews could successfully assimilate; and when they did, this both contributed to and was reflected in their physical and mental health.

In his 1905 study *Materials for the Physical Anthropology of the Eastern European Jews* Fishberg set out to prove that Jews could not be defined along uniform racial lines. Working with comparative statistics on the stature, chest girth, head form, nose shape, and pigmentation of Jews and non-Jews throughout various regions in eastern Europe and among immigrant Jews in the United States, he argued that Jews differed little somatically from the ethnic groups amongst whom they lived. This, he believed, gave the lie to the claim of a Jewish racial uniformity or homogeneity, undermining the ideological foundations of racialist anti-Semites and Zionists. Nonetheless, Fishberg found a consistent

[58] Maurice Fishberg, *Health Problems of the Jewish Poor* (New York, 1903), p. 15. The medicalization of the cheder, as part of a broader critique of traditional "ghetto" life by Jewish reformers, had already emerged in the late eighteenth century. See John Efron, "Images of the Jewish Body: Three Medical Views from the Jewish Enlightenment," *Bulletin of the History of Medicine*, 1995, *69*:349–366. On the cheder as a source of pathology in Russian-Jewish discourse see Steven Rappaport, "Heder and Hygiene: Medicalizing a Social Problem," paper presented at the annual Association of Jewish Studies Conference held in Boston in December 1996.

[59] Fishberg, *Health Problems of the Jewish Poor*, pp. 15–16. On the Zionist idea of a "new man" see Berkowitz, *Zionist Culture and West European Jewry before the First World War* (cit. n. 32).

variation (rather than uniformity) between individual Jewish communities across geographic boundaries. Jews were on average slightly deficient in stature compared with non-Jews. Yet they were always so in ratio to their neighbors, so that in environments where Christians were taller on average, for instance, so too were Jews—even if the Jews remained short relative to their neighbors. This last fact was due in part to the Jews having dwelled for centuries predominantly in towns and cities; yet "the wretched social, economic and sanitary conditions under which the Jews labor in eastern European ghettos will also account for the deficiency they display in bodily height when compared with their gentile neighbors." In Galicia, where conditions were worst, Jews were shortest; in Poland they were somewhat taller; and in southern Russia Jews were "quite tall." In each case physical stature mirrored economic conditions. "Another indirect proof of this theory is the increase of stature of the first generation of Jews in New York City, where the social and economic conditions are much improved. . . . The superior stature of the native American Jews is thus seen to be a result of superior social conditions and environment."[60] Physical traits were environmentally, not racially, determined, and assimilation into a new milieu produced a transformation at the biophysical level.

Fishberg's endorsement of assimilation, however, was by no means unequivocal. He, too, recognized that modernity took its toll upon Jewry. This sort of ambivalent embrace of assimilation was not unusual. There was a sense even among those Jews in the United States who were adamant in approving the integration process that there was a not-inconsiderable price to be paid for collective "health and normalcy." In his 1911 work *The Jews: A Study of Race and Environment* Fishberg offered a more complex reading of Jewish assimilation, modernity, and health than that found in his earlier writings.[61] Fishberg repeated his arguments about the heterogeneity of Jewish bodily shapes and sizes and the evidence this offered about the historical assimilation of Jews into their non-Jewish environments.[62] Yet he also described the detrimental impact of emancipation and modernity on Jewry in clear and forceful terms. "It is evident from all available facts that Judaism thrives best when its faithful sons are isolated from the surrounding people, segregated in Ghettos or Pales of Settlement, excluded from educational institutions frequented by people of the dominant faith, and thus prevented from coming into contact with their non-Jewish neighbors." It was precisely in countries like Russia and Rumania, where Jews were politically, socially, and economically most isolated, that their birth rates were highest and their intermarriage and conversion rates lowest. In the West, where the Jews enjoyed freedom and equality, just the opposite was the case. "Indeed, the western Jews display a striking retrogression and decadence, which is by no means accidental." The only "redeeming factors" for western Jewry were their low mortality rate and eastern European immigration. But neither of these, Fishberg was quick to point out, could offer "indefinite help." Death rates could not realistically be expected to drop much lower, and immigration

[60] Fishberg, *Materials for the Physical Anthropology of the Eastern European Jews* (cit. n. 52), pp. 39–46, on p. 40. "The general rule in Europe seems to be that the urban type is physically degenerate," Fishberg wrote, quoting the prominent American writer William Ripley (p. 39).

[61] Fishberg, *Jews* (cit. n. 57). Fishberg's attack on Zionism in this work is his most sustained denunciation of the movement to be found in his social scientific writings. He did publish an attack in 1906, in Yiddish; the translated title is *The Perils of the Jewish Nationalist Movement.*

[62] Fishberg, *Jews*, pp. 506–513. Fishberg did stress in this work that, while some parts or components of the body, like stature and chest size, were directly influenced by environment, others were almost exclusively the result of heredity. The most stable traits, such as skin pigmentation or eye and hair color, were the product of genetics, not socioeconomic conditions. But this, too, in Fishberg's view, demonstrated his point about the historical possibility and fact of Jewish assimilation—in this case through intermarriage and conversion.

from the East was not an inexhaustible resource. Moreover, eastern European Jewry would surely follow the western pattern of integration and assimilation as political conditions improved. Employing the direst of phrases, Fishberg spoke of "race suicide" among the Jews of both East and West.[63] Invoking the danger of collective disappearance, Fishberg sounded more like a Zionist than an assimilationist. For it was the Zionist social scientists who had demonstrated how the assimilative process worked to dilute Jewish identity and ultimately threatened to destroy the Jewish people.

In the end, however, Fishberg reaffirmed his faith in assimilation as the only "normal" and "natural" course for Jews to take. He denied that the statistics concerning fertility, intermarriage, and conversion indicated any genuine pathology: "We do not want to be misunderstood on this point. In showing the tendency to retrogression of the Jews we do not at all mean to imply that this is a sign of a physical degeneration of a physiological or pathological nature." Indeed, Fishberg proceeded to normalize the condition of modern western Jewry by universalizing it: The Jews "are not unique. The same phenomena are to be observed among all civilized peoples in varying degrees of intensity. 'Prudential foresight' goes hand-in-hand with an improved standard of life, with greater ambition to raise children who are fortified with a proper education, the best weapon in the fierce struggle for existence which has been going on in modern commercial and industrial life."[64]

Moreover, Fishberg by no means softened his hostility to the Jewish nationalists. Zionism, he insisted, was abnormal and pathological. In the penultimate chapter of the book, entitled "Assimilation versus Zionism," Fishberg defined Jewish nationalism as an ideological movement "born of negatives": the "rejection of emancipation" and fear of assimilation; the adoption of "the chauvinistic ideas of [the Jews'] Christian neighbors" in order to "revive a racial and national spirit"; and "apprehension for the future of Judaism." Zionists assert not merely that Jews are "a race," that is, that there is more to collective Jewish identity than religion and history: "To the Zionists, the Jews are a distinct, non-European race which has preserved itself in its original purity in spite of the Jews' wanderings all over the globe. They hold that the Jews can never merge with the European races, and are bound to remain distinct from their Christian or Mohammedan neighbors."[65] This is the source of the Zionists' unwavering belief in the national identity and destiny of Jewry.

Fishberg did not do justice to the Zionist position, which was far more heterogeneous on the question of Jewish racial "purity." It was, for instance, a belief in the lack of any demonstrable "racial antipathy" between Christians and Jews that led many Zionist social scientists to identify intermarriage as the crucial problem facing modern Jewry. In their warnings about dangerous levels of interracial mixture, they referred repeatedly to Rudolf Virchow's study of Jewish and Christian German schoolchildren. In the early 1870s, at the behest of the newly constituted German Anthropological Society, Virchow conducted a study of close to seven million children that measured the color of the eyes, hair, and skin and the shape of the skull. The survey demonstrated, as George Mosse has written, that "there was no such thing as a pure German or a pure Jewish race." Pure races, Virchow concluded, did not exist, and science could not establish any clear and simple correlation between particular bodily features (color of hair or shape of nose) and particular races. For Zionist social scientists, as for many non-Jewish authorities, statistics showing a sur-

[63] *Ibid.*, pp. 466, 520.
[64] *Ibid.*, p. 521.
[65] *Ibid.*, pp. 469, 473, 470.

prisingly high percentage of blond or "mixed type" German Jews were not a cause for rejoicing. While they could take some comfort in the fact that such a study disproved the anti-Semitic claims of a physiological or instinctual antipathy between Christian and Jew, it nonetheless made clear that the assimilation of Jewry—and hence the loss of Jewish particularity (*Eigentümlichkeit*)—was proceeding apace.[66]

For Fishberg, this sort of physical and social intermixture between Jews and surrounding peoples and races was normal; the Jewish past and present both demonstrated that Jewry was a product of just such a phenomenon. Moreover, Jewry—although not necessarily Judaism—was "healthiest" when it acted on these impulses. Isolation, whether imposed by external regimes or by internal ideologies, was an "unnatural," abnormal condition. In the nineteenth century European Jewry proved that "there is nothing within the Jew that keeps him back from assimilating with his neighbors of other creeds; that as soon as the political and civil laws which previously kept him apart from the general population are abrogated, he begins to adapt himself to the new surroundings in a wonderful manner."[67]

CONCLUSION

Jewish social science emerged in Europe and the United States in the first decades of this century for myriad reasons. The Jews had been an object of research and debate within the general scientific community for decades. Jewish social scientists perforce used the methods, ideas, and images of European science, even as they sought to undermine significant elements within it. Anthropological, medical, and demographic studies of Jewry, produced by Jews, were impelled by both apologetic and self-critical, reformist motives. A Jewish social science, it was believed, might serve to counter the scientific anti-Semitism that had developed since the 1870s, even as Jewish experts diagnosed the "ills" of their own people and prescribed cures. On the one hand, Jewish social scientists accepted many of the negative images and ideas about the condition and characteristics of Jews and incorporated these into their work. (One can, it ought to be noted, also find positive images of Jews here, and even assertions of Jewish superiority—most notably the widespread belief in the higher developed intelligence of Jewry.) On the other hand, the environmental determinism adopted by Jewish social scientists offered a foil to the biological determinism posited by anti-Semites.

But Jewish social science was ideologically riven in addition to being ideologically driven. If Jewish scientists were united in their desire to appropriate the discourse that had traditionally evaluated their people in a negative way and in their labor to reverse the condition of Jewry, they were divided over how, more precisely, to interpret the social scientific data and what practical conclusions to draw from it.

Zionist social scientists saw in the anthropological and medical data generated about the Jewish body evidence of a Jewish national and racial identity that transcended geographic boundaries; and they understood both Jewish health and disease as proof that

[66] Mosse, *Toward the Final Solution* (cit. n. 1), pp. 91–92; see also note 6, above. In 1911 Ruppin, for instance, argued that the best evidence against this notion of "instinctual racial antipathy" was "the fact that even the Jews themselves have begun to avoid the choice of marriage with the Jewish type, and prefer the so-called 'Aryan type.' One often sees in Jewish marriage ads explicit calls for blond hair, or the phrase 'nicht jüdisch aussehend.'. . . The only real deep-seated instinctual racial difference, leading to aversion—namely, difference in skin color—does not come into play in relations between Jews and Europeans": Ruppin, *Juden der Gegenwart* (cit. n. 25), p. 170.

[67] Fishberg, *Jews* (cit. n. 57), pp. 479–480.

"modernity," in the guise of emancipation and assimilation into non-Jewish society, posed an unprecedented danger to Jewry. Integrationists shared with Zionist social scientists the desire to counter the negative image of Jewry prevalent in much of the modern social scientific literature. In this sense, there was a unity of purpose behind the analyses of Jewish social scientists that transcended the political commitments that divided them. At the same time, Fishberg, Maretzki, and others engaged with Jewish nationalists at the intellectual and ideological levels, offering a spirited scientific defense of Jewish integration. Both assimilationist and Zionist social scientists, therefore, recognized the reality of Jewish assimilation at the physical as well as the social and cultural level. What they debated were the effects of assimilation and how, ultimatcly, Jewish identity ought to be defined.

The literary critic Ferdinand Brunetière in his home, 1899.

The Solvency of Metaphysics

The Debate over Racial Science and Moral Philosophy in France, 1890–1919

By Jennifer Michael Hecht

ABSTRACT

French nineteenth-century political theory generally held that the empirical scientism associated with the Enlightenment was inextricable from Enlightenment egalitarian ideals. As the century drew to a close, a racist anthropology developed that seemed to confound this idea because it combined scientism with a devotion to human inequality. There were several varieties of this phenomenon, but Georges Vacher de Lapouge's anthroposociology was most prominent. The racist doctrine was effectively combated by the anthropologist Léonce Manouvrier, but it remained troubling because it had demonstrated that science could be used against democracy. Three left-wing social theorists addressed this issue to great effect: Alfred Fouillée, Célestin Bouglé, and Jean Finot. Though each of these men was dedicated to secular rationalism, each came to the conclusion that if society had to choose between Enlightenment ideals and Enlightenment methods it should preserve the ideals. In so doing, Fouillée, Bouglé, and Finot proposed a sort of metaphysical leap of faith that would hold certain basic human values beyond the reach of scientific theories, however persuasive they might seem.

THE RISE OF ANTHROPOLOGICAL THEORIES OF HUMAN INEQUALITY in late nineteenth-century France created a schism between Enlightenment empirical scientism and Enlightenment egalitarian ideals. The relationship between scientific authority and politics began to shift, and the resulting changes contributed to the development of the "new right" and "new left." Broadly, the old right can be said to have relied upon monarchical tradition and revealed religion, while the left looked to science and the concomitant vision of historical progress. Between 1880 and 1914, the right began to employ science and numbers to support its social hierarchies and the left saw that, in some instances, the preservation of their political ideals required a rejection of science. A significant aspect of this reconfiguration was precipitated by antidemocratic biological theories. In France, the most important of these was the "anthroposociology" of Georges Vacher de Lapouge. This essay examines the antiracist and antiscientific doctrines devised in reaction to Lapouge and theorists like him. The primary authors of this late nineteenth-century antiracism were Alfred Fouillée, Célestin Bouglé, and Jean Finot.

It has long been noted that at the end of the nineteenth century the republican political left began to mistrust the ability of science to defend its ideals. Harry Paul's insightful article of 1968 discussed this phenomenon in terms of the "debate over the bankruptcy of science," which he described as arising from a public call, issued in 1895 by the positivist Ferdinand Brunetière, for a return to religion in the service of a moral society.[1] This call was significant on a broad cultural field: it was deplored by angry chemists and government positivists and applauded by clerics. This study will demonstrate that anthropologists and philosophers of the period mocked the answers proffered on both sides of the issues raised by Brunetière. They were engaged in a much more trenchant debate on morality that was generating the new ideas, both horrifying and heroic, that would define the following century. This article takes Brunetière's "bankruptcy of science" debate as a background map of ideological positions at the turn of the century. From opposite sides, the opponents in a new debate were pushing each other off that map and contributing to the creation of a new political order.

Ferdinand Brunetière was a prolific and respected literary critic, well known for his multivolume tomes on lyric poetry and the history of French literature. By the turn of the century he would distinguish himself as a leading lay participant in the Catholic revivalist movement and as an anti-Dreyfusard, but before 1895 he was best known as a rationalist freethinking scholar, *maïtre de conférence* at the École Normale Supérieure, and editor of the *Revue des Deux Mondes.* He used the pages of that journal to discuss Darwinian theory, anticlericalism, and philosophical materialism—all in the name of secular democracy. (See Frontispiece.) When he called for a pragmatic return to Catholicism in his article "Après une visite au Vatican," he was not announcing a new intellectual position regarding the nature of reality but, rather, proclaiming a change in social strategy. He feared that without the morality of religion the republican body politic would fall into chaos. In defense of bourgeois security, he dramatically asserted that he had been mistaken: science could not convince the mass of human beings to be good. When Brunetière spoke of science, he meant positivist secularism as opposed to religiosity. He bemoaned the failure of science to provide a convincing social morality because he believed that without sanctions in the afterlife, no morality could be sufficiently imposing. In one instance he did state that "if we ask Darwinism for lessons in moral behavior, the lessons which it gives us will be abominable," but his main point was that the secular morality of scientists taught the same lessons as religion without the requisite supervision.[2] His leading opponents had no argument with his terms. The chemist and government minister Marcelin Berthelot, for example, argued that positivist republicanism was doing fine and had no need to run back into the arms of a paternalist, dogmatic religion. On both sides of this debate, the same basic model of rationalism prevailed, and both sides had the same conception of good and bad: the scientists were making no claims to have found a new scientific morality, and Brunetière did not insist on spirituality as a rationale for his proposed Catholic revival.

When Vacher de Lapouge spoke of science, however, he meant Darwinism's "abominable lessons." He advocated the breeding of human beings through a process of state selection (based on anti-Semitism and skull measurement), artificial insemination, and

[1] Harry Paul, "The Debate over the Bankruptcy of Science in 1895," *French Historical Studies,* 1968, *3*:299–327.

[2] Ferdinand Brunetière, "Après une visite au Vatican," *Revue des Deux Mondes,* 1895, *127*:97–118, on p. 104.

euthanasia, and he vilified any conception of morality that would hinder his proposals.[3] Secular, positivist republicans who responded to Lapouge's work understood that to counter such doctrines with an uninspired Catholicism would be as ineffective as it would be hypocritical. But they were driven to believe that science could not be left as the authority on human morality. In the hands of Brunetière and his opponents, the debate over morality was rather tame, posing a pragmatic Catholicism against a science that was essentially uninterested in revising morality. Philosophers and anthropologists went further, pitting a new metaphysics of human values against a harsh materialist doctrine of antimoralism and racist law.

ANTHROPOSOCIOLOGY

There was nothing new about mixing biological theory with politics, but the practice had proliferated after 1859, when Darwin published his theory of evolution. "Racial conflicts" were ubiquitous, for the term signified tensions between groups that are racially homogeneous by today's definitions. In the 1860s and 1870s Paul Broca's anthropometrical studies had quantified these purported racial differences. The famed Count Gobineau, an ardent opponent of evolution, invented a theory of history based on the interactions of three human races that corresponded closely to aristocrats (Gobineau's white race), bourgeois (yellow), and workers (black). By the turn of the century many French theorists applied a biological hierarchy to social categories; discussions of degeneration, criminal anthropology, and eugenics were relatively commonplace.[4] Nonetheless, Vacher de Lapouge stood out as having invented a racial doctrine that his contemporaries saw as scientific, erudite, vicious, and revolutionary. He called his science "anthroposociology," and this term, though soon widely used, generally maintained its reference to Lapouge. It would be hard to overestimate Lapouge's industriousness in the service of his science. He kept up a tremendous correspondence with disciples, sympathizers, opponents, journal editors, and colleagues all over the world. In so doing, he actively facilitated the exchange of information, the translation of work, the publication of new studies, and the creation of conferences. He sent fellow travelers his collections of data and his photographs of "types," lent them skulls, annotated their work, wrote prefaces, and even helped them attain university positions. Contemporary discussions of racialist anthropological politics regularly cited him as the leader of the movement in France.[5] Thus, while the defenders of egalitarian

[3] Lapouge's dismissal of good and evil is quite similar to Nietzsche's, though neither seems to have read the other's work. The chief difference between them is that, ultimately, Nietzsche called for the individual to strive, inwardly, to become the "Ubermann," while Lapouge called for a dismissal of individuality in pursuit of racial group progress. At times Nietzsche seems to be advocating very Lapougian constructs, but while the philosopher's work is punctuated with brilliant insights that explode his otherwise racialist ideas, the anthroposociologist's work is punctuated with very clever wit that merely serves to further his racialist convictions.

[4] See Stephen J. Gould, *The Mismeasure of Man* (New York: Norton, 1981); Robert Nye, *Crime, Madness, and Politics in Modern France: The Medical Conception of National Decline* (Princeton, N.J.: Princeton Univ. Press, 1984); Linda Clark, *Social Darwinism in France* (University: Univ. Alabama Press, 1984); J. Edward Chamberlin and Sander Gilman, eds., *Degeneration: The Darker Side of Progress* (New York: Columbia Univ. Press, 1985); Daniel Pick, *Faces of Degeneration: A European Disorder, c. 1848–c. 1918* (Cambridge: Cambridge Univ. Press, 1989); and William Schneider, *Quality and Quantity: The Quest for Biological Regeneration* (Cambridge: Cambridge Univ. Press, 1990).

[5] See, e.g., Léon Winiarski, "L'anthropo-sociologie," *Devenir Social,* 1898, *4*(3):193–232; and Charles Fages, "L'evolution du Darwinisme sociologique," *Humanité Nouvelle,* 1898, *2*:28–42. See Lapouge's extensive collection of correspondence, "Fonds Vacher de Lapouge," housed in the Paul Valéry Library of the University of Montpellier; his government personnel files, located in the Archives Nationales de Paris, especially BB/6(II)419 and F/17/22640; and the Archives of the Société d'Anthropologie de Paris, located at the Musée de l'Homme in Paris.

democracy did not direct their arguments exclusively at Lapouge, it is useful to consider this central target.

Lapouge is also an important subject because, despite the strenuous rejection of his ideas that will be described in this essay, those ideas were profoundly influential—celebrated in Germany and the United States during his lifetime and, after his death, revived in France. Hans Günther, the principal Nazi race scientist, cited Lapouge regularly. The two men corresponded over many years and aided one another in important ways. Lapouge had many other German collaborators, but his influence was not limited to Germany. He corresponded with many Americans and was entertained at the White House, along with other eugenists, under the Harding administration. In 1939 the Vichy regime set up an Institut d'Anthroposociologie that formed the "scientific" arm of a group of institutions dedicated to racist propaganda. The institute was based on Lapouge's ideas, and his son, Claude Vacher de Lapouge, was made its president. In 1940 Pierre Laval authorized a commission to investigate the possibility of putting Lapouge's eugenics program into practice.[6] And yet years earlier, at the turn of the century, Lapouge had been repudiated by the French intelligentsia.

In his two major works, *Les sélections sociales* (1896) and *L'Aryen* (1899), Lapouge described humanity as divided into two races, each of which could be identified by its "cephalic index": the dolichocephalic and the brachycephalic. (See Figure 1.) The index was calculated by comparing the width and breadth of the skull: dolichocephali had long, narrow heads, while brachycephali had round heads. The cephalic index originated with the Swedish anthropologist Anders Retzius; Lapouge's innovation lay in the qualitative characteristics he associated with these head shapes.[7]

"Dolichos"—or Aryans, as Lapouge alternatively called them—were fair skinned with blue eyes and temperamentally creative, adventurous, and refined. More often Protestant than Catholic, they were the majority population in northern Germany, Scandinavia, and England. "Brachies" (also his term) were much darker in complexion, tended to live in mountainous regions, and were more often Catholic than Protestant. They preferred to stay near their homes and chose constancy over change. They were good but mediocre people, lacking imagination and courage. This mild and uninspired group dominated the populations of France, Spain, and Italy, all of Asia, and most of the Slavic countries. Lapouge insisted that the aristocracy of the ancien regime had been dolichocephalic. Because the French Revolution had removed the aristocracy from control in France, the dolichos were overwhelmed by the masses of brachies. Stripped of their legal and financial power, they were intermixing with the brachies and dying out. According to Lapouge, the only dolicho group that was both succeeding in the public world and refusing to breed outside the group

[6] In many of his books, Günther cited Lapouge more frequently than any other author and there were many more general borrowings. See, e.g., Hans Günther, *Kleine Rassenkunde Europas* (Munich, 1925); Günther, *Rassenkunde des Deutschen Volkes* (Munich, 1923); and Günther, *Racial Elements of European History* (London, 1927). For their correspondence see Lapouge/Günther and Lapouge/A. F. DuPont (DuPont often wrote to Lapouge on Günther's behalf), Fonds Vacher de Lapouge. On Lapouge's trip to the White House see Lapouge to Madame Albertine de Lapouge, no. 068-50, 23 Sept. 1921, and [no number] 28 Sept. 1921, Fonds Vacher de Lapouge. On the institute see Michael R. Marrus and Robert O. Paxton, *Vichy et les juifs* (Paris: Calmann-Lévy, 1981), pp. 395, 413, 416. Laval's commission is discussed in Hubert Thomas-Chevalier, *Le racisme française* (Nancy: Thomas, 1943), pp. xi, xix, cited in George L. Mosse, *Toward the Final Solution: A History of European Racism* (Madison: Univ. Wisconsin Press, 1978), pp. 58–62.

[7] Georges Vacher de Lapouge, *Les sélections sociales* (Paris, 1896); and Lapouge, *L'Aryen, son role social* (Paris, 1899). On the history of the cephalic index see Claude Blanckaert, "L'indice céphalique et l'ethnogénie européenne: A. Retzius, P. Broca, F. Pruner-Bey (1840–1870)," *Bulletins et Mémoires de la Société d'Anthropologie de Paris,* 1989, *1*(3–4): 165–202.

Figure 1. *Head types. Lapouge loaned these photographs to William Ripley, who published them in his* Races of Europe *(New York, 1899). Note the cephalic index printed beneath some of the portraits. Lapouge considered an index below 80 to be dolichocephalic, the superior type.*

was the Jews. This was a problem for Lapouge because he was an anti-Semite who saw the Jews as a dangerous, venial, "false aristocracy" that could exploit but not create civilization. Lapouge also recognized many racial subdivisions, mixes, and exceptions to his rules, all of which he explained in great numerical, anecdotal, and linguistic detail. He and his disciples measured many thousands of skulls and heads in order to demonstrate and decipher these relationships.[8]

Lapouge wanted France, redesigned as a "selectionist state," to regulate its citizens' professional and reproductive lives. He argued that the diluted semen of one perfect dolicho man could serve for the impregnation of two hundred thousand women a year, and he proposed a government that would organize the enterprise. "Liberty, Equality, Fraternity,"

[8] Lapouge had many zealous disciples, but he also got help from some unexpected quarters. The poet Paul Valéry, for instance, joined Lapouge in his laboratory, helping him to measure six hundred skulls taken from an old cemetery. The young poet frequented Lapouge's anthroposociology lectures while studying law at Montpellier. Valéry later commented that he did not learn anything useful but that, "among all the things that I learned that were never useful to me, those pointless measurements were not more pointless than the others." Paul Valéry to Henri Bégouen, 1936, quoted in Henri Bégouen, "Vacher de Lapouge—Père de l'Aryenisme," *Journal des Débats Politiques et Littéraires,* 22 Aug. 1936, p. 3.

Lapouge insisted, must be replaced by "Determinism, Inequality, Selection!"[9] He saw the world on the brink of a total revolution based on his theories of heredity, writing that "in the next century people will be slaughtered by the millions for the sake of one or two degrees on the cephalic index." This kind of language was rare with Lapouge and it was more a prediction than a threat. Yet although Lapouge did not ask specifically for extermination, he did call for the death of any moral sentiments that stood in its way—because they also stood in the way of his breeding plans. His invective against morality was explicit: "Here is why I have been speaking to you of the abyss and of a cataclysm. All of morality and all of the ideas which serve as a base for law and for the political sciences, in their present-day conceptions, constitute a series of deductions of which the first term assumes the existence of a personal divinity. . . . Remove all validity from this source and there is nothing left." When he predicted that "the superior races" would soon conquer human groups "retarded in evolution," he added that "the last sentimentalists will witness the copious extermination of entire peoples."[10] As a rule, he showed more anger toward these "last sentimentalists," who continued to believe in God and morality, than toward the human groups "retarded in evolution."

In general, scholars have understood Lapouge as a frustrated aristocrat who added Darwinism to the racist and classist theories of Gobineau and in so doing fashioned one of the first political utopias based on scientific racism. Indeed, in his own memoirs Lapouge described himself as an aristocrat who had always struggled against the republic and its ideals.[11] But this is not accurate. Vacher de Lapouge began his career as a resolute left-wing anticlerical republican; he worked as a magistrate for the young Third Republic and insisted on democratic ideals in his public and private correspondence. He was not very successful in this career—partly in consequence of his boisterous anticlerical zeal—and when he was still a young man he resumed his studies, switching his attention from law to science.[12] Lapouge began taking classes at various Parisian institutions, among them the École d'Anthropologie, which was run by radically anticlerical atheists. These anthropologists were freethinkers who had trained themselves in anthropology in order to use

[9] Georges Vacher de Lapouge, "Préface," in Ernst Haeckel, *Le monisme, lien entre la religion et la science,* trans. Lapouge (Paris, 1897), pp. 1–8. Lapouge claimed to have performed the first "telegenesis," mailing a dose of human sperm from one town to another and there attaining a successful conception: Lapouge, *Sélections sociales* (cit. n. 7), pp. 472–473.

[10] Georges Vacher de Lapouge, "L'anthropologie et la science politique," *Revue d'Anthropologie,* 1887, *16*:136–157, on pp. 151, 142, 151.

[11] Georges Vacher de Lapouge, "Souvenirs," in *Georges Vacher de Lapouge: Essai de bibliographie,* ed. Henri de La Haye Josselin (Paris, 1986), pp. 9–16, on p. 15. On Lapouge see Guy Thuillier, "Un anarchiste positive: Georges Vacher de Lapouge," in *L'idée de race dans la pensée politique française contemporaine,* ed. Pierre Guiral and Emile Témime (Paris: CNRS, 1977), pp. 48–65; and Pierre-André Taguieff, "L'introduction de l'eugénisme en France: Du mot à l'idée," *Mots,* 1991, *26*:23–45. For an article that exhibits a decided sympathy for Lapouge and his ideas see Jean Boissel, "Georges Vacher de Lapouge: Un socialiste revolutionnaire Darwinien," *Nouvelle École,* 1982, *38*:59–84. Insofar as these essays discuss the life of Lapouge, they largely draw on two articles written at the time of his death: Bégouen, "Vacher de Lapouge" (cit. n. 8); and Etienne Patte, "Georges Vacher de Lapouge," *Revue Générale du Centre-Ouest de la France,* 1937, *46*:769–789. For works in English see Clark, *Social Darwinism in France* (cit. n. 4), pp. 143–154; Schneider, *Quality and Quantity* (cit. n. 4), pp. 56–63; and Jennifer Michael Hecht, "Anthropological Utopias and Republican Morality: Political Atheism and the Mind/Body Problem in France, 1880–1914" (Ph.D. diss., Columbia Univ., 1995), Chs. 5, 6.

[12] In one instance, Lapouge apparently took advantage of the fact that a revolutionary law forbidding ecclesiastic dress had never been officially revoked and arrested the first priest he saw passing in the street outside his window; see Bégouen, "Vacher de Lapouge." Lapouge's youthful republicanism is best illustrated in the supervisor's reports written about him during his period in the magistrature: Archives Nationales de Paris, BB/6(II)419 and F/17/22640. Indications can also be found in the personal correspondence in Fonds Vacher de Lapouge. For a history of Lapouge that traces his leftist, republican origins see Hecht, "Anthropological Utopias," Ch. 5.

the young science as a bludgeon against religion. When Brunetière needed a quotation in order to demonstrate the position of total antimetaphysics, he took it from the writings of one of these anthropologists. They were atheists, explicitly antiphilosophical, and committed to scientific materialism, but they were also devoted to principles of equality. The group put forward a number of programs to increase knowledge and social equality and claimed, if often a bit mournfully, that these projects were an adequate replacement for the functions of religion.[13]

In contrast, Lapouge was deeply nihilistic. Even his own project for human breeding seemed to him a poor, though serviceable, replacement for religious purpose. His belief in selection as the evolutionary mechanism surpassed even that of Darwin, who allowed for some effect of the inheritance of acquired characteristics. For Lapouge, control of the breeding of human beings was the only true arena of progress, despite its difficulty. "We are on our way," he wrote, "by new formulas based on social hygiene, toward the elimination of the idea of morality. It is an evolution which has its advantages and its inconveniences, but which the progress of human knowledge renders inevitable." Lapouge believed that only cowardice prevented his contemporaries from adopting his anthroposociological theories—especially those contemporaries who claimed to be freethinkers. Indeed, he said that it was "an act of faith" that enabled them to escape the conclusions of "Darwinian political science."[14] For Lapouge and many of his contemporaries, "faith" was a rather bad word implying a lack of intellectual rigor. It was in opposition to Lapouge and his ilk that a handful of republican theorists rehabilitated this loose notion of faith, insisting that certain human principles were unconditionally true—even if they could not be supported by empirical evidence.

For Lapouge, Brunetière's solution to the crisis of morality and meaning was the epitome of cowardice. "Already liberals, socialists, and anarchists treat Darwinians as barbarians. So be it! The barbarians are coming, the besiegers have come to be besieged, and their last hope of resistance is to lock themselves up in the citadel which they were attacking. The near future will show our sons a curious spectacle: the theoreticians of the false modern democracy constrained to shut themselves back up in the citadel of clericalism." Many contemporaries agreed with Lapouge, lauding his scientific and scholarly abilities and holding him up against "Christians and mystics of all types." In one of the few articles to compare Brunetière and Lapouge directly, an 1897 review of *Les sélections sociales* repeatedly praised Lapouge's dismissal of God and traditional morality and congratulated him for using "the most dependable anthropological methods: measurement of skulls." For

[13] The freethinking anthropologist Brunetière quoted is André Lefèvre. It is worth repeating his extreme statement: "Religions are the purified residues of superstition. The value of a civilization is in inverse proportion to its religious fervor. All intellectual progress corresponds to a diminution of the supernatural in the world. The future is science." André Lefèvre, *La religion* (Paris: Reinwald, 1892), pp. 572–573, cited in Brunetière, "Après une visite au Vatican" (cit. n. 2), p. 98. On the anthropologists see Hecht, "Anthropological Utopias"; Elizabeth Williams, "The Science of Man" (Ph.D. diss., Indiana Univ., 1983); Joy Harvey, "Races Specified, Evolution Transformed" (Ph.D. diss., Harvard Univ., 1983); and Michael Hammond, "Anthropology as a Weapon of Social Combat," *Journal of the History of the Behavioral Sciences,* 1980, *16*:118–132. For their antiphilosophical stance see, e.g., Lefèvre, "La philosophie devant l'anthropologie," *Homme,* 1884, *1*:577–584; and René Fauvelle, "Il faut en finir avec la philosophie," *ibid.,* 1885, *2*:139–146. For more on the anthropologists' programs and their role as replacements for religion see Jennifer Michael Hecht, "French Scientific Materialism and the Liturgy of Death: The Invention of a Secular Version of Catholic Last Rites (1876–1914)," *French Hist. Stud.,* 1997, *20*:703–735.

[14] Lapouge, *L'Aryen* (cit. n. 7), pp. 509, 514. Unlike most French eugenists, who tended toward Lamarckianism, Lapouge believed that educating individuals had no place in an evolutionary program because it would have no real effect on their offspring.

this reviewer and many others, the chief concern was the issue of morality: "Lately—before and after Brunetière—the favorable influence of moral instruction has been put into serious doubt. Vacher de Lapouge goes further." According to Lapouge, the review continued, "it is necessary to finally replace service to a supernatural and chimerical God with service to the species."[15]

Lapouge excoriated any solution to the "crisis of morality" that, having professed a belief in materialism, either prescribed a return to Christian morality or proffered belief in a naturally improving hereditary morality. He asserted that "the socialist, anarchist or so-called democratic journals" were celebrating Darwinism because they saw it "above all" as "an argument with which to oppose the Church." They were happy to use science to disprove religion, Lapouge charged, but when science suggested a course of action that contradicted their own moral universe they refused to acknowledge it. "When I say 'they,' " wrote Lapouge, "I mean the freethinkers, or those who qualify as such, because from the very beginning the churches have seen the consequences of these new theories and have taken steps to denigrate them." He was right. Catholic writers had not had any trouble repudiating his vicious theories. They simply denied that science was a source of truth and laughed at Lapouge's odd construct, usually including a few jokes regarding the idea that humans shared an ancestry with apes.[16]

For scientistic, positivist republicans, on the other hand, countering Lapouge was not so easy. Emile Durkheim, for instance, felt compelled to allow a Lapougian disciple a place in his *Année Sociologique*—editing a section on anthroposociology. Durkheim did not like it, but as long as Lapouge appeared to be a good scientist, Durkheim felt obliged to permit extremely positive reviews of his work to be printed in the journal.[17] Republicans championed fearlessness in the face of science as a cardinal virtue, upholding a belief, at least as old as the Enlightenment, that the world was a conflict between the darkness of familiar, comforting, but infantilizing dogma and the light of new, somewhat harrowing, but ultimately emancipatory lawful truths.[18] To these progressive republicans, ignorance or fear of science seemed to explain the backwardness of conservatives. How could they now turn against what seemed to be a lawful, quantifiable scientific discovery?

The scientistic republicans were lucky to have an excellent anthropologist on their side, in the person of Léonce Manouvrier. Manouvrier had earned his medical degree, and a

[15] *Ibid.,* p. 514; and R. F. [no full name given], "Les sélections sociales," *Revue des Revues,* 1896, *23*(135):375–377, on pp. 376, 377.

[16] Lapouge, *L'Aryen,* p. 513. For jokes at Lapouge's expense see, e.g., J. Rochette, "Compte rendu—Les sélections sociales," *Études Religeuses,* July 1897, pp. 279–281.

[17] The disciple, Henri Muffang, sent all of his reviews to Lapouge for "correction" before submitting them to Durkheim. See Henri Muffang to Georges Vacher de Lapouge, A067-1 through A067-91, Fonds Vacher de Lapouge. Durkheim explains his decision to include anthroposociology when introducing the section in its first appearance: Emile Durkheim, *Année Sociologique,* 1896–1897, *1*:519.

[18] This admittedly abbreviated account of a complicated phenomenon does not address the question of how numbers came to be seen as objective and authoritative. One answer is that the Newtonian description of the world in numerical ratios was so astoundingly powerful (allowing one to predict, explain, and manipulate the physical universe) that it made quantification attractive. Another, related answer is that the complex, unsettling transition to modernity led to a desire to extend the pure truths of mathematics into the world of subjects and objects. Also, historians of statistics have described relationships between the rise of modern states and the rise of the authority of quantitative methodologies; see, e.g., Ian Hacking, *The Taming of Chance* (Cambridge: Cambridge Univ. Press, 1990). Theodore M. Porter's recent work, *Trust in Numbers: The Pursuit of Objectivity in Science and Public Life* (Princeton, N.J.: Princeton Univ. Press, 1995), offers a new, persuasive interpretation, suggesting that the rise of quantification had to do in part with the need to share knowledge at a distance. In private or local situations knowledge of the speaker certifies the information, whereas the modern global community requires a system of detailed universal standards.

lauréate du prix de thèse, from the Paris Faculty of Medicine. He left the profession in 1878, when the famed anthropologist Paul Broca invited him to work in the Paris Laboratory of Anthropology. After Broca died in 1880, Manouvrier spent the rest of his life running the lab (though not always as its titular chief). He was also a professor at the Paris School of Anthropology, served as Secretary General of the Paris Society of Anthropology for twenty-four years, gave frequent, well-attended public lectures, and wrote a number of anthropological articles. Most of these lectures and articles were dedicated to debunking the racist and sexist methodology of contemporary physical anthropology. Manouvrier's strong attack on Lapouge, "L'indice céphalique et la pseudo-sociologie," defended anthropology against anthroposociology, but it was not scientistic. He dismissed the idea that abilities and character traits are determined by biology, positing instead a social philosophy in which environment shapes personality. Yet it was his scientific credentials that gave the article its tremendous weight, and it was cited widely.[19] Durkheim used Manouvrier's critique as an opportunity to cut the "Anthroposociology" section from his journal. The first issue that did not include the section instead ran back-to-back reviews of Manouvrier's article and Lapouge's latest book, *L'Aryen.* Even here, Lapouge was given credit for an "incontestable erudition," but his theories were pronounced scientifically erroneous, according to the "magisterial critique of M. Manouvrier." There were many similar celebrations of Manouvrier's article, all of which conspicuously cited and admired his scientific credentials.[20] The anthropologist's refutation was just what the republican political world was looking for and might have ended the matter, had not several theorists noticed a flaw: the Lapougian threat would always exist if science was held to be the authority on questions of human society.

REPUBLICAN REACTION TO LAPOUGE

Many of Lapouge's contemporaries responded publicly to his theories. Of them, the theorists who most aggressively and significantly addressed the relationship between science and social morality were Alfred Fouillée, Célestin Bouglé, and Jean Finot. Questions of natural morality versus *a priori* moral facts have long occupied philosophers, and many inquiries on such topics substantially antedate the work of Fouillée, Bouglé, and Finot. The significance of these later theorists lies in the fact that their actions were directly political and their concerns immediate. Their work reacted to an increasingly popular theory that described nature as amoral and called upon human beings to be antimoral and murderous. Darwinian evolutionary theory made explicit the progressive possibilities of abandoning traditional morality, and henceforth both morality and amorality could be described as having vital social benefits. The work of Fouillée, Bouglé, and Finot was engendered by the assault on humane and democratic ideals such possibilities occasioned.

ALFRED FOUILLÉE

Solidarism has come to be known, in the words of the historian J. E. S. Hayward, as "the ideology of the Third Republic." The philosopher Alfred Fouillée formulated the doctrine,

[19] Léonce Manouvrier, "L'indice céphalique et la pseudo-sociologie," *Revue de l'École d'Anthropologie,* 1899, *9*:233–259. On Manouvrier's critique of Lapouge and other anthropological excesses see Jennifer Michael Hecht, "A Vigilant Anthropologist: Léonce Manouvrier and the Disappearing Numbers," *J. Hist. Behav. Sci.,* 1997, *33*(3):221–240.

[20] For the two reviews see "Compte rendu—L'indice céphalique et la pseudo-sociologie," *Ann. Sociol.,* 1899–1900, *4*:143–145; and "Compte rendu—L'Aryen, son role social," *ibid.,* pp. 145–146. For another approving citation of Manouvrier's essay see Gustave Rouanet, "Les théories aristocratiques devant la science," *Petite Republique,* 2 Jan. 1900, p. 1.

beginning in the 1870s, and by the 1890s it dominated political discourse. Solidarism's most dedicated political champion, Léon Bourgeois, was made prime minister in 1895, and in the following years he continued to articulate and popularize Solidarist goals.[21] The popularity of Solidarism was due to its concerned moderation: it was a reaction against laissez-faire individualism that stopped short of socialism. William Logue has characterized it as a kind of neoliberalism that asserted that the maximum liberty for all could be attained only through organized social action that ran counter to classic liberalism's concept of freedom. In a different context, Hayward said much the same thing when he cited the growing nineteenth-century notion that "in the inegalitarian economic sphere, it was laissez-faire that oppressed and social intervention that liberated."[22] Neither Hayward nor Logue, however, takes into account the themes of natural history and evolution that dominate nineteenth- and early twentieth-century discussions of Solidarism. It would be difficult to exaggerate the extent to which the theorists of Solidarism discussed political theory in terms of Darwinian evolution. This was not a simple case of scientific jargon borrowed from the anthropologists to increase the doctrine's cultural authority. Rather, Léon Bourgeois and the theorists of Solidarism clearly felt both burdened and blessed to be the first thinkers with access to anthropology's stunning new information. It was with a sense of duty and resolve (and a sense that misinterpretation could be calamitous) that they brought naturalism to bear on the old questions of social contract, human character, general will, and a prepolitical "state of nature."

Solidarism's concern about the political implications of evolutionary theory concentrated firmly on Darwinian, *not* neo-Lamarckian, theory. Unlike French scientists, French politicians and political theorists made frequent reference to Darwin and the "survival of the fittest" but seldom mentioned Lamarck or his work.[23] Darwinian evolution was well known in political circles, and its frightening implications were by no means lost on the French. They understood that civilization's moral goal of taking care of the "unfit" was preventing, even reversing, the work of evolution. And yet that goal could not be abandoned: to return to the political, social, and economic equivalent of the "state of nature"

[21] The relationship between Fouillée's philosophy and Bourgeois's political career has been well established. See J. E. S. Hayward, "Solidarity: The Social History of an Idea in Nineteenth-Century France," *International Review of Social History,* 1959, *4*:261–284; John A. Scott, *Republican Ideas and the Liberal Tradition in France, 1870–1914* (New York: Columbia Univ. Press, 1951); and Clark, *Social Darwinism in France* (cit. n. 4), pp. 54–57, 157–186. For a discussion of Solidarism see Hayward, "The Official Philosophy of the French Third Republic: Léon Bourgeois and Solidarism," *Int. Rev. Soc. Hist.,* 1961, *9*:22–25; Hayward, "Solidarity"; and Célestin Bouglé, *Le solidarism* (Paris, 1907).

[22] William Logue, "Sociologie et politique: Le libéralisme de Célestin Bouglé," *Revue Française de Sociologie,* 1979, *20*:141–161; and Hayward, "Official Philosophy of the French Third Republic," p. 33.

[23] Several studies have demonstrated that French scientists (and in some cases the English as well) held on to Lamarckianism long after Darwinian theory had replaced it elsewhere. See Yvette Conry, *L'introduction du Darwinisme en France au XIXe siècle* (Paris: Vrin, 1974); Thomas F. Glick, ed., *The Comparative Reception of Darwinism* (Austin: Univ. Texas Press, 1974); Robert Nye, "Heredity or Milieu: The Foundations of Modern European Criminological Theory," *Isis,* 1976, *67*:335–355; and Peter Bowler, *Theories of Human Evolution: A Century of Debate, 1844–1944* (Baltimore: Johns Hopkins Univ. Press, 1986). As for the Darwinian orientation of political figures and social theorists, the works examined in this study all consistently ignore Lamarck (though in some cases he would have aided their arguments) and use the terms "Darwinism," "Darwinian," or "struggle for life" ("lutte pour la vie," "lutte pour l'existence") in their references to evolution. See also Léon Bourgeois, *Essai d'une philosophie de la solidarity: Conférences et discussions* (Paris, 1902). This conference was attended by twenty-one prominent French philosophers, academics, lawyers, and politicians. In the discussions no one brought up Lamarck or his "inheritance of acquired characteristics." This is not to say that they discussed Darwinism in detail, either, or that they never assumed a gradual human progress that might be construed as a sort of Lamarckian improvement. My point is that the French wrote copiously about the meaning and consequences of Darwinian evolution. Our discovery of the French romance with Lamarck should not blind us to their obsession with the Darwinian "struggle."

would lead to a brutal world. Solidarism was partly conceived of as a defense of civilization—a humane call for society to remain above nature's base struggle. But it was also born of the notion that the natural world was more just (impartial, uncorrupted) than the world of human society, because human society creates artificial barriers (unequal wealth) and artificial hazards (war, machinery, and "virtuous" celibacy) to the survival and propagation of the most fit. Because of these conflicting impulses, Solidarism was, at first, sometimes described as the policy of a just society working to insure that cruel natural competition was tempered by human reason and sometimes described as an effort to return to a natural condition in which the "fittest" have the opportunity to succeed—regardless of their original social station. We might call these "civil Solidarism" and "natural competition Solidarism." It further confuses matters that some Solidarist theorists saw the natural world as much more cooperative than the world Darwin described. This gave rise to a "natural cooperation Solidarism" that held that mutualism was a natural fact, either because animals were seen to be interdependent or because society was understood as a single organism in which individuals and classes acted as cells and organs, respectively. All three forms had their champions from the beginning, but we can discern a clear shift over time: in the natural Solidarism of the early years modern society was generally held to be the villain (indicted either for promoting artificial rather than natural inequality or for replacing natural cooperation with artificial competition), while the later civil Solidarism assumed that whatever evils could be found in society, nature was worse.

Hayward discussed the dual nature of Solidarism, calling it "simultaneously 'is' and 'ought,' a datum and an imperative"—even showing that it came to rest on the imperative—but he did not give much consideration to the natural history debate from which it stemmed. That debate, however, is crucial to understanding how the duality was eventually resolved. In Hayward's words, by 1908 Bourgeois "had (following Fouillée) recognized that natural solidarity—the fact of interdependence—was a-moral and that it was only through the rational intervention of men that it could be made the foundation of social justice."[24] But to understand why Fouillée came to rest—after much vacillating—on the idea of civil Solidarism, we must consider the battle he was waging against Lapouge. The brutality of nature had once meant a lion killing a zebra and, further, a less apt lion dying of starvation. Now the brutality of nature might mean the erasure of morality from public life, European races dominating or even slaughtering one another, state-determined laboratory pregnancies, the end of the family, the end of democracy, the elevation of the race and the state above all.

By the time Lapouge published his first book, Fouillée had long been engaged in a fight against arguments for social individualism based on natural history—primarily those articulated by Herbert Spencer. Fouillée held that an understanding of biological facts about individuals and human groups was necessary for the creation of an ideal state. Unlike Spencer, however, he believed that these biological facts argued for mutualism as strongly as for individualism and that human reason must, in any case, mitigate the harsh interpretation of "survival of the fittest" prescribed by Spencer and other social Darwinists. Indeed, from his doctoral thesis, "La liberté et le déterminisme," of 1872 through his *Humanitaires et libertaires au point de vue sociologique,* which appeared posthumously in 1914, Fouillée published twenty-seven books that grappled with the relationship between biology and politics. He deliberated extensively on the morality of redirecting or accelerating Darwinian evolution and devoted considerable attention to anthroposociology. Fouillée's accep-

[24] Hayward, "Official Philosophy of the French Third Republic" (cit. n. 21), p. 27.

tance of some degree of social Darwinism made him particularly sensitive to Lapouge's claims. In one of the earliest articles to take up the discussion of Lapouge, Fouillée complained that "the 'struggle for life' between the whites, blacks, and yellows was not enough; some anthropologists have also imagined a struggle for life between blonds and brunettes, long-heads and short-heads." He argued, essentially, that Lapouge had grossly exaggerated the biological aspect of psychological differences between the Germans and the French, and he cited Léonce Manouvrier as an anthropologist who denied the significance of the cephalic index.[25] Though Fouillée objected strongly to Lapouge's "fanaticism," he still quoted Lapouge consistently and, in general, approvingly. Fouillée maintained that, if used prudently, inquiries into the national physiological differences between Europeans could help to establish their psychological differences. He would devote much of his large oeuvre to defining the characteristics of various nationalities. Fouillée is remembered as one of the most important late nineteenth-century French philosophers, but he should also be known as a central figure in what might be termed "fin-de-siècle national character studies." This phenomenon seems best explained by the coincidence of discussions of evolutionary heredity with the moment at which Western Europe became a solid block of nation-states. After 1871 "character studies" of these states proliferated, fetishistically characterizing the natural "likes and dislikes," virtues and failings, friends and enemies of each hitherto-amorphous national group. French works of this sort generally claimed several high virtues as inherently French, but the national character studies were also sites of anxiety and self-doubt. (See Figure 2.) As authors attempted to reimagine the nations of Europe in light of shifting political balances and new anthropological data, they considered not only the strangeness of others but also how strange (or decadent) their own group seemed in others' eyes.[26]

Fouillée was certain that nationalities had biologically determined intellectual and social characteristics, but he firmly objected to many of Lapouge's claims. For example, in his study of the psychology of the French people, Fouillée cited Lapouge's national characterizations but found his pragmatic suggestions distasteful. Wrote Fouillée, "This ethics of breeding-studs founded on naturalist hypotheses and on the dreams of utopians is not really human morality." In any case, Fouillée doubted Lapouge's contention that one could breed any desired psychic and intellectual type and ridiculed the further suggestion that one could create races of naturalists, fishermen, farmers, and blacksmiths. He found this last notion particularly amusing: "A race of naturalists! As if the quality of naturalist follows a cerebral formation distinct from that of a fisherman or a farmer! What audacity it would take to want to intervene in the creation of men, on the basis of information as vague as that of the forms of skulls and of their problematic relationship to mental superiority!" It is surprising that Fouillée found so much fun in this, because the gradations he did endorse were only marginally less fine. Indeed, such characterizations were the whole

[25] Alfred Fouillée, "La psychologie des peuples et l'anthropologie," *Rev. Deux Mondes,* 1895, *128*:365–396, on p. 365. Fouillée quotes Lapouge's warning of "copious exterminations"—without attribution—on the first page of this piece. His reference to Manouvrier predated the latter's specific critique of Lapouge. Fouillée's concerns with the relationship of biology and politics were later combined with analyses of the works of the philosopher Marie-Jean Guyau, who became Fouillée's son-in-law when Fouillée married Augustine Guyau. Her work may have been the best known of the three: under the pseudonym Bruno, she was the author of the famous *Le tour de France par deux enfants,* which sold 7.4 million copies between its publication in 1877 and 1914. Subtitled "Devoir et patrie," it was a primer for civic and moral virtue. On Guyau the philosopher see Geoffrey C. Fidler, "On Jean-Mari Guyau, Immoraliste," *Journal of the History of Ideas,* 1994, *55*(1):75–97.

[26] They even exported this imagined criticism. See, e.g., Alfred Fouillée, "As Others See Us," *Living Age,* 1899, no. 212, pp. 67–72.

Figure 2. *"How the French are seen by others." From the* Revue Encyclopédique *(1899). The flurry of anthropological "national character studies" that appeared in the last decades of the nineteenth century exhibited some anxiety regarding foreign opinions of the French character and style.*

point of his book, and though he intended them to be used to help the nationalities understand one another, he offered scientific explanations as to why any given group was more or less nervous, imaginative, prone to dreaming, sexually energetic, and so on. For this reason, Fouillée was sometimes referred to as an anthroposociologist, though he himself strictly rejected the appellation within his published works, in his correspondence with Lapouge, and in reported conversations with Lapouge's disciples.[27] Fouillée never rejected the idea that national character types were based in heritable biological traits, but there is no doubt that the years he spent arguing against Lapouge's antimoralist naturalism shifted his thinking toward civil Solidarism. He found himself codifying Solidarism as an "ought" rather than an "is" precisely because Lapouge's natural laws were so convincingly nasty.

In 1903 Fouillée published a lengthy attack on anthroposociology. The work was, as he described it, a study of the various psychological profiles of European nations along sociological, historical, and biological lines. However, while the central chapters of the book kept to a somewhat sociobiological agenda, the introduction and conclusion were devoted to combating anthroposociological ideas. "The real law of human societies," asserted Fouillée, "is not natural selection and the struggle for life, but rational choice and cooperation for life."[28] He had argued in the past both that mutualism in human society was scientifically based in natural models and that the competition that did exist in nature was preferable to the corrupted competition of human society. Now he characterized Solidarism

[27] Alfred Fouillée, *Psychologie du peuple français* (Paris, 1898), pp. 281, 281–282. For the correspondence see Muffang to Lapouge; A 91-6, n.d.; and Alfred Fouillée to Lapouge, A 42-1 through A 42-3, Fonds Vacher de Lapouge.

[28] Alfred Fouillée, *Esquisse psychologique des peuples* (Paris, 1903), pp. 529–530.

Figure 3. *Célestin Bouglé. Bouglé was one of the most important of Durkheim's disciples and succeeded to his university post when Durkheim died in 1917. He later became director of the École Normale Supérieure.*

as distinctly human. Modern society might be brutal and amoral, but, if Lapouge was even partially correct, natural forces were even less humane. Human beings must then create a world based neither on religious dogma nor on natural science. Only "rational choice" could serve as the "real law of human societies." In fashioning his notion of Solidarism, Léon Bourgeois drew heavily on the work of Fouillée, increasingly promoting the idea that human logic and morality dictated mutual assistance. Still, though civil Solidarism triumphed, notions of natural cooperation Solidarism and natural competition Solidarism never entirely disappeared from the arguments of Fouillée and Bourgeois. The other of Solidarism's central theorists, Célestin Bouglé, substantially altered the debate by explicitly rejecting science as a viable means of arriving at sociopolitical truths.

CÉLESTIN BOUGLÉ

Célestin Bouglé was one of Durkheim's primary disciples and closest collaborators. (See Figure 3.) Having written a doctoral thesis entitled "Les doctrines égalitaires," he went on to teach at the Faculté des Lettres de Toulouse. In 1901 he began teaching social philosophy at the Sorbonne, and in 1920 he was named the director of the Centre de Documentation Social at the École Normale. In his many works Bouglé expressed a position on egalitarianism and Solidarism that respected the validity of scientific information on humanity but increasingly considered it to be inconsequential to society. In his 1897 article "Anthropologie et démocratie" Bouglé argued that whether or not science could prove the existence of biologically based differences in the capabilities of races or individuals, these differences should have no effect on the philosophical decision to maintain political equality. Bouglé directly attacked anthroposociology and Lapouge, whom he recognized as the French founder and leader of this movement. He would later use the same arguments to refute a wider range of anthropological, racist doctrines. In "Anthropologie et démocratie" Bouglé argued that Lapouge's descriptions of inequality might be factual but should be functionally insignificant to the republic. In fact, he suggested that the republican attachment to notions of natural equality may have been no more than a necessary but transient stage. Thus, "a morality suffused with the idea of Solidarity may not need to consider the idea of equality as anything more than provisional. . . . If it is true that, in declaring men to be equal, we deliver a judgment not on the way that nature made them, but on the way that society must treat them, well then the most precise craniometry could not prove us right or wrong."[29] Although Bouglé questioned whether human capabilities could be deduced from physiological measurements (citing Léonce Manouvrier's scientific argument), his invective against Lapouge was largely based on the assumptions that anthroposociology made concerning the *significance* of the biologically based inequality of human beings.

Even after Manouvrier had dismissed the scientific validity of Lapouge's claims, Bouglé continued to write refutations of anthroposociology. For whether or not brachycephali constituted a distinct race, Lapouge had put forth a profoundly disturbing challenge to liberal democracy and laissez-faire capitalism. A crucial aspect of that challenge was that it provided a way of referring to the nation's less capable, less intelligent members as a distinct group. Bouglé believed that the essence of Lapouge's questions (if not their particular formulation) was in fact extremely important. When anthropologists claimed—through the erroneous method of comparing cephalic indexes—that human beings differed in their capabilities, they were, according to Bouglé, pronouncing "a truth as old as the world."[30] People's abilities differed, whether or not they were commensurate with cephalic indexes. However, Bouglé concluded, that fact should have no bearing on their political, judicial, or economic rights.

Bouglé published "Anthropologie et démocratie" in the *Revue de Métaphysique et de Morale.* This journal had been founded in direct reaction to the incursion of materialist science into philosophy's territory. It also specifically denounced the desperate return to organized religion that was elsewhere advocated as a response to materialist, atheistic, "natural" morality. The *Revue* rejected the "pretensions" of those who looked to science or to a self-conscious Catholicism for social morality. Alphonse Darlu, a philosopher and frequent contributor to the *Revue,* reviewed Brunetière's 1895 article and made the journal's position clear: "We would like to remind him that philosophy exists." According to Darlu, Brunetière's problem was that his education had been excessively shaped by positivism. Like many of his generation, he had been raised on Ernest Renan and Hippolyte Taine, leaders in the cult of reason and science and in the attempt to apply scientific methods to history and morality. Following that, he "read and re-read Darwin, at that young age when one lives for intellect." Darlu argued that Brunetière's experience was by no means unique—such changes of heart were a widespread phenomenon affecting men and women who had embraced positivism and evolutionism with too much faith and then, pulled by their strong senses of morality, eventually swung back to Catholicism with great force. "To stop midway," explained Darlu, "requires a philosophical frame of mind and

[29] Célestin Bouglé, "Anthropologie et démocratie," *Revue de Métaphysique et de Morale,* 1897, 5:443–461, on p. 461. Bouglé correctly named Otto Ammon as Lapouge's German counterpart and referred to other authors as "their disciples" (p. 443 and throughout).

[30] *Ibid.,* p. 457.

very deep moral beliefs." French society, he insisted, was coming apart owing to an imbalance between intellectual and moral thought. In the absence of a strong, credible moral code, some members of society were returning to "a very simple, very sweet, very sad Christianity," while others buried themselves in specialized scientific projects. Meanwhile, warned Darlu, society is falling prey to "blind and terrible forces." The light of reason was lost "between positivism that stops at the facts and mysticism that is driven by superstitions."[31]

This journal was thus a natural home for Bouglé's more theoretical work, but he worried that those who were awed by Lapouge's scientific data would be moved only by a scientific, data-laden rebuttal. He thus straddled several different academic fields as he progressed from his early critique of bad science to a later, more general, indictment of the enterprise as a whole, always remaining true to an antireligious secular republicanism. Despite Bouglé's strong rejection of the pragmatic "return to religion," he gingerly approached Brunetière on the subject of combining their resources in a struggle against anthroposociology. In a private letter Bouglé candidly admitted to Brunetière that he had "combated several of your ideas and methods with all possible vigor." But he went on to say that, given an earlier conversation, he was sure that Brunetière would be eager to fight the "pretensions of anthroposociology." Bouglé proposed to discredit anthroposociology by writing a study of the caste system in India; apparently he hoped that Brunetière would publish sections of it in his *Revue des Deux Mondes.* Bouglé asserted that this work would show that Indian marriage rules had not given the results predicted by anthroposociology and that "it is impossible to find, even in this land, a true parallel between social differences, physical differences, and mental differences." We do not have Brunetière's response, but it seems that he declined the proposal. Bouglé went ahead with his study nonetheless; it was published in sections in the *Année Sociologique* and in the *Grande Revue* and appeared as a book entitled *Essais sur le régime des castes.* This work was explicitly aimed against anthroposociology (especially the section on race) and came to be one of Bouglé's most influential sociological studies.[32]

In 1904 Bouglé, who was now a professor of social philosophy at the Sorbonne, sharply criticized all attempts to describe history as an epiphenomenon of natural history. In *La démocratie devant la science* he divided such attempts into social Darwinism, organicism, and anthroposociology. To the extent that he could, Bouglé countered each of these notions with anthropological, "scientific" arguments, resisting the idea that Enlightenment ideals might have to be divorced from Enlightenment methodology. He was no longer naming Lapouge as his primary opponent, but he still referred to Lapouge when articulating the philosophical problem of science and political equality: "Anthropology, according to M. Vacher de Lapouge, victoriously refutes the errors of the eighteenth century, 'the most fantasy-believing, the most anti-scientific of centuries,' and demonstrates that a democratic regime is 'the worst condition in which to make good [hereditary] selection.' " Denying the scientific validity of antidemocratic anthropology was not a sufficient reaction to La-

[31] Alphonse Darlu, "Réflexions d'un philosophie sur les questions du jour: Science, morale et religion," *Rev. Métaphys. Morale,* 1895, *3*:239–251, on pp. 249, 248. On the reasons for the journal's founding see "Introduction," *ibid.,* 1893, *1*:1–6; see pp. 2–4 for the quotations.

[32] Célestin Bouglé to Ferdinand Brunetière, n.d., Bibliothèque Nationale de Paris, Manuscrit, 25033, fols. 79–80 NAF. There is a sustained discussion of "anthroposociology" in the chapter on race in Célestin Bouglé, *Essais sur le régime des castes* (Paris: Félix Alcan, 1908), pp. 129–156. Of Bouglé's study the historian of sociology Don Martindale writes: "More than any other single study, this essay laid the basis for the modern theory of caste." Don Martindale, *The Nature and Types of Sociological Theory* (Boston: Houghton Mifflin, 1981), p. 265.

pouge's critique. Bouglé made it clear that if he were forced to choose between Enlightenment political ideals and Enlightenment trust in science, he would choose the ideals:

> Even when it is established that solidarity exists within the best organized animal societies, this animal solidarity does not seem to approach the human ideal: respect for the equal dignity of each of society's members. Democratic societies recognize from this that they are attempting to go above and beyond nature. . . . At times acquiescing and at times resisting nature, society seems to say to natural science both "I will apply your laws" and "Your laws do not apply to me."

This disavowal of the authority of science clearly went beyond the mere negation of unpleasant scientific findings. Bouglé insisted that neither naturalism, nor logic, nor rationality would ever manage to make society "lift its smallest finger" toward equality and social cohesion unless they were joined by sentiment. "In other words, the indispensable condition of moral efficacy of these sociological inferences [of solidarity] is the preliminary existence of a 'social spirit.' "[33]

Though Bouglé called for a return to a philosophical justification of democratic ideals that lay beyond scientific discovery, he did occasionally argue that a truly objective science would demonstrate that natural laws dictated a Solidarist society. He was arguing not that science would discover equality but that it would discover that human societies should be run on principles of equality. Bouglé believed that science might be able to discover such laws in the distant future; until then, he mused, France should concentrate on the revival of moral philosophy. Even those who are most dedicated to science, he continued, should stop assaulting philosophy as an unempirical and thus unnecessary discipline, because until science fulfilled its promises philosophy was, in fact, profoundly necessary. Bouglé warned that

> if it is true that the most objective scientific observation cannot yet suffice to demonstrate to human beings that they must work for the coming of a just city, of which the members aid each other to rise, if right up until the new order it will be necessary to come to this by a sort of rational choice, then maybe it would be imprudent (and in a democracy more than in any other society) to denigrate moral philosophy, which is the art of rational choice and of methodically ordering the purpose of a human life in terms of a universal purpose.

He hoped that someday there would be an objective, egalitarian, "scientific morality" and wondered whether it would "relegate all moral philosophy to the frontiers of society as totally useless." Uncertain of the future relationship between science and morality, he was quite sure that contemporary scientific morality was unacceptable. Anthroposociological doctrines were wrong, he argued, because of the society they imagined. "Against these we can propose, according to experience, our firm conclusions. We will henceforth know them by their fruit." Bouglé knew it was very unscientific to dismiss a methodology because it drew unpleasant results. He was uncomfortable with this position and worried that others would disagree with him, arguing that egalitarian principles are impossible to employ and that "it would be dangerous to try; it would be much better to listen to the lessons of nature." Against this imagined response by those who would "exploit the prestige of science against the attraction of democracy," Bouglé admitted that he offered no positive proofs. "Our conclusions," he wrote, "if not imperative, are at least emancipatory.

[33] Célestin Bouglé, *La démocratie devant la science* (Paris, 1904), pp. 18, 288, 301.

Figure 4. *Jean Finot. Painting by L. Lèvy-Dhurmer. Finot founded and edited the journal* Revue des Revues (*which became the* Revue Mondiale), *publishing a wide range of literary and political opinions and highlighting his primary concern: international pacifism. This painting ran in the* Revue Mondiale *when Finot died in 1922.*

They liberate our society from its naturalist obsession. They remind it that no one has the right to discourage the ambitions of the spirit in the name of a so-called scientific morality."[34]

Bouglé's belief that science could make objective discoveries about the nature of humanity led him to fear either that bad science, proving inequality, would be mistaken for good science or, worse yet, that good science would, in fact, prove inequality. This pushed him to the position that he summed up as "*Noli me tangere*"—natural science cannot touch human values. Even if the science is correct, human dignity and the resulting claim to equal treatment must be set above scientific pronouncements of inequality between human groups.[35] Had he been able to discredit not *bad* race science but *all* race science—that is, had he been able to announce that the search for objective natural racial laws was inherently fallacious—he would not have needed this elegant intellectual device.

JEAN FINOT

Jean Finot demonstrated that such a critique of race science was in fact conceivable. Finot, born Finklehaus, was a Polish journalist who became a French citizen in 1897. (See Figure 4.) In his adopted country he founded a journal, *La Revue des Revues,* which he edited and to which he contributed numerous articles. This work put him in close contact with many of his most illustrious contemporaries. His correspondents included writers as diverse as Zola, Tolstoy, Brunetière, and Cesare Lombroso—many of whom frequently published

[34] *Ibid.,* pp. 302, 303.
[35] *Ibid.,* p. 288.

in *La Revue des Revues*. Finot was well known in his time. When he died in 1922, the sociologist René Worms celebrated him as an eminent "philosopher, philanthropist, patriot, hygienist, feminist, and sociologist." Above all, however, Worms praised Finot for having fought against the whole "school" whose doctrine is "generally known under the name anthroposociology."[36] Finot's greatest fame came from his *Le préjugé des races* (1905) and his later *Préjugés et problème des sexes* (1912). Both of these works put forward lively, witty arguments against the existence of innate, biological character traits and intellectual abilities. Finot too drew on Manouvrier's work for its scientific credentials.[37] His own arguments against racism concentrated in part on debunking scientific race theory and in part on considering the sociopolitical origins of perceived racial differences. His indictment of "race prejudice" was broadly conceived, including discussions of American blacks, animosity between the English and the French, and Aryan supremacy. He often cited racial categories that his generation no longer recognized as such as evidence of the transience and historical specificity of supposed racial divisions. Indeed, in the foreword to the 1906 English translation of *Le préjugé des races,* Finot pleaded for the "indulgence" of his readers by reminding them that he had claimed, as early as 1901, that contrary to popular belief there was no innate immutable hatred between the English and the French races. Finot's need to assert that these were not separate races reminds the modern reader of the extent of this racial conception of nations:

> When my first works appeared in 1901 on that subject, mocking voices were raised to show the impossibility of an *entente* between *two races* which were so inherently different and, presumably, antagonistic. . . . The *Times,* in a remarkable article on my efforts in this direction (November 1st, 1902), was right in maintaining that it is often sufficient to breathe on the subjects of our discord to see them vanish. The union of a few men of goodwill has succeeded in overcoming the stupidity of the theory of races and of age-long prejudices![38]

Finot's central interest, and the central interest of his journal, was the promotion of pacifism through internationalism. He held that the differences between human beings were only individual, and though he criticized Fouillée for his racialist thinking he made use of the philosopher's notion of Solidarism in his efforts on behalf of international peace. Because nations were understood as representing different races, the cause of pacifism was well served by an attack on racialism. *Le préjugé des races* begins with a discussion of English and American eugenic theorists; though Finot dismissed them, he did so without anger, explaining that they were simply trying to ameliorate the public health. "In France and Germany," he added, "the gospel of human inequality has taken on even stranger aspects. It is Vacher de Lapouge who is the most authoritative representative of the new doctrine. Loyal to his principles, convinced of their truth, he defends them in his work with a keenness and a talent worthy of esteem." Finot had so much respect for Lapouge's scholarship that he cited him as the quintessential opponent, writing that "in M. Vacher

[36] René Worms, "Jean Finot, sociologue," *Revue Mondiale,* 1922, pp. 228–232, on p. 229. The only figure Worms mentions here is Gobineau, but he speaks at length about Finot's attack on the cephalic index. *La Revue des Revues* was later called *La Revue* and then *La Revue Mondiale.* Finot wrote many of the journal's articles. According to his son, in the few years before he became a French citizen he used ten different pseudonyms in order to write freely: Jean-Louis Finot, "Mon père," *Rev. Mond.,* 1922, pp. 143–150. For Finot's correspondence with these figures see Bibliothèque Nationale de Paris, Manuscrit, 24494 (1) doc. 264, microfilm 2278/NAF 24519, fols. 174–180, NAF 24530, fols. 380–382, NAF 25038, fols. 292–293.

[37] Jean Finot, *Les préjugé des races* (Paris, 1905), p. 103; and Finot, *Préjugés et problème des sexes* (Paris: Alcan, 1912).

[38] Jean Finot, *Race Prejudice,* trans. Florence Wade-Evans (New York, 1906), p. vi.

de Lapouge, the new doctrine finds a defender of the greatest eloquence and it suffices to examine his books for one to know all the weapons that are taken up by his co-religionists, adepts, and students." Finot bemoaned the uncritical acceptance that this theory had found among journalists, politicians, literary writers, artists, and the greater public and noted with disgust that the doctrine was finding its way into manuals of history and pedagogy. He assured his reader that "without doubt, this doctrine will some day take a place of honor in the history of human errors," but he lamented that, for now, "out of any one-thousand educated Europeans, nine-hundred and ninety-nine are persuaded of the authenticity of their Aryan origins."[39]

Finot's critique of racist anthropology rested on the idea that the science began with certain assumptions about racial difference—he described it as "teleological" and thus without value. The anthroposociologists, he claimed, were "hypnotized by their primordial idea," which they supported by bringing together, "without examination, everything that seems propitious to their theory—a theory that is more political than scientific." Finot's conclusions were surprisingly encompassing. He argued that all character traits were specific to individuals and that even if, indeed, a trait could be found in one human group more than in another, that was due to environment and culture. Beauty, he argued, was a purely social convention, and no single standard could be set for all human beings. Neither a language type nor any system of government could be established as an innate capacity of any single racial group. In general, Finot boldly concluded that "the term race is but a product of our mental gymnastics, the workings of our intellect, and outside all reality. Science had need of races as hypothetical groupings, and these products of art . . . have become concrete realities for the vulgar. Races as irreducible categories exist only as fictions of our brains." Drawing on an analysis of the work of Manouvrier as well as his own investigations, Finot argued that "craniological measurements teach us almost nothing concerning the mental capacity and the moral value of peoples." Ridiculing anthropology's "instruments of precision," he asserted that the "fantastical data" they provided was meaningless. There was "literally nothing," he argued, in the whole collection of external differences that appear to divide men, that could authorize their division into superior and inferior beings. "If this division exists in our thought," insisted Finot, "it only came there as the result of inexact observations and false opinions drawn from them." Finot rejected the scientific categorization of his generation in defense of a notion of eternal change that would continue even after his internationalist dream of peace had been achieved. Progress, then, was not aiming toward an end point of happy stasis but, rather, toward an end point wherein human beings could endlessly change in an environment of peace and equality. "The character of a people," wrote Finot, "is thus nothing but an eternal becoming. The qualities of our soul and its aspirations remain as mobile as clouds chased by the wind."[40]

Like Bouglé, Finot struggled his entire life to strike a balance between science, philosophy, and the religious needs of the masses. Later in life he wrote several books on longevity, a theme that grows in significance when understood as an atheist's alternative to eternity. Finot's longevity theories all carried a wistful optimism that he explicitly maintained in the absence of divine grace. It was this balanced optimism that Henri Bergson celebrated when he presented Finot's *Progrès et bonheur* to the Académie Française in 1914. Two months before the beginning of the war, Bergson stood before the Académie praising Finot for discussing morality without erring either on the side of the "abstract

[39] Finot, *Préjugé des races* (cit. n. 37), pp. 505, 27, 356–357.
[40] *Ibid.,* pp. 491, 312, 501, 109, 489–490, 345.

deductions of the old metaphysics" or on the side of "pure empiricism" and, instead, for balancing between the two, "for a knowledge which is clearly of the philosophical order, but, without pretending to embrace the totality of the real, concentrates its attention on human activity." Finot's other major work on longevity, *La science du bonheur,* also struggled against the dogmas of science and of religion. In dealing with the issue of morality, Finot insisted on the priority of rational expectations over religious ideals, writing that he refused to be "intoxicated by the religion of self-sacrifice, of altruism toward all and for all, and especially by that of future existence . . . we have wrapped it carefully in a purple shroud, there where the dead gods sleep." Yet following such comments, issued always with that combination of spite and sadness reserved for absent deities, Finot similarly lamented the lost prestige of science: "Nature, we are told, knows only the species. She neglects and dooms the individual. Nature is calumniated. Science is libeled in the same way."[41]

Finot never abandoned science but constantly reminded his readership of its fallibility. In the struggle between science and religion, of "freethought against the dogmas," he believed that science would and should prevail, but not to the exclusion of a devotion to human solidarity and love. Finot chided scientists who pretentiously dismissed metaphysics: "From the scientific standpoint, nothing authorizes the logic of the sectarian mind violently rejecting everything that is not in harmony with its comprehension." Similarly, he warned spiritualists to stop mocking secular moralists: "Dogmatic religions are also wrong in seeking to struggle against lay morality . . . Social harmony requires their mutual respect. Mankind can exist only upon moral foundations. Why discredit those of science and of experience, if a portion of the nation must live by these latter?" Despite such calls for tolerance, Finot tended to treat the religious as rather backward. He did not believe that one could be a pure materialist atheist, but to argue this he invoked the human need for spiritualism rather than its truth. "The most positive rationalists," he wrote, "now admit the existence of spiritual needs and eternal aspirations toward the infinite." He believed that in the future human beings would experience in a more useful way what will forever be the "same awe of and the same longing for the Infinite." When Finot outlined the reasons for his optimism about the future, they had everything to do with reconciling oneself to a godless world and celebrating the progress of a morality conceived in opposition to biological assumptions about human character.[42]

Finot was a powerful adversary. When Lapouge complained, publicly or privately, about his detractors, he blamed Finot, more than any other nonscientist, for the general repudiation of his theories. In his final work, *Race et milieu social,* Lapouge attacked the intelligence, honesty, and education of, among others, Manouvrier, Bouglé, and Finot, but the last was the target of particular malice: Lapouge revealed Finot's real name, Finklehaus, with much anti-Semitic drama. Lapouge was so angry because Finot was so effective. He made this explicit in a letter to Madison Grant, the famed American racist and author of *The Passing of the Great Race.* The year was 1919, and Lapouge was explaining that he had not written much on race in the past few years: "Jews like Finklehaus (called Jean Finot) have so excited public opinion against the theory of races that it would be as

[41] Henri Bergson, "Rapport sur 'Progrès et bonheur' de J. Finot," in Bergson, *Mélanges,* ed. André Robinet (Paris: Presses Univ. France, 1972), pp. 1090–1094, on p. 1093; and Jean Finot, *The Science of Happiness,* trans. Mary Stafford (New York: Putnams, 1914), pp. 20–21. According to Finot's son, this book was translated into fifteen languages.

[42] Finot, *Science of Happiness,* trans. Stafford, pp. 234–240, 242, 244, 242, 246, 248, 257, 331–333.

dangerous as it would be useless to try to do anything."[43] Lapouge continued his campaign until his death in 1936—but not in France.

GOVERNMENT SCIENTISM AGAINST THE ACADEMY

In 1947 George Sarton published a tribute to the life and work of the recently deceased Marie Tannery, in the course of which he described the "scandal of 1903." This scandal pertained to Marie Tannery's husband, Paul, whose professorial election to the faculty of the Collège de France had been overturned by a scientistic government minister merely because Tannery was Catholic. In Harry Paul's article on Brunetière and the "bankruptcy of science" debate, this incident is effectively used to describe the lag between the scholarly turn away from purely materialist science and ideology and the government's continued interpretation of all metaphysics as inimical to the republic.[44] Described as such, the scandal offers a window into tensions between Catholicism and positivism, but it deserves reevaluation within the context of anthropology and philosophy.

The creation of a history of science chair at the Collège de France had been requested by Auguste Comte in 1832 and became a reality for Pierre Laffitte, one of his disciples. When Laffitte died, the applicants for the chair included Léonce Manouvrier, Paul Tannery, and Grégoire Wyrouboff; the final choice was between Tannery and Wyrouboff. Tannery, a widely known historian of science, was elected by a large majority. It was almost unprecedented when the minister of public instruction, Joseph Chaumié, overturned the professors' vote and gave the position to Wyrouboff instead. Tannery belonged to the Catholic Scientific Society of Brussels (as did Pasteur and Pierre Duhem), which was dedicated to demonstrating that science and Christian faith were by no means mutually exclusive.[45] His work, however, betrayed no sign of his religious beliefs; indeed, he had been much influenced by Comte. Tannery's work on the history of science was internationally acclaimed and stood as a model long after his death. Wyrouboff, on the other hand, was not known as a historian of science. He was a competent scientist and arguably the foremost living representative of the positivist school, but he had little relevant experience. Indeed, ac-

[43] Lapouge to Madison Grant, A 068-80, 23 Mar. 1919, Fonds Vacher de Lapouge. For the books see Georges Vacher de Lapouge, *Race et milieu social* (Paris: Rivière, 1909), pp. xi, xx; and Madison Grant, *The Passing of the Great Race* (New York: Scribners, 1916).

[44] George Sarton, "Paul, Jules, and Marie Tannery (with a note on Grégoire Wyrouboff)," *Isis,* 1947, *38*:33–51; for more on Tannery see René Taton, "Paul Tannery," in *Dictionary of Scientific Biography,* ed. Charles Coulston Gillispie, Vol. 13 (New York: Scribners, 1976), pp. 251–256. Paul, "Debate over the Bankruptcy of Science" (cit. n. 1), pp. 325–326; see also Paul's study of the origins of twentieth-century history of science through an analysis of the Collège de France chair: Harry Paul, "Scholarship and Ideology: The Chair of the General History of Science at the Collège de France, 1892–1913," *Isis,* 1976, *67*:376–387. There are several recent contributions to the history of the science and religion debates in the Third Republic. Fritz Ringer has explored the question in terms of educational practice and ideology; see *Fields of Knowledge: French Academic Culture in Comparative Perspective, 1890–1920* (Cambridge: Cambridge Univ. Press, 1992), esp. pp. 207–225. Also important is Herman Lebovics, *True France: The Wars over Cultural Identity, 1900–1945* (Ithaca, N.Y.: Cornell Univ. Press, 1992). Chapter 1 considers the anthropologist and politician Louis Marin in a study of the activity and eventual decline of the "old right" vision of France in the twentieth century (pp. 12–50).

[45] Chaumié was part of the extremely anticlerical Combes government of 1902–1905. See Sarton, "Paul, Jules, and Marie Tannery"; Paul, "Debate over the Bankruptcy of Science," pp. 325–327; and Pierre Duhem, *Paul Tannery* (Montligen, 1905), p. 14. On the Catholic Scientific Society of Brussels see Paul, "Debate over the Bankruptcy of Science," p. 322; and Antonin Eymieu, *La part des croyants dans les progrès de la science au XIXe siècle* (Paris, 1910).

cording to Sarton, Wyrouboff "was not a trained historian of science and contributed nothing whatsoever to the subject, neither before his election nor after."[46]

This episode dramatically demonstrates that the French government understood its fundamental ideological standpoint to be at odds with the new ideas emanating from philosophy and social theory, but Sarton's account tells only part of the story. Wyrouboff did get the appointment, but he was in very poor health; in fact, in the academic years 1909/1910 and 1912/1913 he was too ill to teach. His replacement was Léonce Manouvrier. In applying to teach at the Collège, Manouvrier had reminded the board of his well-known scientific critique of Lapouge and explicitly represented his work as a bulwark against the advancement of political anthropology. Manouvrier rejected the "bankruptcy of science" proposition, but he did not dispute Brunetière's assessment of the dangers of looking to Darwinism for lessons in moral behavior. He simply believed that a philosophical anthropologist was the best safeguard against the anthroposociologists. When he asked the Collège professors "What's all this talk about the failure of science?" it was with the knowledge that for years he had been winning a battle against the moral "abominations" Brunetière feared. Offering himself as a necessary moral guard, Manouvrier wrote that "a plethora of real aberrations" had arisen from the incompetent use of anthropology, producing "a social and moral movement . . . that gains ground every day, carried by a multitude of books, journals, learned societies, and congresses." Manouvrier concluded: "It is not to be doubted that, from this point of view, the teaching proposed here responds to an urgent necessity. The movement of which I have just spoken could be fertile, but in the absence of a critique that is both scientific and philosophical, it risks becoming nothing more than a sterile agitation that is more of a retardant than a boon to the progress of morality."[47] Manouvrier had come to the same conclusion as Fouillée, Bouglé, and Finot: the great questions of existence, those pertaining to morality (and, lurking behind it, mortality), should no longer be approached through religion or positivism but, rather, through a "critique that is both scientific and philosophical."

CONCLUSION

In the late nineteenth century French theorists had to revise their understanding of science and republicanism as innately joined in the struggle against authoritarianism and dogma. Lapouge's vision of a scientifically engineered society jolted republicans into a realization that science had the potential to be extraordinarily antirepublican. Fouillée, Bouglé, and Finot each struggled to maintain the connection between republicanism and naturalist

[46] Sarton, "Paul, Jules, and Marie Tannery," p. 40. Paul, "Debate over the Bankruptcy of Science," corrects Sarton's exaggerated claim that only Tannery was suited for the chair. Still, it is widely agreed that Tannery was much better qualified than Wyrouboff. In 1867 Wyrouboff founded *Philosophie Positive* with Émile Littré, and for seventeen years he edited and wrote for this journal. His scientific work was in the field of crystallography and physicochemistry.

[47] Manouvrier twice applied for posts at the Collège de France—the first application was for the chair in philosophy. His first essay of application is in the Collège de France Archives, G.IV.f—45F (1899), "Considèrations présentées par M. L. Manouvrier à l'appui de sa candidature." The quotations are from this application. For his second effort see G.IV.g—13Q (1903), "Considèrations présentées par M. L. Manouvrier à l'appui de sa candidature pour la Chaire d'Histoire générale des Sciences." Paul sees Manouvrier as a Comtian positivist and argues that the professors rejected him for the permanent position because of this; see the subsection "Wyrouboff and Manouvrier: A Triumph for Positivism and Science," pp. 384–390, in "Debate over the Bankruptcy of Science." Though Paul takes Manouvrier at his word regarding his work in debunking anthropological racism, sexism, etc., he does not seem to be aware of the grand scale of Manouvrier's war against bad science. See Hecht, "Vigilant Anthropologist" (cit. n. 19).

scientism. They devoted large sections of their many works to arguing that anthroposociology was bad science. What is extraordinary, however, is that they each managed to step outside the scientific argument and question its relevance. Touting a mixture of natural cooperation Solidarism and natural competition Solidarism, Fouillée described zoological and anthropological data as natural fact models for civilization. But when anthroposociology made him wary of natural lessons he abandoned the project, insisting that society should not be modeled on animal truth but, rather, distinguished from it. Following him, civil Solidarism took hold in the political world. Bouglé tended to believe that true natural science was inherently emancipatory and moral but that its contemporary manifestations were dangerous and degrading. He thus concluded that any cherished aspect of human life must be independent of scientific proof. His switch from empiricism to idealism was more self-conscious and troubled than was that of Fouillée and is thus rather more compelling. As for Finot, his humanistic relativism led him to conclude that even the most advanced science was an act of interpretation and art. Good or bad, science could never dictate the tenets of moral philosophy. All of these positions required a certain philosophical bravery. For a long time, republicans had based their political ideologies and their public rhetoric on the conviction that scientistic empiricism was the sole road to truth. In the battle against religious and political dogma, they had used this empirical conviction as both weapon and shield: it gave mettle to their public polemics and supported them in their private existential malaise. Abandoning a commitment to empiricism without returning to religion meant a nerve-racking subjection to relativism and uncertainty. This experience defined a generation of theorists and deserves further attention from historians.

The common idea that it took Nazi eugenics to silence racialist genetic science may have to be reconsidered. That interpretation harbors the notion that, were it not for those cataclysmic excesses, genetic science now would be unencumbered by the burdens of politics. The ability of these late nineteenth-century thinkers to dethrone science and insist that it be treated as a mere servant of moral philosophy is deeply significant to this question. Tensions between present-day scientists and the academic left are also illuminated by this history. Given the tenor of late twentieth-century debates on the nature of scientific truth, it is useful to note that the relativism at which these earlier theorists arrived was not an abandonment of objective truth in favor of a valueless universe. It was, instead, an appreciation that scientific theories of humanity are, inherently, in eternal flux. These theorists argued that intuitive moral wisdom must be held above science. They gave themselves license simply to "know" what is right and to assert that some scientific proclamations are wrong, despite indexes, bell curves, Latin names, and calibrated tools. Henceforth, they hoped, we should know them by their fruit.

The American Scientist as Social Activist

Franz Boas, Burt G. Wilder, and the Cause of Racial Justice, 1900–1915

By Edward H. Beardsley

ACCORDING TO TRADITIONAL INTERPRETATION American scientists who joined in the public debate over the Negro question in the early twentieth century entered the dispute on the side of those favoring greater discrimination and repression. To be sure, some scientists opposed racist ideas and practices, but supposedly they did so in passive and private fashion, leaving public involvement to their more bigoted colleagues. Eventually, of course, American science became an active force for racial egalitarianism, but allegedly the shift began only in the late 1920s, reaching its peak in the 1930s, when Nazi brutalities against European Jewry made the inherent dangers of racism more clear. In sum, American scientists were Johnny-come-latelies in advocating racial justice for Negroes.[1]

Like many of the time-honored historical generalizations concerning the Negro, however, this one also suffers from some distortion. Far from being bigots, a number of influential scientists of the early twentieth century strongly supported the concept of racial justice by their research and in more visible ways. Preliminary investigation of the racial attitudes of American scientists indicates that among that group were anthropologist Franz Boas, psychologists E. B. Titchener and Livingston Farrand,

I am grateful to the National Endowment for the Humanities for a 1969 summer grant which supported the research for this article.

[1] For examples of what I have termed the traditional view see Oscar Handlin, *Race and Nationality in American Life* (Garden City, N.Y.: Doubleday, 1957), pp. 136, 141, 145, 149; Arnold Rose, *The Negro in America* (Boston: Beacon Press, 1962), p. 34; I. A. Newby, *The Development of Segregationist Thought* (Homewood, Ill.:Dorsey Press, 1968), p. 4; I. A. Newby, *Jim Crow's Defense. Anti-Negro Thought in America, 1900–1930* (Baton Rouge, La.:LSU Press, 1965), pp. xi, 20, 50. August Meier, in *Negro Thought in America, 1880–1915* (Ann Arbor:University of Michigan Press, 1963), pp. 162, 183, does identify Boas as actively involved in the early civil rights movement but does not mention Wilder. T. F. Gossett in *Race: The History of an Idea in America* (Dallas:SMU Press, 1963), pp. 81–83, 94, 424, noted defections from scientific racism as early as the late nineteenth century, but he did not find scientists taking a public stand against racism until the 1920s, One important exception to the traditional viewpoint is George Stocking's *Race, Culture, and Evolution* (New York: The Free Press, 1968), which argues (pp. 261–267) that the behavioral sciences were in a transitional stage in the early twentieth century, characterized by a shift away from a biological explanation of race and toward a cultural one. By about 1917 the shift was virtually completed (at the level of theory) for anthropology and sociology, Focusing on the development of ideas, Stocking's book does not, however, deal with scientists' involvement with public aspects of race. The implication is that this came later.

anatomists Burt Green Wilder and F. P. Mall, and sociologists Frances Kellor and R. E. Park.[2]

Although further study is needed to establish the depth of involvement of most of those men, at least two—Boas and Wilder—were deeply and publicly committed to the ideal of justice for black Americans. While Boas' and Wilder's activism was not necessarily typical of that of other scientists sympathetic to the Negro cause, the vigor of their involvement, along with the fact that theirs was a shared outlook, suggests a need for revising the traditional racist image of American scientists of the early twentieth century.

Boas and Wilder, in many ways very dissimilar men, shared two important characteristics. Each was solidly grounded in his particular science and each exhibited an independence of mind that kept him in perpetual tension in relation to surrounding society. Wilder, the elder of the two, was born in Boston in 1841. As a student at the Lawrence Scientific School, on the eve of the Civil War, he studied under Louis Agassiz and Jeffries Wyman and proved himself an exceptionally able pupil. Agassiz, in fact, considered Wilder his best student in the ante-bellum period. While Wilder reciprocated the feelings of admiration and was greatly influenced by his teacher, he did not share Agassiz's scientific outlook on race. On that question the more egalitarian views of anatomist Wyman had far greater impact on his thinking.[3]

When the war broke out, Wilder enlisted as a medical cadet and later served as a surgeon in the all-Negro Massachusetts Fifty-fifth Infantry Volunteers. Afterward, following the completion of medical studies at Harvard, he joined the faculty of the newly opened Cornell University as professor of comparative anatomy and zoology, a position he held until his retirement in 1910. At Cornell he soon established himself as a leading scholar and something of a character besides. A perpetual and energetic crusader, he was a mainstay in the national campaign against smoking, an ardent enthusiast for temperance, and along with Theodore Roosevelt gave lifelong allegiance to a more simplified spelling. Most undergraduates remembered him, however, for his annual lectures on hygiene and sex, presentations so graphic that coeds were known to faint from shock. A thorough-going moralist, Wilder exhorted freshmen to

[2] Farrand was active in the formation of the N.A.A.C.P. See *The New York Times*, May 31, 1909, p. 3. On Titchener's views see E. B. Titchener to B. G. Wilder, May 22, 1915, B. G. Wilder Papers, Regional History and University Archives, Cornell University, Ithaca, N.Y.; an example of Kellor's outlook is Francis Kellor, "The Criminal Negro," *The Arena*, 1901, *26*. For Park's view see R. E. Park, "Negro Home Life and Standards of Living," *Annals of the American Academy of Political and Social Science*, 1913, *49*: 147–163. Mall, in addition to Boas and Wilder, is discussed below. To say that these men were concerned with what we would today term "racial justice" does not mean that they had what we would call a modern outlook. All were, to varying degrees, influenced by the intellectual climate of their time. Park, for example, apparently came to his egalitarianism by way of a Lamarckian analysis of racial development (see Stocking, *Race, Culture, and Evolution*, p. 247). For all his commitment to equality, Boas believed it possible that the Negro race, on the average, was slightly (but inherently) inferior to the white in mental ability. See, e.g., Franz Boas, "The Negro and the Demands of Modern Life," *Charities and the Commons*, 1905, *15*: 85–86.

[3] Morris Bishop, *A History of Cornell* (Ithaca, N.Y.: Cornell University Press, 1962), pp. 110–111. Wilder later commented on his two teachers and their racial views. See his article on Wyman in David Starr Jordan, ed., *Leading American Men of Science* (New York: Henry Holt, 1910), pp. 190–192, and "What We Owe to Agassiz," *Popular Science Monthly*, July 1907, *71*: 10. While Wilder conceded Agassiz's scientific racism, he insisted that Agassiz never wavered from his support of civil rights for Negroes.

shun all evil, whether in the form of hazing parties, cigarettes, coffee, narcotics, alcohol, or women of pleasure.

But his scholarship attracted as much attention as his causes, with the result that his laboratory drew many of the ablest students at Cornell. One of them was Theobald Smith, a future leader in bacteriology, who later told Wilder: "'It was a fortunate thing for us that your laboratories were so small and crowded, because all of your work was done in the presence of your pupils, and we could not very well escape the infection of your enthusiasm'"[4] Wilder first earned his reputation in the field of zoology, but toward the end of his career he shifted his research interest to neurology and became a specialist on the morphology of the brain.[5]

Boas came from a different background. Born in Germany in 1858, the son of a prosperous and politically conscious middle-class family, he was also a product of the German university tradition. Following work at Heidelberg and Bonn he took his Ph.D. at Kiel University in physics and mathematics. But even as a graduate student his interests began to veer in other directions. Shifting first to geography, by 1890 he was firmly committed to cultural anthropology as a result of opportunities he had had to study North American Indians. While he was defining his scholarly interests, he was also laying down roots in the United States. In 1899, after two museum appointments, a period at Clark University, and a stint as an editor of *Science*, Boas accepted a permanent position as head of anthropology at Columbia University, a post he would hold until 1937.[6]

As an anthropologist his chief contribution to American scholarship, in addition to his elaboration of the idea of culture as independent of race, was the introduction of the rigorous methods of the physical sciences into a field previously characterized by bias and unsupported generalization. Boas' approach to anthropology, one student recalled, was one of "'icy enthusiasm,'" which meant an "ardent, unsparing drive for understanding, completely controlled by every critical check."[7] As a teacher he was brusque, demanding, and supremely self-confident. He inspired awe from colleagues and fear from most students, although those who could work independently found him a superb mentor. And yet he had great compassion for those students with severe personal problems. One student remembered that he worried over them "like an anxious father or grandfather."[8]

What catalyzed the social activism of Boas and Wilder, however, was not their education or their independent-mindedness. The key to understanding their involvement with the Negro question lay in certain experiences which each had as mature men. The pivotal influence on Wilder was his war service. He accepted the assignment as a regimental surgeon to black troops largely in hopes of furthering his scientific study of Negro anatomy, an interest which he inherited from his teacher, Wyman. Ultimately that interest won official sponsorship when B. A. Gould of the United

[4] Bishop, *Cornell*, pp. 110–111; see also *Who Was Who in America*, Vol. I (Chicago: Marquis, 1950), p. 1345. The lecture remarks are from B. G. Wilder, "Hygiene and Morality," Miscellaneous Pamphlets (1899–1901), Wilder Papers.

[5] J. H. Comstock, "Burt Green Wilder," *Science*, May 22, 1925, *61*:533.

[6] A. L. Kroeber, "Franz Boas: The Man," *Memoirs of the American Anthropological Association*, 1943, *61*:5–13.

[7] *Ibid.*, pp. 22, 26.

[8] Margaret Mead, *An Anthropologist at Work. The Writings of Ruth Benedict* (Boston: Houghton Mifflin, 1959), p. 346.

States Sanitary Commission put him in charge of gathering physical data on Negro soldiers and supplied him with instruments to do the work.[9] That project, plus his duties as surgeon, brought him into intimate contact with the troops of the Massachusetts Fifty-fifth. As a result the young scientist soon formed a number of strong friendships with his men, several of whom he taught to read and write.[10]

Besides friends, his military experience also gave him a new view of the race. One thing that impressed Wilder was the Negro's devotion to principle. When Massachusetts authorized the Fifty-fifth the state set the Negroes' pay at the level of laborers ($10 per month) rather than soldiers ($13). The black troops responded by refusing to accept the discriminatory wage, even though the state declared that that was all they would get. In all, the Fifty-fifth served for over a year without pay before they finally won their point. More remarkable to Wilder was the fighting spirit of the regiment during that difficult period. In June 1864 the Fifty-fifth played a crucial role in capturing a Confederate fortification on James Island, South Carolina. The impressive thing about the Negroes' participation was its apparent spontaneity. From what Wilder could learn no white officer ordered the blacks into battle; they simply took themselves in when they saw the attack floundering.[11]

Wilder's Civil War experiences were so central to his later involvement in racial matters that even if his research had not pointed up the essential equality of the Negro, he would perhaps have still played an activist role. As he remarked years afterward, his "personal observation" had convinced him "that the [Negroes'] title to . . . rights and opportunities was earned during the Civil War by the general conduct of soldiers of African descent, by their valor, by their initiative, and by their deliberate self-sacrifice for the sake of a principle."[12] In responding as actively as he did to the growing pattern of discrimination in American life after the 1890s Wilder was merely honoring a basic rule of friendship which said that when comrades were unfairly attacked one came to their aid.

If Wilder's commitment had its basis in personal experience, Boas was an activist for what were essentially professional reasons. A cautious and skillful investigator who deplored hasty and unfounded generalization, Boas took strong exception to American scientists' readiness to embrace what he regarded as unsupported speculation about the Negro race.

When Boas assumed his Columbia University post in the late 1890s the concept of evolutionism was still the fashion in anthropology. A doctrine that arranged all the races of man in a sort of evolutionary line-up from the savage at the rear to the most civilized races up front, this theory provided not only a system for rating achievements of contemporary cultures but also a description of the pathway the higher races had

[9] Jeffries Wyman to Wilder, July 25, 1864, box 1, Wilder Papers; also see Wyman to Wilder, Oct. 12, 1864, box 1, and Mary A. Q. Gould to Wilder, Apr. 14, 1868, box 1, Wilder Papers.

[10] Wilder Diary, Apr. 6, 1865, p. 215, box 13, Wilder Papers.

[11] B. G. Wilder, "Two Examples of the Negro's Courage, Physical and Moral," *Alexander's Magazine*, Jan. 1906, *1*: 22–25.

[12] B. G. Wilder, "The Brain of the American Negro," *Proceedings of the First National Negro Conference* (n.d.), p. 23. Certainly it would be wrong to claim for Wilder the outlook of the modern-day civil rights activist. Like most racial liberals of his time, Wilder was to some degree a captive of prevailing racial ideoogy, although he struggled against it. In one publication he noted that it was common knowledge that the white race had, on the whole, evolved to a higher position than had the black. See Wilder in Jordan, ed., *Leading American Men of Science*, p. 192.

followed in getting to the top. But the way up was clearly not open to all. To the evolutionist the term "civilized race" carried the additional connotations of superior and white (more precisely, Anglo-Saxon), while "savage" and "barbarian" implied inferior and non-white. Furthermore, certain of the non-white peoples assigned to the lower orders of being (the Negroes, for example) were destined to remain there. The belief was that evolution had simply stopped working for them.[13]

By the mid-1890s Boas strongly rejected such thinking, largely because anthropologists had never put their evolutionary model to the test, had never paused in their zeal for sweeping generalization to make those detailed analyses of specific cultures upon which generalization had to rest. Boas in that period was as interested as the evolutionist to find laws of cultural development, but he knew that scientists must first "understand the process by which the individual culture grew before we can undertake to lay down the laws by which the culture of all mankind grew."[14]

But if evolutionism with its belief in superior races was yet unproven, available evidence, Boas believed, also suggested that it was erroneous. A common misconception about primitive people was that they had a lower order of intelligence than white men, rendering them fickle, excitable, and unable to concentrate at length. As one contemporary scientific account put it, even short conversations wearied the native, "'particularly if questions are asked that require efforts of thought or memory. . . .'"[15] Boas' research led him to quite different conclusions. Willing to study primitive peoples on their own cultural terms, he realized early that the alleged weakness of the savage mind was merely a mirage which owed its existence to anthropologists' insistence on viewing native mores and actions from the vantage of white cultural values. If natives lost interest in conversing with an investigator, it was usually because his questions seemed trifling and were often posed in the white man's language besides. "I can assure you," Boas reported to a group of colleagues, "that the interest of . . . natives can easily be raised to a high pitch and that I have often been the one who was wearied out first."[16]

Fundamentally, then, Boas' public involvement on the issue of race sprang from a scientist's desire to rid his discipline of prevailing amateurism and to substitute the

[13] On the connection between evolutionism and the idea of racial inferiority see George Stocking, "American Social Scientists and Race Theory, 1890–1915" (unpublished Ph.D. dissertation, University of Pennsylvania, 1960), pp. 466–467. The theory that Negroes lay outside the evolutionary process is treated in John Haller, *Outcasts from Evolution. Scientific Attitudes of Racial Inferiority, 1859–1900* (Urbana: University of Illinois Press, 1971), pp. ix, 100, 198–199.

[14] Franz Boas, "The Growth of Indian Mythologies," *Journal of American Folklore*, 1896, *9*:11. On Boas' rejection of evolutionism (by the 1890s) see Stocking, "American Social Scientists and Race Theory," pp. 522–523, 526–527. Boas was not the first anthropologist to question linear evolution or to emphasize the empirical approach to research. The German anthropologist Rudolph Virchow argued the inherent equality of the races, and his influence on Boas is believed to have been substantial. Also, American anthropologist Horatio Hale, who supervised Boas' British Columbia research, stressed not only the need for an empirical approach but also skepticism of the ideas of the evolutionists. (On Virchow see Clyde Kluckhohn and Olaf Prufer, "Influences During the Formative Years," *The American Anthropologist*, 1959, *61* (No. 89): 21–22. On Hale, see Jacob Gruber, "Horatio Hale and the Development of American Anthropology," *Proceedings of the American Philosophical Society*, 1967, *111*:18–19, 25–33. Perhaps Hale's influence was more a reinforcement than a catalyst to Boas' thinking, for Gruber himself notes that Boas had already begun to question prevailing ideas before he met up with Hale.)

[15] Quoted by Boas in "Human Faculty as Determined by Race," *Proceedings of the American Association for the Advancement of Science*, 1894, *43*:320–321.

[16] *Ibid.*

rigorous methods of science for facile generalization. The broad issue at hand was the matter of sovereignty in science; and those professionals, like Boas, who regarded their approach as the more scientific were vitally concerned to educate not only colleagues but also the more enlightened part of the public to distinguish between kings and pretenders. When Boas began work at Columbia, anthropology was still largely the preserve of those he regarded as amateurs in their approach. As an aggressive professional, holding strong views about the need for his discipline to become more truly scientific, he felt obliged to oppose that amateurism wherever he found it. In segregationist America there was no lack of targets.[17]

Although Boas and Wilder had differing perspectives on the race issue, their response to problems of discrimination and oppression was similar. Each played essentially an educative role, whose goal was to advance more rational thinking on race among both whites and Negroes. With whites—both scientists and laymen—their purpose was to liberalize opinion. With blacks, the aim was to dissolve inbred feelings of inferiority and encourage greater self-esteem. Both men proved equally skillful in cutting through the inconsistencies and strained reasoning of the segregationist defense.

In dealing with racism among whites Wilder's approach was more personal than Boas' and had more of the earmarks of a moral crusade. While not a belligerent man, he was argumentative and often blunt. His targets tended to be individuals rather than groups or institutions, and usually people with some influence on public opinion. One of those targets was the novelist Owen Wister, a Harvard-trained Philadelphia lawyer, who first achieved popularity with *The Virginian* (1902). In 1905 the *Saturday Evening Post* began serializing a new book, *Lady Baltimore*. The story of a young Northerner on a genealogical mission in contemporary South Carolina, the novel was in large part a literary endorsement of both the Southern view of Reconstruction and the national decision to return the management of Negroes to the former master class. Annoyed by Wister's patrician racism, Wilder was particularly angered by the author's attempt to make science serve his white supremacist notions. One passage in the book had the main character comparing skulls of a gorilla, an Aryan, and a South Carolina Negro. The Negro and Caucasian skulls were easily distinguishable, but between the ape's and the black man's the hero could see only a"'kinship which stares you in the face.'"[18]

[17] Certainly there was more impelling Boas on the question of Negro inferiority than just his concern for scientific professional standards. His own students and colleagues testify to this. In describing the motive forces in his life, they cite not only his dedication to facts and the scientific method but also his liberal outlook on social and political questions, his great affection for the primitives he studied, his sense of an obligation to use his research in the interest of freedom, and his passionate opposition to anti-Semitism. While all were factors in his public involvement on the Negro question, his commitment to scientific objectivity and reliability still seems the most basic and fundamental explanation. His involvement on the anti-Semitism issue supports this. Although from a Jewish background and a foe of anti-Semitism since his youth, Boas did not become actively and publicly involved on that issue until the 1920s, when Nazi racists made a major effort to enlist science in support of their views. Boas also never involved himself with the Indian's plight as he did with the Negro's or Jew's. And it is suggestive, I think, to note here that the idea of Indian inferiority was never a major tenet of scientific racism. See Melville Herskovits, *Franz Boas. The Science of Man in the Making* (New York: Charles Scribner's, 1953), pp. 113–114. Also see the comments of other students and colleagues in *Amer. Anthropol.*, 1943, *45* (memoir 61): 8, 14, 23, and 1959, *61* (memoir 89): vi, 2, 6, 21–22, 30.

[18] Wilder, "The Brain of the American Negro," p. 26; for a sketch of Wister see *Who Was Who in America*, Vol. I, p. 1370; also see Owen Wister, *Lady Baltimore* (New York: The MacMillan Co., 1906), pp. 289–300, *passim*.

Frankly astonished that a "liberally educated writer who had every opportunity for ascertaining the facts" could harbor such ideas, Wilder was even more disturbed by the probable influence of Wister's story on its readers. So he wrote the novelist, pointing out his errors and inquiring about the source of his racial notions. Through his secretary Wister replied crisply that his anatomical views merely reflected common knowledge, that kind "'found in any museum of anatomy or academy of natural sciences.'"[19] With that, Wilder dropped his cover of amiability. "Pardon my persistence," he wrote, "but there is more to be said." Reputable scientific opinion had long held—and Wilder spoke, he said, from the vantage of forty-five years in science—that the differences between Negro and white crania were insignificant compared to the vast gap between Negroes and the higher apes. Even a child could distinguish skulls of various human races from those of primates. What Wister needed to do was to re-visit his local museum for a closer look. But more than that, he should immediately and publicly retract his inaccuracies. If he did not, Wilder assured him that "unwelcome as the task may be" he would have to try to "arrest further diffusion of the scientific error and the political venom that characterize the passages in question."[20]

Ultimately the dispute ended in a standoff. Wister promised to "'set the matter right'" if further inquiry proved his statements incorrect, but when the book version appeared, the passage still left the impression of an affinity between apes and Negroes. Meantime, Wilder published his own correction in *Alexander's Magazine*, for he had decided that even full atonement by Wister would be inadequate. The novelist never offered that, for he regarded Wilder's demand as unreasonable. Wilder, of course, disagreed: from his standpoint a man who should have known better had involved himself in the sorry business of spreading racism and should be called to account.[21]

That rule also applied to women. In 1915 Margaret Deland, a well-born Easterner who had attained a fair measure of popularity and wealth from her writing, published a story entitled "A Black Drop." In it she argued that racial intermarriage was not only harmful to the Caucasian race but instinctively and inherently repulsive to whites. If a white man did not feel repulsed, "'there is something wrong with him,'" the authoress insisted.[22] What Wilder basically objected to was Deland's claim that racial aversion was instinctive. He was certain that it was merely the result of prejudice or social convention. "No natural instinct," he wrote her, "has prevented the existence of millions of mulattoes."[23] Southern whites might proclaim their inherent repulsion in public, but that merely reflected their hypocrisy; in private they obviously observed a different set of ethics. Indeed, if there were only some way to identify kinship through the blood, Wilder felt certain that "very few of the proud Southern families would fail of representation in the veins of their despised poor relations."[24]

Deland was unconvinced. There were undeniably millions of mulattoes, but that was no argument against her theory; it merely showed that "there is something wrong with

[19] Wilder, "The Brain of the American Negro," p. 26.

[20] *Ibid.*, pp. 26–29.

[21] *Ibid.*, pp. 29–30; also see Wister, *Lady Baltimore*, p. 171, for the author's revision of offensive passages. Wilder's attack clearly put Wister on the defensive, for in the prefatory remarks in his book he insisted that he bore no malice toward the colored race, and he expressed surprise that anyone could have thought otherwise.

[22] Quoted in Wilder to Margaret Deland, May 13, 1915, box 2, Wilder Papers; on Deland see *Who Was Who*, Vol. II (1950), p. 151.

[23] Wilder to Deland, May 13, 1915, box 2, Wilder Papers.

[24] Wilder notation, "Negro Ravia" file, n.d., box 13, Wilder Papers.

a vast number of white men!"[25] But Wilder should not think her prejudiced: Booker T. Washington, she enthused, was "one of the most splendid people in the World" and superior to a large proportion of whites, even though, racially considered, he was "nearer the primitive man than is the Anglo-Saxon."[26]

Such encounters were surely frustrating. But Wilder, an old hand at the aggravating business of reform, was undeterred. In fact, at the same time he was battling literary figures he was also keeping watch on the scientific community. One colleague whom he sought to expose was the Virginia anatomist Robert B. Bean. In 1906 Bean published three articles on the Negro brain. One, a highly technical and fairly cautious treatment, appeared in a professional journal of anatomy and was the product of his doctoral research under the anatomist Franklin P. Mall. Its central point was that the Negro brain was smaller than the white in the region controlling the higher intellectual faculties. The other two articles, published in a popular periodical and less restricted by the demands of experimental evidence, emphasized what Bean saw as the social implications of his research. The first argued that Negroes, because of the size and shape of their brain, not only lacked the white man's intellectual capacity but were also creatures of uncontrollable passion. The second concluded that since the Negro mind was essentially unredeemable, schoolmen should discourage college education for blacks and train them instead for work more in line with their ability, such as farming and the manual trades.[27]

Wilder, who regarded even the technical article as a challenge to his standing as a neurologist, did not discover the two popularized treatments (and the extent of Bean's departure from his evidence) until 1909. When he did, he quickly decided that he must do more than merely answer Bean in a scientific journal; he must expose him publicly.[28] And he had best do as thorough a job as possible, for it appeared that Bean's ideas were already having an impact. One man who read the articles was the editor of *American Medicine*, who shortly afterward urged disfranchisement of all Negroes on the ground that they were "'an electorate without brains.'"[29] To lay a firm basis for his rejoinder Wilder redid much of Bean's research himself, using his own sizeable collection of Negro and Caucasian brains. By spring 1909 he had his results in hand. He also had a forum from which to present them: a national civil rights group, planning a conference in New York, invited Wilder to address them on the Negro brain. In that speech he devoted major attention to Robert Bean.[30]

The basic flaw in Bean's claim about Negro intellectual potential, Wilder argued, was that it rested on too small a sampling of Negro brains. Indeed, at one point Bean even tried to generalize from "*a* Negro and *a* Caucasian [brain] . . . as if they represented a constant racial difference. . . ."[31] Invidious comparisons, Wilder objected,

[25] Deland to Wilder, May 19, 1915, box 2, Wilder Papers.

[26] *Ibid.*

[27] On Robert B. Bean see *Who Was Who*, Vol. II, p. 51; the technical article appeared in *American Journal of Anatomy*, 1906, *6*; the two popularized treatments are Robert B. Bean, "The Negro Brain," *The Century Magazine*, Sept. 1906, *50*, and "The Training of the Negro," *ibid.*, Oct. 1906, *50*.

[28] Wilder had read the *Journal of Anatomy* and was preparing to comment on it in a paper to be delivered to a meeting of the American Philosophical Society in April 1909. He had to ask Bean for the references to his *Century* articles, however. See Wilder to Robert B. Bean, Feb. 22, 1909, box 2, Wilder Papers.

[29] Wilder commented on this in "Brain of the Negro," p. 36.

[30] Wilder's brain collection is noted in J. H, Comstock, "Burt Green Wilder," *Science*, May 22, 1925, *41*:532–533; also see Wilder, "Brain of the Negro," pp. 22, 35–39.

[31] "Brain of the Negro," p. 38.

collection a large brain of a onetime murderer and a smaller one once belonging to an esteemed judge. Surely he was not to conclude that the criminal class was superior to "those who pass on their misdeeds."[32] But more was involved than faulty methodology, Wilder suspected. Bean's findings also reflected his own personal bias against Negroes. One man who shared his suspicions, Wilder noted, was Bean's teacher, Mall, who had recently accused Bean of shaping his evidence to support his anti-Negro views.[33]

Although the speech later appeared in print as part of the published proceedings of the conference, Wilder knew that such a document would give his views on the Negro brain and on Bean only limited exposure among those white readers he wanted to reach. So he decided to circulate additional copies of the article himself. By summer 1910 he had spent $500 for printing and distribution and had put his article into the hands of a thousand additional readers.[34]

In Bean's case Wilder apparently did not know the man he attacked, which made it easier to say unpleasant things. But even when personal acquaintance was involved, Wilder did not back away, as was clear in his encounter with anatomist Robert W. Shufeldt, a former student who had spent a number of years in scientific work in Washington, D.C. In 1915 Shufeldt published a book titled *America's Greatest Problem: The Negro*, which represented a high point in a long career of Negro-baiting. Wilder was thoroughly offended by it and wrote a lengthy and devastating review which he sent to the editors of *Science*. Just afterward he told a friend that "it has been no easy or pleasant task, and will probably cost me his regard; he was an old pupil, but I think we never have met since and I have not been attracted by a certain flavor in his writings which also characterizes the present work; some of it is really disgusting."[35] Ultimately Wilder had to settle for publication of a much shorter review than he wanted. *Science* editor James McKeen Cattell not only found the article too long but also thought it improper to "take up in *Science* the question of the social relations, etc. of the Negro."[36] Still, even the abbreviated review was stinging: in it Wilder ridiculed Shufeldt's obsession with the supposed sexuality of Negroes.[37]

Franz Boas also tried to check the more unfounded assertions of his colleagues. His approach, however, was less rancorous, largely because he addressed his criticisms not to individuals but to all his colleagues. One of his platforms was the annual meeting of the American Association for the Advancement of Science, whose subsection of anthropology provided the first national home for American workers in the field. In 1894 and again in 1908 Boas served the subsection as its presiding officer and thus had the responsibility for delivering a state-of-the-profession address. Both times he used that speech to call for more rigorous standards in the study of the Negro race.

In 1894 his subject was the racial bias of many African studies. Too often, he said, such investigations began with the assumptions that the white race represented the highest perfection in man and that deviation from whiteness was a measure of race in-

[32] *Ibid.*

[33] *Ibid.*, p. 39.

[34] Wilder to G. Spiller, Aug. 15, 1910, box 2, Wilder Papers.

[35] Wilder to Dr. Lamb, Aug. 8, 1915, "Shufeldt" file, box 13, Wilder Papers; see Shufeldt, Biographical Sketch, Apr. 1902, Collection 239, Academy of Natural Sciences, Philadelphia; Shufeldt's racist attitudes are well expressed in Robert W. Shufeldt to Wilder, Aug. 17, 1915, box 13, Wilder Papers.

[36] James McKeen Cattell to Wilder, Aug. 31, 1915, box 13, Wilder Papers.

[37] *Science*, Nov. 26, 1915, N.S. *42*:768.

feriority. Such ideas were patent myths, Boas insisted, which owed their existence partly to the anthropologists' tendency to confuse actual cultural achievement with aptitude for achievement. Because black Africans did not then exhibit high culture, anthropologists concluded that they lacked capacity for it. That view was not warranted, for in earlier times certain African regions which had encountered Mohammedan influence had developed sophisticated civilizations.[38]

A second fundamental error was the belief that achievements of various races were the result of biology. Actually, history, far more than inheritance, caused one race to dominate over others. For example, Boas noted, many favorable conditions quite apart from race aided the rapid growth of civilization in Europe. Later, when European culture began to spread to other continents the peoples there, again for nonracial reasons, "were not equally favorably situated. . . ." As a result, European culture had smothered "all [the] promising beginnings" which had appeared elsewhere.[39] But that was the story only for the modern era. Once, history stood on the side of certain colored races: then they played the lead roles, while white men were mere primitives.[40]

A decade and a half later, when Boas made his second address, racial segregation had entered its most intense phase. Those clamoring for still greater repression were raising the specter of racial interbreeding and arguing that the Negro posed a threat not only to the biological integrity of the white race but to civilization itself. In Boas' mind that situation made any racial bias of scientists more regrettable than ever, for racial fear-mongers were then looking to science to substantiate their claims.

Matters were grave enough, he felt, to call for a moratorium on unfounded scientific statements on race mixing. In his 1908 address Boas issued that call: "I feel that it behooves us to be most cautious and particularly to refrain from all sensational formulations of the problem, that are liable to add to the prevalent lack of calmness . . .; the more so since the answer to these questions concerns the welfare of millions of people."[41] But scientists could do more, Boas suggested, than merely keep quiet. Since there was a lack of even the "most elementary facts" about interbreeding, one important service they could perform was to investigate thoroughly the whole question of the fitness of mulattoes.[42] Boas did not know what conclusions might result, but he suspected they would remove much of the fear of amalgamation. His own studies had convinced him that all so-called pure races—even the Anglo-Saxon—were products of considerable mingling of diverse people, and such mixing had not proved a serious barrier to high cultural attainment.[43]

Focusing on the scientific community might do much to alter the racial views of his colleagues, but it was a poor way, Boas recognized, to change popular thinking. Greatly concerned to widen the influence of his views, in the early twentieth century Boas began an assault on middle-class white racism that was extraordinary for a man so heavily engaged in strictly professional tasks. His ideas were presented in various forms, in newspaper and periodical articles, book reviews, and a major book of his own. The periodical pieces appeared or were noted in such journals as *Everybody's Magazine*, *Van Norden's Magazine*, *Charities and the Commons*, and *The Century*, and

[38] Boas, "Human Faculty," pp. 301–306.

[39] *Ibid.*, pp. 307–308.

[40] *Ibid.*, p. 303.

[41] Franz Boas, "Race Problems in America," *Science*, May 28, 1909, *29*: 845–846.

[42] *Ibid.*

[43] *Ibid.*, pp. 848–849, 840–841.

his articles in *The New York Sunday Times* extended his range even further. Altogether he argued his case at least a dozen times in the decade before World War I.[44]

His emphasis was chiefly the African origins of the Negro. As he explained to Booker T. Washington, "I am particularly anxious to bring home to the American people the fact that the African race in its own continent has achieved advancements which have been of importance in the development of civilization. . . ."[45] In fact, he told a Columbia colleague, the evidence from Africa "is such that we must class the Negro among the most highly endowed human races."[46] Awareness of that, he reasoned, would force fair-minded men at least to give up the notion that licentiousness, shiftlessness, and criminality were racial characteristics of black men.[47]

But winning wide acceptance—even consideration—of such heretical ideas was clearly beyond the power of a single individual. If prevailing racist attitudes were to be substantially changed, a large-scale, collective assault was needed. Accordingly in 1906 he threw his energy behind the establishment of an African museum, an institution which would combine public exhibits with facilities for extensive scholarly research on the Negro. In its main aim of substituting truth for error in the public mind, the proposed museum illustrated well the almost boundless faith in reason of that generation of scientists. Downplaying the force of irrationality, Boas seemed convinced that people instinctively opted for truth when presented with a choice. As he told one colleague, the African museum would "serve eminently well to counteract the prejudices which hinder the advancement of the Negro race."[48] He even foresaw basic alterations in national policy toward Negroes as a result of the institution's work.[49]

As Boas conceived it, the museum would have two primary functions. On the one hand it would attempt to persuade the public by means of exhibitions and publications that Africa had once been a seat of highly developed civilizations. Its second function, and the one with greatest potential for influencing government policy, would be a comprehensive investigation of Negro anatomy. Scientists had previously written a great deal on such questions as the Negro brain, the deleterious effects of race mixing, and the arrested development of Negro children. But all those accounts were written "entirely from a partisan point of view."[50] Objective investigation not only promised to elevate white concepts of the Negro but also would enable government to utilize "unprejudiced scientific investigation" in deciding "what policy should be pursued."[51]

The major obstacle before the project, not surprisingly, was cost. Physical facilities alone would require about $500,000, and operating expenses demanded an additional $670,000 endowment. In the first decade of the twentieth century such funding was almost unavailable. Certainly the federal government offered little avenue of hope. Washington might spend millions of dollars annually to combat livestock and plant diseases, but support of social research lay far in the future. Private sources were like-

[44] For a full listing of Boas' publications in this period see "Bibliography of Franz Boas," in *Amer. Anthropol.*, 1943, *45* (No. 3): 82–92; also see Wilder to Boas, May 11, 1909, Boas Correspondence, American Philosophical Library, Philadelphia; for the *Times* articles see *N. Y. Times*, Sept. 24 and Oct. 1, 1911.

[45] Boas to Booker T. Washington, Nov. 8, 1908, Boas Correspondence.

[46] Boas to Felix Adler, Oct. 30, 1906, Boas Correspondence.

[47] Boas, "The Negro and the Demands of Modern Life," pp. 86–87.

[48] Boas to Adler, Oct. 30, 1906, Boas Correspondence.

[49] *Ibid.*

[50] Boas to Lucius P. Brown, June 5, 1909, Boas Correspondence.

[51] Boas to Starr Murphy, Nov. 23, 1906, Boas Correspondence.

wise unpromising. Large scale philanthropy was of recent origin and up to then had gone mostly to noncontroversial projects in such fields as public health, medicine, and graduate education.

But Boas refused to concede the odds and confidently undertook to enlist the country's major givers behind his effort. Included in his solicitation were John D. Rockefeller and Andrew Carnegie, who had recently endowed institutes bearing their names, and two supporters of Southern Negro education, Robert C. Ogden and George F. Peabody. In each appeal Boas stressed the irrationality and social danger of racism and the value of the African museum idea as a vehicle for dissolving such tension. The worth he himself placed on the project was evident in his appeal to the head of the Carnegie Institution: "You are aware that I am exceedingly anxious to interest you in other kinds of anthropological research. However, if I were given the choice, I do not know whether I should not think the Negro work more important on account of its practical bearings at the present time."[52]

If Boas seriously hoped for support, he was disappointed. None of the wealthy givers whom he solicited regarded the project as sufficiently important to merit a contribution. Carnegie merely referred Boas to the director of his Institution, who in turn referred him to the federal government. Apparently all found the proposal too radical and needlessly upsetting to white sensibilities. Peabody for one preferred a more gradual approach to race problems: "What you have in mind will be a most desireable thing to do, but there are special developments now underway in the South . . . which will make it perhaps more advantageous to have some delay in the carrying out of any such program. . . ."[53]

Rebuffed by private sources Boas turned to the one public agency which might aid him, the Smithsonian Institution's Bureau of Ethnology. Seeking to capitalize on growing national interest in the "new" immigration, he redesigned his project and in 1909 proposed an investigation of the relative influence of heredity and environment which would focus on Negroes, mulattoes, and ethnics. The investigation was tailor-made for the Bureau, he argued, and certain to strengthen its relationship with Congress, which was then vitally interested in the so-called European races.[54] But the Bureau would have none of it. Such research would only arouse the "race feeling" of Congress and jeopardize Bureau appropriations.[55] Incensed by such timidity Boas complained to a colleague that the policy of the Smithsonian "has been for years one of hesitation when scientific questions of considerable importance have come up. . . ."[56]

In the absence of substantial public and private support Boas saw no recourse but to try to create a program of Negro studies on his own at Columbia. His ability to coax modest sums from his many institutional and individual contacts (Peabody ultimately

[52] Boas to R. S. Woodward, Dec. 4, 1906; also Boas to Andrew Carnegie, Dec. 8, 1906; Boas to Starr Murphy, Nov. 23, 1906; Boas to Robert C. Ogden, Feb. 11, 1907; Boas to George Peabody, Feb. 11, 1907, all in Boas Correspondence. Ogden and Peabody were deeply involved in efforts to provide technical and vocational education for Negroes. Both served as trustees for Hampton and Tuskegee institutes and both were officers of the Southern Educational Board. See *Dictionary of American Biography*, Vol. VIII, pp. 641–642 (Ogden) and Supplement, Vol. II, pp. 520–526 (Peabody).

[53] Peabody to Boas, Feb. 13, 1907, Boas Correspondence.

[54] Boas to Charles D. Walcott, May 4, 1909, Boas Correspondence.

[55] J. W. Hodge to Boas, Mar. 18, 1910, Boas Correspondence.

[56] Boas to J. W. Jenks, Mar. 21, 1910, Boas Correspondence.

became one of his staunchest supporters) and his success in attracting a number of capable students, several of them Negroes, combined to produce at least a scale-model version of the museum idea. By the mid-1920s he could report a variety of projects: he was sending one student to Africa, and several others were studying the American Negro. One was gathering folk tales in the Caribbean, another was measuring Harlem blacks, and a third was studying the structure of a Negro community in Virginia.[57]

What made Boas' and Wilder's involvement particularly exceptional for that day, however, was not that they did something *for* Negroes, but that they did something *with* Negroes—especially with a certain group of them. The first decade of the century witnessed the rise of a vigorous civil rights movement in America. Led by outspoken individuals of both races who accepted no compromise with white supremacy, it demanded immediate justice for at least the talented minority of blacks. Whereas the vast majority of whites sympathetic to the Negro's plight recoiled from the new militance, Boas and Wilder did not hesitate to stand with it.

One reflection of their commitment was their many appearances before civil rights conferences in that period. Wilder's appeal stemmed only in part from his special knowledge of anatomy; he was also what that generation called an inspirational speaker. His zeal for the Negro's cause and his ability to puncture white pretensions were amply illustrated by a 1905 address at the Boston A.M.E. Zion Church on the occasion of the William Lloyd Garrison Centennial. His major theme that day was the bravery of the black Massachusetts soldiers during the Civil War. By all measurements, Wilder noted, the Negroes had served as ably as any troop of whites. But where was the justice, he asked, of always imposing the white race as the standard? Certainly history gave no endorsement of white supremacy. It sometimes argued the opposite: "Who are the leaders of Tammany Hall?" Wilder asked. "Who are the mismanagers of insurance companies? Who compose the lynching mobs and the gangs of college hazers? Whites." But the so-called superior race had not just ridiculed the law. It had also evaded basic responsibilities: "Who were the importers of the first Negro slaves, and who were the progenitors of multitudes of their descendents? White men again."[58]

Less a spellbinder than Wilder, Boas nonetheless had something important to contribute to the budding civil rights movement. One Negro who recognized that was W. E. B. DuBois, then teaching at Atlanta University, Southern center of the new black activism. In 1906 he invited Boas to deliver the university's commencement address and take part in its annual Negro conference.[59] Boas' message to Atlanta Negroes was the one he characteristically gave to black people: despite insistent claims of whites, there was no proof of the racial inferiority of Negroes. To the contrary, the significant achievements of Negroes in pre-modern Africa pointed to the essential equality of the black race and justified its playing as active a role in American life as any other people. Of course, white-dominated society, however unfairly, would stiffly resist that par-

[57] For a review of projects for 1926 and 1927 see Boas to G. F. Peabody, June 25, 1926; also see Boas to Zora Hurston, May 3, 1937; Melville Herskovits, Memorandum to Boas, "End of December, 1927" file, all in Boas Correspondence. On Boas' relationship with Peabody see Boas to Peabody, June 25, 1926, and Aug. 4, 1926, and Peabody to Boas, Aug. 5, 1926.

[58] B. G. Wilder, "Two Examples of the Negro's Courage," *Alexander's Mag.*, Jan. 1906, *1*:25; see also pp. 22–25; on Wilder's oratorical skills see Bishop, *Cornell*, p. 111.

[59] W. E. B. DuBois to Boas, Mar. 31, 1906, Boas Correspondence.

ticipation, which meant that Negroes must develop both patience and perseverence in their search for fulfillment. But if Boas warned against expecting the impossible and against unruly agitation, he did not advise letting white society set the timetable for change. Instead, he urged the Atlanta students to press continuously for "full opportunities for your powers."[60]

The most notable landmark in Boas' and Wilder's emergence as social activists was their participation in the 1909 and 1910 conferences of the National Negro Committee. Convened by a biracial group who saw need for an immediate and forthright attack on Southern segregation, the conferences proved to be staging sessions for the creation in 1910 of the National Association for the Advancement of Colored People. In supporting those meetings Boas and Wilder were out of step with the great majority of whites claiming to be liberals on race. Most reacted as did William James and Andrew Carnegie, who boycotted the New York City sessions on the ground that they would do the Negro no good and only fan the animosity of white racists.[61]

Wilder, one of about a dozen whites on the 1909 program, spoke on his speciality, the Negro brain. His opening remarks revealed again that his was not merely an academic interest. "The American Negro is on trial, not for his life but for the recognition of his status, his rights, and his opportunities. At this, as at most other trials, experts disagree." But Wilder claimed to be one witness who could testify impartially, even though he was a white man. For one thing he believed in evolution and therefore minimized not just differences between men but even those between man and the apes. His own past was his second qualification: "during both my army and university experiences, there have been occasions when I was tempted to exclaim, 'Yes, a white man is as worthy as a colored man—provided he behaves himself as well.'"[62] After roasting again his old adversary Wister, and after dealing with the scientific missteps of Bean, Wilder then carefully weighed the evidence for and against the Negro in the matter of innate mental capacity. He concluded that "as yet there has been found no constant feature by which the Negro brain may be certainly distinguished from that of a Caucasian."[63]

The next year Boas appeared before the conference. His topic, "The Anthropological Position of the Negro," promised little more than a review of past research. What he actually delivered, however, was a veiled endorsement of racial intermarriage as the best long-run solution to problems of race.[64] Most white Americans, Boas knew, recoiled from such a proposition, for they viewed Negroes as inferiors, and mulattoes—who supposedly inherited the worst traits of both parent races—as even worse. Yet available evidence belied both beliefs. Historians and anthropologists had found, for example, that race mixing not only had been occurring since prehistoric times but also had produced many peoples of marked capability and achievement.[65]

[60] Franz Boas, "Commencement Address at Atlanta University," in *Race and Democratic Society* (New York: J. J. Augustin, 1945), p. 69; also see pp. 63–66.

[61] For a discussion of the origins of the N.A.A.C.P. and the reactions to the Negro conferences see Elliott M. Rudwick, *W. E. B. DuBois: Propagandist of the Negro Protest* (New York: Atheneum, 1968), pp. 120–131.

[62] Wilder, "The Brain of the Negro," p. 22.

[63] *Ibid.*

[64] On arrangements for Boas' speech, see Oswald Garrison Villard to Boas, Mar. 16, 1910; and Boas to Villard, Mar. 17, 1910, both in Boas Correspondence. Boas' speech was reprinted in *The Crisis*, Dec. 1910, *1*:22–25.

[65] *Crisis*, pp. 22–24.

Although Boas did not say it explicitly, his meaning was clear: white Americans had no reason to fear large-scale intermarriage between the races. At any rate, such mixing was inevitable, he proclaimed. White and black lived side by side in America, and outside additions to the population were almost exclusively European. Those two facts "must necessarily lead to the result that the relative number of pure Negroes will become less and less. . . . The gradual process of elimination may be retarded by legislation, but it cannot possibly be avoided."[66] Of course such a development lay far in the future, but once extreme physical differences were eliminated, existing race tensions would appreciably lessen.[67] Boas' solution, while surely repugnant to the vast majority of Americans, left no doubt about his allegiance to the principle of equality.

A fundamental reason why Boas and Wilder made themselves available to the emerging civil rights movement was their conviction that Negroes, as much as whites, needed to learn about the potential of the black race. Both understood, as so many of the white "friends" of the Negro did not, that slavery and segregation had convinced even Negroes of their inferiority. Even among the "best classes of the Negro," Boas had noted a strong "feeling of despondency. . . ."[68] Until the black man gained confidence in himself he could never play the role in national affairs that he ought. One way to foster race pride, both scientists were convinced, was to convey to the Negro a sense of the achievements of his forebears.[69]

Corroboration of that reasoning and a tribute to the work of Boas and Wilder came in 1911 from a young New York Negro. Willis Huggins was an industrial arts student at Columbia University. After reading two articles by Boas on native African culture, he wrote the scientist to thank him for the "timely information you gave out" on the achievements of the Negro race. Huggins had re-read the pieces to a meeting of young blacks at a local YMCA, and "to a man those present voiced appreciation of your careful research and . . . fair play." Indeed, Huggins concluded, the whole "Army of Negro youth" who were seeking to advance their race had "much to be grateful for in the products of your research."[70]

Like the student Willis Huggins, later historians would generously applaud the humanism of Franz Boas, but unlike the New York Negro most would recall only his efforts in the period after World War I. Burt Green Wilder they would neglect completely.

Overlooked by scholars, the activism of Boas and Wilder in the early twentieth century deserved much better. For one thing, their influence within their respective disciplines surely created a pressure on colleagues to proceed with caution in applying their research to contemporary racial problems. Reputable scientists from other disciplines who also opposed such distortions in that early period likely exerted a similar influence in their fields. To the degree that professional scientists were responsive to that kind of pressure—or example—from their leaders (and the authoritarian nature of scientific organizations would seem to make this likely), then the meaning of Boas'

[66] *Ibid.*, p. 25.

[67] *Ibid.* Radical as these ideas were for the time, Boas' solution would be regarded by some today as a racist scheme, for it would solve the race issue by a gradual purge of an identifiable black race and culture.

[68] Boas to Starr Murphy, Nov. 23, 1906, Boas Correspondence.

[69] *Ibid.*

[70] Willis N. Huggins to Boas, Oct. 16, 1911, Boas Correspondence.

and Wilder's involvement is that early-twentieth-century American science was less racist than commonly portrayed and actually featured a vigorous egalitarian tradition.[71]

In the realm of contemporary popular attitudes Boas' and Wilder's impact was much more limited. But they did exert some influence. Along with a small band of courageous men and women of both races, they at least succeeded in keeping a surging American racism off balance during those bleak decades after 1890. Outside the South, racist doctrine never gained full ascendancy, partly because of the kind of resistance they posed. Historian Oscar Handlin once commented on the value of such holding actions: "Always the 'glittering generalities' of the Revolutionary period have been remembered by some, and the respect due them remained an impediment in the way of those who wished to deny the equality of man—not enough of an impediment to prevent the development of racism, but enough to keep alive a sense of discordance with tradition."[72]

If they failed to win many converts from racism in that early day, Boas' and Wilder's influence on later Americans was far greater. One product of their labors (and that of colleagues sharing their views) was a countervailing ideology of brotherhood, which a recent generation of activists has used with telling effect aginst white supremacist ideas. Indeed, many of the key arguments of the modern civil rights movement were elaborated by Boas and Wilder in the early twentieth century. When later spokesmen talked of the impact on the Negro of a hostile social environment, his need for race pride, the contributions of Africa to culture and civilization, the essential equality of the races, and the insignificance of some racial differences in the face of larger individual variations, they were merely voicing notions advanced a half-century before by scholars like Wilder and Boas.

[71] See the beginning of this article for a listing of other scientists who apparently shared Boas' and Wilder's outlook. To be sure, there were reputable scholars in the period who were committed to the concept of inferior races. E. A Ross in sociology and C. B. Davenport from genetics are notable examples, although the historian of the eugenics movement describes Davenport as a careless and manipulative scientist, whose methods were often faulty. See Mark Heller, *Eugenics. Hereditarian Attitudes in American Thought* (New Brunswick: Rutgers University Press, 1963), pp. 67–68. An alternative way of viewing the racial attitudes of American scientists would be to focus on academic disciplines rather than on individuals. Doing this reveals what appears to be a common pattern in the development of many of the behavioral and biological sciences. Over time, an increased commitment to inductive scientific procedures came to characterize these disciplines. As that happened, an earlier attachment to sweeping generalization gave way to a more critical approach emphasizing empirical data and de-emphasizing grand theory. One result of the shift was a reconsideration of the presumed inferiority of the Negro. These developments occurred at different times for various disciplines, but it seems apparent that anatomy had adopted a more cautious, empirical approach by the end of the nineteenth century, anthropology had reached that stage before World War I, genetics and sociology shifted over by the 1920s, and psychology was reconsidering some of its earlier racial theories by the 1930s. Discussions of this phenomenon for particular disciplines are found in Heller, *Eugenics*, pp. 66–67, 163, 167; Roscoe C. and Gisela J. Hinkle, *The Development of Modern Sociology* (Garden City, N.Y.: Doubleday, 1943), p. 22; and Ruth Benedict, *Race, Science and Politics* (New York: Viking Press, 1968), pp. 71–78, which looks at changing racial views in one area of psychology. Stocking considers this process of professionalization and its impact on race thinking in a different way. Focusing on the behavioral sciences, Stocking argues that what caused the decline of the idea of racial determinism within those disciplines was the adoption of the culture concept from anthropology. Although assimilation proceeded at varying rates for the several disciplines, by the 1930s the idea of a cultural explanation for human behavior, he says, had achieved the status of a paradigm throughout the behavioral sciences. See *Race, Culture, and Evolution*, pp. 303–305.

[72] Handlin, *Race and Nationality in American Life*, p. 146.

Finally, their efforts also had significance for the social history of science in the United States in that their fight for racial justice set one of the historic precedents for the recent and widespread social activism of American scientists. Following World War II scholars in a variety of fields came to regard such involvement as a normal professional role: nuclear physicists compaigned publicly for disarmament and control of atomic weapons; biologists fought against pollution and population explosion; psychiatrists and biologists moved in the vanguard of the civil rights movement. Not since the period of the American Revolution, when "philosophers" such as Thomas Jefferson, David Rittenhouse, Benjamin Rush, and Benjamin Franklin assumed leading roles in the political and social movements of their day, had American scientists so eagerly sought to relate their studies to the reform needs of their time. Franz Boas and Burt Green Wilder, as much as any other scholars of their day, helped revive the idea that a scientist owed more to society than mere pursuit of knowledge and its economic application.

Blind Law and Powerless Science

The American Jewish Congress, the NAACP, and the Scientific Case against Discrimination, 1945–1950

By John P. Jackson, Jr.

ABSTRACT

This essay examines how the American Jewish Congress (AJC) designed a legal attack on discrimination based on social science. This campaign led to the creation in 1945 of two new AJC commissions, the Commission on Community Interrelations and the Commission on Law and Social Action. The AJC's attack on discrimination highlights the difficulties and potentialities of using social science research in the law. On the one hand, the relationship between the communities of law and social science was inherently unstable, with constant conflicts over their mission, work styles, timetables, audiences, and standards for success. On the other hand, the AJC's commissions generated new types of arguments and new types of evidence that were powerful tools against racial discrimination and eventually the key to toppling legal segregation. The essay concludes that while the processes by which lawyers and scientists produce their arguments are very different, the products of science can be very useful in the courtroom.

ON 17 MAY 1954, in one of the most important civil rights cases of the century, the U.S. Supreme Court declared segregation unconstitutional in *Brown v. Board of Education.* Social science received prominent mention in a footnote in Chief Justice Earl Warren's opinion. Soon after the decision, Warren's citation of social science became a source of controversy, as scholars tried to sort out its importance. New York University professor of jurisprudence Edmond Cahn expressed concern that Warren's opinion rested on the "flimsy foundation" of social science rather than on solid legal reasoning. Cahn believed that in the *Brown* case social scientists had overstepped the bounds of proper

An earlier version of this essay was delivered at the 1996 meeting of the Southern Conference on Afro-American Studies, Tallahassee, Florida. This research was supported by the National Science Foundation's Program in Law and Social Science (NSF Award Number SBR-9421729). I am grateful to Arthur L. Norberg, Leila Zenderland, Benjamin Harris, Stephen Berger, Will Maslow, Margaret Rossiter, and three anonymous referees for their patience and expert editorial advice.

science and ventured into the realm of advocacy. Given the undeveloped state of social psychology, Cahn thought it unable to provide truly objective scientific evidence. Hence social psychology posed a danger to the law. It was imperative, Cahn argued, for judges to "learn where objective science ends and advocacy begins." At present, he insisted, it was still possible for the social psychologist to "hoodwink a judge who is not over wise."[1]

Kenneth B. Clark, a psychologist who had served as the chief liaison between the National Association for the Advancement of Colored People (NAACP) and the social science community during the *Brown* litigation, was quick to defend his work against Cahn's allegation that he had been an "advocate" rather than an objective scientific advisor. How could this be, Clark asked, when the "primary research studies were conducted ten years before these cases were heard on the trial level"? Clark concluded that "one would have to be gifted with the power of a seer in order to prepare himself for the role of advocate in these specific cases ten years in advance."[2]

Clark was certainly correct that the social scientific community had no crystal ball to predict the course of the *Brown* litigation. Yet it is also true that social scientists had been working to use their expertise to dismantle legal segregation in the years immediately following the end of World War II. The most famous social scientific evidence used in the *Brown* litigation was Kenneth and Mamie Clark's projective tests using black and white dolls to measure psychological damage caused by segregation. The "doll tests" have become so well known that commentators on the case often focus on them to the exclusion of any other social science testimony. Because of his work in *Brown,* Kenneth Clark became the exemplar of the socially conscious social scientist in the postwar era, but his doll tests were not the only social scientific evidence used by the NAACP, nor even the most important. A group of social scientists centered at the American Jewish Congress (AJC) had mounted a social scientific attack on discrimination that was expressly designed to be integrated with a legal attack. Although Clark worked briefly for the AJC in the 1940s, this group carried out its research program without his involvement. The AJC social scientists exemplified what Sheila Jasanoff has called "scientific subcultures that . . . coalesced in and around the processes of adjudication."[3]

[1] *Brown v. Board of Education of Topeka,* 347 U.S. 483 (1954); and Edmond Cahn, "Jurisprudence," *New York University Law Review,* 1955, *30*:150–169, on pp. 157–158.

[2] Kenneth B. Clark, *Prejudice and Your Child* (Boston: Beacon, 1963), p. 191.

[3] Sheila Jasanoff, *Science at the Bar* (Cambridge, Mass.: Harvard Univ. Press, 1995), p. 8. The perception that the Clarks' doll tests were central to *Brown* is incorrect. Many authors, however, treat them as the only empirical evidence in *Brown.* A sampling of this work includes Hadley Arkes, "The Problem of Kenneth Clark," *Commentary,* Nov. 1974, *58*:37–46; Harold Cruse, *Plural But Equal: Blacks and Minorities in America's Plural Society* (New York: Morrow, 1987), pp. 72–74; William H. Tucker, *The Science and Politics of Racial Research* (Urbana: Univ. Illinois Press, 1994), pp. 144–146; Roy L. Brooks, *Integration or Separation? A Strategy for Racial Equality* (Cambridge, Mass.: Harvard Univ. Press, 1996), pp. 12–16; and Ernest van den Haag, "Social Science Testimony in the Desegregation Cases: A Reply to Professor Kenneth Clark," *Villanova Law Review,* Sept. 1960, *60*:69–79. In the doll tests, Kenneth and Mamie Clark presented young children with two dolls identical in every way except skin color. The children were asked to identify the dolls as white or black. Then the children were asked to identify one doll with themselves and to identify one doll as "good" or "bad." The Clarks argued that the identification of a black doll as "bad" by African American children was a sign of psychological damage. A complete description of the doll tests can be found in William E. Cross, *Shades of Black: Diversity in African-American Identity* (Philadelphia: Temple Univ. Press, 1991). On the Clarks' work more generally see *ibid.;* Ben Keppel, *The Work of Democracy: Ralph Bunche, Kenneth B. Clark, Lorraine Hansberry, and the Cultural Politics of Race* (Cambridge, Mass.: Harvard Univ. Press, 1995); and Darryl Michael Scott, *Contempt and Pity: Social Policy and the Image of the Damaged Black Psyche, 1880–1996* (Chapel Hill: Univ. North Carolina Press, 1997). A work that situates Clark as a "public psychological expert" is Ellen Herman, *The Romance of American Psychology: Political Culture in the Age of Experts* (Berkeley: Univ. California Press, 1995). On Clark's brief tenure at the AJC see Gerald Markowitz and David Rosner, *Children, Race, and Power: Kenneth and Mamie Clark's Northside Center* (Charlottesville: Univ. Virginia Press, 1996), p. 81.

This essay will examine how the AJC designed an attack on discrimination that was founded on the assumption that law and social science could be merged. This assumption led to the creation of two new AJC commissions, the Commission on Community Interrelations (CCI) and the Commission on Law and Social Action (CLSA). The AJC's attack on discrimination highlights the difficulties and potentialities of combining social science research and the law. On the one hand, the relationship between the communities of law and social science was inherently unstable, with constant conflicts over their mission, work styles, timetables, audiences, and standards for success. On the other hand, the AJC's merger generated new types of arguments and new types of evidence that were powerful tools against racial discrimination and eventually the key to toppling legal segregation. In short, while the AJC failed to institutionalize any long-term working relationship between lawyers and social scientists, it did succeed in creating a powerful attack on racial discrimination.

The first section of the essay looks at the postwar context of social science research on race prejudice and at the AJC's place in that context. The second section considers the design of each commission within the AJC. Next, I examine the work each commission undertook in order to carry out the attack on discrimination. Of particular importance here is how AJC attorneys developed new legal arguments that required new kinds of social scientific evidence and how social scientists stepped up to provide that evidence. The fourth section outlines how the NAACP began to use social scientific data in ongoing litigation against segregated schools and how it began to turn to the AJC to supply that data. The final section of the essay examines the end of the collaboration of social scientists and lawyers at the American Jewish Congress and considers the ramifications for our understanding of the interplay between the law and science.

THE POSTWAR FIGHT AGAINST RACE PREJUDICE

World War II had a profound effect on minority/majority group relationships in the United States. After the war was won, many Americans turned their attention to the elimination of racism and prejudice at home. Social science was one tool that could be enrolled against race prejudice. (See Figure 1.) Even before the war, most social scientists had abandoned notions of innate racial traits and had begun to study intergroup relations and race prejudice.[4] But the war transformed the way social scientists approached their work, giving a new urgency to the social scientific study of race prejudice.

There had always been two strands of social scientific thought in the United States. On the one hand, the cultural authority of science depends on the image of scientists as detached, impartial experts who are immune from political or moral concerns. On the other hand, social science makes a claim to social utility by offering to solve numerous political and moral problems.[5] The second strand of social scientific thought was exemplified by

[4] Philip Gleason, "Americans All: World War II and the Shaping of American Identity," *Review of Politics,* 1981, *43*:483–518; David Southern, *Gunnar Myrdal and Black-White Relations: The Use and Abuse of* An American Dilemma, *1944–1969* (Baton Rouge: Louisiana State Univ. Press, 1987); Elazar Barkan, *The Retreat of Scientific Racism* (Cambridge: Cambridge Univ. Press, 1992), pp. 343–346; Carl N. Degler, *In Search of Human Nature* (New York: Oxford Univ. Press, 1991), pp. 187–211; and Franz Samelson, "From 'Race Psychology' to 'Studies in Prejudice,' " *Journal of the History of the Behavioral Sciences,* 1978, *14*:265–278.

[5] There is an extensive literature on this "paradox" of social science in the United States. Major monographs include Joanne Brown, *Definition of a Profession: The Authority Metaphor in the History of Psychological Testing, 1890–1930* (Princeton, N.J.: Princeton Univ. Press, 1992); Mary O. Furner, *Advocacy and Objectivity: A Crisis in the Professionalization of American Social Science, 1865–1905* (Lexington: Univ. Kentucky Press, 1975); John M. O'Donnell, *The Origins of Behaviorism: American Psychology, 1870–1920* (New York: New York Univ. Press, 1985); and Mark Smith, *Social Science in the Crucible: The American Debate over Objectivity and Purpose, 1918–1941* (Durham, N.C.: Duke Univ. Press, 1994).

Figure 1. *Social science arguments as anti-racist propaganda in the postwar United States. (From the Collections of the Manuscript Division, Library of Congress, Washington, D.C.)*

the Society for the Psychological Study of Social Issues (SPSSI), a group of activist social psychologists that came together in 1936.

SPSSI was born in the Great Depression, a time of profound social and economic distress. The social psychologists of SPSSI hoped to create a new form of social psychology, one that could empower their research subjects and be used for positive social change. Members championed a nonreductivist social psychology that moved out of the laboratory and into the lives of real Americans. They rejected what they saw as positivist, experimentalist psychology and sought to bring forth an unapologetic political psychology that would make American society more democratic.[6]

SPSSI's project was not without contradictions. While professing an ideology that could be used to empower workers and create a true socialist democracy, SPSSI was also a tool affording its members professional advancement within the psychological discipline. Within a year after its founding, SPSSI was a force to be reckoned with, at least within the world of psychology, as one of every six American Psychological Association members was also a member of SPSSI. Soon SPSSI began publishing a "SPSSI Bulletin" as part of the *Journal of Social Psychology* and produced *Industrial Conflict,* a book on labor conflict. World War II gave many of these social psychologists unprecedented opportunities to work closely with the government; they were involved in strategic bombing surveying, propaganda analysis, psychological warfare, studies of civilian and enemy morale, and surveys of public opinion.[7] In all of these efforts, psychologists were convinced that their special expertise was necessary to direct governmental power in the most efficient manner. War work led social psychologists to reconceptualize their mission as social engineers in two important ways.

First, social scientists began to focus on the government as an agent of change in society. In the 1930s many SPSSI members had considered themselves "radicals" and "outsiders" and believed that society was best remade by the people, or the workers, rather than by a powerful elite represented by government or state action. After World War II, however, many SPSSI members began to emphasize the role of government, and of legal change, in altering society. Their experiences in World War II linked social psychologists to government to such an extent that even dedicated leftists began writing about how the government could undertake social engineering for the benefit of the people.[8]

Second, after the war social scientists began to focus intently on race relations and race prejudice. A common theme in much of the research done on race prejudice during the war was that social science needed not only to understand race prejudice but to work to eliminate that prejudice. Hitler's rise to power, the struggle against Nazi ideology, and the perceived need to unify the nation behind the war effort transformed the study of prejudice into a struggle against totalitarianism. This viewpoint was epitomized by Gunnar Myrdal's *An American Dilemma,* which laid the blueprint for two decades of social engineering focused on the elimination of race prejudice. At the close of World War II two social psychologists who would soon be active within the AJC program, Ronald Lippitt and Marian Radke, declared that "the need for an understanding of the dynamics of prejudice has no equivalent in importance in the social sciences. In no other aspect of interpersonal and intergroup relationships is there a more urgent need for social sciences to 'get out and do something.' "[9] For Lippitt and Radke, as for many others, race prejudice was the chief

[6] Lorenz Finison, "Unemployment, Politics, and the History of Organized Psychology," *American Psychologist,* 1976, *31*:747–755; and Finison, "Unemployment, Politics, and the History of Organized Psychology, II: The Psychologists League, the WPA, and the National Health Program," *ibid.,* 1978, *33*:471–477.

[7] On SPSSI see Benjamin Harris, "Reviewing Fifty Years of the Psychology of Social Issues," *Journal of Social Issues,* 1986, *42*:1–20. On the membership see Finison, "Unemployment, Politics, and the History of Organized Psychology," p. 753. SPSSI's publishing ventures are discussed in Lorenz Finison, "The Psychological Insurgency: 1936–1945," *J. Soc. Issues,* 1986, *42*:21–34, on pp. 28–29. On the war work of social psychologists see James Capshew, *Psychology on the March* (Cambridge: Cambridge Univ. Press, 1998); and Blair T. Johnson and Diana R. Nichols, "Social Psychologists' Expertise in the Public Interest: Civilian Morale Research during World War II," *J. Soc. Issues,* 1998, *54*:53–78.

[8] Finison, "Psychological Insurgency," pp. 31–33.

[9] Ronald Lippitt and Marian Radke, "New Trends in the Investigation of Prejudice," *Annals of the American*

threat to the democratic way of life. Such a threat demanded more than dispassionate study; it demanded eradication. This understanding contrasted sharply with prewar views that saw race prejudice as irrational but seldom portrayed it as a danger to society.

The change in how social scientists viewed race prejudice coincided with a change in Jewish leadership. After World War II a host of middle-class American Jews of Eastern European descent with socialist-labor backgrounds began to take over Jewish leadership roles. In contrast to the older generation of leaders, these men were professional civil rights workers who were not interested in placating the WASP elite but were eager and willing to join public battle for Jewish rights.[10]

The new militancy in the Jewish leadership was owed in part to the horrors of Nazi Germany. The Holocaust united the American Jewish community against anti-Semitism with the goal of eliminating it forever. New funds poured into the coffers of Jewish organizations dedicated to fighting anti-Semitism, including the decades-old American Jewish Congress, the American Jewish Committee, and the Anti-Defamation League of B'nai B'rith. Unlike other organizations, however, the American Jewish Congress was not willing to rely on moral pleading and the process of education to eliminate anti-Semitism. It preferred a more direct approach through litigation and lobbying for legal change.[11]

The AJC's belief that legal changes were necessary to eliminate prejudice dovetailed with the interests of activist social scientists who wanted to use governmental power to create a more just and fair society. The next section examines how lawyers and social scientists came together at the AJC.

TWO BLUEPRINTS FOR MERGING SCIENCE AND THE LAW

Founded in 1918, the American Jewish Congress was one of the more militant and confrontational Jewish organizations. From the beginning—and in contrast to the older and more staid American Jewish Committee—the American Jewish Congress protested, confronted, and publicized anti-Semitism. After World War II, when the full horror of the Nazi regime became public, and fearing rampant anti-Semitism in the United States, the AJC redoubled its efforts. Drawing on the insights of two innovative émigré scholars, Kurt Lewin and Alexander Pekelis, it launched a "comprehensive program of legal, legislative and social action which would protect and safeguard the rights of Americans . . . by outlawing every form of discrimination on grounds of race, creed, color or national origin."[12] At the heart of the program was the belief that education and moral exhortation against

Academy of Political and Social Science, 1946, *244*:167–176, on p. 167. On the transformation to a struggle against totalitarianism see Herman, *Romance of American Psychology* (cit. n. 3), pp. 174–207. For Myrdal's book see Gunnar Myrdal, *An American Dilemma* (New York: Harper, 1944); on the immense impact of Myrdal's work in creating a "liberal orthodoxy" concerning race relations see Walter Jackson, *Gunnar Myrdal and America's Conscience* (Chapel Hill: Univ. North Carolina Press, 1990).

[10] On this change in Jewish leadership generally see Lenora E. Berson, *The Negroes and the Jews* (New York: Random House, 1971), pp. 98–107.

[11] Edward S. Shapiro, *A Time for Healing: American Jewry since World War II* (Baltimore: Johns Hopkins Univ. Press, 1992), pp. 16–17; and Stuart Svonkin, *Jews against Prejudice: American Jews and the Fight for Civil Liberties* (New York: Columbia Univ. Press, 1997), pp. 79–112.

[12] David Petegorsky, "Report of the Executive Director to the Biennial National Convention of the American Jewish Congress," 31 Mar.–5 Apr. 1948, p. 10, American Jewish Congress Papers, Box 19, American Jewish Historical Society, Waltham, Massachusetts (hereafter cited as **AJC Papers**). Accounts of the formation of the AJC can be found in Morris Frommer, "The American Jewish Congress: A History, 1914–1950" (Ph.D. diss., Ohio State Univ., 1978); and Henry L. Feingold, *A Time for Searching: Entering the Mainstream, 1920–1945* (Baltimore: Johns Hopkins Univ. Press, 1992).

prejudice would always fail if official discrimination continued. Official discrimination—whether in the form of quotas determining the number of Jews allowed into medical school or the more blatant segregation statutes of the South—would have to be eliminated to eliminate prejudice. Only then could education to combat prejudice have some hope of success. In short: attacking discrimination was the key to attacking prejudice.

To attack official discrimination the AJC created two new divisions, both concerned with using both the law and social science. The Commission on Community Interrelations (CCI) hoped to translate social science research into social action in order to lead a "scientific attack on anti-Semitism and other minority problems in the United States."[13] CCI recognized that the law could play a valuable role in its work. The Commission on Law and Social Action (CLSA) was to do for Jewish Americans what the NAACP had been doing for African Americans—litigate to protect their rights. CLSA attorneys viewed social science as a valuable resource for the creation of new laws against discrimination and segregation.

Commission on Community Interrelations

The Commission on Community Interrelations was the brainchild of one of the most influential social psychologists of the time, Kurt Lewin. Educated at the University of Berlin, Lewin made several fundamental contributions to social psychology in Europe, concentrating on worker education and job satisfaction. In 1934 he fled Europe for the United States. Finding himself in a strange new country, Lewin turned his attention to issues surrounding group identification, prejudice, aggression, and Jewish identity. After two years at Cornell University, he spent 1935–1945 at the Child Welfare Research Station at the University of Iowa.[14]

In 1944 some of Lewin's publications came to the attention of AJC president Rabbi Stephen Wise. The AJC had earmarked $1 million for the creation of a research center for the study of intergroup relations, and Wise thought Lewin should lead it. Lewin saw an opportunity to carry out scientifically informed social engineering. He held that social engineering was parallel to industrial engineering: "as industrial plants have found out that physical research pays, social organization will soon find out that social research pays."[15] The AJC initiative seemed tailor-made for Lewin's vision of a social engineering organization, and he quickly agreed to undertake the creation of the new commission.

In July 1944 Lewin submitted a plan for the new organization, "Memorandum for the Commission on Anti-Semitism," to the American Jewish Congress. It outlined a vision of lawyers and social scientists working side by side to fight for democracy. Lewin argued that many existing programs were based on an insufficient understanding of the causes of and cures for anti-Semitism. His program would be based on two criteria. First, "it has to be objective, i.e. it has to uncover the essential facts in an unbiased scientific manner."

[13] "New Scientific Attack on Anti-Semitism Launched by American Jewish Congress," 28 June 1945 (press release), Alfred J. Marrow Papers, Box 20, Folder: "CCI, Pamphlets, Public Relations," Archive for the History of American Psychology, Akron, Ohio (hereafter cited as **Marrow Papers**).

[14] Mitchell Ash, "Cultural Contexts and Scientific Change in Psychology: Kurt Lewin in Iowa," *Amer. Psychol.*, 1992, *47*:198–207.

[15] Kurt Lewin, "The Place of the Commission on Community Interrelations within the Work of Jewish Organizations," address to the National Community Relations Advisory Council, 16 Nov. 1944, Kurt Lewin Papers, Folder 19: "American Jewish Congress, CCI," Archive for the History of American Psychology (hereafter cited as **Lewin Papers**). See also Alfred J. Marrow, *The Practical Theorist: The Life and Work of Kurt Lewin* (New York: Basic, 1969), pp. 161–164.

Second, the program "has to be practical, i.e. it should lead to coordinated significant actions." Lewin envisioned an organization with two main divisions: a research division that would consist of community sociologists, opinion analysts, group psychologists, individual psychologists, and statisticians; and an "operational division" that would combine two existing AJC commissions that included numerous lawyers: the Commission on Economic Discrimination and the Commission on Law and Legislation. In the new commission, research workers and operational personnel would be equal participants. As Lewin wrote, "The need for action determines the content of the research; scientific requirements determine its technique." Lawyers and other legal personnel, he believed, should work hand in hand with social science researchers. Fundamental to CCI's views on prejudice was the notion that "if we can break down the social segregation and discrimination which defines a racial or religious group as a sanctioned target for prejudice and scape-goating, time will take care of the individual prejudice."[16]

CCI's focus on official discrimination rather than the attitude of prejudice is reflected in one of its first projects: a general survey of the state of knowledge about prejudice and discrimination undertaken by Goodwin Watson. As one of the founding members of SPSSI, Watson had long been concerned with using psychology to create a more democratic and just world. Watson had made one of the first attempts to measure race prejudice in his Ph.D. dissertation at Columbia's Teacher's College in 1925. His survey for CCI was part of his struggle to become, as he put it, a "social engineer" in the postwar world and also reflects the postwar emphasis on government action in the creation of a just society.[17]

In his report for CCI Watson surveyed different techniques for fighting prejudice, including education, moral exhortation, and other methods. He concluded that "*it is more constructive to attack segregation than it is to attack prejudice.*" Segregation was amenable to public control, unlike people's private prejudices. Moreover, education would fail to end prejudice unless official "caste barriers" were torn down because "habits built around those barriers will silently undo anything we accomplish."[18]

CCI believed that law could do what education could not: break through the irrational attitude of race prejudice. Thus the law became an essential component in fighting discrimination and prejudice. Stuart W. Cook, who would later play a pivotal role in the *Brown* litigation, came to CCI in 1948 to serve as co-director with Lewin. A University of Minnesota Ph.D. (1938), Cook had long been interested in the effects of intergroup contact on attitude change. Soon after joining the staff of CCI, he set forth the commission's philosophy regarding the relationship between education and legal change:

> Educational programs aimed at reducing discrimination are likely to make slow headway when there are no anti-discrimination laws. But passage of a law changes the atmosphere in which education is carried on. Once legislation exists, an educational program can draw support from the law-abiding tradition of most citizens. After a law is passed, an educational campaign designed to explain the law's purpose and to encourage compliance with it is no longer an

[16] Kurt Lewin, "Memorandum on Program for the Commission on Anti-Semitism of the American Jewish Congress," pp. 1, 9, Lewin Papers, Box M946, Folder 22: "CCI"; and "Some Basic Issues," 12 Dec. 1944, Marrow Papers, Box M1938, Folder 21: "CCI Papers."

[17] Watson laid out his blueprint for psychology's role in the postwar world in Goodwin Watson, "How Social Engineers Came to Be," *Journal of Social Psychology,* 1945, *21*:135–141. For details of Watson's life and work see Ian A. M. Nicholson, "The Politics of Scientific Social Reform, 1936–1960: Goodwin Watson and the Society for the Psychological Study of Social Issues," *J. Hist. Behav. Sci.,* 1997, *33*:39–60; and Nicholson, " 'The Approved Bureaucratic Torpor': Goodwin Watson, Critical Psychology, and the Dilemmas of Expertise, 1930–1945," *J. Soc. Issues,* 1998, *54*:29–52.

[18] Goodwin Watson, *Action for Unity* (New York: Harper, 1947), p. 64.

inefficient technique but is in a position to produce a great return for a relatively small investment.[19]

Cook was presenting the key idea for these researchers: that the law is not an alternative to education against group prejudice but is best used in conjunction with such education.

Isidor Chein, CCI's research director, made a similar point. Chein, like Kenneth B. Clark, had received his Ph.D. in psychology (1939) from Columbia, one of the centers of activist social psychology in the 1930s. Before joining CCI Chein worked with one of the first organizations in New York City dedicated to the study of intergroup relations, the Committee on Unity established in the wake of the 1943 Harlem riots. In 1946 he wrote that legal changes "constitute virtually the only means of breaking into the vicious circle [of prejudice and discrimination]; legislation, in all areas to which it may be applied, against discriminatory practices in employment, education, housing, and so on." Hence, Chein was quite receptive to the AJC's program of direct legal action against discrimination. He joined CCI as a research associate and became director of research in 1950. At CCI Chein continued to maintain that "education and the law . . . go hand in hand; each approach helps to bring out the best potentialities of the other. Either, alone, is apt to be fruitless."[20]

The particular method Lewin developed for CCI was "action research." Lewin envisioned social scientists researching a problem and "actionists" implementing a reform program. "Actionists" included not just lawyers but also community workers, local leaders, religious personnel, and others who lived and worked in the community under study. From the beginning, CCI personnel anticipated difficulties with these two groups of people working side by side. Ronald Lippitt, a student of Lewin's from the Iowa days, made the issues clear in an internal memorandum. Social scientists, he wrote, were "likely to feel that action personnel have no appreciation of the problems and requirements of data collection" and were "likely to be put in situations where [their] previous sources of satisfaction—recognition for competence in technical publication, theorizing, etc., are not relevant." By contrast, action personnel were "likely to feel that social research isn't practical enough yet to make improvements on the great mass of experience which has led to certain more or less institutionalized practices" and that "data collection procedures are an unwarranted nuisance when they begin to call for certain modifications of action plans."[21] As we will see, Lippitt's argument would prove prescient, for while CCI pursued valuable research that would later become important in the *Brown* litigation, social scientists and lawyers would coexist only uneasily within the institutionalized context of the AJC.

[19] Stuart Cook to Will Maslow, 5 Mar. 1947, Marrow Papers, Box M1938, Folder 14: "1946–1947, CCI Correspondence."

[20] Isidor Chein, "Some Considerations in Combating Intergroup Prejudice," *Journal of Educational Sociology,* 1946, *19*:412–419, on p. 416; and Isidor Chein to Garner Roney, 9 Feb. 1949, Marrow Papers, Box M1938, Folder 15: "1948–1949, CCI Correspondence." A brief biography of Chein can be found in "Awards for Distinguished Contributions to Psychology in the Public Interest: 1980," *Amer. Psychol.,* 1981, *36*:67–70. On the Committee on Unity see Gerald Benjamin, *Race Relations and the New York City Commission on Human Rights* (Ithaca, N.Y.: Cornell Univ. Press, 1972), pp. 38–70. On activist social psychology in the 1930s see Katherine Pandora, *Rebels within the Ranks: Psychologists' Critique of Scientific Authority and Democratic Realities in New Deal America* (Cambridge: Cambridge Univ. Press, 1997).

[21] Ronald Lippitt, "Action Research—Idea and Method," 17 July 1945, Marrow Papers, Box M1938, Folder 13: "1943–1945, CCI Correspondence." For a more complete description of CCI's program of "action research" see Frances Cherry and Catherine Borshuk, "Social Action Research and the Commission on Community Interrelations," *J. Soc. Issues,* 1998, *54*:119–142.

In 1945, however, despite reservations, most CCI social scientists felt that the problems linked with fusing research and action could and would be worked out. At a February 1945 meeting, Stephen Wise announced that the AJC would fund Lewin's "Commission of Community Interrelations" for five years and that its purpose was to investigate scientifically the causes and cures of anti-Semitism and race prejudice. The expansion of CCI's mission to include forms of race prejudice beyond anti-Semitism was in keeping with the larger agenda of the AJC. Shad Polier, an attorney and Wise's son-in-law, put the matter this way:

> We make no claim that our activities are compounded wholly of altruism, or that they are entirely divorced from our more partisan and proximate objective of enhancing the security of Jews in America. On the contrary, since we believe Jewish interests to be inseparable from the interests of justice we have always contended that for the Jewish community there is an unfailing advantage to be derived from performance of the principled act.[22]

The focus on all forms of discrimination rather than on anti-Semitism alone was mirrored in other branches of the AJC. As the work of CCI was getting under way, the AJC was also forming the Commission on Law and Social Action; it too began with great plans for merging social scientific research and legal action.

Commission on Law and Social Action

Although Lewin had hoped that CCI would incorporate the existing Commission on Law and Legislation and Commission on Economic Discrimination, in November the AJC instead merged them into a new Commission on Law and Social Action under the leadership of Will Maslow.

Maslow had come to the United States when he was three years old. He had studied economics at Cornell but switched to law at Columbia when he discovered that universities hesitated to hire Jewish professors. Unfortunately, most large law firms were equally hesitant to hire Jewish attorneys. The newly formed bureaucracies of the New Deal, however, had no such strictures, and Maslow worked as a trial attorney for the National Labor Relations Board and later supervised the Fair Employment Practices Commission.[23]

In 1945, when Maslow was hired to head the newly formed CLSA, he found Alexander Pekelis waiting for him. Pekelis had been the director of the Commission on Law and Legislation until it was collapsed into CLSA. Born in Russia, Pekelis spent most of his adult life in Italy and France as a professor of jurisprudence. He had fled to the United States in 1941, just before the Nazis entered France. He took a position on the Graduate Faculty of the New School for Social Research and in 1942 entered the Columbia School of Law, where he became editor of the *Columbia Law Review.* Because he was not yet a member of the bar, Pekelis could not serve as a staff attorney for CLSA, but Maslow created a special position for him—"Chief Consultant."[24] Pekelis's first job for CLSA was

[22] H. Epstein, " 'Forward,' Accent on Action: A New Approach to Minority Group Problems in America," 1945, Stuart W. Cook Papers, Box M2337, Folder 2: "Commission on Community Interrelations," Archive for the History of American Psychology (hereafter cited as **Cook Papers**); and Shad Polier, "Why Jews Must Fight for Minorities," 4 Nov. 1949, pp. 2–3, Shad Polier Papers, Box 16, Folder: "Polier/AJC on Civil Liberties," American Jewish Historical Society.

[23] Murray Friedman, *What Went Wrong? The Creation and Collapse of the Black–Jewish Alliance* (New York: Free Press, 1995), pp. 133–134.

[24] M. R. Konvitz, "Introduction," in *Law and Social Action: Selected Essays of Alexander H. Pekelis,* ed. Konvitz (Ithaca, N.Y.: Cornell Univ. Press, 1950), pp. v–vii, on p. vii.

to prepare a memorandum that would set out the commission's organization and philosophy, just as Kurt Lewin was doing for CCI.

Pekelis based his understanding of the law on an older American legal tradition, "legal realism." There are as many definitions of legal realism as there were legal realists. Generally the term referred to "that body of legal thought produced for the most part by law professors at Columbia and Yale Law Schools during the 1920s and 1930s."[25] These law professors were following the lead of progressive theorists who rejected nineteenth-century legal reasoning.

At the end of the nineteenth century American jurisprudence was dominated by the belief that judges discovered, rather than made, law. According to this view, law was a process of deductive reasoning that took legal rules and case precedents as its major premises and the facts of a particular case as its minor premises. The judge could make a proper ruling in each case by following a formal procedure that inevitably led to the correct decision. The case method of legal education, pioneered by Christopher C. Langdell at Harvard, perpetuated this formalistic system by teaching that legal precedent, combined with proper reasoning, would lead to uniform law.[26]

In the twentieth century the formal legal system came under increasing attack, as legal theorists and practicing lawyers began to demand that the law pay more attention to how the world actually worked. The Columbia law professor Karl Llewellyn was the first to attempt a definition of the new jurisprudence in a 1930 law review article, "A Realistic Jurisprudence: The Next Step." Llewellyn called for the findings of the behavioral sciences to be used by lawyers and judges to bring jurisprudence more into line with the way the world truly operated. The legal realists, as they came to be known, attacked two major concepts. First, they denied that judges "discovered" law. Rather, argued the realists, judges *made* law, and the law either helped or hindered certain social policies. Second, the realists argued for new sources of information to replace the abstractions of nineteenth-century jurisprudence. New sources of information, including social science, began flooding the legal system with facts about how the law operated in society.[27]

At least some of those that could be called legal realists were interested in using social science in legal proceedings. Progressive lawyers such as Charles E. Clark and William O. Douglas attempted to use social science research to further their reforms. Unfortunately, social research initiated for use in litigation could not be completed in time to be useful in the courtroom. In addition, if the results of the research did not suit the needs of the lawyers it was ignored.[28] The same issues would soon haunt CCI researchers and CLSA attorneys in their attempt to integrate social science and the law.

Pekelis had entered Columbia as a mature scholar; law school for him was almost a formality, a credential he needed to gain access to the American bar. The central notions of realism—that judges make rather than discover law and that information about the "real" world should play a central role in the creation of law—clearly resonated with him. "Similar theories had been developed in Europe long before legal realism became popular

[25] Morton J. Horowitz, *The Transformation of American Law, 1870–1960: The Crisis of Legal Orthodoxy* (New York: Oxford Univ. Press, 1992), p. 169.

[26] Edward A. Purcell, *The Crisis of Democratic Theory: Scientific Naturalism and the Problem of Value* (Lexington: Univ. Kentucky Press, 1973), pp. 74–75.

[27] Karl Llewellyn, "A Realistic Jurisprudence: The Next Step," *Columbia Law Review,* 1930, *30*:431–465. A detailed treatment of how social science made inroads into the legal culture is John W. Johnson, *American Legal Culture, 1908–1940* (Westport, Conn.: Greenwood, 1981).

[28] See, e.g., John Henry Schlegel, *American Legal Realism and Empirical Social Science* (Chapel Hill: Univ. North Carolina Press, 1995).

here," he wrote in 1943.[29] Pekelis wrote the central doctrines of legal realism into his memorandum for the AJC leadership.

At the heart of Pekelis's proposal was the unique nature of anti-Semitism in the United States. In contrast to European anti-Semitism, which was a function of official government action, American anti-Semitism "comes from the forces of society itself. Anti-Semitism here is private or communal, not public or governmental in nature." These social forces—in the form of unofficial quotas on Jews in professional schools, for instance—would be much harder to detect than anything so blatant as a law. Hence, to combat American anti-Semitism, Jews needed the sorts of data that could only be provided by social science. "Contemporary experience," Pekelis wrote, "has shown that no political, social, administrative, or legal action can be conducted efficiently unless means are found to narrow the gap between those who devote themselves to the study of social reality and those who, in legislative communities and courts, shape the law of the community. . . . Law without a knowledge of society is blind; sociology without a knowledge of law, powerless." Pekelis wanted to insure successful cooperation between social scientists and lawyers by having them work in the "same functional unit." In a line that Lewin might have written, Pekelis urged "a close-knit integration of projects. . . . The same type of integration must be achieved between social and legal action and between research and operational activities."[30]

Although Pekelis did not realize his dream of social scientists working within the legal department, his ideas resonated with the working attorneys of CLSA. Just as Pekelis predicted, CLSA attorneys found social science materials necessary because of the special problems of American anti-Semitism. Maslow and his colleagues recognized that social science data would be necessary for much of their litigation to be successful. A February 1946 memorandum argued that "legal skills, social science training and the capacity for social action must be joined if specific tasks are to be defined intelligently and pursued successfully." In 1949, contrasting the sorts of problems confronted by CLSA, which wanted to eliminate anti-Semitism, with those confronted by the NAACP, which wanted to eliminate discrimination against African Americans, a "Note" in the *Yale Law Journal* observed that "discrimination against Jews in the United States is usually non-governmental, non-violent, and extremely subtle" and that "an organization concerned with Jewish problems must employ sociological research to expose the more subtle discrimination to which Jews are subjected."[31]

Pekelis, like Lewin, had set out a blueprint for collaboration between social scientists and lawyers in the fight against discrimination. Both argued not only that merging the law and social science would be advantageous in the battle against discrimination but that it was vital for lawyers and social scientists actually to plan and carry out their projects together. What they envisioned went beyond the citation of a few research results in a legal brief, involving lawyers' partnership with social scientists in the creation of that brief. Moreover, social scientists would not merely study the effects of particular legal changes but would be active agents in creating those changes. Looking beyond mere "cooperation,"

[29] Alexander H. Pekelis, "The Case for a Jurisprudence of Welfare" (1943), in *Law and Social Action*, ed. Konvitz (cit. n. 24), pp. 1–41, on p. 3.

[30] Alexander H. Pekelis, "Full Equality in a Free Society: A Program for Jewish Action" (1945), in *Law and Social Action,* ed. Konvitz, pp. 218–259, on pp. 256–257.

[31] CLSA Memorandum, Feb. 1946, AJC Papers, Box 33, Folder: "CLSA Memorandum"; and "Private Attorneys-General: Group Action in the Fight for Civil Liberties" (note), *Yale Law Journal,* 1949, *58*:574–598, on pp. 589, 594.

Lewin and Pekelis wanted lawyers and social scientists to work in an almost symbiotic relationship. In reality, however, the lawyers and social scientists would have trouble working as closely as Lewin and Pekelis had hoped. While social scientists did try to design research that would be of use in the legal arena, and while lawyers designed arguments that required social scientific data, the close collaboration Lewin and Pekelis anticipated never emerged.

GENERATING RESEARCH FOR THE LEGAL ARENA

The research sponsored by CCI demonstrated social scientists' belief that education and the law could be partners. CCI researchers attempted to show that a change in the actual physical circumstances of society could precede a change in attitudes. In other words, the law did not have to wait for a change in social climate to be effective; it could itself change attitudes. Two lines of research illustrate how CCI attempted to merge social science with an attack on discrimination: research conducted on the separation of attitudes from behavior and research on the effects of interracial contact.

On the separation of attitudes from behavior, CCI's social scientists built on the work of Richard T. Lapiere. In the 1930s Lapiere traveled through the United States with a Chinese couple. They stayed in hotels or auto camps and ate in a total of 184 restaurants. Except in one hotel, they were served without incident. Six months later, Lapiere sent a questionnaire to these establishments asking if they served "members of the Chinese race." Over 90 percent of those responding indicated they would not, despite the fact that they had done just that six months earlier.[32]

At CCI Bernard Kutner successfully duplicated Lapiere's research when he sent two white women into New York restaurants. They were later joined by an African American woman, who was seated without incident. When Kutner inquired as to the policies of the restaurants, however, he was informed that they did not serve African Americans. Gerhardt Saenger conducted a similar study on the integration of sales personnel, discovering that customers who, moments earlier, had been assisted by African American salesclerks at a large New York department store would tell a pollster that they would never trade at a store that employed African Americans.[33]

Of more lasting influence than the research on attitudes and behavior was that on interracial contact. CCI social scientists envisioned a vicious circle of discrimination and prejudice: because discrimination seemed to teach people that minority groups were inferior, it led to prejudicial attitudes; these attitudes, in turn, led to the erection of more discriminatory barriers preventing minorities from fully entering society. CCI saw interracial contact as the point at which the circle of discrimination and prejudice could be broken.

As a program of scientific study, CCI researchers attempted to discover the circumstances under which contact between individuals of different races tended to decrease prejudice. The "contact hypothesis"—that contact between different races decreased prejudice—was an open question in the late 1940s. While some unsystematic evidence indicated that interracial contact decreased prejudice, there was no firm consensus, and, indeed, some evidence suggested that contact increased prejudice. In a 1948 survey of various

[32] Richard T. Lapiere, "Attitudes vs. Actions," *Social Forces,* 1934, *13*:230–237.

[33] Bernard Kutner, C. Wilkins, and P. R. Yarrow, "Verbal Attitudes and Overt Behavior Involving Racial Prejudice," *Journal of Abnormal and Social Psychology,* 1952, *47*:649–652; and Gerhardt Saenger and E. Golbert, "Customer Reactions to the Integration of Negro Sales Personnel," *International Journal of Opinion and Attitude Research,* 1950, *4*:57–76.

programs designed to decrease prejudice, the Cornell social scientist Robin M. Williams called for further research into interracial contact. Noting that "establishment of the effects of segregation per se will be an extraordinarily difficult task," Williams argued that social scientists should study the effects of segregated and nonsegregated situations in order to answer questions such as, "Where is friction greatest? Where are the areas of high and low intensity and incidence of verbal prejudice? How do stable areas of intermingling compare with shifting areas and 'invasion' points?"[34]

Soon CCI researchers were attempting to answer the sorts of questions posed by Williams. In 1948 CCI research director Stuart Cook wrote that "insofar as successful action against discriminatory practices brings about a decrease in segregation this will mean increased contact between persons from different backgrounds that will, under favorable circumstances, create a reduction in prejudice. Such a reduction in prejudice should, of course, have the consequences of further reduction of discriminatory practices and hostile behavior."[35]

As soon as he arrived at CCI in 1947, Cook was testifying before city commissions that contact between the races served to decrease racial tensions. Before the City Commission of Jersey City Cook stated that "joint occupancy of the same housing community by Negroes and whites has consistently worked out: Initial, mistaken ideas are soon corrected through day-to-day contact and, in many places, members of the different races have come to share identical responsibilities in a completely democratic way."[36]

Building on earlier work on the desegregation of the armed forces, CCI's researchers concentrated on interracial public housing and employment. These situations grew out of the migration of African Americans into New York and the newly desegregated employment and public housing opportunities for them. CCI seized the research opportunities offered by these real-life situations to study interracial contact in a series of what can best be described as field studies.

In one of the first studies of interracial housing, for instance, two CCI staffers conducted interviews with families living in two desegregated and two segregated housing projects. The researchers found that white prejudice was much higher in the segregated projects. They posited that the contact possible in integrated neighborhoods gave individuals the opportunity to realize that their prejudices had no basis in reality. By contrast, in the segregated neighborhoods no opportunity existed for prejudiced individuals to overcome their stereotypes about African Americans. A parallel set of studies explored the effects of interracial workplaces. John Harding and Russell Hogrefe polled the white workers on a newly integrated sales floor. They found that while the basic attitude of the white workers toward their African American co-workers may not have changed significantly, they could nonetheless work peacefully side by side. What CCI and other researchers on interracial contact attempted to discover were the specific conditions under which interracial contact would decrease prejudice. In the early 1950s the Harvard psychologist Gordon Allport summarized what social science had learned about conditions necessary for contact to reduce prejudice: "first, that the contact must be one of equal status; and second, that the members must have objective interests in common." In order to achieve these conditions,

[34] Robin M. Williams, Jr., *The Reduction of Intergroup Tensions* (New York: Social Science Research Council, 1948), p. 91.

[35] Stuart W. Cook, "The Program of CCI," 17 Dec. 1948, Marrow Papers, Box M1938, Folder 15: "1948–1949, CCI Correspondence."

[36] Stuart W. Cook, "Some Psychological and Sociological Considerations Related to Interracial Housing," Nov. 1947, p. 2, AJC Papers, Box 21, Folder: "CCI, 5/20/49."

Allport argued, "artificial segregation should be abolished. Until it is abolished equal status contacts cannot take place. And until they take place cooperative projects of joint concern cannot arise. And until this condition is fulfilled we may not expect widespread resolution of intergroup tensions. Hence, nearly all the investigators agree that the attack on segregation must continue."[37] This would be a key point for the social scientists when they became involved in the litigation campaign: that the abolition of segregation was a *necessary* rather than a *sufficient* step toward bettering race relations. In terms of the "contact hypothesis," for example, social scientists were arguing not that all that was required to reduce prejudice was to eliminate legal segregation but, rather, that *nothing* could be done to reduce prejudice until legal segregation was eliminated. The elimination of segregation was not the end of the journey, but the beginning.

The work on the contact hypothesis, like the work on the separation of attitudes and behavior, was designed to show that social change could be enacted through legislation or court order. What CCI social scientists were arguing was a reversal of William Graham Sumner's famous dictum, "Stateways cannot change folkways." On the contrary, the CCI social scientists insisted, stateways could indeed change folkways. Eliminating discrimination—for example, through a Fair Employment Practices Act or through unsegregated housing projects—would go a long way toward reducing prejudicial attitudes. Pointing to the body of research that CCI had amassed, the sociologist Arnold Rose, Gunnar Myrdal's collaborator on *An American Dilemma*, argued that segregation statutes had always had the effect of increasing prejudice and that the latest research demonstrated that the law could similarly decrease prejudice. Rose declared:

> It has been thus demonstrable for a long time that law and power could create or increase attitudes of prejudice, it should not be surprising from newly available evidence that law and power would also decrease prejudice. Yet the latter conclusion has been contrary to most experts' opinions for a long while. Perhaps the sociologists have misled us with their notions of "mores," "folkways," and the "inevitable" slowness of social change. . . . Now we know that law and authority can reduce prejudice.[38]

THE LEGAL USES OF SOCIAL SCIENCE

While CCI was building the scientific case for legal change, CLSA was taking action against discrimination, drafting model Fair Employment Practices statutes and campaigning heavily for a federal statute. It brought four test cases before the New York State Commission against Discrimination, sued Columbia University Medical School for its "unofficial" quota on Jewish applicants, filed six cases in three states against discrimination

[37] Morton Deutsch and Mary Evans Collins, "Intergroup Relations in Interracial Public Housing: Occupancy Patterns and Racial Attitudes," *Journal of Housing,* 1950, *7*:127–129; Deutsch and Collins, *Interracial Housing: A Psychological Evaluation of a Social Experiment* (Minneapolis: Univ. Minnesota Press, 1951); Daniel M. Wilner, R. P. Walkley, and S. W. Cook, "Residential Proximity and Intergroup Relations in Public Housing Projects," *J. Soc. Issues,* 1952, *8*:45–69; Wilner, Walkley, and Cook, *Human Relations in Interracial Housing: A Study of the Contact Hypothesis* (Minneapolis: Univ. Minnesota Press, 1955); John Harding and Russell Hogrefe, "Attitudes of White Department Store Employees toward Negro Co-workers," *J. Soc. Issues,* 1952, *8*:18–28; and Gordon Allport, *The Resolution of Intergroup Tensions* (New York: National Conference of Christians and Jews, 1952), pp. 22–23.

[38] Arnold Rose, "The Influence of Legislation on Prejudice" (1949), rpt. in *Race Prejudice and Discrimination: Readings in Intergroup Relations in the United States,* ed. Rose (New York: Knopf, 1951), pp. 545–555, on p. 554.

in housing, and fought the granting of a radio operator's license to the *New York Daily News* on the grounds that it was prejudiced against Jews and African Americans.[39]

Social science played an important role in many of these cases. In the cases before the New York State Commission against Discrimination, for example, CLSA established statistical data as proof of discrimination. In the case against the *Daily News,* CLSA established content analysis as a technique with probative value. However, while these cases made use of social science evidence, that evidence came from research conducted by CLSA itself rather than by CCI. For example, in the *Daily News* case, Pekelis and his staff performed the content analysis themselves rather than having CCI undertake the task.[40] There would be one case, however, that would lead to a close relationship between the arguments generated by the CLSA attorneys and research conducted by CCI researchers.

Westminster v. Mendez

A case against segregated education provided the first opportunity for a true collaboration between CCI and CLSA. The case was *Westminster v. Mendez,* a California school segregation case involving Mexican Americans. Segregation of Mexican American schoolchildren had been a common practice in California since the 1920s, when immigration made Mexicans the state's largest minority. Although the practice was not sanctioned by California law, the white or "Anglo" populations of many communities insisted that their local school boards create separate schools for Mexican American children, ostensibly because of language problems.

On 2 March 1945 five Mexican American parents filed a suit in Federal District Court to enjoin the segregation of their children on the grounds that such segregation constituted a violation of Fifth and Fourteenth Amendment equal protection guarantees. The school districts argued that such segregation was not based on race but served sound educational purposes involving bilingual education. Moreover, they argued that control of the schools was a local matter and hence outside the jurisdiction of the federal courts. Finally, they claimed that the separate facilities were in any case equal and therefore constitutional under the "separate but equal" doctrine of *Plessy v. Ferguson,* the 1896 U.S. Supreme Court case that had entrenched segregated facilities in constitutional law.

The Federal District Court's ruling came nearly a year later. Because California had no segregation statutes, the court ruled that there was indeed a violation of the equal protection clauses of the Constitution. The court found that, in the absence of such segregation statutes, the "separate but equal" doctrine did not apply. Finally, it found no sound educational reason for the segregation of Spanish-speaking pupils; in fact, the court argued, assimilation would proceed more rapidly in integrated schools. The school district quickly appealed the case to the Ninth Circuit Court of Appeals.[41] During this appeal, the case came to the attention of the American Jewish Congress.

The *Westminster* case, though not originated by CLSA, provided a unique opportunity. The case was "low risk"—that is, very few AJC resources were expended in the litigation

[39] Petegorsky, "Report of the Executive Director" (cit. n. 12).

[40] "Content Analysis—A New Evidentiary Technique" (note), *University of Chicago Law Review,* 1948, *15*:910–924; and *Memorandum in the Nature of Proposed Findings Submitted at the Direction of the Federal Communications Commission by the American Jewish Congress,* AJC Papers, Box 189, Folder: "Brief, Nov. 12, 1946, Federal Communications Commission."

[41] This account of the case is taken from Charles Wollenberg, *All Deliberate Speed: Segregation and Exclusion in California Schools, 1855–1975* (Berkeley: Univ. California Press, 1976), pp. 110–121.

equal" doctrine—one of the first chances for CLSA to present its views on racial segregation and discrimination.

Although not a member of the bar, Alexander Pekelis wrote the CLSA brief for the *Westminster* case. In his introduction he noted that CLSA firmly agreed with the central point that the NAACP had made in its own *amicus* brief: "If facilities were really duplicated, financial ruin of the local bodies of the states would ensue. If financial disaster is to be avoided, the facilities granted to minorities are bound to be physically inferior." However, so as not to duplicate the argument made by the NAACP, Pekelis predicated his arguments on the assumption that facilities were "identical" rather than attempting to show that those provided to the minority students were inferior. To prove that even identical facilities were inherently unequal, he relied heavily on sociological and psychological data.[42]

As all attorneys arguing against segregation were obliged to do, Pekelis had to deal with the *Plessy* precedent. Pekelis's strategy was to accept the legal doctrine, propounded in *Plessy,* that separate but equal facilities were, in fact, constitutional. However, he analyzed the findings of the 1896 case as firmly anchored in "factual" rather than "legal" grounds. He noted that *Plessy* found segregated railroad cars constitutional because "it proceeded on the factual and sociological assumption that such segregation did '*not necessarily imply the inferiority of either race to the other.*' " Pekelis then argued that the legal "fiction" of *Plessy*—that segregation does not connote the inferiority of one race to another—was contradicted by nearly all social scientific knowledge available. Pekelis asked starkly: "Will any court today, in the light of the sociological and psychological findings made in the last fifty years, prove so lacking in candor and so blind to realities as to subscribe to the fiction of benevolent segregation on which *Plessy v. Ferguson* relies? That is the issue. Not the legal doctrine of *Plessy v. Ferguson* is in question but the factual fallacy on which it rests." To prove his point that *Plessy* was grounded in a "factual fallacy," Pekelis argued that

> whenever a group, considered "inferior" by the prevailing standards of a community, is segregated by official action from the socially dominant group, the very fact of official segregation, whether or not "equal" physical facilities are being furnished to both groups, is a humiliating and discriminatory denial of equality to the group considered "inferior" and a violation of the Constitution of the United States and of treaties duly entered into under its authority.

Pekelis first noted that all parties agreed that separate facilities had to be equal if they were to pass the constitutional test of *Plessy.* Second, he maintained that "equality" was defined not by "mere identity of physical facilities" but by "identity of substantial similarity of their *values.*" Pekelis then defined "values" by their "social significance and psychological context, or in short, on the community judgment attached to them." For example, Pekelis maintained that a probate court would not hold two physically identical houses equal if one were in a slum and the other in a fashionable neighborhood.[43]

Having established that the law recognized the reality of what he called "social inequality," Pekelis then showed how the segregation of a group previously deemed socially

[42] *Westminster School District v. Mendez,* Brief for the American Jewish Congress, *amicus curiae,* 1946, p. 3, Case Files, Ninth Circuit Court of Appeals, Record Group 276, Box 4464, Folder 11310, National Archives—Pacific Sierra Region, San Bruno, California (hereafter cited as **CLSA Brief**).

[43] CLSA Brief, pp. 21, 22, 4, 5. See also *Plessy v. Ferguson,* 163 U.S. 537 (1896).

inferior in and of itself constituted a legal inequality. The act of segregation, he insisted, was tantamount to an official declaration of the inferiority of the segregated group. Once the "legal inferiority" of segregation is adopted, it reinforces and intensifies the "social inequality," leading to a vicious circle of discrimination and prejudice.

Such problems were especially acute in segregated education, according to Pekelis: "The official imposition of a segregated pattern based on notions of inferiority and superiority produces its deepest and most lasting social and psychological evil results when applied to children." He argued:

> Since segregation reinforces group isolation and social distance it helps to create conditions in which unhealthy racial attitudes may flourish. By giving official sanction to group separation based upon the assumption of inferiority it helps to perpetuate racial prejudice and contributes to the degradation and humiliation of the minority child. The crippling psychological effects of such segregation are in essence a denial of equality of treatment. In this sense segregation is burdensome and oppressive and comes within the constitutional prohibition.[44]

Pekelis argued that the internal psychology of schoolchildren should be a constitutional issue, regardless of the effect of segregation on the larger society. This is the "damage" argument that would underpin the *Brown* decision eight years later. Pekelis supported the argument with a wide variety of social science materials on segregation and discrimination.

On 14 April 1947 the Ninth Circuit Court of Appeals ruled unanimously that the segregation of Mexican American schoolchildren in California schools violated the Fourteenth Amendment. The basis of the opinion was not the far-reaching legal case made by CLSA but a narrow legal ground: segregation was unconstitutional because California law had no provisions for the segregation of Mexican American schoolchildren. Governor Earl Warren made any further litigation moot when, on 14 June 1947, he signed a repeal of California's education segregation statutes that ended *de jure* segregation in the state.[45]

This case showed that the existing social science literature was not adequate to support Pekelis's argument. Given the dearth of psychological studies specifically on segregated education, Pekelis had relied on studies of general childhood development and general studies of segregation. Except for an article by Howard Hale Long, these studies did not directly address segregated education.[46] Neither did the social science literature isolate segregation as required by law as a separate variable from segregation that arose by custom. Finally, there was no social science that could distinguish the psychological effects of segregation in general from the psychological effects of being segregated to inferior facilities. CCI researchers could have undertaken research projects to fill these holes in the social science literature. Such projects, however, would have been tremendously difficult to design, and CLSA might have had to wait months or years for the results to be reported. In order to make a more authoritative pronouncement, CLSA turned to CCI's social scientists, who found a way to enroll scientific authority without undertaking additional studies.

Survey of Social Science Opinion

Isidor Chein, after consultation with CLSA, decided to poll a wide range of social scientists on the issue of the psychological damage of segregation.[47] The results of this survey,

[44] CLSA Brief, pp. 14–15.

[45] Wollenberg, *All Deliberate Speed* (cit. n. 41), p. 132.

[46] The article was Howard Hale Long, "Some Psychogenic Hazards of Segregated Education of Negroes," *Journal of Negro Education,* 1935, *4*:336–350.

[47] Tracy S. Kendler, "Contributions of the Psychologist to Constitutional Law," *Amer. Psychol.,* 1950, *10*:505–510.

designed explicitly for use in legal proceedings, would become one of the most cited articles of social science in subsequent briefs by CLSA and by the NAACP in cases against segregation.

The survey was mailed to 849 social scientists in May 1947. Respondents were asked if they agreed that "enforced segregation has detrimental psychological effects on members of racial and religious groups which are segregated, even if equal facilities were provided." A parallel series of questions asked about psychological effects on "groups which enforce the segregation." The respondents were also asked to state the basis of their opinions; they could choose four options: "My own research findings. Research findings of other social scientists. My own professional experience. Professional experiences of other social scientists which have been made available to me." The cover letter was explicit as to the motivations for the survey: "For the purpose of providing legislative bodies, courts and the general public with a consensus of responsible scientific opinion, we are asking social scientists to indicate their position in this issue."[48]

Chein presented the results at the 1948 Eastern Psychological Association meeting. He reported that 90 percent of the respondents believed segregation psychologically damaging to the segregated group and 83 percent believed it also damaged the enforcing group. All but 10 percent of the respondents checked one of the four alternatives offered as the basis for their opinion. "All in all, then," Chein concluded, "we may not only say that there is widespread agreement among social scientists that enforced segregation is psychologically detrimental despite equal facilities, but we may add that the majority of these social scientists believe that there is a factual basis for such agreement." Chein noted that there were problems with the existing research regarding the questions he posed in his survey. "Since equal facilities are in fact not provided," he explained, "the proposition that enforced segregation does have detrimental psychological effects, even under conditions of equal facilities, seems impossible to prove." Indeed, Chein found "virtually nothing in the published literature that is explicitly devoted to this problem." That did not mean that his respondents had no basis for their opinions, however, for there was a large literature on segregation "from which one may cull a great deal of information pertinent to [the problem]."[49] Chein called for social scientists to turn their attention to research designed to prove psychological damage even given equal facilities.

Chein's survey was a curious piece of social scientific research. As a document for a legal trial, however, it would be tremendously useful. Attorneys fighting segregation could now make part of the legal record the fact that the vast majority of social scientists found segregation to be psychologically damaging. That there was no study that specifically isolated *de jure* segregated education with equal facilities would not be immediately relevant in a court of law. An expert witness is often called to provide an opinion; the reasons for the opinion are seldom queried. Hence the opinions of social scientists, as discovered

[48] The cover letter and the survey are reprinted in Max Deutscher and Isidor Chein, "The Psychological Effects of Enforced Segregation: A Survey of Social Science Opinion," *Journal of Psychology,* 1948, *26*:286–287. Deutscher was a statistician who helped Chein crunch the numbers in interpreting his survey.

[49] Isidor Chein, "What Are the Psychological Effects of Segregation under Conditions of Equal Facilities?" paper read at the Nineteenth Annual Meeting of the Eastern Psychological Association, 16–17 Apr. 1948, Kenneth B. Clark Papers, Box 22, Folder: "American Jewish Congress Reports, 1946–49," Manuscript Division, Library of Congress, Washington, D.C. (hereafter cited as **Clark Papers**). A version of this paper, which includes more commentary than the essay cited in note 48, above, was published: Chein, "What Are the Psychological Effects of Segregation under Conditions of Equal Facilities?" *Int. J. Opinion Attitude Res.,* 1949, *3*:229–234; the quotations are from pp. 232, 259.

by Chein's survey, would be relevant. How they came to those opinions was simply not that important for the lawyers.[50]

The results of Chein's survey were published in May 1948. Chein, with his colleague Max Deutscher, explained that the survey was inspired by the recent Ninth Circuit Court of Appeals decision in *Westminster.* "According to the Court," they announced, "the basic evidence for this decision consisted of studies in race relations made by anthropologists, psychologists, and sociologists, demonstrating that legal segregation does in reality imply the inferiority of one 'race' to the other and implements such status." The characterization of the Ninth Circuit decision as primarily based on social science material was, in fact, completely inaccurate. As noted earlier, *Westminster* was decided on a relatively narrow point of law. It did not address either the larger question of the constitutionality of separate but equal facilities or the sociological and psychological questions raised in CLSA's *amicus* brief. Nevertheless, Deutscher and Chein announced that the present study was designed for use in the legal arena. They continued: "Final decision on the legality of enforced segregation, regardless of equal facilities, has not yet been rendered by the Supreme Court. Here too, social science may be a significant if not crucial factor. For social scientists interested in 'social engineering' this represents a concrete opportunity to apply the relevant findings of social science data."[51]

More than any other single work, Chein's survey represented a confluence of interests between social scientists and attorneys. For CLSA attorneys, the survey represented a terse, nontechnical presentation of social science opinion that could add authority to legal briefs. For the CCI social scientists, it represented an opportunity for research to be meaningfully translated into legal action. The social science survey was particularly well timed for use in the battle against legalized segregation. By 1948, CCI social scientists were beginning to attract the attention of the NAACP, just as it was entering a new phase of litigation against segregated education. To understand the NAACP's use for social science, it is necessary to understand the shape of its campaign against segregated education.

THE NAACP AND THE USE OF SOCIAL SCIENCE

The roots of the NAACP campaign against segregated education can be traced to 1930, when the association hired a young lawyer, Nathan Margold, to draft a legal strategy that could be used to secure adequate educational facilities for African American children. Margold counseled against a futile campaign of litigation aimed at the equalization of facilities in the "separate but equal" world of the South. Such a campaign would entail a separate lawsuit for each of the thousands of southern school districts to prove that facilities in each of them were unequal, quickly depleting the NAACP's meager resources. Margold argued, however, that if the NAACP "boldly challenge[d] the constitutional validity of segregation if and when accompanied irremediably by discrimination," it could eliminate segregated schooling in one stroke and thus guarantee African American children adequate educational facilities.[52]

[50] See, e.g., "Rogers on Expert Testimony," NAACP Papers, Series IIB, Box 138, Folder: "Schools, Kansas, Topeka, Brown v. Board of Education, Expert Witnesses, 1951," Manuscript Division, Library of Congress (hereafter cited as **NAACP Papers**).

[51] Deutscher and Chein, "Psychological Effects of Enforced Segregation" (cit. n. 48), p. 259.

[52] Mark V. Tushnet, *The NAACP's Legal Strategy against Segregated Education, 1925–1950* (Chapel Hill: Univ. North Carolina Press, 1987), p. 27.

The Margold report was the blueprint for the NAACP's challenge to segregated education, providing a simple but powerful strategy for the elimination of Jim Crow schools. Margold's legal strategy was well timed. It arrived just as the NAACP's legal staff was undergoing significant changes, the most important of which was the arrival of Charles Hamilton Houston, dean of the Howard University School of Law. Houston would be responsible for the planning and execution of the litigation envisioned by Margold.

Houston was a 1922 graduate of the Harvard University School of Law, where he was an exceptional student who earned a doctorate in juridical science rather than the more conventional J.D. He came to the attention of two of the most famous and demanding members of Harvard's faculty, Felix Frankfurter and Roscoe Pound. Frankfurter, one of Harvard's legal realist professors, became Houston's doctoral advisor. Frankfurter taught Houston that the law was a tool of social engineering and that an attorney needed an understanding of the law's social setting in order to be successful. Pound, one of the founders of the earlier school of "sociological jurisprudence," believed that the law must include an understanding of the social sciences in order to operate effectively. As dean of Howard Law School, Houston brought the realist emphasis on the use of social science to an entire generation of African American attorneys who would lead the fight for civil rights.[53] The NAACP's later reliance on social science during the school segregation cases might have been expected, given that much of the legal staff received its training at Howard, an important center of legal realist thinking and its concomitant reliance on "real world" factual data.

In 1933 Houston began the NAACP litigation campaign against segregation. The NAACP focused on suits to eliminate segregated graduate and professional schools, because Houston felt that these were most vulnerable to a direct attack on segregation.[54] Segregated states seldom made a pretense of offering equal opportunities for African Americans seeking a graduate education. Seventeen of the nineteen states that required segregated education had no graduate schools whatsoever for African Americans. Only three states—West Virginia, Missouri, and Maryland—offered out-of-state scholarships that would pay the tuition for African American students who wished to pursue graduate education. Houston felt that this near-absolute absence of opportunities for graduate education would enable him to demonstrate easily the inequality of segregation.

Five years after the Margold report, Houston was joined in his effort by a former pupil, Thurgood Marshall. Marshall was born in Baltimore, which he described as the "most segregated city in the United States," and raised in the heart of its African American middle-class community. He went to Lincoln University in Pennsylvania in 1925, and then to the newly-accredited Howard Law School. Marshall excelled under the strict hand of Houston at Howard and, after graduation, moved into what turned out to be a brief private practice. In 1936 he joined the NAACP staff, with an initial commitment from the organization to pay his salary for six months. Marshall quickly moved up in the hierarchy, however. When Houston left the NAACP in 1938 to return to private practice Marshall took charge of the association's legal activities.[55]

Under Marshall's direction, the NAACP won a major victory in 1938 when the Supreme Court decided that the state of Missouri had a constitutional obligation to provide a legal

[53] Genna Rae McNeil, *Groundwork: Charles Hamilton Houston and the Struggle for Civil Rights* (Philadelphia: Univ. Pennsylvania Press, 1983), pp. 49–56, 76–85; and Tushnet, *NAACP's Legal Strategy,* p. 118.

[54] Tushnet, *NAACP's Legal Strategy*, p. 34; and McNeil, *Groundwork,* pp. 116–117.

[55] Mark V. Tushnet, *Making Civil Rights Law: Thurgood Marshall and the Supreme Court, 1936–1961* (New York: Oxford Univ. Press, 1994), pp. 28–29.

education to Lloyd Gaines, a young African American represented by Marshall. However, the Court left open the question whether Missouri could have fulfilled its obligation if it had a separate law school for African Americans.[56]

World War II interrupted the litigation campaign, and further suits were suspended "for the duration." Thus, as the war drew to a close in 1945, the NAACP had one Supreme Court victory to its credit: *Gaines.* That was a limited victory, however, since the Court left open the possibility that separate graduate education programs for African Americans would be constitutionally permissible. Southern states quickly went about setting up such programs. After the war, the task of the NAACP would be to prove that these hastily assembled programs could not possibly be equal to their long-established counterparts for white students, a task that the organization assumed would be relatively simple. In fact, proving any two schools unequal was extraordinarily difficult.[57] One reason the NAACP turned to social science was to prove that separate systems of education were necessarily unequal and, hence, discriminatory.

In a 1947 case to desegregate the University of Texas Law School, *Sweatt v. Painter*, the NAACP sought social science data and social scientists who could serve as expert witnesses. The NAACP turned to CLSA for the names of social scientists who could testify at the *Sweatt* trial. Will Maslow sent a copy of the social science survey to Thurgood Marshall, even though it had not yet been published, and asked if "it can be used in any form in the University of Texas suit." The survey would be of some use, but what the NAACP really needed was a prominent social scientist willing to testify to the evils of segregation. If possible the social scientist should be at a southern institution, minimizing the appearance of northern hostility toward the South. Unfortunately, Will Maslow had to inform Marshall, "CCI believes there are no nationally known psychologists in the South."[58]

The NAACP did have the University of Chicago anthropologist Robert Redfield testify in Texas. Redfield was a rare resource for the NAACP. The son of a prominent Chicago attorney, he had received his J.D. from the University of Chicago Law School in 1921. After a brief practice, Redfield found himself dissatisfied with the law and returned to the university, receiving his Ph.D. in anthropology in 1928. His conversion from the law to social science may have been encouraged by his wife, Margaret Lucy Park, the daughter of the founder of the Chicago school of sociology, Robert Park. In any event, Redfield was much happier as an anthropologist than as a lawyer and published several influential books on Latin American folk culture.[59]

If his almost unique credentials of degrees in both the law and anthropology were not enough to bring Redfield to the attention of the NAACP, then his concern for racial justice would have been. During World War II Redfield proclaimed that African Americans were the victims of a society that treated them as "half-citizen[s]." The problems of African Americans, he declared, were in the "mythology of the modern man" that proclaimed the natural inferiority of the African American race. Redfield argued that this was nonsense and that "a child of one skin color starts even with a child of any other skin color, if you let him. We don't let him, and we entertain a false biology which seems to justify us. I say again that race is of consequence because of what men think and feel about it and not

[56] *Missouri ex re. Gaines v. Canada*, 305 U.S. 337 (1938).

[57] Tushnet, *NAACP's Legal Strategy* (cit. n. 52), pp. 87–88.

[58] Maslow to Thurgood Marshall, 14 Apr. 1947, 28 Apr. 1947, NAACP Papers, Series IIB, Box 204, Folder: "University of Texas, Sweatt v. Painter, Correspondence, Jan.–June 1948."

[59] George W. Stocking, Jr., "Robert Redfield," in *Dictionary of American Biography,* Suppl. 6: *1956–1960*, ed. John A. Garraty (New York: Scribner's, 1980), pp. 532–534.

because of anything that race is of itself." Between 1947 and 1950 Redfield served as the director of the American Council on Race Relations (ACRR), an organization founded in 1944 to "bring about full democracy in race relations through the advancement of the knowledge concerning race relations." The NAACP had been working extensively with two ACRR sociologists, Robert Weaver and Louis Wirth, to amass sociological data on the effects of racially restrictive housing covenants. It is undoubtedly through this work that Redfield came to the attention of the NAACP. While Redfield had no formal connection to CCI, he was aware of the research on integrated sales counters and interracial housing.[60]

Redfield's testimony demonstrates how the social scientific ideas that were at the heart of CCI's research program could be used in a court of law to argue against segregation. Many of Redfield's arguments were precisely those that CCI's researchers were attempting to make in the social science literature. Redfield testified that segregation was inimical to "public security and general welfare." Drawing on his experiences at the desegregated University of Chicago, he argued that "segregation policy, and the stigma which segregation attaches to the segregated increases prejudice, mutual suspicion between Negroes and Whites and contributes to the divisiveness and disorder of the national community, contributing to crime and violence." Redfield's third and final point was that desegregation could be expected to proceed smoothly, that the "abolition of segregation in education, is likely to be accomplished with beneficial results to public order and the general welfare."[61]

When cross-examining Redfield, attorney general Price Daniel turned to an issue that would prove to be one of the dominant themes in litigation concerning educational segregation: whether desegregation could be "imposed" on a community that did not desire it. In other words: Could legal change precede attitude change in race relations? Daniel pressed Redfield to admit that it was "impossible to force the abolition of segregation upon a community that has had it for a long number of years." Redfield refused to admit the point, arguing that "segregation in itself is a matter of law, and that law can be changed at once." Daniel refused to give up and questioned Redfield about the speed of attitude change within the community—wasn't it true that such change could not be forced on a community? Redfield then admitted that, depending on circumstances, the attitudes of the community could resist desegregation; nonetheless, he insisted, "in every community there is some segregation that can be changed at once, and the area of higher education is the most favorable for making the change."[62]

The arguments that Redfield made in Texas underscore the importance of the research undertaken by CCI. The idea that segregation increases racial tension and that integration would decrease it was at the heart of CCI's research into the contact hypothesis. Moreover, the empirical studies that CCI undertook would add to the credibility of the witnesses in

[60] Robert Redfield, "Race and Human Nature: An Anthropologist's View" (1944), rpt. in *The Social Uses of Social Science: The Papers of Robert Redfield,* Vol. 2, ed. Margaret Park Redfield (Chicago: Univ. Chicago Press, 1963), pp. 137–145, on p. 142; and Louis Wirth, memorandum, 1 May 1947, Louis Wirth Papers, University of Chicago Archives, Regenstein Library, Chicago, Illinois. On the NAACP's work with ACRR see Clement E. Vose, *Caucasians Only: The Supreme Court, the NAACP, and the Restrictive Covenant Cases* (Berkeley: Univ. California Press, 1959), pp. 159–163. Redfield's pretrial notes for the *Briggs* trial a few years later demonstrate his familiarity with CCI research: Robert Redfield Papers, Box 23, Folder 4, University of Chicago Archives (hereafter cited as **Redfield Papers**).

[61] Pretrial notes, Redfield Papers, Box 33, Folder 7.

[62] Redfield's testimony in *Sweatt* was part of the record of *Briggs v. Elliot,* 98 F. Supp. 529 (1951). All references are to Redfield's testimony as printed in the *Briggs* transcript. See Redfield testimony, pp. 166–167, copy in Tom C. Clark Papers, Tarleton Law Library, University of Texas, Austin, Texas.

Brown who would argue, as Redfield had in the Texas case, that desegregation could be imposed on an unwilling populace.[63]

Just a few months after Redfield's testimony in Texas, the NAACP persuaded CCI director Stuart Cook to present the results of the Chein survey in an Oklahoma lawsuit against segregated graduate education. Unfortunately, because of the way the legal issues were framed Cook's testimony was disallowed by the trial judge. Nonetheless, Cook reported to American Jewish Congress executive director David Petegorsky that the NAACP was "very enthusiastic about the potential effect of our survey . . . on the psychological effects of segregation."[64]

The Texas and Oklahoma trials demonstrate that the NAACP was beginning to show considerable interest in the legal use of social science data. A good summary of the NAACP's position on social science was provided by Annette H. Peyser in 1948. Peyser was a young staffer with a sociology degree from New York University who had come to work for the NAACP in 1945 when the political scientist Harold Lasswell recommended her as a propaganda analyst. Soon after arriving she began to work with the legal staff to assemble sociological materials relating to the NAACP campaign against restrictive covenants in housing. The lone social scientist on the NAACP staff, Peyser began a "public relations attempt at compensating for the fact that the NAACP does not have a counterpart to the Commission on Law and Social Action, such as CCI." The announced purpose of her paper, which was presented at a legal conference, was an effort to "effect a better relationship between the legal expert and the social scientist." Peyser recounted the use of social scientific data and expert witnesses in *Westminster*, the graduate school cases, and other cases. She noted that the NAACP had emphasized sociological data to prove that segregation leads to inequalities in facilities because there simply were no studies regarding psychological damage. The NAACP had, "in the absence of specific scientific studies relating to the psychological effects of segregation, emphasized the factual aspects of segregation." Peyser looked to the AJC to bridge this gap. While the NAACP had neither the funds nor the personnel to undertake original research, the same was not true of the AJC, with its two divisions of social scientists and lawyers: "The CCI works independently of but in cooperation with the Commission on Law and Social Action of the American Jewish Congress. It may be that AJC, because of its physical structure, will be able to perform some of the necessary research on the psychological effects of segregation." Peyser also noted that CCI had already been an invaluable source of social scientific materials for the NAACP. In an obvious reference to the social science survey, she claimed that the "American Jewish Congress has been instrumental in reprinting articles that have appeared in scientific journals as well as preparing and writing these articles for the purpose of eventually having them 'planted' in such journals." Peyser closed her address with a call for further cooperation between social scientists and lawyers.[65]

[63] See, e.g., the trial testimony of M. Brewster Smith in *Davis et al. v. County School Board of Prince Edward County,* 1952, Case File 1333, Box 126, Vol. 2, pp. 292–293, Civil Case Files, 1938–1958, Richmond Division, Records of the U.S. District Court to the Eastern District of Virginia, Record Group 21, National Archives—Mid Atlantic Region, Philadelphia, Pennsylvania.

[64] Cook to David Petegorsky, 24 May 1948, AJC Papers, Box 19, Folder: "CCI, 1948–49."

[65] Annette H. Peyser to M. Brewster Smith, 25 Mar. 1952, M. Brewster Smith Papers, Box M605, Folder: "NAACP," Archive for the History of American Psychology (hereafter cited as **Smith Papers**); and Annette H. Peyser, "The Use of Sociological Data to Indicate the Unconstitutionality of Racial Segregation," Clark Papers, Box 63, Folder: "Background Reports, Undated." This document is undated, but internal evidence suggests that it is the 1948 address. On Peyser's work more generally see Jack Greenberg, *Crusaders in the Courts* (New York: Basic, 1994), p. 35; and Tushnet, *Making Civil Rights Law* (cit. n. 55), p. 89.

The NAACP's need for social scientists would intensify after 1950, when the Supreme Court ruled that segregated graduate education was unconstitutional. In both the Texas and Oklahoma graduate school cases, the Court based its decision on ineffable and intangible factors, benefits denied African Americans by segregation. In the case of Heman Sweatt, such an intangible was a thing like the opportunity to attend a law school with the reputation of the University of Texas. In the case of George McLaurin in Oklahoma, it was the opportunity to exchange ideas with his fellow graduate students. While the Supreme Court stopped short of directly overturning the "separate but equal" doctrine, the door was now opened wide to do so.[66] Moreover, the way through that door could be through pointing out "intangible factors." When they turned their attention to elementary and secondary education in the cases that eventually were decided in *Brown,* the attorneys of the NAACP would attempt to use social science to help them articulate some of those factors. And social scientists were waiting for them, armed with studies identifying lines of evidence that supported the notion that segregation was psychologically damaging. The NAACP could not merely turn to CCI to find social scientists, however, for on the eve of the campaign against elementary and public school segregation the social scientists at the AJC were dispirited and in disarray.

CONCEDING DEFEAT: DISSOLVING THE SCIENCE/LAW COLLABORATION

Civil rights attorneys and socially minded social scientists were dedicated to the same end: the elimination of prejudice and discrimination. Their styles and the methods they employed were fundamentally different, however. The law worked on a strict timetable, and lawyers had to adhere to that timetable. To be persuasive to judges and juries, a good legal argument had to be forceful and unambiguous, leaving no room for other interpretations. On the other hand, social science proceeded at a much more leisurely pace, since professional publications usually enforced no strict deadlines. To be persuasive to social scientific peers, a good social scientific argument had to be provisional and filled with careful caveats. The possibility of other interpretations would often be left open. The differences in the processes by which each community produced its arguments against discrimination eventually doomed the institutionalized merger of law and social science at the AJC.

Isidor Chein's survey of social science opinion was by far the most successful example of social science research designed to be used as a legal argument. More typically, CCI's social scientific research did not fit in with the AJC's larger campaign against discrimination. Despite the fact that the CCI social scientists quite deliberately aimed their research priorities in a manner that would be useful to CLSA and other activists at the AJC, social science never made a significant contribution to the AJC's struggle against discrimination.

A tension always existed between CCI and the rest of the American Jewish Congress. On the one hand, CCI was to be an objective, scientific research agency. On the other hand, it was funded by and worked within the American Jewish Congress—an avowedly political and activist organization. This tension was recognized by Stuart Cook, who told a colleague that "first, CCI must be an active functioning participant in the general [American Jewish] Congress program. Second, within the framework of [the American Jewish] Congress—which to the outsider's eye is a partisan, political organization—it must be a scientific research group holding the complete confidence of non-Congress organizations

[66] *McLaurin v. Oklahoma State Regents for Higher Education,* 339 U.S. 637 (1950); *Sweatt v. Painter,* 339 U.S. 629 (1950); and Tushnet, *NAACP's Legal Strategy* (cit. n. 52), pp. 130–132.

and individuals. As you know, we have waged an up-hill fight for this dual objective." Despite the optimistic tone of much that was written by CCI social scientists, the organization never fit comfortably within the AJC framework. The concerns raised by Ronald Lippitt in his 1945 memorandum came to haunt CCI in the course of the next five years: the production of social scientific materials was too slow to meet the needs of the other branches—the real world, it was argued, moved too fast for the social scientists of CCI. The result was that in 1948, soon after Cook assumed control, he faced a series of budget and staff cuts that eventually decimated CCI. Cook tried desperately to hold the organization together, but by 1950 he had had enough and accepted an offer from New York University. In his resignation letter he wrote that he was unhappy with the AJC leadership. "What made me accept [the NYU offer] was that no meaningful assurances were really possible; that [CCI's budget of] $240,000 had not really shrunk to a stable $78,000 but rather that no one could really know where the end of the path was to be," he explained.[67]

With Cook's resignation, Isidor Chein became CCI director. He faced further staff reductions and budget cuts. By 1952 the situation had come to a "crisis," and Marie Jahoda, a distinguished social psychologist, was called in to take stock. Jahoda surveyed both CCI staffers and American Jewish Congress leaders in an attempt to discover the source of the friction. CCI staff members argued that the AJC neither understood nor respected what they were attempting to accomplish. One, John Harding, complained that the AJC's position on any issue was decided "ideologically." That is to say, the AJC had its agenda; and if social science happened to agree with that agenda, well and good. But social science could never actually guide the AJC's position. "Empirical research *is* seen by Congress leadership as serving a useful function in providing evidence from time to time of the correctness of the Congress stand on various issues," Harding wrote to Jahoda. "However this use of research is a dispensable luxury, since the correctness of the Congress position is always clearly evident . . . before the research is done."[68]

Chein took the opportunity provided by Jahoda's assessment to announce that the situation at CCI was "inherently unstable and similar crises, with all the incident demoralization, must inevitably recur. In other words, regardless of the outcome, I shall be looking for another position."[69] Soon Chein would join Cook at NYU.

Jahoda's survey revealed that, while CCI was roundly criticized by the other branches of the American Jewish Congress, there was no clear consensus as to what exactly was wrong. In her final report, she noted that the criticisms leveled at CCI by the other branches often were contradictory: "CCI's scientific standards were said by some to be too high, by others too low. On the one hand, CCI was presented as being too perfectionist and too much concerned with meeting scientific requirements when the needs of Congress might have been satisfied with a less thorough job. On the other hand, jobs whose usefulness were recognized . . . were criticized as not really representing a scientific contribution." All these criticisms could be true, she argued, because "*there exists . . . no generally recognized and defined standard against which the functioning of CCI could be measured and judged adequate or inadequate. CCI's function as a social science department within Congress is undefined.*"[70]

[67] Cook to Alfred Marrow, 5 June 1947, Marrow Papers, Box M1938, Folder 14: "1946–1947 CCI Correspondence"; and Cook to Marrow, 11 Aug. 1950, AJC Papers, Box 70, Folder: "Dr. Stuart Cook."

[68] John Harding to Marie Jahoda, 9 Apr. 1952, Marrow Papers, Box M1938, Folder 17: "1952–1953 CCI Correspondence."

[69] Chein to Jahoda, 9 Apr. 1952, Marrow Papers, Box M1938, Folder 17: "1952–1953 CCI Correspondence."

[70] Marie Jahoda, "The Commission on Community Interrelations of the American Jewish Congress," 1952, pp. 14, 15, Marrow Papers, Box M1938, Folder 21: "CCI Papers."

The fundamental problem confronting CCI, Jahoda claimed, was that "science proceeds at a lower speed, with less flexibility in tackling new problems and with different standards of success than [other] organization activities." Social scientists had little to contribute to an attack on discrimination until they had completed a study on the specific problem in question. Until then, all a social scientist could report was that "the analysis of the data continues." Jahoda noted that such a report could be regarded as "unsatisfactory and even annoying," but she also noted: "A more exact content report . . . might have been even more annoying: it might well have read thus: 'a code was developed and applied. Reliability checks proved that it was inadequate. The code was revised with better reliability checks.' Many of the processes which enter into research are . . . boring and unrevealing."[71]

Jahoda recommended that the American Jewish Congress continue to fund CCI but define more clearly what it expected from social scientific work. CCI did continue, as a shadow of its former self—it conducted no more serious research into intergroup relations. Many of CCI's social scientists followed Cook and Chein to NYU, where they continued their research in a more academic setting.

CONCLUSION

The tension between CCI and the larger AJC demonstrated the difficulties of combining the disparate cultures of social science and legal action. Social scientists were definitely interested in being part of the activist AJC, and the AJC was interested in receiving whatever aid the social scientists could offer. But even given the socially conscious nature of the social science work done at CCI, the merging of the two groups worked only poorly. CCI's social scientists realized that they had to conduct careful, scholarly research in order to be useful, but the activism of the AJC could not endure the natural pace of scientific research. Nor was the AJC happy with the provisional nature of social scientific research—the uncertain conclusions, the meticulous caveats, and the calls for further study that are the hallmarks of careful scientific work. All of these characteristics limited the usefulness of social science for the larger AJC. The processes by which social science argued against discrimination and prejudice were incompatible with the processes by which the law attacked the same foes. Hence the AJC's attempt to combine social science research and legal action can be viewed as a failure.

And yet, in another sense, the merger was a triumph, for the work done by CCI was appreciated from afar. The NAACP greatly appreciated the social scientific research generated by the American Jewish Congress. CCI had, for all intents and purposes, come to an end by 1952, just as the NAACP was beginning to litigate against segregated education in public elementary and secondary schools. Yet CCI left behind a body of research that would prove useful. Its research on the separation of attitudes and behavior and on the contact hypothesis would emerge time and again in the *Brown* litigation. Moreover, the social scientists who had worked at CCI, most notably Cook, Chein, and Saenger, would become key players in the later litigation campaign. Chein's survey of social science opinion would be cited twice by Earl Warren in the *Brown* opinion.[72] Despite Kenneth Clark's protestations to the contrary, the social scientific community had indeed prepared for the legal battle of *Brown* for a decade.

[71] *Ibid.,* pp. 18, 20.

[72] *Brown v. Board of Education of Topeka,* 347 U.S. 483 (1954). See also John P. Jackson, Jr., "Creating a Consensus: Psychologists, the Supreme Court, and School Desegregation, 1952–1955," *J. Soc. Issues,* 1998, *54*:143–177.

We are left, then, with what appears to be a paradox: CCI, the very organization that failed to integrate law and science under the auspices of the American Jewish Congress, was the organization responsible for the research that played a central role in the dismantling of legal segregation. In other words, the products of CCI's research proved valuable in the legal battles against segregation, though the processes by which those studies were produced were only an annoyance to the lawyers of the CLSA. Maurice Rosenberg wrote that "in general any meeting [between the legal and scientific styles of thought] that occurs will take place after the scientists have gathered their data. . . . The methods are far apart until the scientist has made his findings."[73]

The studies produced by CCI were very useful in the courtroom; however, lawyers did not need—and did not really desire—to be involved in their creation. In fact, the attempt at close collaboration between social scientists and lawyers produced frustration for both groups. The dream of Kurt Lewin and Alexander Pekelis that lawyers and social scientists would work side by side to battle prejudice and discrimination did not come to fruition. But the attempt led to the dismantling of legalized segregation and forever changed the landscape of the United States.

[73] Maurice Rosenberg, "Comments," *Journal of Legal Education,* 1970, *23*:199–204, on p. 204.

German Eugenics between Science and Politics

By Peter Weingart

THE HISTORY OF EUGENICS is one of a reciprocal involvement of science and politics. Simply put, that history can be characterized as beginning with two scientific theories, evolutionary theory and its complement, the theory of human heredity. Taken up by political discourse, these theories helped to create or crystallize concerns about the hereditary quality of the human stock. They delineated and helped to focus political concerns about population policy and control. Scientists used eugenics as a vehicle for their political convictions and social biases, just as politicians used its scientific framework, sketchy as it was, to advance their particular causes.

It was only in Germany, however, that the small community of race hygienists (as eugenicists called themselves there), seeking status and recognition, formed a coalition with politicians of the conservative and radical right. This coalition enhanced the political influence of race hygienists and aided the implementation of many of their ideas. The subsequent moral catastrophe had repercussions far beyond the German community, forcing eugenics-minded geneticists all over the world to strive for a sharper demarcation between science and politics. Their efforts eventually resulted in a fundamental change in their original program: human genetics thereafter renounced all claims to improving the gene pool of the human race as a whole and instead concentrated on the cure of genetic diseases on an individual basis.

This study does not encompass the entire development of eugenics in Germany. It focuses on the crucial role of an institute in Berlin, the Kaiser Wilhelm Institute (KWI) for Anthropology, Human Heredity and Eugenics (Kaiser-Wilhelm-Institut für Anthropologie, Menschliche Erblehre und Eugenik), which became the single most important institution for race hygiene in Germany. Although there were a number of other important centers devoted to various aspects of eugenics and race hygiene—such as the Institute for Genealogy and Demography under Ernst Rüdin in Munich, which was part of the German Research Agency for Psychiatry and in 1924 became an institute in the Kaiser-Wilhelm-Gesellschaft (KWG)—the institute in Berlin was clearly the leading institution in the field between its foundation in 1927 and 1945. Given that the most eminent representatives of race hygiene, Eugen Fischer, Fritz Lenz, and Otmar

This article is part of a larger project on eugenics and race hygiene in Germany on which a book-length study has also appeared: Peter Weingart, Jürgen Kroll, Kurt Bayertz, *Rasse, Blut und Gene: Geschichte der Eugenik und Rassenhygiene in Deutschland* (Frankfurt: Suhrkamp, 1988). Some passages in this essay are taken from an unpublished paper, "Eugenics under the Nazis," by P. Weingart and H. Kranz, presented at the XVIIth International Congress of History of Science, Berkeley, California, August 1985.

"hereditary" scientists, took part. The eventual plan—to use the proposed *Reichsanstalt* to coordinate research in different parts of the country and to model it on the Kaiser Wilhelm Institute for Physics—met with financial and administrative approval.[8]

Two and a half years later, and most likely in some continuity with these plans, the president of the Kaiser-Wilhelm-Gesellschaft, Adolf von Harnack, and its secretary, Friedrich Glum, submitted a plan to the senate of the KWG arguing that it had proved "absolutely necessary to create a scientific center for anthropology, human heredity and eugenics in Germany, since Sweden, the United States, and England have gone ahead with work in this area, in *particular because these inadequate and dilettante efforts in this area have to be countered.*"[9] The designated director of the new institute was to be Eugen Fischer, then the leading anthropologist in the country, best known for his 1913 study on the bastard population in Rehoboth, South-West Africa, in which he was the first to apply Mendelian genetics to the study of racial mixture, thus founding what came to be known as the "anthropobiological" school. The KWG was determined to move rapidly. The projected cost of the new institute was RM (reichmarks) 600,000 for construction of the building and the director's private home and for funding his position. The Reich was to appropriate RM 300,000; Prussia, RM 100,000, the director's post, and the lot in Dahlem. The remaining RM 200,000 was to be raised from private sources. Harnack pointed out at the meeting that it was important that the inauguration of the new institute take place a year later, in 1927, when the International Congress of Genetics would convene at the KWI for Biology.

In a programmatic article published in August of 1926 Fischer spelled out the purpose and direction of the institute. He described anthropology as the science dealing with the complex relations between heredity and the environment. One issue that seemed particularly interesting to him and "anthropologically beyond all tendentious attitudes" was the "question of the Jewish population living among a non-Jewish one." Race biology to him was the biology of man in general, which included the study of race crossing. Turning to the study of heredity, Fischer called for extensive research on normal and pathological lines of heredity in the population, pointing out how little was known and how ill defined the line was between the normal and the pathological. Believing that "pure anthropology does not pass value judgments," he moved cautiously to what he termed "social anthropology," which in his view involved the question of whether certain social associations could determine not only the life of the individual but also his hereditary lines. This was Fischer's careful and distanced way of adapting the research questions of social hygiene to those of eugenics: the most important issue within this realm was the impact of membership in certain social associations on reproductive behavior, or bluntly put, the worry that the better strata in society reproduce at a lower rate. The inverse problem—whether the qualities of certain populations have an impact on their fate as a social group—led to the question of purity of race and the consequences of race crossing. But Fischer took great

[8] "Vermerk über die am 22. Januar 1923 im Reichsministerium des Innern abgehaltene Besprechung betreffend Errichtung einer Reichsanstalt für menschliche Vererbungslehre und Bevölkerungskunde," *ibid.*, pp. 236–238.

[9] Minutes (emphasis added), Senate of the KWG, 19 June 1926, KWG Generalversammlung, KWI für Anthropologie und Psychiatrie, Archiv der Max Planck Gesellschaft, Berlin (hereafter **KWI-Anthropologie, MPG Archiv**), Sign. A1/62–69, p. 86.

pains to distance himself from the exaggerated claims of that part, called "race hygiene," of "our entire science." He clarified that the term race in this context did not mean "systemic race," but the "favorable form with respect to its entire constitution and hereditary strength for a particular population." Fischer left no doubt, however, that he considered the control and direction of human reproduction, both by way of ethical persuasion and by social and legislative measures, the overriding task of human culture. To preserve hereditary lines, to study and care for them, was the "actual and ultimate objective inherent in all this research."[10]

When the institute opened as planned just in time for the International Congress of Genetics, Fischer stressed once more that the problems subsumed under the "problematic concept" of *Rassenkunde* (race studies) would "naturally be researched on a strictly scientific basis and free of any other kind of thoughts." The new KWI was to be a "purely theoretical institute for research into the nature of man" where "the question of human races and racial differences" would be researched "purely scientifically without regard for political and other tendencies." Setting the institute aside from most others in the KWG, Fischer asserted that it would not seek applicable results, a surprising claim given that a year earlier he had stated that the ultimate objective of race hygiene research was to gain control over the biological "wheel of fortune."[11]

The institutionalization of race hygiene under the umbrella of the highly reputed KWG nurtured the hopes and assuaged the fears of those who looked with suspicion upon some of the movement's right-wing pronouncements. Julius Moses, representative of the Social Democratic Party in the Reichstag, expressed this sentiment in a session of the appropriations committee in March 1927 in which RM 300,000 was set aside for the new institute. Moses hailed the plan and pointed out that the working class had grown to distrust science in recent years because it had shown nationalistic tendencies. Especially in the area of race hygiene, nationalism had run rampant during and after the war. Moses believed that the new institute would pursue research "from a strictly scientific basis, with no regard for nationalistic or nonnationalistic, national or international results." In response to protests from other members of the committee and from Harnack, Moses supported his assertions, validating working-class suspicion of nationalistic science by quoting from *Einführung in die Rassen- und Gesellschaft-Physiologies,* by a Professor Adolf Basler, which described the proletariat as the sewer into which all those would sink who were unusable and detrimental for human society.[12] Friedrich Glum, the secretary of the KWG, later reminisced in his autobiography that the foundation of the institute—"against some resistance"—was explicitly designed to "put up something scientifically tenable against the nonsense about race problems and eugenics circulated by the rising National Socialism."[13]

To some extent this motive also reflected developments within the eugenics

[10] Eugen Fischer, "Aufgaben der Anthropologie, menschlichen Erblichkeitslehre und Eugenik," *Naturwissenschaften,* 1926, *32*:749–755.

[11] Eugen Fischer, "Von wissenschaftlichen Instituten," *Kultur und Leben,* 1927, *4*:315.

[12] See Adolf Basler, *Einführung in die Rassen- und Gesellschafts-Physiologie* (Introduction to physiology of race and society) (Stuttgart: Franckh'sche, 1925); see also Verhandlungen des Reichstags, III. Wahlperiode 1924/27, Ausschuß für den Reichshaushalt, 229. Sitzung, 4 Mar. 1927.

[13] Friedrich Glum, *Zwischen Wissenschaft, Wirtschaft und Politik* (Bonn: Bouvier, 1964), p. 371.

movement itself. The conflict between the more moderate "Berlin" and the conservative, race hygiene–oriented "Munich" factions had erupted again in 1925 with the establishment of the German Federation for Population Betterment and Heredity (Deutscher Bund für Volksaufartung und Erbkunde). The new association was an initiative of the Reich Association of Registrars in Germany, an organization numbering about 7,000 registrars of births, marriages, and deaths. This group was pressing for the dissemination of knowledge on human heredity and eugenic ideas. It had originally planned to join forces with the German Society for Race Hygiene, but this did not materialize. Now the German Federation, although still inclined to limit its activities to the dissemination and popularization of eugenic programs so as to develop activities "complementary" to those of the Society for Race Hygiene, naturally developed into a competitor. The race hygienists had received most of their financial support from the Prussian administration, and this money now shifted to the German Federation, primarily because that body represented the interests of many individuals in the Prussian ministries of health and welfare.

Members of the German Federation, especially its leadership, were closely linked to the political administration. Examples include its first chairman, Arthur Ostermann, a *Medizinalrat* (medical official) in the Prussian Ministry of Public Welfare, and his cochairman, Hermann Muckermann. Both were close to the predominantly Roman Catholic Center party, which formed a coalition with the Social Democrats in Prussia and controlled the Ministry of Public Welfare. Muckermann owed much of his influence to this connection, because as a Jesuit priest, amateur biologist, and self-styled eugenicist, he did not rank high among established scientists. In fact, he was instrumental in setting up the KWI for Anthropology, Human Heredity and Eugenics.

Although the German Federation sought friendly relations with the German Society for Race Hygiene, the race hygienists were skeptical of their new competitor. But eventually they had to acknowledge its success in attracting both public attention and government support. It was against this background that in 1930 members of the Berlin Society for Race Hygiene unanimously agreed to change its name, dropping the term "race hygiene" in favor of "eugenics" in order to avoid "manifold misunderstandings" associated with the former. In 1931 the German Society for Race Hygiene followed suit, renaming itself the German Society for Race Hygiene (Eugenics) and reformulating its constitution. By focusing on the "eugenic composition [*Gestaltung*] of the family and the people," it distanced itself from its earlier emphasis on race and opened the way to the eventual fusion of the two national societies in 1931.[14]

Thus the less ideological, more practical, and more scientifically sound faction within the movement had prevailed. At the turn of the decade it seemed, therefore, that the eugenics movement had been institutionalized both in the organization of basic research in the country and in the medical and public welfare administration and was ready to take on tasks deemed important in the area of population policy and public hygiene. If eugenics was inherently oriented to medical and hygienic practice and thus could hardly be considered nonpolitical, it

[14] "Dokumente aus der eugenischen Bewegung," *Das Kommende Geschlecht,* 1930, *5*(6):20; "Aus der Gesellschaft für Rassenhygiene (Eugenik)," *Arch. Rass. Ges.-Biol.*, 1932, *26*:95; and "Satzungen der Deutschen Gesellschaft für Rassenhygiene (Eugenik)," *ibid.*, p. 101.

seemed at least that the moderate forces had been able to put a leash on the radical right-wing race hygienists and their nationalistic racism. Fischer's institute may be seen as an important element in this strategy. As it turned out, however, the compromises struck between the opposing political factions proved too tenuous to survive in the general political arena as well as in the small sector of science we are examining here.

II. THE BERLIN EUGENICISTS

The inextricable connections between science and politics in eugenics and race hygiene are exemplified by the switch from claims to pure science before 1933 to promises of service to the "new state" after the Nazis came to power. This change was rooted in the intellectual and political convictions of the scientists who were instrumental in setting up the KWI for Anthropology or central to its activities. These scientists demonstrated ideological affinities with that political movement to which most of them adapted so willingly once it had come to power.

Eugen Fischer (1874–1967), the director of the institute, headed its department of anthropology. He had been called to Berlin from Freiburg, where he had held a chair at the university. In 1913 Fischer had published a book based on his studies of the descendants of German colonialists and the Hottentot population in South-West Africa. The book became a classic in anthropology because it was the first to deal with the problem of race crossing in terms of modern genetics, and it made Fischer the recognized leader of German anthropology. Although the book marked a definite step toward linking race biology to genetics, it also reveals a conflation of scientific and political or ideological tenets. Thus although his study showed no negative effects of race crossing, Fischer concluded: "We still do not know very much about the effects of race crossing. But one thing we know for certain: without exception every European people . . . which has assimilated the blood of inferior races—and only dreamers can deny that Negroes, Hottentots, and many others are inferior—had paid for this assimilation of inferior elements with intellectual and cultural decline."[15]

With his new approach Fischer was instrumental in bringing about a shift in Germany from physical anthropology, which had focused on measuring techniques (craniometry, anthropometric methods), to race anthropology, which pursued research on physical and mental (*seelische*) phenomena from a genetic perspective of heredity and selection. His "anthropobiology" shared with race hygiene and eugenics the assumption of a superindividual entity, the race or the people (*Volk*), based on a static and typological concept of race. The composition of this entity, whether conceived of as race or *Volk,* was regarded as having its own natural characteristics, which had to be protected. Inherent in these views was the normative evaluation of these entities and the practical orientation to political action. Although there was no basis in eugenics for explicit or implicit racism—for example, the prevention of race crossing—and in fact some eugenicists even opposed it; a logically equivalent "social racism" directed against social outcasts (*Asoziale, gemeinschaftsunfähige Minusvarianten*) was nonethe-

[15] Eugen Fischer, *Die Rehobother Bastards und das Bastardisierungsproblem beim Menschen* (Graz: Akademische Druck- und Verlagsanstalt, 1913), p. 302. The book was republished in 1961!

government's reluctance to meet their demands that led some race hygienists to look for political allies who would be more receptive. The most promising such ally seemed the National Socialists.

Most explicit in his evaluation of the Nazis as a political force that could advance the cause of race hygiene was Fritz Lenz. He had read Hitler's *Mein Kampf* carefully and given it a long review in the *Archiv für Rassen- und Gesellschaftsbiologie*. Although Lenz did not share Hitler's radical anti-Semitism, he applauded him as the "first politician of great influence who has recognized race hygiene as the central task of politics and who wants to throw his full weight behind it."[21] Lenz saw a spiritual relationship between National Socialism and race hygiene and hoped to be able to win the support of the political movement for race hygienic reforms. When he republished his 1917 essay "Zur Erneuerung der Ethik" (On the reform of ethics), in 1933 under the title *Die Rasse als Wertprinzip* (Race as a value principle), he claimed that it contained all the basic elements of the *Weltanschauung* of National Socialism and may even have contributed to its emergence.[22]

The reaction of Lenz's colleagues to National Socialism was not quite as enthusiastic, primarily because of political prudence, since the Nazis had not yet assumed power. Roughly six months before that happened, the Nazi party, in its turn, approached the race hygiene community in a letter addressed to several of the leading figures in the field. Referring to "Professor Lenz," the letter declared that within the Public Health Section of the party's national organization, a subgroup on race hygiene was being formed that was intended to "prepare those measures that had to be taken in order to secure for the future the seriously threatened existence of the German people quantitatively and qualitatively." It suggested that "loose working groups" be set up in the various subfields of race hygiene. The goal was to win experts, and the party considered the objective so important that political differences could be overlooked. Thus the invitation was extended even to those who actually "stood at a distance" from the National Socialist party.[23]

Fischer responded to this letter favorably in principle, pointing out that he followed the National Socialists' activities with great interest and satisfaction and could subscribe to most of the party's eugenics program. He declined to cooperate beyond giving personal advice, however, because he feared for the political independence of his institute.[24] (This fear was obviously superfluous after 30 January 1933, when the Nazis came to power.)

The invitation was also extended to Muckermann, which aroused the objections of hardliners—primarily of Theo Lang, a collaborator of Rüdin's. Hermann Boehm defended his invitation to Muckermann as a tactical move made in order to anticipate and neutralize possible criticism from the Center party, provided Muckermann's support could be gained. Muckermann himself had responded that he was "willing in principle to cooperate wherever there are efforts to solve

[21] Fritz Lenz, "Die Stellung des Nationalsozialismus zur Rassenhygiene," *Arch. Rass. Ges.-Biol., 1931, 25*:300–308, on p. 308.

[22] Fritz Lenz, *Menschliche Auslese und Rassenhygiene (Eugenik)* (cit. n. 19), 3rd. ed. (Munich: Lehmann, 1931), pp. 415, 418; and Lenz, *Die Rasse als Wertprinzip: Zur Erneuerung der Ethik,* 2nd ed. (Munich: Lehmann, 1933), p. 7.

[23] Document Center, Berlin, File Boehm.

[24] Copy of Fischer's answer, *ibid.*

this existential problem of the German people." Even though he did not agree with all the demands of the National Socialist party, he saw a common goal in "overcoming the degeneration of the biological heredity of the forefathers and in the preservation and augmentation of hereditarily healthy families in all occupational groups of our people."[25] Despite this favorable response, in the end the conflict within the Nazi party forced Boehm to withdraw the invitation to Muckermann.

The tactically motivated caution with which Boehm had proceeded was not needed after Hitler was named *Reichskanzler*. The party could now enact the race hygienic measures it deemed fit.

III. THE ADAPTATION OF THE INSTITUTE TO NATIONAL SOCIALISM

At the end of May 1933 the new government took control of the German Society for Race Hygiene (Eugenics). This affected the KWI for Anthropology directly, since Fischer and Muckermann were in the society's directorate, together with Ostermann. All three had to step down, the former name of the society was reinstated, and the society was transferred back to Munich, where Rüdin was named its *Reichskommissar,* now directly responsible to the minister of the interior. Fischer also lost control of the *Archiv für Rassen- und Gesellschaftsbiologie*. Verschuer had to step down from his position in the Berlin Society for Race Hygiene. Thus the Nazis restored the more radical faction of race hygienists to a position of influence within the movement and neutralized Fischer's institute.

Although this seemed to be an unfriendly act committed by a government that was shortly before considered their ally, the race hygienists had reasons to overlook it—perhaps as a necessary measure to enforce the "new order"—and to welcome the realization of the policies they had longed for. There is no evidence that any of the scientists were pressured to give the sort of eulogies to the new government of which Rüdin's was neither an isolated nor atypical example. In 1934 he wrote: "The importance of race hygiene has become evident to all aware Germans only through the political work of Adolf Hitler, and only through him has our more than thirty-year-old dream become reality: to be able to put race hygiene into action."[26]

State authorization and the expansion of professional control and power go hand in hand, and race hygiene was no exception to the rule. In May 1933 race hygiene assumed the role of a science serving to legitimate politics. With the formation of an "expert committee for population and race policy at the Ministry of the Interior," scientific expertise was utilized in a politically functionalized manner. A case in point is the draft of a sterilization law, prepared by the previous government, that had to be completed in one day. This entailed changing the provision for voluntary sterilization into a compulsory measure and setting up an elaborate legal apparatus, the "hereditary health courts" (*Erbgesundheitsgerichte*), which were to oversee the legality of sterilization measures and hear appeals. Fischer, Lenz, Rüdin, Baur, and Verschuer, among others, were called upon to support the implementation of the law. Verschuer became a member of

[25] Copy of Muckermann's letter to Boehm, *ibid.*

[26] Ernst Rüdin, "Aufgaben und Ziele der Deutschen Gesellschaft für Rassenhygiene," *Arch. Rass. Ges.-Biol.*, 1934, *28*:228–231.

mittees, the Agency for Hereditary Health, and the hereditary courts. For these agencies numerous reports had to be prepared."[35]

After Verschuer moved to Frankfurt in 1935, the institute's departments were reorganized. Verschuer's department for human heredity was dissolved and divided between Fischer and Lenz, who replaced Verschuer. The reason given for this change was twofold: first, a competent successor, so it seemed, could not be found; second, and more important, the expansion of human heredity had led to its merger with anthropology and race hygiene (eugenics) with respect to both subject matter and research methods.[36]

In April 1935 a new department for hereditary psychology opened under Kurt Gottschaldt, a *Privatdozent,* which was devoted to research on the inheritance of normal mental traits. The "fact of the heredity of mental traits" was assumed; the department was to prove it. This research was probably the result of Verschuer's work on a hundred pairs of twins in which he had found a significantly greater similarity in intelligence between monozygotic than between dizygotic twins. Gottschaldt's own research strategy was to organize "twin camps" ("particularly important for hereditary psychology research") during the summer months to observe psychological details and personalities as a whole.[37] In 1936 collaborative work with the psychological division of the Reich War Ministry began in which twin analyses were carried out in reviews of military agencies.

In line with Fischer's commitment, repeated in his 1936 report for the third time at least, the number of "requested reports increased permanently": sixty reports on racial purity for the Reich Agency for Pedigree Research by Fischer and Wolfgang Abel, twenty-eight paternity reports for various courts, and twenty reports by Fischer and Lenz on sterilization.[38] Meanwhile, the Nuremberg laws had been passed in September 1935. Fischer, having overcome his initial hesitation, thanked the Führer in 1936 for enabling "the hereditary scientists to put their research results to the practical service of the entire people."[39]

It is evident, then, that given the demands of hereditary health practice and the teaching of courses for the public health service, the theoretical orientation of the institute receded more and more to the background. How strongly did the Nazis influence the willingness of the scientists to collaborate with the new government? In contrast to most other Kaiser Wilhelm Institutes, Fischer's was in a special situation. The coincidence of the party's ideology and political objectives with the institute's disciplinary delineation, as well as with the scientific and professional interests of the race hygienists, makes it difficult to isolate the effect of Nazi pressure. Race hygiene was as much an applied science as it was politics and ideology. Service to the state was as much a nationalistic duty as it was a fulfillment of the scientists' self-interests. In this context political infringements on scientific autonomy were hardly viewed as threats. When Muckermann was

[35] Excerpt from the report on activities of the KWI for Anthropology, 13 May 1935 (cit. n. 29), pp. 32–33; and meetings of the *Kuratorium,* 10 May 1935 and 20 June 1936, Bundesarchiv Koblenz R 36/1366.

[36] *Naturwissenschaften,* 1936, *24*:27.

[37] *Ibid.,* 1937, *25*:379.

[38] Report on activities of the KWI for Anthropology, 22 June 1936, KWG Generalversammlung, KWI-Anthropologie, MPG Archiv, Sign. A1/2399, pp. 73, 74.

[39] Eugen Fischer, cited in *Allgemeine Zeitung,* No. 211, 6 May 1936.

forced to leave the institute he had been instrumental in setting up, Fischer, from all we know, did not put up any of the resistance he had vowed a year earlier.[40]

IV. RISE AND FALL

The sterilization law of 1933 (which went into effect in 1934) had given the race hygienists their first quasi-professional recognition and had expanded their licensed activities, much along the lines they had fought for throughout the years of the Weimar Republic. Subsequent years brought even more recognition and professionalization. In September 1935, on the occasion of the Nuremberg party convention of the Nazi party, two additional race hygienic laws were proclaimed. The first was the Law for the Protection of German Blood and German Honor, which was supposed to protect the purity of the race and put sexual intercourse and marriage between German citizens and Jews under penalty. This law fulfilled an older demand of the radical, racist wing of the race hygiene movement, which had been supported by the Nazis before their seizure of power. Now it provided the second substantial opportunity for the professionalization of race hygiene, since the determination of ancestry (interpreted as the degree of Jewish ancestry) now became an important task (and, a few years later, a matter of life and death). In 1939 it was resolved that only experts could be licensed to call themselves "physicians of hereditary biology."[41]

The second law, the Law for the Protection of the Hereditary Health of the German People, made marriage counseling mandatory and stipulated that marriage permits could be granted only after a health certificate had been issued. This law reflected demands of eugenicists dating back to before World War I.

The devotion of the regime to the eugenic credo reached even further. In addition to the sterilization and marriage controls, other measures that had been discussed by eugenicists for many years before the Nazis' rise to power were now put into law: the standardization of the public health system (1934), a law providing for the inheritance of farms and settlements (1933), and the provision of financial support for young couples set by the number of children they had. All these measures, together with many parallel organizations in the party, amounted to an encompassing system of "efforts to keep the race pure and for the convalescence of our people."[42]

The astonishing growth and development of race hygiene in Germany were envied by eugenicists abroad.[43] To some extent this politically induced "professionalization" established structures that bore the characteristics of the other great profession with which race hygiene had much in common: medicine. A canon of systematized knowledge was put into textbooks; recruitment processes and educational curricula were at least partly formalized; and, most important, practical application of that knowledge was monopolized, as in ancestry statements, paternity suits, and statements to the heredity courts.

[40] Otmar von Verschuer, "Die Eugenik und Hermann Muckermann," *Deutschlands Erneuerung,* 1932, *16*:463–466, on p. 465.

[41] Horst Seidler and Andreas Rett, *Das Reichssippenamt entscheidet: Rassenbiologie im Nationalsozialismus* (Vienna: Jugend & Volk, 1982), p. 176.

[42] Max Fischer, "Adolf Hitler und die Rassenhygiene," *Psychiatrisch-neurologische Zeitschrift,* 1939, *41*:176–178, on p. 178.

[43] See *Eugenical News,* 1934, *19*:140.

Race hygiene owed its sudden surge of professional power and privilege to the new state, but it paid the price of the loss of scientific autonomy. The political control that was extended to all scientific organizations and societies did not bypass the Society for Race Hygiene. As mentioned above, Frick, the minister of the interior, had removed the board of directors of the society and named Ernst Rüdin as the new *Reichskommissar*. Although the society now assumed official functions for the state, it also became subject to the decisions of the ministry. Its task, together with that of the National Socialist Federation of Physicians and the Bureau for Race Policy of the Party, was to combine scientific and political information and propaganda. According to a decree of the Bureau for Race Policy of 17 May 1934, the positions of its regional directors and the chairmen of the local societies for race hygiene were to be filled with the same persons if at all possible, a measure obviously designed to assure *Gleichschaltung,* or "bringing into line with party control." In similar fashion the bureau collaborated with the German Society for Research on Race, the former Anthropological Society, and other important race hygiene organizations. The chairman of the society was named special adviser in the science department of the bureau. Thus, as in other spheres of politics and society, the National Socialists created the typical dual structure: governmental institutions replicated political or party organizations. This structure also contained the seeds for the eventual bifurcation in the development of race hygiene.

All these developments were reflected, to an extent, in the growth of the KWI for Anthropology. As early as 1934 Fischer had requested RM 207,000 for expansion of the institute. This sum was granted, and in 1936 he could report that the successful construction of the new building would soon lead to "nice results."[44] By 1937 the institute had expanded by one third. It was the most prestigious institution of its kind and prided itself on having visitors from Norway, the Netherlands, Switzerland, Italy, Hungary, India, Japan, China, Brazil, and many other countries well into the war years, as well as having representation at major scientific congresses in Paris, Copenhagen, London, and Edinburgh. A contemporary observer diagnosed an increasing dogmatism, which became apparent at the 1939 Genetics Congress at Edinburgh when Fischer reacted very emotionally to criticism, a reaction that might be explained by a feeling of superiority on the part of Fischer and members of his institute because of their eminent and powerful position in Germany.

The expansion continued without interruption, even after war began. In 1938 Karl Diehl was made head of an external department at Sommerfeld for hereditary research on tuberculosis. Wolfgang Abel, one of Fischer's assistants, became head of a department for studies on race, or *Rassenkunde*. This department's task was to expand knowledge on racial traits; research was undertaken in different parts of Germany ("also on Jews") by Walter Dornfeldt and on Gypsies in Romania and Scotland by Abel. *Rassenkunde* was the continuation of heredity (*Erblehre*) as a science; in Fischer's words, "all *Rassenkunde* and then naturally all race policy are based on the proof of the hereditary nature of racial traits." The new results that had been gained through research on race crossing both at home (through Abel's studies of the so-called Rheinland bastards) and

[44] Report on activities of the KWI for Anthropology, 1 Apr. 1935–31 Mar. 1936, KWG Generalversammlung, KWI-Anthropologie, MPG Archiv, Sign. A1/2399, p. 77.

abroad (in Chile, Trinidad, and South-West Africa) "are immediately useful for race policy."[45] What this meant in concrete terms was that the institute subscribed, although it did not actively contribute, to the forced sterilization of the Rheinland bastards, descendants of German women and African soldiers who had come to the Rheinland with the French army. The first incidents of illegal sterilization of Jews took place in connection with this campaign. By 1940 the institute prided itself on contributing at least indirectly to the "scientific" legitimation of the selection and destruction of Jews.

The bifurcation between race hygiene and what later gradually emerged as human genetics also became apparent in another development. In his report submitted to the *Kuratorium* in January 1941, Fischer presented plans for the more distant future. "The coming victorious end of the war and the monumental extension of the 'Greater German Empire' also present our research agencies with great, new tasks." Fischer claimed two fundamental objectives for research: "In times like these it has to serve the immediate interests of the people, the war, and politics; but second, it must orient itself to the future as well as the present, for one can never know what practical effects pure scientific research might have in the future." "No one could imagine that Gregor Mendel's studies on peas would provide the basis for a hereditary health legislation," Fischer wrote, "or that my study on bastards of 1908 could support race legislation."[46]

Fischer's words signaled a shift in the political justification of what he envisioned as the further development of his institute in particular and of race hygiene in general. The strategy he pursued was clever, and it seems to have unfolded on two levels: with regard both to the institute's relations with the political authorities and to its reorganization along lines suited to future research. Fischer must have had a sense of the growing separation between the ever-more-radical race policy of the Nazis and the direction of development of his science. Wholly aside from his own participation in anti-Semitic propaganda, he was likely to have anticipated greater difficulties in legitimating the kind of research he felt was a logical continuation of the research program of the field. In addition, he obviously had problems with Lenz, for as early as November 1938 he indicated to the administration of the KWG that he wanted Verschuer as his successor. At that time he envisioned Lenz as being named his substitute director, but in 1940 he made it explicit that he wanted to prevent Lenz from getting that position too. Instead, he suggested that Lenz be made head of a separate institute for race hygiene within the KWI for Anthropology and be given the title of director.[47]

In his report on the activities of the institute, Fischer gave a lengthy description of Lenz's work in which he wrote that "as a representative of race hygiene [Lenz is] a scholar of unusual, unique character in his field." But for the organization of the institute, his "particular kind of research as a race hygienic thinker" did not come into consideration. He felt that Lenz needed and should receive total freedom from administrative and other activity in the interest of his theoretical work. Thus he wanted Lenz to have a separate institute for race hygiene. The department of eugenics had been renamed the race hygienic department when Lenz had become its head.

[45] Report on activities of the KWI for Anthropology, 1935–1936, *ibid.*, Sign. 2404.

[46] Report on activities of the KWI for Anthropology, 1936–1940, Anlage 3, *ibid.*

[47] Memorandum, 21 Nov. 1938, by Telschow of an 18 Nov. conversation with Fischer; and memorandum by Telschow of an 18 Oct. 1940 conversation with Fischer; both in Hauptakten, *ibid.*, Sign. A1/2399, pp. 130, 141.

The "new task" for research Fischer saw elsewhere: in phenogenetics. He believed the study of human heredity had advanced so far that it could describe the essential characteristics of normal and pathological hereditary traits, relate phenomenal appearance to these traits, and offer an approximate idea about the impact of environmental effects on them. But the dark spot was the question of how a given hereditary disposition unfolded and "operated" in acquiring the phenomenal appearance. "The path from a given hereditary disposition [*Erbanlage*] to the readily configured hereditary trait [*Erbmerkmal*] in the individual is still unknown," Fischer remarked. The most important problem in this connection was, in his view, the phenogenetic study of diseases and malformations. The way to get results was by using animal experiments. Fischer suggested creating a department for experimental hereditary pathology and named Hans Nachtsheim as head.[48]

Verschuer, who returned to Berlin as Fischer's successor as director of the institute in May 1942, specialized in hereditary pathology. Hans Nachtsheim, one of two members of the institute who had no ties to the Nazi party, later became the founder of human genetics in postwar Germany. Lenz, although he accepted a chair at the University of Göttingen, never again played any important role in this field. This was the significance of Fischer's reorganization of the institute in 1940.

It seems that with this move Fischer acknowledged an apparently growing trend toward a differentiation within race biology (*Rassenbiologie*), a category that encompassed, primarily for political motives, *Rassenkunde,* race hygiene, human heredity, and racial anthropology. Originally the term *Rassenbiologie* had been introduced to give some unity to the heterogeneity of the field. Institutes for heredity and research on race (*Vererbungswissenschaft und Rassenforschung*) existed alongside others for hereditary and race care (*Erb- und Rassenpflege*) or for anthropology and race science (*Anthropologie und Rassenkunde*). Toward the end of the 1930s and in the early 1940s most of these institutes were renamed race biological institutes (*rassenbiologische Institute*). Physical anthropology had always played a strong role in the race hygiene movement, and there was still no clear delineation between it and the human genetics side of race hygiene. But the differentiation between these two subjects did begin under the Nazi regime, in spite of the political pressures that were later cited as having supposedly prevented it.

The common term race biology encompassing these various efforts could not conceal the underlying problem that had beset the grand design to institutionalize race hygiene in the universities: a clearly discernible tendency on the part of scientists to return to their original disciplines. It was the beginning of a long-overdue process of differentiation that had been delayed by the politicization of a field that had entered the Third Reich with too many loose ends and vague delineations. Race scientists (*Rassenkundler*) turned back to anthropology; hereditary pathology (*Erbpathologie*) oriented itself to medicine and was to become the cradle of human genetics in Germany; and psychiatric heredity looked to psychiatry. In 1944 the anthropologist Eberhardt Geyer saw race biology as having arrived at a "crossroads." For "a science that essentially has no common orientation, whose representatives come from the most diverse disciplines and departments, that knows no organized and systematic division of labor," he saw the

[48] Report on activities of the KWI for Anthropology, 1936–1940 (cit. n. 46), pp. 5–8.

only feasible solution in the establishment of race hygienic institutes in the university medical departments for the transmission of a certified canon of knowledge on hereditary pathology, and of anthropological institutes in philosophy departments for the more speculative basic research on such topics as race and people.[49]

This development was reflected to some extent in the internal politics of the KWI. In 1940, when Fischer planned for Verschuer to take his place and for Lenz to head his institute for race hygiene within the KWI, Fischer also suggested to the KWG that the KWI be renamed the KWI for Hereditary and Race Studies (*Erb- und Rassenkunde*).[50] (This plan was never realized, and in fact, on Fischer's seventieth birthday the institute was renamed in his honor the Eugen Fischer Institute, a name devoid of any political connotation.) Fischer's decision to establish a department for experimental hereditary pathology did not imply that he was to turn away from his notion of anthropobiology. In seeing a central task in phenogenetics he continued to search for "race genes." He considered "further research on the differential distribution of respective genes in different races" essential.[51] But in calling to the institute Hans Nachtsheim, who had previously worked in the KWI for Heredity and Breeding Research, Fischer acknowledged the importance of experimental genetics. In other words, even though Fischer himself still tried to retain the unity of a race biology that had gained political support from Nazi ideology, he was forced to recognize the differentiation of the field, a differentiation that also implied its depoliticization. In giving his reasons for the establishment of the new department, he remarked that the new structure of the institute would serve his successor, who, unlike Fischer himself, could not have gathered the rich experience of forty years of research activity in the overall administration of the institute.[52]

Verschuer's activities as the new director of the institute reflected this development in a different way: he represented what was to become the core of postwar human genetics, namely, the study of the genetic basis of diseases. With his notions of hereditary pathology and the hereditary physician, he had pointed the way to a strictly medical orientation of race hygienc as early as 1934. (It is interesting to note that the first "hereditary clinic" offering genetic counseling opened in the United States in 1940.) This enabled him to become the foremost human geneticist in Germany after the war. At the same time, Verschuer was a political opportunist par excellence, if one accepts the judgment of his more critical colleagues, such as Nachtsheim, and weighs available evidence. His opportunism must be interpreted as the same morally ruthless pursuit of scientific and professional interests that motivated so many of the race hygienists. According to Nachtsheim, Verschuer, because of his services to National Socialism, brought human genetics into disrepute at home and abroad as no other geneticist had done.[53] In fact, through Verschuer the institute was to become directly con-

[49] E. Geyer, "Wissenschaft am Scheideweg," *Arch. Rass. Ges.-Biol.*, 1943, *37*:1–6, on p. 2.

[50] Memorandum of conversation between Telschow and Fischer, 18 Oct. 1940 (cit. n. 47). Fischer did not repeat his suggestion at the meeting of the *Kuratorium* in Dec. 1940, for unknown reasons.

[51] Eugen Fischer, "Versuch einer Phänogenetik der normalen körperlichen Eigenschaften des Menschen," *Zeitschrift für Induktive Abstammungs- und Vererbunglehre,* 1939, *76*:47–117, on p. 73.

[52] Report on activities of the KWI for Anthropology, 1936–1940 (cit. n. 46), p. 9.

[53] Nachtsheim in a letter to the newspaper *Die Welt,* 21 Sept. 1960, cited in Karl Saller, *Die Rassenlehre des Nationalsozialismus in Wissenschaft und Propoganda* (Darmstadt: Progress Verlag, 1961), p. 140.

nected with the murderous "experiments on humans" at Auschwitz. Even though this connection was never substantiated in a court of law, evidence accumulated over the years leaves little doubt.

Verschuer was a devout Protestant who before 1933 could have been considered a moderate eugenicist. But as soon as the new government seized power, he began to pay tribute to it. He identified eugenics with race hygiene, and even though his basically modern concept of the hereditary physician did not demand it, he declared he owed his ideas to the *weltanschauliche* revolution of 1933 and the National Socialists' insight that "individualism as the principle of medical practice" was an erroneous development that had to be overcome.[54] In 1936 he became the expert biologist in the Research Department on the Jewish Question in the Reich Institute for the History of the New Germany and contributed to its publications. From 1938 on he belonged to its council of experts and provided anthropological reports on descent. His 1941 review in his own journal, *Erbarzt,* of the first issue of the institute's journal *Weltkampf* (published since 1924 by Arthur Rosenberg, and appearing in 1941 as a "scientific quarterly"), attributed to it "an important role in our fight against Jewry, which is truly a global fight," thus revealing his aggressive anti-Semitism.[55]

In the KWI for Anthropology Verschuer continued to support Fischer's research program on phenogenetics. In 1943 he established a new department for embryology, and during 1943 and 1944 he received substantial grants from the German Research Foundation for projects on comparative hereditary pathology and on hereditary research on tuberculosis. In the fall of 1943 he also received from the Reich Research Council through the Research Foundation nine commissioned projects, among them two (on specific albuminous matter and on eye color) in which his former assistant Josef Mengele was involved.

Mengele had written his medical dissertation under Verschuer in Frankfurt in 1938 and worked in Verschuer's institute until he had to join the Wehrmacht in 1940. He served in the Waffen-SS until he was wounded in 1942 and commissioned to the office of the Reich Physician of the SS and Police in Berlin, which oversaw the "medical experiments" in the concentration camps.[56] Here his path again crossed that of Verschuer, who apparently let Mengele work "on the side" in the KWI.[57] It is likely that Verschuer motivated Mengele to ask for a commission to go to Auschwitz to use this "unique opportunity" for race biological research.[58] From May 1943 on, Mengele was in Auschwitz as camp physician. His scientific experiments on humans were carried out on his own responsibility, in his "free time."[59] When he had killed "his" twins, he dissected them and sent the "material" (blood samples, pairs of eyes, etc.), which was labeled "Urgent! important war material," back to the KWI in Berlin to the attention of his teacher

[54] Otmar von Verschuer, introduction to the new journal *Der Erbarzt,* 1934, *1*:1–2, on p. 1.

[55] Otmar von Verschuer, "Weltkampf," *Der Erbarzt,* 1941, *9*:264. See also Helmut Heiber, *Walter Frank und sein Reichsinstitut für Geschichte des neuen Deutschland* (Stuttgart: Deutsche Verlagsanstalt, 1966), pp. 421, 606.

[56] Zdenek Zofka, "Der KZ-Arzt Josef Mengele: Zur Typologie eines NS-Verbrechers," *Vierteljahreshefte für Zeitgeschichte,* 1986, *34*:245–267, on p. 254.

[57] Verschuer *Nachlass,* University of Münster, cited in Benno Müller-Hill, *Tödliche Wissenschaft* (Reinbek: Rowoholt, 1984), p. 112.

[58] See F. K. Kaul, *Ärzte in Auschwitz* ([East] Berlin: Volk & Gesundheit, 1968), p. 90; and Zofka, "KZ-Arzt Mengele" (cit. n. 56), p. 254.

[59] Evaluation by Dr. Wirths in Mengele File, Document Center, Berlin.

Verschuer.[60] As Dr. Mikos Nyiszli, Mengele's prisoner pathologist assistant, reported later, the "directors of the Berlin-Dahlem institute . . . always thanked Dr. Mengele for the rare and valuable material." Hermann Langbein, in his account, quotes Verschuer in an interview as admitting that "Mengele had sent enormously interesting preparations of pairs of eyes with different pigmentation to this institute."[61] It was never resolved with certainty whether Verschuer knew the origin of the preparations or the circumstances under which they were obtained, but the evidence is overwhelming that he must have known.

Analyses of Mengele's character and motivations continue to accumulate, and because they can never be proved, become ever more complex and run the risk of mystifying him. Leaving aside legitimate psychological explanations, I think it is more revealing here to see Mengele and Verschuer together. Robert J. Lifton quotes one of his source's recollection of Mengele's saying that "not to utilize the possibilities Auschwitz offered would be 'a sin, a crime' and 'totally irresponsible' toward science." It must be remembered that the most severe limitation of research in race hygiene, and to some extent still in modern human genetics, is the inadmissibility of experimentation on humans. When professional fervor is excessive to the point that moral limitations are seen as mere obstacles to the pursuit of knowledge, the ideologically legitimated eradication of these limitations, along with the accessibility of human "material," must create the very context in which the "human capacity to convert healing into killing" is set free.[62]

V. CONCLUSION

Mengele's role in Auschwitz, and even more his connection to Verschuer and to what at one time was the largest and perhaps most respected scientific institution in the field of anthropology, human heredity, and eugenics in Germany, were successfully suppressed, rendered an episodic "aberration" in the history of science, the charge of "abuse" and "suppression" of science resting on the politicians. The institutional inertia of science was not in the least disturbed by its political and moral corruption. Only a few scientists, among them Werner Heisenberg, saw the connection between the ideology of an apolitical science and its utility to and corruptibility for immoral political goals. Fischer's institute in Berlin embodied, probably more than any other research institute in the country, the seeming paradox that what started out as avowed "pure research," undertaken with no regard to political or any goals other than knowledge, ended with at least parts of that research being involved in mass murder.

When it became clear that Berlin would fall, the chief scientific protagonists of

[60] H. Langbein, *Menschen in Auschwitz* (Vienna: Europaverlag, 1972), p. 383.

[61] Mikos Nyiszli, *Auschwitz: A Doctor's Eyewitness Account* (New York: Frederick Fell, 1960), p. 63; and Langbein, *Menschen in Auschwitz*, p. 384. Langbein also reports Verschuer's apparent surprise when he asked him if he knew about the source and circumstances of the "material." In a report to the German Research Foundation in March 1944 Verschuer acknowledged that Mengele worked on the "protein body project for him in Auschwitz, where with the permission of Himmler, anthropological studies of different races are carried out and blood samples are being sent to his [Verschuer's] laboratory." Verschuer to S. Breuer of the German Research Foundation, 20 Mar. 1944, Bundesarchiv Koblenz R 73/15345.

[62] Robert J. Lifton, *The Nazi Doctors* (New York: Basic Books, 1986), pp. 367, 383. Lifton's account and psychological interpretation of Mengele's actions and motives is the most penetrating to date, apart from the eyewitness accounts by Langbein and Nyiszli.

race hygiene left the capital and headed west. Lenz had seen his hopes of 1931 disappointed. In 1944 he had reintroduced the term *eugenics,* pointing to the danger "that measures are carried out in the name of race hygiene that are not, in fact, race hygienic and may even have adverse effects."[63] He opposed the overemphasis on hereditary pathology and clung to the conception that research on counterselection and differential reproduction had to remain the central issues of race hygiene. Lenz remained a race hygienist who steadfastly clung to a Social Darwinist perspective and thus to a holistic conception of race hygiene in which anthropology and hereditary biology would guide human reproduction as well as the organization of society according to the prerequisites of selection. In a way, with this position Lenz represented the core conception of race hygiene, which had lent itself so blatantly to the politicization of the field and its utilization by the ruthless Nazi regime. But the feebleness of his criticism should not lead one to overlook completely that from the vantage point of his notion of race hygiene, he was able to judge National Socialist race hygienic policies critically. When he finally got a new chair for genetics at the University of Göttingen in 1947, Nachtsheim congratulated him, adding that the "value of heredity was obviously considered higher in the West than here," in Berlin.[64] Lenz's race hygienic orientation prevented him from playing a role in postwar human genetics, however.

Verschuer had all of the institute's material that seemed interesting to him put on two trucks and transported to his hometown, Solz, near Bebra, where he waited for better times. In May 1946 the self-declared president of the Kaiser-Wilhelm-Gesellschaft, Robert Havemann, a physicist, published an article revealing the Mengele-Verschuer connection and attacking Verschuer for having indirectly participated in the killings of camp prisoners to obtain research material. Nachtsheim, in a conversation with Havemann regarding the article, criticized him for leveling false or unproved allegations against Verschuer instead of concentrating on "facts" such as Verschuer's party membership, cooperation with the SS and fanatic National Socialists, and so on. Havemann's reply that he considered Verschuer responsible for Mengele's crimes met with Nachtsheim's question: "Do you consider use of material from a concentration camp as such a crime? . . . One could blame von Verschuer only if he knew that the individuals whose eyes he had received had not died a natural death. . . . One could not deny V. good faith."[65] In August an unofficial KWG commission was set up to look into the allegations, and although the commission came to the unanimous opinion that Verschuer had acted against "human and scientific ethics," that judgment was never passed officially because of Nachtsheim's insistence on the same formalistic reasoning that he had used against Havemann. In contrast to Lenz's position, Verschuer's line of research turned out to be more central to the direction that human genetics was to take in the following years. After coming out of denazification procedures virtually unscathed, Verschuer embarked on a successful postwar career in 1951 with a new chair for genetics at the University of Münster.

Nachtsheim's department of experimental hereditary pathology was the only one in the institute that survived the war. The equipment and material of the

[63] Fritz Lenz, "Gedanken zur Rassenhygiene (Eugenik)," *Arch. Rass. Ges.-Biol.*, 1943, *37*:83–109, on p. 106.

[64] Nachtsheim to Lenz, 13 Jan. 1947, MPG Archiv, Sign. N18.

[65] Nachtsheim, notes on a conversation with Havemann, 9 May 1946, *ibid.*, Sign. N47.

other departments had been taken west by Verschuer, who, after the surrender, refused to hand them back to Nachtsheim in Berlin. An air raid in the final days of the war had destroyed seventeen of nineteen experimental strains of rabbits, some more were taken by Russian soldiers, and Nachtsheim offered the remaining ones to colleagues in England because he could not feed them anymore. As the KWG, which had moved its seat to Göttingen, could no longer fund the institute in Berlin, it was taken over by the German Academy of Sciences and renamed the Institute for Comparative Hereditary Biology and Hereditary Pathology. When in 1948 Berlin was split in two and the Academy came under the influence of the Communist government in the east, plans to move the institute to Berlin-Buch were dropped; and after a short interim period, it was taken over by the successor of the KWG, the Max Planck Gesellschaft, with Hans Nachtsheim as its director.[66] Experimental hereditary pathology became the line of research that provided the continuity between eugenics or race hygiene and human genetics. Race hygiene having fallen into political and moral disrepute in the wake of the collapse of the Fascist state and its crimes, and with no governmental demands on that science or ideological support for it, human genetics began to launch attempts at professionalization as a depoliticized science.

[66] Hans Nachtsheim and Herbert Lüers, "Das Max Planck Institut für vergleichende Erbbiologie und Erbpathologie in Berlin-Dahlem," *Münchener Medizinische Wochenschrift*, 1954, *43* (unfoliated page).

The Race Hygiene Movement in Germany

By Sheila Faith Weiss

THE HISTORIES of all national eugenics movements raise difficult and controversial questions, but the case of German eugenics or "race hygiene" (*Rassenhygiene*) is by far the most troubling.[1] For many people the term *German eugenics* immediately brings to mind visions of the Nazi death camps and the "final solution." This presumed connection between German eugenics and the racial policies of the Third Reich makes a sophisticated analysis of German eugenics especially urgent, and especially problematic. Although much progress has been made in recent years toward correcting the tendency simply to subordinate race hygiene to the larger themes of either the history of European racism or the development of *völkisch* thought, the current historiography of German eugenics is far from satisfactory.[2]

Research for this article was supported in part by a Mellon Foundation research grant from Clarkson University and by a National Science Foundation Summer Grant. Special thanks go to Mark Adams and Michael Neufeld.

[1] The German term *Rassenhygiene* (race hygiene) had a broader scope than the English word *eugenics*. It included not only all attempts aimed at "improving" the hereditary quality of a population but also measures directed toward an absolute increase in population. Despite these differences I will employ the two terms interchangeably throughout the essay. Even when German eugenicists limited themselves to measures that fall under the more limited term *Eugenik* (the Germanized form of the English word), they almost always used the term *Rassenhygiene*.

[2] For examples of this tendency to view German eugenics as part of a "larger story" of either the history of European racism or *völkisch* thought see David Gasman, *The Scientific Origins of National Socialism: Social Darwinism and the German Monist League* (New York: American Elsevier, 1971); Leon Poliakov, *The Aryan Myth: A History of Racist and Nationalist Ideas in Europe,* trans. E. Howard (New York: New American Library, 1977); and George L. Mosse, *Toward the Final Solution: A History of European Racism* (London: J. M. Dent, 1978). More recently West German historians have begun to treat eugenics as a subject worthy of study in its own right. See, e.g., Gunter Mann, "Rassenhygiene—Sozialdarwinismus," in *Biologismus im 19. Jahrhundert in Deutschland,* ed. Mann (Stuttgart: Ferdinand Enke, 1973), pp. 73–93; Werner Doeleke, *Alfred Ploetz (1860–1940): Sozialdarwinist und Gesellschaftsbiologe,* med. Diss., Frankfurt, 1975 (Tübingen: privately printed, 1975); Renate Rissom, *Fritz Lenz und die Rassenhygiene,* med. Diss., Mainz, 1982 (Husum: Matthiesen, 1983); and Georg Lilienthal, "Rassenhygiene im dritten Reich: Krise und Wende," *Medizinhistorische Journal,* 1979, *14*:114–133. During the past few years there have been several excellent studies on specific eugenic practices and institutions. See Gerhard Baader and Ulrich Schultz, eds., *Medizin und Nationalsozialismus: Tabuisierte Vergangenheit—Ungebrochene Tradition?* (Dokumentation des Gesundheitstages Berlin 1980, 1) (West Berlin: Verlagsgesellschaft Gesundheit, 1980); Gisela Bock, *Zwangssterilisation und Nationalsozialismus. Studien zur Rassenpolitik und Frauenpolitik* (Schriften des Zentralinstituts für sozialwissenschaftliche Forschung der Freien Universität Berlin, 48) (Opladen: Westdeutscher Verlag, 1986); Georg Lilienthal, *Der "Lebensborn e.V.": Ein Instrument nationalsozialistischer Rassenpolitik* (Forschungen zur neueren Medizin- und Biologiegeschichte, 1) (Stuttgart/New York: Gustav Fischer, 1985); Benno Müller-Hill, *Tödliche Wissenschaft: Die Aussonderung von Juden, Zigeunern und Geisteskranken 1933–1945* (Hamburg: Rowohlt, 1984); and Paul Weindling, "Weimar Eugenics: The Kaiser Wilhelm Institute for Anthropology, Human Heredity and Eugenics in Social Context," *Annals of Science,* 1985, *42*:303–318. Despite this new research, many topics have yet to be investigated.

Looking only at developments during the Third Reich, it would be easy to come to the false conclusion that race hygiene was always a right-wing movement. German eugenics, however, was far more heterogeneous in its politics and ideology than is generally assumed. Although its advocates were overwhelmingly recruited from the ranks of the *Bildungsbürgertum* (educated middle classes), they embraced no single political outlook. Until Hitler's seizure of power in 1933 precluded the expression of political diversity within the movement, German eugenics captured the interest of individuals whose allegiance spanned the breadth of the Wilhelmine and Weimar political spectrum. While there were few committed Communists associated with the movement, the important position played by the socialist Alfred Grotjahn and the large number of members affiliated with the Weimar left-center eugenics society, the Deutscher Bund für Volksaufartung und Erbkunde (German Alliance for National Regeneration and the Study of Heredity), render it impossible to view German race hygiene as solely or even primarily a right-wing phenomenon. About the only unanimity discernible between such men as Grotjahn and the political conservative Fritz Lenz was on the question of laissez-faire capitalism: like the vast majority of other German race hygienists, both men viewed it as dysgenic. Their consistent critique of capitalism should make us suspicious of interpretations that see race hygiene as just another intellectual prop of corporate capital.[3]

Just as German eugenicists varied greatly in their political orientation, they differed in the degree to which they accepted and promoted racist ideologies. Like the great majority of educated whites in Europe and North America of their day, all race hygienists accepted the racial and cultural superiority of Caucasians as a matter of course. From today's vantage point all German eugenicists would be considered racist; however, since this type of racism was shared by most eugenicists everywhere, emphasizing it in the case of Germany, where the population was relatively homogeneous, tells us very little. The situation is more complicated with regard to ideologies of Aryan or Nordic supremacy. It is undeniable that many race hygienists, including several in the vanguard of the movement such as Alfred Ploetz (1860–1940), Max von Gruber (1853–1927), Ernst Rüdin (1874–1952), and Fritz Lenz (1887–1976), were Aryan enthusiasts. Indeed, among the prominent Aryan-minded eugenicists there were those who were—sometimes secretly, sometimes openly—in favor of using race hygiene to promote the so-called Nordic race. However, extreme caution must be taken not to equate the pro-Aryan sentiments of a handful of German eugenicists with the aims of the movement as a whole. Many of Germany's leading eugenicists, such as Wilhelm Schallmayer (1857–1919), Hermann Muckermann (1877–1962), Artur Ostermann (1876–?), and Alfred Grotjahn (1869–1931), were uncompromising in their critique of Aryan ideologies. Together with large segments of the Deutsche Gesellschaft für Rassenhygiene (German Society for Race Hygiene), and virtually all of the members of the Deutscher Bund, they rejected out of hand the desirability of a "Nordic race hygiene." In addition, anyone who examines the content of the two major Wilhelmine and Weimar eugenics journals and looks at

[3] This point is clearly presented in Loren Graham's insightful article "Science and Values: The Eugenics Movement in Germany and Russia in the 1920's," *American Historical Review*, 1977, *82*:1133–1164, rpt. in Graham, *Between Science and Values* (New York: Columbia Univ. Press, 1981), pp. 217–256.

the platform of the Deutsche Gesellschaft will be struck by the relative lack of space devoted to *völkisch* ideologies or "Nordic eugenics" compared with other issues. On the whole this is true even of the writings of those eugenicists who embraced the Aryan mystique.

What both Aryan apologists and those eugenicists rejecting Aryanism *did* stress were strategies designed to increase the number of Germany's "fitter" elements and eliminate the army of the "unfit"—fitness being defined in terms of social and cultural productivity. The eugenicists' equation of fitness with productivity and achievement, and of degeneracy with asocial behavior and the inability to contribute to society, reflected their own middle-class prejudices. In sum, then, German eugenics before (and to some extent even after) 1933 was not primarily concerned with replenishing and improving the Nordic stock of Europe; occasional public displays of pro-Aryan sentiment notwithstanding, *Rassenhygiene* was more preoccupied with class than with race. Prior to Hitler's seizure of power the concerns of German race hygiene were not fundamentally different from those of many other Western eugenics movements.

Despite this unity of class bias, the diversity of political outlooks and racial attitudes among German eugenicists would appear to preclude their having had any single goal in common. At first glance it is easy to see differences and conflicts of interest—so much so that it is tempting to try to divide the movement into right-wing racist and left-wing nonracist camps as a first attempt at organizational clarity. While the institutional development of Weimar race hygiene, to be discussed later in this essay, does offer some justification for this general classification, the story is more complex. Viewing the movement as an uneasy union of two separate and competing "camps" obscures the underlying rationale and logic of German eugenics. Whatever additional reasons may have motivated them, all German race hygienists embraced eugenics as a means to create a healthier, more productive, and hence more powerful nation. Race hygiene was, however, quite unlike the usual political and economic strategies designed by those in power for the same purpose. Eugenics embodied a technocratic, managerial logic—the idea that *power* was a product of the rational management of *population*. For its practitioners race hygiene was a sometimes conscious, often unconscious strategy to buttress the supposedly declining cultural and political hegemony of Germany and the West through the rational management and control of the reproductive capacities of various groups and classes. Such a rational administration of human resources, the eugenicists believed, would ensure the level of hereditary fitness thought necessary for the long-term survival of Germany and Western Europe and the allegedly superior cultural traditions they embodied. This logic constituted the common bond that united all German eugenicists.

THE ORIGINS OF GERMAN EUGENICS, 1890–1903

Social, Professional, and Intellectual Contexts

German eugenics cannot be understood without examining the conjunction of circumstances that collectively account for its origin as a movement. Three contexts stand out as being particularly significant in shaping the early development of race hygiene: the *social* problems resulting from Germany's rapid and thoroughgoing industrialization; the *professional* traditions of the German medical community; and the *intellectual* currency of the "selectionist" variant of social Darwinism then fashionable among certain German biologists and self-styled social theorists. These three contexts will be dealt with in turn.

During the last quarter of the nineteenth century the newly unified German Empire was transformed from an agricultural into an industrial society. The industrialization and urbanization process, expeditious and thorough as it was, produced profound changes in the social and economic structure of the young Reich, engendering a myriad of serious social tensions and problems.[4] Had Imperial Germany not possessed a rigidly authoritarian political structure shaped primarily by the self-interest of preindustrial elites and their allies in heavy industry, the social dislocations precipitated by industrialization would not have appeared so threatening to the stability of the state and the social order. But the *Kaiserreich* was certainly no democracy, and, given the stranglehold that the landed aristocracy, the military, the barons of industry, and high-ranking members of the bureaucracy had on politics, these tensions and problems could not be effectively remedied.[5]

Foremost among the problems afflicting the Reich as a result of this combination of political immobility and rapid social change was the rise of a radical labor movement. The growing number of strikes, lockouts, and other forms of labor unrest, coupled with the growing success of the officially Marxist Social Democratic Party at the polls, provoked fear and anxiety among many middle- and upper-class Germans regarding the seemingly hostile, uncontrollable, and ever-increasing industrial proletariat.[6] In addition, there were other social problems that were viewed by Germany's *Bildungsbürgertum* as posing a threat to the proper functioning of the state. These included an increase in various types of criminal activity; a rise in prostitution, suicides, alcohol consumption, and alcoholism; and a heightened awareness of the existence of large numbers of insane and feebleminded individuals. This latter group, the so-called mental defectives, was singled out by both medical and lay observers as an especially grave social and financial liability for the new Reich.[7]

[4] Walther G. Hoffmann, "The Take-Off in Germany," in *The Economics of Take-Off into Sustained Growth,* ed. W. W. Rostow (London: Macmillan, 1963), pp. 95–118; Hans-Ulrich Wehler, *Das deutsche Kaiserreich* (Göttingen: Vandenhoeck & Ruprecht, 1977), pp. 24, 41–59; and Wolfgang Köllmann, "The Process of Urbanization in Germany at the Height of the Industrialization Period," *Journal of Contemporary History,* 1969, *4*:59–72, on p. 62.

[5] For a discussion of this issue see Wehler, *Das deutsche Kaiserreich,* pp. 60–140; Richard J. Evans, ed., *Society and Politics in Wilhelmine Germany* (New York: Barnes & Noble, 1978), pp. 16–22; and Wolfgang Mock, "Manipulation von oben oder Selbstorganisation an der Basis? Einige neuere Ansätze in der englischen Historiographie zur Geschichte des deutschen Kaiserreichs," *Historische Zeitschrift,* 1981, *232*:358–375.

[6] For a brief discussion of the middle-class fear of the proletariat see Fritz Ringer, *The Decline of German Mandarins: The German Academic Community, 1890–1933* (Cambridge, Mass.: Harvard Univ. Press, 1969), p. 129; Fritz Stern, "The Political Consequences of the Unpolitical German," in *The Failure of Illiberalism: Essays on the Political Culture of Modern Germany* (Chicago: Univ. Chicago Press, 1975), p. 15; and Guenther Roth, *The Social Democrats in Imperial Germany* (1963; New York: Arno Press, 1979), pp. 85–101.

[7] See Vincent E. McHale and Eric A. Johnson, "Urbanization, Industrialization and Crime in Imperial Germany," *Social Science History,* 1976–1977, *1*:45–78, 210–247, on pp. 212–214; Eduard O. Mönkemöller, "Kriminalität," in *Handwörterbuch der sozialen Hygiene,* ed. Alfred Grotjahn and J. Kaup (Leipzig: F. C. W. Vogel, 1912), Vol. I, pp. 687–688; Mönkemöller, "Selbstmord," *ibid.,* Vol. II, p. 376; E. Fuld, "Das rückfälliger Verbrechertum," *Deutsche Zeit- und Streit-Fragen,* 1885, *14*:453–484; Richard Evans, "Prostitution, State and Society in Imperial Germany," *Past and Present,* 1976, No. 70, pp. 106–129, on pp. 106–108; James S. Roberts, "Der Alkoholkonsum

These problems were hotly debated by many of Germany's academic social scientists and reform-minded religious leaders under the rubric of the *soziale Frage*—a term referring to the social and political consequences of unbridled economic liberalism and the industrialization process.[8] Although those discussing the "social question" embraced different economic and political ideals, all agreed that some kind of *Sozialpolitik* (social policy) was necessary to integrate Germany's proletariat (and asocial subproletariat) into the Reich, thereby preventing the collapse of the state. Like most educated middle-class Germans, the early eugenicists were keenly aware of this debate and were fully cognizant of the serious social problems that plagued the Reich as a result of the industrial revolution.[9] The increased visibility of a number of asocial, nonproductive types—an important component of the much-debated "social question"—was the problem they set out to tackle using a new form of *Sozialpolitik*: race hygiene.

That these race hygienists would be inclined to offer a biomedical solution for social and political problems can be attributed to the second major influence that shaped their eugenics: the distinctive social, political, and intellectual traditions of the German medical community. All of the movement's important leaders were physicians by training and had studied medicine before turning their attention to eugenics. Moreover, fully a third of those affiliated with the Deutsche Gesellschaft during its early years were medically trained. As physicians, the founders of German eugenics not only shared the prejudices and posture of the *Bildungsbürgertum* as a whole, but were also heir to a well-defined set of assumptions about the hereditary nature of disease and the role of medical professionals in safeguarding the health of the nation.

The medical professionals' perception of themselves as custodians of national health, and hence of national wealth and efficiency, has a long history. In Germany it dates at least as far back as the mid-nineteenth century, when German physicians demonstrated their responsibility to the state during the so-called health reform movement.[10] Later, during the third quarter of the nineteenth century, the rise of scientific medicine and hygiene bestowed upon academic physicians, and medical professionals in general, an unprecedented level of social

---

deutscher Arbeiter im 19. Jahrhundert," *Geschichte und Gesellschaft,* 1980, *6*:220–242, on pp. 226, 232, 237; Alfred Grotjahn, "Alkoholismus," in *Handwörterbuch der sozialen Hygiene,* Vol. I, p. 14; Grotjahn, "Krankenhauswesen," *ibid.,* Vol. II, p. 643; Ludwig Meyer, "Die Zunahme der Geisteskranken," *Deutsche Rundschau,* 1885, p. 83; Alexander von Oettingen, *Die Moralstatistik in ihrer Bedeutung für eine Sozialethik* (3rd ed., Erlangen: A. Deichert, 1882), p. 671; and Arthur von Fircks, *Bevölkerung und Bevölkerungspolitik* (Leipzig: C. L. Hirschfeld, 1898), p. 116.

[8] Albert Müssigang, *Die soziale Frage in historischen Schule der deutschen Nationalökonomie* (Tübingen: J. C. B. Mohr, 1968), p. 4; Adolf Wagner, *Rede über die soziale Frage* (Berlin: Wiegandt & Grieben, 1872); Gustav Schmoller, "Die soziale Frage und der preussische Staat," in *Quellen zur Geschichte der sozialen Frage in Deutschland,* ed. Ernst Schraepler (2nd ed., Göttingen: Musterschmidt, 1964), Vol. II, pp. 62–66; Friedrich Naumann, "Christlich-Sozial," *ibid.,* pp. 79–84; Ringer, *Decline of the German Mandarins* (cit. n. 6), pp. 145–147.

[9] For a discussion of the attitudes and prejudices of the *Bildungsbürgertum* see Klaus Vondung, ed., *Das Wilhelminische Bildungsbürgertum: Zur Sozialgeschichte seiner Ideen* (Göttingen: Vandenhoeck & Ruprecht, 1976); and Stern, "Political Consequences of the Unpolitical German" (cit. n. 6), *passim.*

[10] Erwin H. Ackerknecht, "Beiträge zur Geschichte der Medizinalreform von 1848," *Archiv für Geschichte der Medizin,* 1932, *25*:61–109, 112–183; and George Rosen, "Die Entwicklung der sozialen Medizin," in *Seminar: Medizin, Gesellschaft, Geschichte,* ed. H.-U. Deppe and M. Regus (Frankfurt am Main: Suhrkamp, 1975), pp. 99–102.

esteem and, indirectly, political importance.[11] At this time many young medical professionals eager to make a contribution to national health turned their attention to bacteriology; others, like some of Germany's future eugenicists, adopted a different approach. Their exposure to fields of medicine that emphasized the role of heredity in the etiology of disease (e.g., neurology and psychiatry) led them to question the efficacy of concentrating solely on pathogens. Instead, they were convinced that serious disorders such as mental illness, feeblemindedness, criminality, epilepsy, hysteria, and the tendency to tuberculosis were often inherited and could quite frequently be traced back to a "hereditary diseased constitution."[12] Many medically trained race hygienists argued that the surest way to improve the general level of national health was to upgrade the bodily constitution of all individuals in society—a task to be accomplished by means of an energetic eugenics program.

In addition to the social question and the German medical tradition there was a third influence that greatly shaped the early development of the movement: the "selectionist" variety of social Darwinism popularized by Germany's most outspoken biologist, Ernst Haeckel (1834–1919), and later legitimated by the scientific writings of the Freiburg embryologist August Weismann (1834–1914).

Haeckel went far beyond Darwin in his attempt to flesh out the larger philosophical and social meaning of the evolutionary theory. Although, like Darwin, he believed in the inheritance of acquired characteristics, Haeckel always stressed Darwin's selection principle as the most important engine of forward-directed organic change; indeed, for Haeckel, Darwinism was synonymous with selection.[13] Weismann, who came to reject the possibility of the inheritance of acquired characteristics through his work on heredity, afforded Darwin's principle of natural selection an even greater role in organic and social evolution than did the author of the *Origin of Species* himself. His famous mechanism of heredity, "the continuity of the germ plasm," first articulated in 1883, challenged the basic tenets of the more optimistic first-generation social Darwinists who assumed that new characteristics acquired by an organism as a result of environmental change would be transmitted to future generations.[14] As one German social Darwinist and eugenicist expressed it,

[11] George Rosen, *A History of Public Health* (New York: M.D. Publications, 1958), p. 44; Eduard Seidler, "Der politische Standort des Artzes im zweiten Kaiserreich," in *Medizin, Naturwissenschaft, Technik und das zweite Kaiserreich,* ed. Gunter Mann and Rolf Winau (Göttingen: Vandenhoeck & Ruprecht, 1977), pp. 91–92; and Hans-Heinz Eulner, "Hygiene als akademisches Fach," in *Städte-, Wohnungs- und Kleidungshygiene des 19. Jahrhunderts in Deutschland,* ed. Walter Artelt et al. (Stuttgart: Ferdinand Enke, 1969), p. 18.

[12] Erwin H. Ackerknecht, *A Short History of Psychiatry,* 2nd ed., trans. S. Wolff (New York/London: Hafner, 1968), p. 82; K. Grassman, "Kritische Ueberblick über die gegenwärtige Lehre von der Erblichkeit der Psychosen," *Allgemeine Zeitschrift für Psychiatrie,* 1896, *52*:960–1022; Oscar Aronson, *Ueber Heredität bei Epilepsie* (Berlin: Wilhelm Axt, 1894); M. Wahl, "Uber den gegenwärtigen Stand der Erblichkeitsfrage in der Lehre von der Tuberculose," *Deutsche Medizinische Wochenschrift,* 1885, No. 1, pp. 3–5, No. 3, pp. 36–38, No. 4, pp. 34–36, No. 5, pp. 69–71, No. 6, pp. 88–90; and "Über die Vererbung von Geisteskrankheiten nach Beobachtung in preussischen Irrenanstalten," *Jahrbuch für Psychiatrie und Neurologie,* 1879, *1*:65–66.

[13] Ernst Haeckel, *The History of Creation,* trans. and rev. by E. Ray Lankester (London: Henry S. King, 1876), Vol. I, p. 120.

[14] August Weismann, "On Heredity," in *Essays on Heredity and Kindred Problems,* ed. Edward Poulton, Selmar Schönland, and Arthur Shipley (Oxford: Clarendon Press, 1889); and Frederick Churchill, "August Weismann and a Break from Tradition," *Journal of the History of Biology,* 1968, *1*:91–112.

> It was Weismann's teaching regarding the separation of the germ plasm from the soma, the hereditary stuff from the body of the individual, that first allowed us to recognize the importance of Darwin's principle of selection. Only then did we comprehend that it is impossible to improve our progeny's condition by means of physical and mental training. Apart from the direct manipulation of the nucleus, only selection can preserve and improve the race.[15]

Indeed, for those who accepted Weismann's views with respect to both heredity and the "all-supremacy" of selection, eugenics was the only practical strategy to ensure racial progress and avert racial decline.

If the ideas of Haeckel and Weismann encouraged many contemporaries to view natural selection as the sole agent of all organic and social progress, the writings of the two biologists also emphasized that progress was not inevitable. Under certain conditions the "unfit" might prosper, thereby posing a challenge to further evolutionary development. This "selectionist" perspective and language provided Germany's future eugenicists with novel tools of analysis that enabled them to come to grips with the social question by transforming it into a scientific problem: the asocial individuals created by industrialization became for them the biologically and medically unfit. The only way to eliminate this group from the population was through a policy of "rational selection," or race hygiene.

The significance of the three contexts is nowhere more clearly visible than in the intellectual backgrounds and early writings of Alfred Ploetz and Wilhelm Schallmayer. Working largely independently of one another during the prehistory of the movement (1890–1903), both men laid the foundations for the future course of race hygiene in their country. Ploetz's organizational talents and charismatic personality allowed him to create the institutional basis for the young movement almost single-handedly. Schallmayer's treatises on eugenics defined the significant theoretical and practical problems that would occupy German eugenicists for decades.

Alfred Ploetz

Ploetz was born in 1860 into an upper-middle-class family in Swinemünde on the Baltic Sea. Although details of his early life remain sketchy, he had become acquainted with the works of Darwin and Haeckel while still at the *Gymnasium*. Even before he began to study economics at the University of Breslau in 1884, he developed a strong interest in the *soziale Frage*; during his student days at Breslau he became increasingly sympathetic to some forms of socialism. Indeed, he transferred to Zurich in 1885 in order to become better acquainted with the various brands of socialist theory. There he not only attended lectures on socialism but became personally acquainted with August Bebel, the leader of the German Social Democratic Party, and other socialists in exile from Bismarck's oppressive Anti-Socialist Law.[16]

Ploetz's interest in economic theory, particularly socialism, was not merely

[15] Hermann W. Siemens, *Die biologischen Grundlagen der Rassenhygiene und der Bevölkerungspolitik* (Munich: J. G. Lehmann, 1917), p. 10. This and all translations are my own.

[16] Doeleke, *Alfred Ploetz* (cit. n. 2), pp. 4, 18; and Hermann Muckermann, "Alfred Ploetz und sein Werk," *Eugenik,* Erblehre, Erbpflege, 1931, *1*:261.

theoretical: he was determined to establish a kind of pan-Germanic utopian commune. As he stated in his memoirs, the popular novels of Felix Dahn, professor of early German history, as well as works of other enthusiasts of Germany's Teutonic past, awakened his interest in the Germanic race. Indeed, Ploetz became so obsessed with the glories of the old Teutonic tribes that he and several friends took an oath under an oak tree to do everything in their power to elevate the Germanic race to the level it had allegedly attained a thousand years earlier. Thanks to this passionate concern, he chose to study economics rather than his first love, biology, believing it would prove to be more useful in helping him accomplish his goal. While at the University of Breslau, Ploetz and a small circle of friends—including the writers Carl and Gerhart Hauptmann—formed a society with the expressed intention of establishing a colony or socialist cooperative in a country where a large part of the population was of Germanic stock. The American Pacific Northwest was chosen as the best possible site.[17]

Ploetz traveled to the United States to familiarize himself with the social and economic conditions of the region and to experience, at first hand, life in one of the already established utopian socialist colonies. He spent six months in Iowa living and working in the cooperative known as Icarus. Appalled at the amount of fighting, laziness, egotism, and infidelity he observed in a community whose economic organization was supposed to eradicate such behavior, Ploetz came to an unusual conclusion: "The unity of such colonies, especially those offering a large amount of individual freedom, cannot be maintained owing to the average [quality] of human material at present. . . . I came to the conclusion that the plans we wished to execute would be destroyed as a result of the [low] *quality of human beings*." His reaction to this revelation was no less surprising: "For this reason I must direct my efforts *not merely toward preserving the race but also toward improving it*. . . . My views . . . immediately led me to the field of medicine—which appeared to be relevant to the biological transformation of human beings."[18]

After his six-month stay in the United States, Ploetz returned to Zurich to begin his medical studies; later, under the direction of the psychiatrist and future eugenicist August Forel, he began an internship in a Swiss mental hospital. Although he had harbored certain proeugenic sentiments even before he began his medical training, Ploetz moved a step closer to articulating the need for race hygiene as a result of it. His experiences in the psychiatric hospital acquainted him with the so-called mental defectives and focused his attention on one cause of the problem: alcoholism. Largely as a result of his discussions with Forel on the subject of alcohol and heredity, Ploetz took an oath of abstinence. From that point on, the counterselective effects of alcohol consumption became one of his major eugenic concerns.[19]

After taking specialized medical courses in Paris in 1890, Ploetz returned to the United States and opened a medical practice in Springfield, Massachusetts. However, he found this work very disappointing. In addition to resenting the time it took from his study of eugenic problems, Ploetz was dismayed by the limitations of therapeutic medicine. Having read Haeckel, he could not have

[17] Muckermann, "Ploetz und sein Werk," p. 261; and Doeleke, *Alfred Ploetz,* pp. 5, 6.
[18] Doeleke, *Alfred Ploetz,* p. 13; emphasis in original.
[19] *Ibid.*, pp. 17–18.

overlooked the attack on medical science in his popular work, *The History of Creation.* By this time Ploetz had recognized the need for a separate discipline dedicated to the hereditary improvement of the race—a discipline more effective in eliminating disease than the thankless "Sisyphean labor" carried out by modern therapeutic medicine.[20]

Influenced by this constellation of social, professional, and intellectual contexts, Ploetz completed *Die Tüchtigkeit unsrer Rasse und der Schutz der Schwachen* (The fitness of our race and the protection of the weak) in 1895.[21] Although initially his book did not generate much public interest, it raised the broad biological, social, and ethical problems that created the need for race hygiene in the first place. It also revealed the technocratic logic underlying eugenic thought.

The major thrust of Ploetz's argument recalls Darwin's personal dilemma in the *Descent of Man*: How can human beings reconcile the inevitable conflict between the humanitarian ideals and practices of the noblest part of our nature with the interest of the race, whose biological efficiency is allegedly impaired by those very ideals and practices? Translated into concrete economic and political terms, Ploetz viewed the problem as follows: Should the state continue to expand the social net and regulate various aspects of economic life in order to lessen the hardship of the weak and economically underprivileged, at the risk of undermining the overall biological fitness of its citizens? Would not health, accident, and old-age insurance invariably lead to an increase in the number of unfit, perhaps at the expense of the fittest members of society?[22]

Ploetz was not oblivious to the serious moral and social issues raised by this alleged conflict. As important as preventing *Entartung* (degeneration) was for him, Ploetz did not believe in ignoring the needs of the present generation; the danger of *Entartung* was not a signal for Germany to abandon health and welfare legislation, despite its counterselective effects. Nor did it mean that one must embrace capitalism, the seemingly most "proselective" economic system, and relinquish all hope of creating a humane, socialist society. The solution to these pressing conflicts was the substitution of a humane and scientific policy of "rational selection" for the inhumane and inefficient process of natural selection. Unlike the existing personal hygiene movement, with its concern for the health of the individual, the new hygiene would direct its attention to improving the hereditary fitness of the human race. Ploetz named it *Rassenhygiene.*[23]

Considering Ploetz's own enthusiasm for all things Teutonic and the heated controversy that later ensued over the use of the word *Rassenhygiene* as a synonym for *eugenics,* it is worth examining what he meant by the term. His definition of *Rasse* is ambiguous and difficult to translate into English: "einer durch Generationen lebenden Gesammtheit von Menschen in Hinblick auf ihre körperlichen und geistigen Eigenschaften." Roughly speaking, Ploetz seems to view as a *Rasse* any interbreeding human population that, over the course of generations,

[20] *Ibid.,* p. 19.

[21] Alfred Ploetz, *Die Tüchtigkeit unsrer Rasse und der Schutz der Schwachen. Ein Versuch über Rassenhygiene und ihr Verhältnis zu den humanen Idealen, besonders zum Socialismus* (Berlin: Fischer, 1895). The work was intended to be the first part of a two-part study entitled *Grundlagen einer Rassenhygiene* (Foundations of race hygiene). The other section was never completed.

[22] Alfred Ploetz, "Rassentüchtigkeit und Sozialismus," *Neue Deutsche Rundschau,* 1894, 5:989–997; and Ploetz, *Die Tüchtigkeit unsrer Rasse,* esp. Ch. 3.

[23] Ploetz, *Die Tüchtigkeit unsrer Rasse,* p. 5.

continues to demonstrate similar physical and mental traits. This imprecise term could denote any small ethnic community, a nation, an anthropological race, or the entire human race.[24] Ploetz's use of his newly coined term *Rassenhygiene* is equally broad and vague, denoting the hygiene of any and all of these groups. Somewhat later he described race hygiene as the measures needed to ensure "the optimal preservation and development of a race." Hence the term referred to the hereditary improvement of such disparate populations as the Jews, the Germans, the "Aryans," and all humanity. Lenz later suggested that Ploetz was not familiar with Francis Galton's term *eugenics* when he wrote his book and simply chose the word *Rassenhygiene*. In later publications Ploetz explained that *Rassenhygiene* had a much larger scope than the English term *eugenics,* embracing not only those measures designed to improve the hereditary quality of a population but also those aimed at achieving its so-called optimal size.[25] As alternatives, he could later have used either the Germanized form of the word *eugenics, Eugenik,* or employed *Rassehygiene* (hygiene of the human race), as Schallmayer suggested.

What Ploetz actually thought about race is most clearly revealed in *Die Tüchtigkeit unsrer Rasse,* the one published source in which he devotes a significant amount of discussion to these questions: "The hygiene of the entire human race converges with that of the Aryan race, which apart from a few small races, like the Jewish race—itself quite probably overwhelmingly Aryan in composition—is the cultural race *par excellence*. To advance it is tantamount to the advancement of all humanity." Although he states elsewhere that Germanic stock probably represents the best portion of the "Aryan race," he is primarily concerned here with whites in general. His views regarding the alleged cultural superiority of white people, however outrageously chauvinistic, were not fundamentally different from those of Schallmayer (who was vehemently opposed to the Aryan mystique) or, indeed, from those of most European intellectuals of his time. Nor was his pro-Aryan sentiment in any way anti-Semitic. Ploetz was, if anything, pro-Semitic at the time he wrote his book (although his views later appear to have changed.)[26] Not only did he stress the significant role played by Jews in the intellectual history of humanity, placing them on the same level as the Aryans in terms of their cultural capacity, but he also opposed all attempts to ghettoize or otherwise separate the former from the latter. He was strongly in favor of marriages between Jews and Aryans on the grounds that these would be both socially and *biologically* advantageous. Ploetz wrote his treatise at a time when economic anti-Semitism was making a strong comeback in Germany. His favorable discussion of the Jews, he stated, was included in his work partly in order to combat the new trend. He had, at least at the time, little patience with Jew-haters. "All anti-Semitism is a pointless pursuit—a pursuit whose support will slowly recede

[24] *Ibid.,* p. 2; see also Lilienthal, "Rassenhygiene im dritten Reich" (cit. n. 2), pp. 115–116.

[25] Alfred Ploetz, "Die Begriffe Rasse und Gesellschaft und die davon abgeleiteten Disziplinen," *Archiv für Rassen- und Gesellschafts-Biologie,* 1904, *1*:2–26, on p. 11. For Ploetz's discussion of the numerous subdivisions of *Rassenhygiene* see his "Zur Abgrenzung und Einteilung des Begriffs Rassenhygiene," *ibid.,* 1906, 3:864–866, on p. 865.

[26] Ploetz, *Die Tüchtigkeit unsrer Rasse* (cit. n. 21), p. 5. Although Ploetz does not discuss his views on the Jews, his letters to his lifelong friend Gerhart Hauptmann, written during the Third Reich, show evidence of at least tacit support for the Nazi regime. Anyone who views himself as a "Mithelfer des dritten Reiches" could hardly have been pro-Semitic; see Ploetz to Hauptmann, 12 Oct. 1936, GH Br NL A. Ploetz, Staatsbibliothek Preussischer Kulturbesitz Berlin.

with the tide of scientific knowledge and humane democracy," as he said in *Die Tüchkigkeit.*[27] Hence, the ideas expressed in this book could later be incorporated into Nazi racial policy only by misrepresenting the views of the author.

Although Ploetz discussed the merits of the Aryan race at length and defined the terms *Rasse* and *Rassenhygiene,* the major purpose of his book lay elsewhere. Above all, Ploetz sought to reconcile the inherent conflict between the Darwinian world view and the humanitarian-socialist ideal through a conscious policy of "control over variation." He had in mind a utopian vision of pushing selection back to the prefertilization stage—a form of germ plasm selection. According to this plan, the genetically best germ cells of all married couples would be chosen as the hereditary endowment for the next generation. As a result, inhumane social measures and economic systems previously deemed necessary to avert biological decline would become superfluous. "The more we can prevent the production of inferior variations," Ploetz asserted, "the less we need the struggle for existence to eliminate them."[28] Although Ploetz's particular solution to the "degeneration problem" was unfeasible and was never seriously entertained by any of Germany's race hygienists, it embodied the view, shared by Schallmayer and later by other eugenicists, that population was a resource amenable to "rational management." As such, it was a biomedical solution to sociopolitical problems: eugenics experts, armed with their knowledge of evolutionary theory and the laws of heredity, would solve the social question with the aid of science.

Although Ploetz's work was the first to employ the term *Rassenhygiene,* it was not the first treatise on eugenics to be published in Germany. The author of the earliest such tract was the Bavarian physician Wilhelm Schallmayer.

Wilhelm Schallmayer

Schallmayer's intellectual biography and early career closely parallel those of Ploetz. Schallmayer was born in Mindelheim, Bavaria. Like Ploetz, he enjoyed the comforts of middle-class life. His father was the owner of a prosperous carriage and wagon business. Before turning to medicine, Schallmayer studied economics, sociology, and socialist theory for two years at the University of Leipzig; he found the works of Karl Marx and the German economist and social theorist Alfred Schaeffle especially interesting.[29] Yet despite the interest he shared with Ploetz in the *soziale Frage* and socialist theory, Schallmayer's concern was purely theoretical—at least until he realized that a faulty social and economic system could have grave eugenic consequences. Unlike Ploetz, he never harbored utopian dreams of building a socialist Germanic community in the New World. Indeed, throughout his life he refused to embrace those ideologies of Aryan supremacy so important to Ploetz and several other eugenicists.

In 1881 Schallmayer enrolled in medicine at Munich. Upon completion of his degree he secured an internship in the university hospital's psychiatric clinic, where he worked under Bernhard von Gudden (1824–1886). It is possible only to

[27] Ploetz, *Die Tüchtigkeit unsrer Rasse,* p. 142.

[28] *Ibid.,* pp. 226–239; see Graham, "Science and Values" (cit. n. 3), p. 1137.

[29] Max von Gruber, "Wilhelm Schallmayer," *Arch. Ras. Ges.-Biol.,* 1922, 14:52–56, on p. 55; and Fritz Lenz, "Wilhelm Schallmayer," *Münchener Medizinische Wochenschrift,* 1919, 66:1294–1296, on p. 1295.

speculate on how Schallmayer's internship might have influenced his later eugenic thought. He undoubtedly witnessed some of the most severe forms of insanity, thus becoming directly acquainted with many "mental defectives." Regarding the treatment and care of the insane and retarded, the young physician probably came away with the views of his teacher, which, as one obituary of von Gudden reported, amounted to a "near complete resignation regarding the effectiveness of medical intervention."[30]

Precisely when Schallmayer became interested in eugenics is unknown. It seems likely, however, that his work in the psychiatric clinic led him to doubt the value of medicine for improving the health of the race. A self-proclaimed social Darwinist and an admirer of Haeckel, he could not have failed to see the connection between his clinical experience and the articulation of the counterselective effects of medicine presented in *The History of Creation.* Schallmayer's own experiences working with "mental defectives," coupled with his "selectionist" outlook, accounted for his own indictment of therapeutic medicine in his first eugenic treatise, a short work entitled *Über die drohende körperliche Entartung der Kulturmenschheit* (Concerning the threatening physical degeneration of civilized humanity). Published in 1891, and reprinted under a slightly different title in 1895, it was Germany's first eugenic tract.[31]

Although Schallmayer's slim volume attracted even less attention than Ploetz's treatise would four years later, it touched on the social, economic, and political justifications for eugenics, and it offered such practical proposals as the creation of medical genealogies and health passports and the introduction of marriage restrictions.[32] Schallmayer's book also stressed the role of physicians and the importance of education and propaganda as the most effective means of achieving eugenic goals—two hallmarks of German race hygiene policy until 1933. Most important, however, Schallmayer's treatise emphasized the technocratic logic and the cost-benefit analysis that later so colored the race hygiene movement.

Schallmayer's frustrations over the limitations of therapeutic medicine also shaped his personal career. Soon after he began work as a general practitioner, he decided to specialize in urology and gynecology. At least in this area, Schallmayer thought, the prevention and treatment of disease would benefit future generations as well as the individual. Ultimately, however, he found even this work disappointing. Like Ploetz, Schallmayer was eager to devote all his time to the cause of eugenics. So, in 1897, after he had acquired sufficient means to give up his lucrative medical practice in Düsseldorf, Schallmayer settled down as *Privatgelehrter.*[33] His newly won freedom afforded him the time to compose a second eugenic treatise with the specific intention of submitting it to the Krupp *Preisausschreiben.*

In 1900 Friedrich Krupp, son of Essen's munitions baron Alfred Krupp, set aside 30,000 marks to be used in a contest to answer the question: What can we learn from the theory of evolution about internal political development and state legislation? It seems likely that Krupp, an amateur biologist, greatly resented

[30] Emil Kraepelin, "Berhard von Gudden," *Münch. Med. Wochensch.*, 1886, *33*:577–607, on p. 607; see also Gruber, "Wilhelm Schallmayer" (cit. n. 29), p. 55.

[31] Wilhelm Schallmayer, *Die drohende physische Entartung der Culturvölker* (2nd ed., Neuwied: Heuser, 1895). I have been unable to obtain a copy of the first edition of Schallmayer's book. From the reviews of the first edition, it seems that the second edition varied little, if at all.

[32] *Ibid.*, pp. 23–32.

[33] Lenz, "Wilhelm Schallmayer" (cit. n. 29), p. 1295.

Social Democratic attempts to use Darwin's theory in support of socialism. Wishing to remain anonymous, Krupp delegated most of the responsibility for the contest to politically sympathetic scholars, with Ernst Haeckel brought in as a figurehead.[34]

On 7 March 1903 the prize committee announced that first prize in the contest was awarded to Schallmayer's *Vererbung und Auslese im Lebenslauf der Völker* (Heredity and selection in the life process of nations), a dense 381-page treatise representing, in Heinrich Ernst Ziegler's appraisal, a "hygienic-sociological" approach to the question. Schallmayer certainly saw the practical and political aims of his book as timely. Whereas the nineteenth century had been concerned with Darwin's evolutionary hypothesis on a purely theoretical level, "the twentieth century," argued Schallmayer, "is called upon to apply the theory of descent to everyday life."[35]

The book's central theme was the rational management of national efficiency. The real political lesson to be learned from Darwin's theory was that long-term state power depended upon the biological vitality of the nation; neglect of the hereditary fitness of its population, such as might result from unenlightened laws and customs, was "bad politics" and would inevitably result in the downfall of the state. Hence the wise politician "would recognize that the future of his nation is dependent on the *good management* of its human resources." In the interest of self-preservation, he argued, it was imperative that Germany take an active part in regulating the overall biological efficiency of its citizens by embarking on a political program that would encourage the biologically best elements in society to reproduce more than those with objectionable hereditary traits. Eugenics, or *Vererbungshygiene* (hereditary hygiene), as he still called it, was the perfect tool to ensure a strong and healthy state; it went hand in hand with his political ideal—a meritocracy.[36]

Schallmayer also presented his readers with a series of eugenics reforms, but he was very cautious in the area of negative eugenics. Although he clearly believed that marriage restrictions for the insane, the feebleminded, the chronic alcoholic, and other defectives were in the best interest of the state and the race, he refrained from openly supporting state legislation as a means to this end. Until such time as more exact information regarding the laws of heredity was known, and enough detailed genealogies could be amassed, he felt that eugenicists would have to concentrate on voluntary measures. He emphasized positive eugenics: convincing the "fitter" groups in society to increase their fertility rate. The question of course remained, Which groups were, biologically speaking, the "fittest"? Schallmayer assumed that biology would one day decide the question objectively. "In the meantime," he argued, "it would not be incorrect to view highly

[34] Heinrich Ernst Ziegler, "Einleitung zu dem Sammelwerke Natur und Staat," in *Natur und Staat: Beiträge zur naturwissenschaftlichen Gesellschaftslehre* (Jena: Gustav Fischer, 1903), pp. 1–4; Ziegler to Haeckel, 4 Oct. 1899, Ernst-Haeckel-Haus Jena, Best. A-Abt. 1 No n 00005; Krupp "*Niederschrift,*" Jan. 1900, Krupp-Archiv Essen, IX-d-244, and correspondence in III-D-159. I want to thank Jeffrey Johnson for his notes from the Krupp-Archiv. For a fuller discussion see Sheila Faith Weiss, *Race Hygiene and National Efficiency: The Eugenics of Wilhelm Schallmayer* (Berkeley/Los Angeles: Univ. California Press, forthcoming 1987), Ch. 3.

[35] Wilhelm Schallmayer, *Vererbung und Auslese im Lebenslauf der Völker: Eine staatswissenschaftliche Studie auf Grund der neueren Biologie* (Jena: Gustav Fischer, 1903), p. x. The treatise went through two revised editions (1910 and 1918).

[36] *Ibid.*, pp. 380–381 (quotation: emphasis in original), 354, 373. Schallmayer's clearest position regarding meritocracy can be found in "Rassehygiene und Sozialismus," *Die Neue Zeit*, 1907, *25*:731–740, on p. 735.

socially productive individuals, especially the better educated, as being, on the average, more biologically valuable." Civil servants, officers, and teachers were encouraged to marry as early as possible. Those who chose to remain single should suffer some sort of financial disadvantage. To encourage civil servants to have larger families, Schallmayer suggested they be given a bonus for each school-aged child.[37] The class bias implicit in Schallmayer's criteria could hardly be more blatant; his own social group, the *Bildungsbürgertum,* turns out to be the "fittest." Other eugenicists shared this orientation.

THE WILHELMINE RACE HYGIENE MOVEMENT, 1904–1918

The Journal and the Society

The Krupp competition marked a turning point both in Schallmayer's personal career and in the attention paid to eugenics in Germany. Prior to this time Schallmayer and Ploetz were virtually lone prophets in their eugenics crusade. To be sure, there were undoubtedly other Germans concerned with similar issues; yet insofar as they were not personal friends of either of the two cofounders, they remained in complete intellectual isolation. Only in the years immediately following the publication of *Vererbung und Auslese* in 1903 were the first institutional steps undertaken to transform an idea into a movement: the creation of Germany's most respected eugenics journal and the foundation of a race hygiene society.

The *Archiv für Rassen- und Gesellschafts-Biologie,* the first journal in the world dedicated to eugenics, was founded by Ploetz in 1904. Although there is no direct evidence linking the creation of the journal with the results of the Krupp contest, it seems likely that the scientific recognition and public attention given eugenics in the immediate aftermath of the *Preisausschreiben* at least suggested to Ploetz and his two assistant editors, the sociologist and economist Anastasius Nordenholz and the zoologist Ludwig Plate, that a more organized and "strictly scientific" manner of discourse on the subject was possible. Without openly admitting it, the editors of the *Archiv* sought to establish a more clearly focused and academically prestigious form of the *Politisch-anthropologische Revue*—a journal that occasionally carried articles on eugenics-related issues but was not taken seriously by most professionals because of its unmistakably *völkisch* tone. During the first four years the *Archiv* was financed by the publishers themselves. By 1908 the journal had proved marketable enough to convince a publishing company to underwrite the cost of its production; whether Ploetz, who was independently wealthy owing to his marriage to Nordenholz's sister, continued to help finance it is unknown.[38]

The *Archiv* sought to attract a wide variety of articles bearing on the "optimal preservation and development of the race."[39] It included entries not only by

[37] Schallmayer, *Vererbung und Auslese,* 1st ed. (1903), p. 338.

[38] Ploetz married Anita Nordenholz after divorcing Ernst Rüdin's sister. The Nordenholzes were the children of a wealthy merchant family from Bremen. This marriage enabled Ploetz to retire from his medical practice and devote his time exclusively to eugenics. He settled with his wife on their estate in Herrsching, near Munich; see Heinrich Reichel, "Alfred Ploetz und die rassenhygienische Bewegung der Gegenwart," *Wienischer Klinische Wochenschrift,* 1931, *44*:1–9, on p. 6.

[39] Alfred Ploetz, Anastasius Nordenholz, and Ludwig Plate, "Vorwort," *Arch. Ras. Ges.-Biol.,* 1904, *1*:iii–vi, on p. iv.

Germany's prominent race hygienists but also from individuals who in no sense considered themselves eugenicists. Most of the articles appearing in the journal during the Wilhelmine period fall into one of five categories: technical articles dealing with genetics and evolution by such leading biologists as Weismann, Plate, Ziegler, Richard Semon, Carl Correns, Hugo deVries, Erich von Tschermak, and Wilhelm Johannsen; entries concerned with so-called degenerative phenomena (insanity, alcoholism, homosexuality, etc.); articles preoccupied with the alleged dysgenic effects of certain social institutions and practices (medicine, welfare, etc.) and the social and economic costs of "protecting the weak"; studies pertaining to the need for population increase and the hazards of neo-Malthusianism; and a potpourri of anthropological contributions, including many racialist, but not always racist, articles as well as high-quality entries from the eminent anthropologist Franz Boas.

Besides publishing rather specialized and lengthy articles, the *Archiv* also tried to keep its readers abreast of developments in eugenics through its numerous book reviews and announcements. Its volumes were substantial indeed—the four quarterly issues together often totaled more than six hundred pages. Although most educated middle-class Germans could have "plowed through" the *Archiv*, the long, dry, and technical articles made neither enjoyable nor easy reading. Its national and international reputation as a highly respected scholarly publication notwithstanding, the journal did little to spread the eugenics gospel in Germany beyond the small group of professionals already committed to the new discipline.

The second institutional development was the formation of the Gesellschaft für Rassenhygiene (Society for Race Hygiene)—the world's first professional eugenics organization. Founded in Berlin on 22 June 1905 by Ploetz, Nordenholz, the psychiatrist Ernst Rüdin (who was also Ploetz's former brother-in-law), and the ethnologist Richard Thurnwald, the society had as its aim "the study of the relationship of selection and elimination among individuals as well as the inheritance and variability of their physical and mental traits."[40] Although there is some confusion as to the exact title of the organization during the first two years of its history, there is little doubt that Ploetz always intended the society, which had begun with only twenty-four members, to be international. Since the word *Rasse* was frequently used by Ploetz as a synonym for "white race," any race hygiene society worthy of the name had to transcend national boundaries and embrace individuals from all white "civilized" nations. Yet it was not until 1907 that the Gesellschaft was able to attract anyone from other countries, at which time it became the Internationale Gesellschaft für Rassenhygiene. Two local groups of the Gesellschaft, in Berlin and Munich, were formed soon after.[41]

The society wished not merely to spread the eugenics gospel but also, and perhaps more important, to serve as a model for what rational selection could accomplish. By offering membership only to those white individuals who were "ethically, intellectually, and physically fit" and from whom "economic prosperity could be expected," the society proposed to demonstrate from statistics collected on both the members and their progeny "how much better the vital statistics, the military fitness, and physical and intellectual efficiency are, compared to

[40] Alfred Ploetz, "Denkschrift über die Gründung der Internationale Gesellschaft für Rassenhygiene," 1907, Ploetz family archives, Herrsching, West Germany, p. 3.

[41] "Zweite Bericht der Internationalen Gesellschaft für Rassenhygiene," 1907, Ploetz family archive, pp. 1, 17.

the population at large . . . and how much more efficient the population of a state would necessarily be that followed race hygiene principles."[42] The society's understanding of "fitness" thus mirrored Schallmayer's own definition of the term: the most important criteria for eligibility were material success and social usefulness.

The Internationale Gesellschaft did not articulate any specific social policy or proposals. However, at a meeting of public hygienists in 1910, Ploetz set forth a list of practical goals that included the following:

> (a) Opposition to the two-child system, fostering "fit" families with large numbers of children, combatting luxury, reestablishment of the motherhood ideal, strengthening the commitment to the family;
> (b) Establishment of a counterbalance to the protection of the weak by means of isolation, marriage restrictions, etc., designed to prevent the reproduction of the inferior; support of the reproduction of the fit through economic measures designed to make early marriages and large families possible (especially in the higher classes);
> (c) Opposition to all germ-plasm poisons, especially syphilis, tuberculosis, and alcohol;
> (d) Protection against inferior immigrants and the settlement of fit population groups in those areas presently occupied by inferior elements—to be accomplished, if need be, through the expropriation laws;
> (e) Preservation and increase of the peasant class;
> (f) Introduction of favorable hygienic conditions for the industrial and urban population;
> (g) Preservation of the military capabilities of the civilized nations;
> (h) Extension of the reigning ideal of brotherly love by an ideal of modern chivalry, which combines the protection of the weak with the elevation of the moral and physical strength and fitness (*Tüchtigkeit*) of the individual.[43]

These proposals reflect the international orientation of Ploetz and the early movement while simultaneously demonstrating their concern with national efficiency. The explicit statement regarding the encouragement of marriages and large families among members of the upper class once again shows the tendency to equate fitness with class. Belonging to the so-called Nordic or Germanic race, interestingly enough, was not a criterion for *Tüchtigkeit*; indeed, the terms *Nordic* and *Germanic* do not appear in the list of tasks and programs. Having mentioned in his speech the lack of any general consensus regarding what constitutes the best race, as well as the rarity of finding pure races anywhere in Europe, Ploetz was indeed reluctant to make special claims for the Nordic population. Since, as he argued, "all these races [Alpine, Jewish, etc.] are seldom found pure here, it is best . . . to rely on fitness as a guide. This is because fitness—both individual and social—is the true guiding star. What particular colors or shapes are attached to [fitness] will come to light in the future."[44] Although Ploetz may have had a particular interest in the Nordic race, his position does suggest that fitness, as defined in terms of social and cultural productivity, was the true measure of the worth of both individuals and races. This belief remained the cornerstone of both his eugenic policy and that of the movement before 1933.

In 1910, the year in which Ploetz presented these proposals, the individual

[42] *Ibid.*, p. 3; Ploetz, "Denkschrift über die Gründung" (cit. n. 40), p. 5.
[43] Alfred Ploetz, "Ziele und Aufgaben der Rassenhygiene," *Vierteljahresschrift für öffentliche Gesundheitspflege*, 1911, *43*:164–191, on p. 165.
[44] *Ibid.*, p. 190.

Table 1. Occupational Composition of the Society for Race Hygiene

Occupation	1907 (International Society) Number	1907 (International Society) Percent	1913 (German Society) Number	1913 (German Society) Percent
Physicians and medical students	27	32.5	136	33.4
Nonmedical academics	14	16.9	76	18.7
Writers and artists	10	12.1	22	5.4
Civil servants and teachers	3	3.6	29	7.1
Miscellaneous	7	8.4	78	19.2
Wives*	22	26.5	66	16.2
Totals	83	100.0	407	100.0

SOURCES: Zweite Bericht der Internationale Gesellschaft für Rassenhygiene, 1907, and Mitgliederliste der Deutsche Gesellschaft für Rassenhygiene, 1913, Ploetz family archives, Herrsching, West Germany.

* Includes only those women, listed with their husbands, who had no other occupation.

German *Ortsgruppen* (chapters) were brought under the banner of the Deutsche Gesellschaft für Rassenhygiene, which was initially a sort of national subdivision of the Internationale Gesellschaft für Rassenhygiene. Yet the international society was not to last. By 1916, in the wake of both World War I and the creation of numerous national eugenics societies in Europe and the United States, Ploetz was forced to give up his dream of a single "intellectual center" for the preservation of the race. At this time the German society officially supplanted the international society; it had long since done so in practice.[45]

In the meantime the total membership remained small, but also grew steadily, and the occupational and class backgrounds of the members of the two societies continued to mirror those of its founders and leaders (Table 1). In both the international and the German society *Bildungsbürger* dominated the membership. Table 1 also reveals that medical professionals made up the single largest group in both organizations, accounting for approximately one third of those affiliated with the two societies. It seems likely that the self-image of German physicians as custodians of the nation's health had much to do with the disproportionate number of prominent physicians, hygienists, and professors of medicine in the early movement. Of the academics from other fields enrolled in the two societies, most were professors of zoology and anthropology. In addition to Ernst Haeckel and August Weismann, who as honorary members probably did not participate much in its activities, the Deutsche Gesellschaft included such distinguished biologists as Ludwig Plate, Heinrich Ernst Ziegler, and Erwin Baur. The two societies also included members of virtually all German political parties; moreover, Jews, as well as Protestants and Catholics, were among the members.[46] The only specific qualifications mentioned in the statutes were that members be both white and "fit."

[45] Deutsche Gesellschaft für Rassenhygiene, 1910 Satzungen, Ploetz family archives; Eugen Fischer, "Aus der Geschichte der Deutschen Gesellschaft für Rassenhygiene," *Arch. Ras. Ges.-Biol.*, 1930, *24*:1–5, on p. 3.

[46] Alfred Ploetz, "Gesellschaften mit rassenhygienischen Zwecken," *Arch. Ras. Ges.-Biol.*, 1909, *6*:277–281, on p. 278.

The Intellectual Development of Wilhelmine Eugenics

The writings of Wilhelmine Germany's race hygienists exhibit some common themes and concerns. The primary intellectual preoccupation of the early movement was with collecting and analyzing data on degeneration. A study of the celebrated Family Zero—a kind of Swiss counterpart to the legendary American Jukes family—was undertaken to demonstrate that central Europe had its own share of degenerate stock. The psychiatrist Ernst Rüdin wrote numerous studies dealing with the inheritance of insanity—emphasizing the Mendelian nature of the transmission of various kinds of mental disorders. Agnes Bluhm, Germany's only prominent female eugenicist, concentrated on proving the degenerative effects of alcohol on future generations and studying the alleged decreased ability of German women to breast-feed their infants.[47] Other eugenicists reported on such topics as the increase in venereal disease in large cities and its impact on the race, the degenerative effects of homosexuality, and the need to reform Germany's penal code along eugenic lines.[48] By and large the tone of these studies was scientific, not popular; they seem to have been written less to stir people to action than to communicate abstract information.

Like eugenicists in the United States and Britain, the Germans also analyzed the cost of maintaining the army of the unfit. The word most often used to describe these individuals was *Minderwertigen*—a term that literally means "the less valuable" and was frequently employed as a synonym for nonproductive people. Certainly the *Umschau*, a popular science journal, used the word in this way when, in 1911, it sponsored a written contest entitled: "What do the inferior elements [*Minderwertigen*] cost the state and society?" Accepting the premise that "all efforts to improve the environment break down in the face of hereditary sickness and inferiority," the sponsors of the competition suggested to potential contestants that only a reduction in the number of "minus variants" would allow society to continue to preserve the life of all those living. Only five contestants applied for the prize, however, and the problem did not yet generate the great concern that it would during Weimar.[49]

In his commentary on the cost of the unfit, Ignaz Kaup, professor of hygiene and member of the Deutsche Gesellschaft, reported on the results of a seminar held to discuss the subject. Since he doubted that the German people were ready to accept American-style sterilization methods as a means of alleviating the problem, some way of physically separating the "unfit" from the rest of society was necessary. False humanitarian considerations were not appropriate, since "all forward-striving nations had the duty to ease the burden of the cost of the inferior as much as possible." Recognizing that the *Minderwertigen* were a financial burden to the state who "despite the expenditure paid out on their behalf are almost never in the position during their working lives to repay the money spent on them," Kaup recommended the creation of work colonies where they could

[47] J. Jörger, "Die Familie Zero," *Arch. Ras. Ges.-Biol.*, 1905, *2*:494–559; and Alfred Ploetz, "Dr. Agnes Bluhm 70 Jahre am 9. Januar 1932," *ibid.*, 1932, *26*:63.

[48] See, e.g., Fritz Lenz, "Über die Verbreitung der Lues, speziell in Berlin, und ihre Bedeutung als Faktor des Rassentodes," *Arch. Ras. Ges.-Biol.*, 1910, *7*:306–327; Ernst Rüdin, "Zur Rolle der Homosexuellen im Lebensprozess der Rasse," *Arch. Ras. Ges.-Biol.*, 1904, *1*:99–109; and H. von Hentig, *Strafrecht und Auslese* (Berlin: Springer, 1914).

[49] Ignaz Kaup, "Was kosten die minderwertigen Elemente dem Staat und der Gesellschaft," *Arch. Ras. Ges. Biol.*, 1913, *10*:723–748, on p. 723.

be prevented from having inferior children and could be made to earn their keep at the same time.[50] At this time, however, most German eugenicists would have been satisfied with some form of permanent institutionalization.

As World War I approached, a third emphasis of the Wilhelmine eugenics movement came to the fore: *Bevölkerungspolitik* (population policy). While Germany's eugenicists did, of course, aim at instituting a eugenically healthy qualitative population policy, there was a marked tendency throughout the last years of the Empire to view the prevention of a decline in population growth as an important measure in its own right. As early as 1904 Alfred Grotjahn spoke of the "growth of population quantity" as the "*conditio sine qua non* of a rational prophylaxis against degeneration." Later the issue had become more pressing: as Schallmayer put it in 1915, arresting population decline was nothing short of "a matter of survival for the German nation."[51]

In order to understand why German eugenicists became obsessed with the population question, it is worth discussing briefly the prewar demographic, social, and political changes in Germany that colored their intellectual perspective. On the surface there seemed little cause for alarm. Wilhelmine Germany was the second most populous country in Europe; it had also witnessed a very substantial population increase of twenty-four million people between 1871 and 1910. Yet this healthy population growth owed far more to the dramatic decline in the death rate, particularly the infant mortality rate, than to a growth in fertility. Indeed Germany, like all Western industrialized nations, experienced a steady birthrate decline during the last third of the nineteenth and first third of the twentieth centuries. Between 1902 and 1914, for example, the Reich suffered an 8.3 per thousand drop in the number of live births.[52] This and the steady decline in the excess of births over deaths after 1902 gave statisticians and eugenicists cause to expect an eventual population standstill or even a decline. Many sought to account for Germany's declining birthrate; however much their explanations differed, all investigators agreed on two points: that the actual decline in population growth was less frightening than the prospect that Germany's situation might soon begin to mirror French demographic realities, and that the drop in the birthrate was deliberate and was directly related to the practice of birth control methods advanced by supporters of neo-Malthusianism.[53]

Considered by Rüdin and the president of the Deutsche Gesellschaft, Gruber, to be even more dangerous than the "relative increase of the unfit,"[54] German neo-Malthusianism encouraged birth control as a means of eliminating poverty and its attendant social problems. Much work remains to be done on neo-Malthusianism in Germany, but it seems likely that the German movement received

[50] *Ibid.,* p. 747, 748.

[51] Alfred Grotjahn, "Soziale Hygiene und Entartungsproblem," in *Handbuch der Hygiene,* ed. Th. Weyl, Suppl. Vol. IV (Jena: Gustav Fischer, 1904), pp. 727–789, on p. 761; and Wilhelm Schallmayer, "Zur Bevölkerungspolitik gegenüber dem durch den Krieg verursachten Frauenüberschuß," *Arch. Ras. Ges.-Biol.,* 1914/1915, *11*:713–737, on p. 729.

[52] Gerd Horhorst, Jürgen Kocka, and Gerhard A. Ritter, *Sozialgeschichtliches Arbeitsbuch. Materialien zur Statistik des Kaiserreiches 1870–1914* (Munich: C. H. Beck, 1974), pp. 15, 29–30.

[53] Lujo Brentano, "Die Malthusische Lehre und die Bevölkerungsbewegung der letzten Dezennien," *Abhandlungen der Akademie der Wissenschaften, 1908–1909,* pp. 567–625; Julius Wolf, *Der Geburtenrückgang* (Jena: Gustav Fischer, 1912); and Reinhold Seeberg, *Der Geburtenrückgang in Deutschland* (Leipzig: A. Deichert, 1913).

[54] Max von Gruber and Ernst Rüdin, *Fortflanzung, Vererbung, Rassenhygiene* (Munich: J. F. Lehmann, 1911), p. 158.

hygiene displayed for the first time by some government officials, not one eugenics-related law was passed during the Wilhelmine period.

There were undoubtedly many reasons for this state of affairs. Among them were the reluctance of German eugenicists to push for sterilization laws or other forms of negative eugenics and their emphasis on a rather abstract and diffuse set of positive eugenics proposals that would have been difficult to translate into concrete statutes. Initially, those in the vanguard of the movement had been content to educate the public as to the social and political need for eugenics. Having an exaggerated sense of their own importance as intellectual leaders of the nation, German race hygienists overestimated the power of their well-manicured public utterances. However, the social, political, and economic disaster brought on by the war both encouraged the growth of the movement and stimulated a bolder approach.

EUGENICS IN THE WEIMAR REPUBLIC, 1918–1933

The three major concerns of Wilhelmine race hygiene—degenerative phenomena, analysis of the burden of the *Minderwertigen,* and population policy—continued to preoccupy the second generation of the movement. The Weimar years, however, witnessed an increased emphasis upon reducing the social cost of the unproductive. Whereas eugenicists had earlier spoken in very abstract terms about improving the "race"—however differently that term was understood by individual practitioners—race hygiene during the Republic was far more concerned with preventing the decline of the German *Volk* and state. This does not mean that the movement lost its international orientation entirely. German eugenicists continued to correspond with their English and American colleagues and, after the early 1920s, participated in international eugenics and genetics conferences. One still finds talk about saving "civilized nations" from degeneration. Yet on balance German eugenicists were absorbed with the problems besetting their own country. Especially during the early Weimar years, eugenicists saw the fatherland as engaged in a life-or-death geopolitical and economic struggle for survival with its Western European and Russian enemies.[64] Oppressed by the economic and psychological impact of the Versailles treaty and inflation, forced—as the geneticist Erwin Baur put it—to suffer foreign domination by people "culturally beneath them," and consigned to live under an unstable and for the most part unloved republic, race hygienists realized that improving Germany's biological and national efficiency was no longer of merely intellectual interest.[65]

Fritz Lenz

The one person who did more than any other to spell out the importance of eugenics during the Weimar years was Fritz Lenz. A decidedly complex individual, Lenz became Weimar Germany's most prominent and, in many ways, most

[64] Heinrich Ernst Ziegler, rev. of Oskar Hertwig, *Zur Abwehr des ethischen, des sozialen, des politischen Darwinismus, Arch. Ras. Ges.-Biol.,* 1922, *14*:212–218, on pp. 217–218.

[65] Baur to the American geneticist and eugenicist A. F. Blakeslee, 1921, printed in Bentley Glass, "A Hidden Chapter of German Eugenics between the Two World Wars," *Proceedings of the American Philosophical Society,* 1981, *125*:357–367, on p. 364.

controversial eugenicist. The death of Schallmayer in 1919, and Ploetz's growing reluctance to shoulder the burdens of discipline building, left Lenz as the acknowledged leader of the Munich chapter of the society—much to the dismay of his less conservative, non-Aryan-supremacist colleagues in Berlin. Lenz viewed himself as a student of Ploetz; he shared his mentor's enthusiasm for the Nordic race—an enthusiasm undoubtedly strengthened through his contact with the anthropologist Eugen Fischer (1874–1967), whom he met while enrolled as a medical student at the University of Freiburg. During his medical studies at Freiburg from 1906 until 1912, Lenz also attended the lectures of August Weismann. From his own account, Weismann made a lasting impression on him and was probably responsible for his lifelong interest in the inheritance of hereditary diseases and intelligence. Lenz's training made him particularly receptive to the ideas of Ploetz, whom he first met in 1909. From that time on he sought to devote his life to the "practical" application of the study of human heredity; even his medical dissertation, completed in 1912, stressed eugenic concerns.[66]

Although Lenz was active in the Munich chapter of the society before the war, he first came to the attention of the international eugenics community in 1921 as coauthor of *Grundriß der menschlichen Erblichkeitslehre und Rassenhygiene* (Principles of human heredity and race hygiene). The treatise comprised two volumes. The first had a theoretical orientation and contained chapters by Erwin Baur on the principles of heredity, Eugen Fischer on the world's racial groups, and Lenz on human inheritance. The second volume, composed entirely by Lenz, dealt exclusively with race hygiene. Such respected American geneticists as Raymond Pearl and H. J. Muller considered the section written by Baur to be a clear and objective state-of-the-art summary; portions of Fischer's and Lenz's contributions, as they stood in the 1931 American edition of the text, were thought by Muller to be less so.[67] Even discounting the current prevalence of typological thinking about race, there can be little doubt that Fischer's and Lenz's discussions were largely a collection of personal and social prejudices masquerading as science. Given the important position Lenz held in the movement and the subsequent outcome of Nazi eugenics, it is worth examining his views on this subject further.

Like Galton and many other non-German eugenicists, Lenz believed in the reality of physical and mental racial traits. He understood these traits to be hereditary in the way that other traits common to all humans are hereditary. As such, according to Lenz, their relative frequency in a population was not static but was influenced by an all-powerful and ubiquitous selection process. Although he fully recognized physical differences between the world's races, he found these uninteresting in themselves and sometimes unreliable when it came to assessing an individual's racial type. Lenz concentrated almost exclusively on what he called the *seelische* (spiritual) differences, by which he meant the sum total of all nonphysical qualities of the major races. He clearly believed in a hierarchy of races, despite his comments to the contrary. Moreover, all talk of a transcendental "racial principle" aside, Western culture was the yardstick by which Lenz measured the "fitness" of races. Those races seen as having a high level of culture—by which he meant European and, in particular, German culture—were

[66] Rissom, *Fritz Lenz* (cit. n. 2), pp. 15–17.
[67] Reported in Glass, "Hidden Chapter of German Eugenics" (cit. n. 65), p. 357.

fitter and hence more worthy of preservation than others. Not surprisingly, the Negroid race stood at the bottom of the scale; the Nordics and the Jews (the latter comprising two main races, the Near Eastern and Oriental) were the most culturally productive. According to Lenz Nordic *man* was future-directed, steadfast, and prudent and hence able to subordinate sensual pleasure to more long-term goals; *he* was not only the religious and philosophical man *par excellence*—always searching but never finding what he needs—but he also exceeded all others in objectivity. Of Nordic *woman* he has less to say except that she, like women of other races, was on the average less objective then men. That, however, was no great problem, "since women have an entirely different mission to fulfill in the life of the race."[68]

Of coursc, Lenz did not recognize his sexism and the almost laughable manner in which he projected the values of the German educated middle class onto "Nordic man" as prejudices. Above all, Lenz thought of himself as an objective scientist who arrived at his conclusions after careful consideration of the facts. He found all demagoguery and emotionalism essentially "un-Aryan." Indeed, in his critique of the "emotional" anti-Semitism found in Theodor Fritsch's *Handbuch der Judenfrage,* Lenz accused the work of being too "Jewish" and insufficiently Germanic in its lack of "absolute objectivity."[69] Given his temperament he could never have written an inflammatory book such as Madison Grant's *The Passing of a Great Race,* though he was not reluctant to discuss its merits. This desire to remain *sachlich* (objective) undoubtedly colored his attitude toward Jews. Lenz's anti-Semitism was of the subdued variety commonly found among conservative German academics. However, insofar as he believed in the reality of racial types and was hence forced to describe the "spiritual" elements of the Jewish race, his stereotypical caricature of Jews has occasionally led people to see him incorrectly as an intellectual forerunner of Hitler. Although hardly pro-Semitic, he considered the Jews to possess many of the admirable qualities that Nordics, as well as others, did not possess to the same degree—much to the dismay of Germany's numerous rabid anti-Semites. Indeed, he felt that Nordics and Jews were more similar than dissimilar. What he did not like about the Jews (e.g., their preoccupation with making money and their liberal politics), he of course also projected onto their list of racial qualities, which he then attempted to relate to his reader in a cool objective manner. It is revealing of the degree to which typological thinking about race was generally accepted that Lenz's book was praised even by Jewish authors. Lenz was proud that the respected Jewish sexologist Max Marcuse, a specialist in the area of venereal disease and prostitution, had apparently accepted his "very unprejudiced and purely objective depiction of the racial condition and psychic constitution of the Jews as compared to that of the Germans."[70] As contemptible as Lenz was (especially for his willingness to cooperate with the Nazis after it was clear to him what *their* policies were), he seems to have believed that the promotion of the Nordic race need not go hand in hand with anti-Semitism. Although he later saw Hitler as the only political leader who truly embraced eugenic ideals, and as a result was favorably

[68] Erwin Baur, Eugen Fischer, and Fritz Lenz, *Grundriß der menschlichen Erblichkeitslehre und Rassenhygiene,* 2 vols. (2nd ed., Munich: Lehmann, 1923), Vol. I, pp. 406, 409, 417–427, esp. pp. 419, 422 (quotation).

[69] Fritz Lenz, rev. of Theodor Fritsch, *Handbuch der Judenfrage, Arch. Ras. Ges.-Biol.,* 1923, *15*:428–432, on p. 431.

[70] Müller-Hill, *Tödliche Wissenschaft* (cit. n. 2), pp. 37–38, 426 (quotation).

disposed toward him as early as 1931, he found his maniacal anti-Semitism too extreme.[71]

While Lenz's acceptance of ideologies of Nordic supremacy was clearly evident in virtually everything he wrote, it should be pointed out that, of the more than six hundred pages he contributed to *Menschliche Erblichkeitslehre,* only about fifty dealt with the race question. The bulk of his work was concerned with topics such as the transmission of hereditary diseases, the inheritance of intelligence and talent, the methodology of genetic research, and the theoretical principles and practical teachings of race hygiene. In his discussion of the inheritance of disease and talent Lenz sometimes cited the work and methodology of British and American geneticists and eugenicists, most frequently Galton. A convinced Mendelian, Lenz sought to demonstrate the Mendelian pattern of inheritance for various pathological traits; when focusing his attention on "metrical" traits such as intelligence, he naturally used the statistical tools developed by the British biometricians. Having at least some training in genetics, he was far more knowledgeable than most German race hygienists about the newest developments in the field. In general, however, the technicalities of genetics were important to Lenz only insofar as they could be used to support and legitimize his eugenic views.

Lenz's major eugenic aim was the preservation of his own class, the *Bildungs bürgertum,* from biological extinction. Perhaps more than anyone else he viewed eugenics as a means of boosting Germany's level of cultural productivity. Although virtually all German eugenicists equated the "fit" with the educated and socially useful elements in society, nobody was crasser in his class prejudices than Lenz. "Productivity and success in social life," Lenz affirmed in his textbook, "serve as a measure of the worth of individuals and families." Indeed, for him, *Entartung* was virtually synonymous with a lack of culture. Lenz, even more than Schallmayer, saw the real threat of degeneration not in the marginal increase in the number of those with serious hereditary diseases but rather in the low birthrate of the educated middle class and the "extinction of highly talented and otherwise distinguished families." Contemptuous of the value of manual labor compared with that of "mental labor," Lenz was particularly dismayed at the drop in the standard of living of academics during the early years of the Republic, as well as by the supposedly preferential treatment shown to workers after 1918: "The German revolution had an overwhelmingly unfavorable selective effect. As a result of the one-sided promotion of the interest of the manual workers, those who work with their brains are forced into a terrible struggle for survival. . . . If one views German society as a whole, there can be little doubt that the results of the revolution will lead to the extinction of educated families—the primary standardbearers of German culture."[72]

Yet for Lenz the events of 1918–1919 were not only dysgenic but also politically distasteful. A conservative academic, he belonged to the far-right-wing German National People's Party and had little tolerance for the Republic and its allegedly untalented leaders. He viewed the German revolution with horror, attributing it to the "extermination" of a large number of "socially minded" individuals on the battlefield. He found the new democratic order, with its promise

[71] Fritz Lenz, "Die Stellung des Nationalsozialismus zur Rassenhygiene," *Arch. Ras. Ges.-Biol.,* 1931, *25*:300–308, on p. 302.

[72] Baur et al., *Grundriß* (cit. n. 68), Vol. II, pp. 206, 192, 63.

It is not clear exactly why the Bund was formed. While the pro-Aryan sympathies that alienated non-Aryanists in the society may have also contributed to the creation of the Bund, there was undoubtedly a more important reason for its foundation. As was mentioned earlier, the Deutsche Gesellschaft made virtually no effort to reach out to all classes in society. Its rhetoric notwithstanding, the society had done little beyond attracting a relatively small number of medical professionals and academics to the movement; indeed, during the early 1920s it probably did not have more than a thousand members. It seems as if the leadership of the society did not quite know how to draw large numbers to their fold without compromising their "scientific integrity." The civil servants who formed the Bund, on the other hand, wanted first and foremost to popularize eugenics—to bring the problem of degeneration and the possibility of "national regeneration" to the awareness of the largest number of people possible. Although not without class prejudices, those involved in the Bund were at once less elitist in their view of the hereditary fitness of the working classes and more willing to write in a style that all Germans could understand. This is especially evident in the Bund's journals, the *Zeitschrift für Volksaufartung und Erbkunde* (1926–1927) and its successor, *Volksaufartung, Erblehre, Eheberatung* (1928–1930). Edited by a high-ranking public health official in the Prussian Ministry of Welfare, Artur Ostermann, the two journals published short, nontechnical articles that were decidedly different in style and tone from those found in the *Archiv*. Besides its popular style the Bund had something else the society lacked: real influence in government circles. Members of the society had direct links to the Association of German Registry Officials and close ties with the German Ministry of Welfare, the German Ministry of the Interior, and the Prussian Ministry of Welfare. All of these agencies contributed money to the Bund and its journals.[80]

Though lacking technical, scientific articles, the *Zeitschrift* and *Volksaufartung* voiced many of the same concerns as the older and more established *Archiv*. Both publications continued to warn about the dangers of birthrate decline and the tendency of the fitter classes to have fewer children; both also lamented the slow progress made in bringing genetics and eugenics into the high school classroom.[81] Yet if there was overlap, there were also important differences in the two eugenics journals. While carrying racist articles only infrequently, the *Archiv* devoted space to reviews (often written by Lenz) of blatantly racist publications; the two Bund publications were free of pro-Aryan and anti-Semitic sentiment. Perhaps more importantly, however, the Berlin journals saw as one of

---

7. Mai," *Zeitschrift für Volksaufartung und Erbkunde*, 1927, *2*:57–58. On these developments in general see Otto Krohne, "Position of Eugenics in Germany," *Eugenics Review*, 1925–1926, *17*:143–146, on p. 144.

[80] Fritz Lenz, "Ein 'Deutscher Bund für Volksaufartung und Erbkunde,'" *Arch. Ras. Ges.-Biol.*, 1925, *17*:349–350; and von Behr-Pinnow, "Jahresversammlung," p. 58.

[81] Regarding population policy see, e.g., Fritz Lenz, "Kinderreichtum und Rassenhygiene," *Z. Volks. Erbk.*, 1927, *2*:104–105; Lenz, "Die bevölkerungspolitische Lage und das Gebot der Stunde," *Arch. Ras. Ges.-Biol.*, 1929, *21*:241–253; Hermann Muckermann, "Differenzierte Fortpflanzung," *ibid.*, 1930, *24*:269–289; Fritz Brüggemann, "Bevölkerungspolitische Probleme," *Z. Volks. Erbk.*, 1927, *2*:18–19; Isch, "Ist Geburtensteigerung bei den Intellektuellen möglich?" *ibid.*, pp. 141–142; and Friedrich Burgdörfer, "Eugenik und Bevölkerungspolitik," *Volksaufartung, Erblehre, Eheberatung*, 1928, *3*:248–262. With respect to the teaching of genetics and eugenics see Spiler, "Vererbungslehre und Rassenhygiene im biologischen Unterricht der höheren Schulen," *Arch. Ras. Ges.-Biol., 1927, 19*:63–69; and Karl von Behr-Pinnow, "Vererbungslehre und Eugenik in den Schulen," *Volks. Erb. Ehe.*, 1928, *3*:73–80.

their major missions the popularization of Prussia's recently instituted *Eheberatungstellen* (marriage counseling centers). Created by a 1926 decree of the Prussian Ministry of Welfare, these centers may have been a concession to those eugenicists and government health officials who had pleaded, without success, for a compulsory exchange of health certificates for couples prior to marriage. Although launched with good intentions, the more than one hundred *Eheberatungstellen* were plagued from the beginning by a shortage of funding and the lack of a unified purpose. Established primarily for genetic counseling, they were not heavily frequented by prospective couples. The Dresden marriage counseling center, the oldest in Germany, had only sixty-four customers between 1911 and 1915.[82]

The attempt to popularize eugenics during the Weimar period was also accompanied by substantial institutional expansion. Before 1920 Germany lacked any institutional center for eugenics and could boast only a few isolated university courses in race hygiene. In 1923 a university chair for race hygiene (held by Lenz) was founded in Munich, and by 1932 over forty eugenics lecture courses were given at various German universities—many, if not all, in faculties of medicine. Two research centers were also established. The German Research Institute for Psychiatry was founded in Munich in 1918 and, with funding and aid from the Rockefeller Foundation, was made a Kaiser Wilhelm Institute in 1924. It was directed by Rüdin after 1931. The Kaiser Wilhelm Institute for Anthropology, Human Heredity, and Eugenics in Berlin was founded in 1927 and directed by Eugen Fischer.[83]

In addition, the movement was becoming increasingly visible both at home and abroad. In 1926, at the Great Exhibition for Health Care, Social Welfare, and Physical Training held in Düsseldorf, several members of the executive committee of the Deutsche Gesellschaft chose the exhibits for health care. Two years later, Munich was to host the International Alliance of Eugenic Organizations, at which time the German eugenicists' foreign colleagues had the opportunity to visit Ploetz's private research laboratory at Herrsching and were given a guided tour of the Kaiser Wilhelm Institute for Psychiatry headed by Rüdin.[84]

Throughout the 1920s and early 1930s the society also continued to grow, reaching a membership of nearly 1,100 by 1931. At a national meeting of the Deutsche Gesellschaft in Munich on 18 September 1931, it merged with the Deutscher Bund für Volksaufartung. The name was changed to the Deutsche Gesellschaft für Rassenhygiene (Eugenik)—the word *Eugenik* was included to demonstrate that the term *Rassenhygiene* was merely its German equivalent. The executive committee was strengthened and given more power, a change that resulted in the Deutsche Gesellschaft becoming more centralized than it had previously been. In addition, members could now join at large.[85]

[82] Artur Ostermann, "Eheberatungstellen," *Volks. Erb. Ehe.*, 1928, *3*:293–298, on p. 295; and F. K. Scheumann, "Sinn und Wesen der Eheberatung," *ibid.*, pp. 19–22, on p. 22.

[83] Maria Günther, *Die Institutionalisierung der Rassenhygiene an den deutschen Hochschulen vor 1933*, med. Diss., Mainz, 1982 (Mainz: privately printed, 1982), p. 61; and "Ein deutsches Forschungsinstitut für Anthropologie, menschliche Erblehre und Eugenik, *Arch. Ras. Ges.-Biol.*, 1927, *19*:457–458.

[84] Hans-Peter Kröner, "Die Eugenik in Deutschland von 1891 bis 1934," med. Diss., Münster, 1980, pp. 84–85.

[85] Hermann Muckermann, "Aus der Hauptversammlung der Deutschen Gesellschaft für Rassenhygiene (Eugenik) zu München am 18. September 1931," *Arch. Ras. Ges.-Biol.*, *26*:94–101, on pp. 94–95; Kröner, "Eugenik in Deutschland," p. 87.

The net effect of these changes was to create a larger, more popular, and more influential society that, as Hermann Muckermann put it, was true to "the historical line" of the movement. For Muckermann, a former Jesuit active in the Berlin chapter of the Deutsche Gesellschaft and a zealous popularizer of race hygiene during Weimar, that meant a non-Aryan-supremacist eugenics movement. Although Lenz, Rüdin, Ploetz, and a handful of others certainly never gave up their pro-Aryan sentiment, they were willing to put that in the background in the interest of a unified movement. Ironically, their influence over the movement was never weaker than it was at the end of Weimar—at a time when the Nazis were gaining strength daily. Despite the fact that the Munich chapter contained many of the prominent leaders of the Deutsche Gesellschaft, it did not grow to the same extent as the Berlin chapter; indeed, by 1931 even the Stuttgart chapter was larger. Nearly all of the numerous new local chapters that sprang up during the late Weimar years employed the term *Eugenik* rather than *Rassenhygiene* in their names, largely owing to the influence of Muckermann.[86] Had the Nazis not forced a drastic change in course in 1933, there is every reason to believe that the movement would have become even more similar to its counterpart in Britain.

Depression and Sterilization

The depression that began in 1929 not only eventually made more than six million people unemployed, but also forced a reexamination of the continued expansion of the welfare state. Calls were heard from industrial circles to trim Germany's welfare budget; "social policy must be limited by the productivity of the economy," it was argued. Although such cries lamenting Germany's economic inefficiency and high welfare costs were not new, they were taken quite seriously by the half-dictatorial, half-parliamentary government of Heinrich Brüning (1930–1932). By 1931 Germany's *Sozialpolitik* had become, at least in the eyes of some, too high "an insurance premium against Bolshevism."[87]

The critique of burgeoning social costs and the desire, even on the part of left-wing politicians, to allocate Germany's dwindling resources in the most cost-effective manner possible did not go unnoticed by race hygienists. This is clearly visible in the more substantial journal *Eugenik,* which superseded the earlier Bund publications in 1930. Edited, like its predecessors, by Ostermann and boasting a circulation of over five thousand, it included both racist society members (Lenz and Rüdin) and non-Aryanists (Muckermann) on its editorial board. It was not formally affiliated with the Bund.[88] Although never as well known internationally as the Archiv, *Eugenik* expressed the trends of the movement during Weimar's financially and politically troubled final years.

One concern mirrored in the journal was the problem of crime.[89] If much of

[86] Hermann Muckermann, "Hauptversammlung der Deutschen Gesellschaft für Rassenhygiene (Eugenik), München, 18. September 1931," *Eugenik,* 1931, *2*:47–48.

[87] David Abraham, "Corporatist Compromise and the Re-emergence of the Labor/Capital Conflict in Weimar Germany," *Political Power and Social Theory,* 1981, *2*:59–109, on p. 84.

[88] Fritz Lenz, "Eugenik, Erblehre, Erbpflege," *Arch. Ras. Ges.-Biol.,* 1931, *23*:451–452; Hermann Muckermann, "Aus der Hauptversammlung" (cit. n. 85), p. 99.

[89] See. e.g., Dr. Finke, "Biologische Aufgaben in der Kriminalpolitik," *Eugenik,* 1930, *1*:55–58; Johannes Lange, "Verbrechen und Vererbung," *ibid.,* 1931, *1*:165–173; and Hermann Muckermann, "Eugenik und Strafrecht," *ibid.,* 1932, *2*:104–109.

Germany's growing crime problem was a manifestation of bad germ plasm, then the millions spent yearly to detain criminals could be saved through an active race hygiene policy. In addition, *Eugenik* carried numerous articles that sought to demonstrate that eugenics was one of the best ways of eliminating waste in the welfare budget. According to one report, entitled "Marriage Counseling and Social Insurance," if more people had used the marriage counseling centers, Germany's hereditary defectives—who allegedly accounted for between eight and ten percent of all those between the ages of sixteen and forty-five—would not constitute such a "heavy burden on our expenditures." More explicit was a statement made by Muckermann in an article on welfare and eugenics. Complaining that 3.45 marks was needed daily to support one institutionalized mental defective, and pointing out that this saddled Germany with a financial burden of over 185 million marks a year at a time when there was barely enough money to keep healthy individuals from starving, Muckermann presented his readers with a sensible solution to the problem of the Reich's overtaxed social net:

> If one compares the money given out for defectives with the amount which a healthy family has at its disposal, one quickly comes to the conclusion that in the future everything must be done to reduce the number of hereditarily diseased individuals—a task that can be achieved by means of eugenics. Besides that, a clear differentiation must be made in the entire welfare system such that the means available are first appropriated for preventive care, and only then given out to people who cannot be brought back to work and life.[90]

Of course Muckermann never suggested that Germany's nonproductive elements shoud be treated in an inhumane fashion, but in hard economic times they would become second-class citizens who would receive from the state only the minimum required to maintain their existence.

Muckermann's cost-benefit analysis reflected Weimar Germany's preoccupation with rationalization and economic efficiency. During the 1920s industrialists sought ways to make Germany competitive on the world market—ways that included the elimination of inefficient facilities, the introduction of better methods of cost accounting and administration, the reorganization of factory work along the lines advocated by Henry Ford and Frederick Taylor, and the amalgamation of operations and firms into more efficient corporations and cartels.[91] Although not connected to the industrialists introducing such innovations, eugenicists nonetheless saw the relationship between race hygiene and the various forms of rationalization. As one eugenics supporter succinctly put it:

> We can protect our position in the world and ensure a high level of culture for our people only through a wise human economy (*weise Menschenökonomie*). Its goal must be an increase in those capabilities of the people who create a larger living space—that is, we must strengthen with respect to procreation, education and employment all those who achieve high quality manual and intellectual work. . . . At the same time it is absolutely essential . . . to limit the number of those who consume

[90] "Eheberatung und Sozialversicherung," *Eugenik,* 1930, *1*:182; and Hermann Muckermann, "Illustrationen zu der Frage: Wohlfahrtspflege und Eugenik," *ibid.,* 1931, *2*:41–42, on p. 42.

[91] Robert A. Brady, *The Rationalization Movement in German Industry* (Berkeley: Univ. California Press, 1933).

> more than they produce, who make the struggle for survival of our people difficult, and who depress [the people's] standard of living.[92]

Thus people became a manipulable resource to be administered in the interest of a healthy and culturally productive nation.

Perhaps nowhere, however, was the true nature of race hygiene better depicted than in the preface to *Eugenik*:

> Civilization has eliminated natural selection. Public welfare and social assistance contribute, as an undesired side effect of a necessary duty, to the preservation and further reproduction of hereditarily diseased individuals. A crushing and ever-growing burden of useless individuals unworthy of life are maintained and taken care of in institutions at the expense of the healthy—of whom a hundred thousand are today without their own place to live and millions of whom starve from lack of work. Does not today's predicament cry out strongly enough for a "planned economy," i.e., eugenics, in health policy?[93]

The devastating financial crisis of the late Weimar period only brought to the fore the logic implicit in eugenics from its very inception: it was a strategy to manage national efficiency rationally in order to preserve Germany's and the West's political and cultural hegemony.

The need to cut welfare costs, together with the constant pressure exerted by Ostermann, Muckermann, and others with influence, finally forced the Prussian government to take action. On 20 January 1932 the Prussian Upper House received and approved a resolution by one of its representatives, a Dr. Struve, to recognize eugenics and popularize it in every way possible and to decrease immediately the amount of money given out for the care of the defective to "a level that can be supported by a completely impoverished people." On 2 July, the Committee for Population Policy and Eugenics of the Prussian Health Council heard talks by Muckermann and three others on the topic of "Eugenics in the Service of National Welfare" and consequently adopted several eugenic proposals, including a draft for a sterilization law.[94]

The drafting of a sterilization law in Germany was a long time coming. Prominent eugenicists had carefully monitored events in the United States, where sterilization was legally practiced after 1907. During the Empire leading members of the Deutsche Gesellschaft did not push even for the voluntary sterilization of hereditary defectives, largely because they were certain that the country would find such a practice abhorrent. By the early Weimar years, however, their attitude had changed. Although by and large still opposed to mandatory sterilization, most members of the Deutsche Gesellschaft were open to voluntary sterilization; however, they still seemed to place more emphasis on institutionalization and work colonies as a means of preventing the unfit from reproducing.[95] During the 1920s a few obscure physicians did exploit the ambiguities in paragraph 224 of the Reich's legal code in order to carry out sterilizations on eugenic grounds.

[92] W. F. Winkler, "Bevölkerungspolitische Zukunftsfragen Europas," *Volks. Erb. Ehe.*, 1928, *3*:169–173, on p. 173.

[93] "Geleitwort," *Eugenik*, 1930, *1*:n.p.

[94] The resolution was cited in *Eugenik*, 1931–1932, *2*:109; see also "Eugenische Tagung des preussischen Landesgesundheitsrates," *Eugenik*, 1931–1932, *2*:187–189.

[95] "Aus der rassenhygienischen Bewegung," *Arch. Ras. Ges.-Biol.*, 1922, *14*:374.

One doctor, Gerhart Boeters, not only bragged about the sixty-three sterilizations he had performed but also sought to encourage other physicians, as custodians of the nation's health, to do likewise.[96] Although Boeters would later lose his civil service position as district physician in Saxony as a result of his boldness, his pleas—published in many of Germany's leading medical journals—ensured that the issue would be discussed. After prolonged debate among members in the Deutsche Gesellschaft and the medical community, in 1932 the Prussian Health Council drafted a sterilization law that permitted the *voluntary* sterilization of certain classes of hereditarily defective individuals and required that proof be given that the defective traits were in fact genetic. There was no mention of sterilization on either racial or social grounds. In addition, the committee that proposed the bill rejected out of hand the use of euthanasia for eugenic purposes.[97]

These proposals were embraced by several medical organizations both inside and outside of Prussia just weeks before the National Socialist takeover in 1933. In general physicians responded positively to the proposed law. Even in Protestant church circles the bill had its supporters. Only the Catholic Church, following the 1930 papal encyclical "Casti conubii," condemned the practice of sterilization.[98] However, owing to the political chaos following the deposition of the Prussian government by the Reich in July 1932, the sterilization draft never became law under the Republic, although it would later serve as the basis of the Nazi mandatory sterilization law of July 1933.

Thus throughout the Weimar years, as during the Empire, the movement was concerned first and foremost with boosting Germany's national efficiency and cultural productivity. Despite ideological differences among its advocates, race hygiene appealed to Aryan supremacists and their critics alike as a scientific means of solving social problems. Especially during the last troubled years of the Republic, more and more people of all political persuasions turned to the new discipline as one of the only effective ways of reducing the welfare budget and ensuring that Germany maintain its rightful position among the "cultured nations." Late Weimar eugenics expressed even more clearly the managerial logic implicit in German eugenics from its earliest days: population could and should be scientifically manipulated in the interest of power.

EUGENICS UNDER THE SWASTIKA, 1933–1945

Although the Weimar years witnessed the gradual adoption of a "eugenic outlook" on the part of certain government officials, prior to 1933 the cause of race

[96] Joachim Müller, "Sterilisation und Gesetzgebung bis 1933" (paper delivered at the Institute for the History of Medicine, Mainz, 7 Nov. 1978), pp. 14–16.

[97] "Eugenische Tagung" (cit. n. 94), p. 187. During the 1920s Germany witnessed a medical and legal discussion over "euthanasia"; Karl Heinz Hafner and Rolf Winau, " 'Die Freigabe der Vernichtung lebensunwerten Lebens': Eine Untersuchung zu der Schrift von Karl Binding und Alfred Hoche," *Medizinhist. J.*, 1974, 9:227–254.

[98] Lilienthal, "Rassenhygiene im dritten Reich" (cit. n. 2), p. 120; and Gerhard Baader, "Das 'Gesetz zur Verhütung erbkranken Nachwuchses'—Versuch einer kritischen Deutung," in *Zusammenhang: Festschrift für Marielene Putscher*, ed. Otto Baur and Otto Glandien (Cologne: Wienand, 1984), p. 869. For a detailed discussion of the attitudes of the Catholic and Protestant churches to the 1932 sterilization bill see Kurt Nowak, *"Euthanasie" und Sterilisierung im dritten Reich: Die Konfrontation der evangelischen und katholischen Kirche mit dem "Gesetz zur Verhütung erbkranken Nachwuchses" und der "Euthanasie"-Aktion* (3rd ed., Göttingen: Vandenhoeck & Ruprecht, 1984).

hygiene was advanced by a relatively small group of intellectuals, primarily medically trained professionals, within the confines of the Deutsche Gesellschaft. The Nazi seizure of power changed this drastically. Now heading the Reich was a man for whom race hygiene represented a key element in a much larger "biological" and racial world view—a world view to which the entire nation would be pledged and ultimately sacrificed. Hitler's maniacal obsession with the Aryan race as the motive force of world history assured that anything useful to the preservation of "Nordic blood" would become a cornerstone of national policy and the subject of intense government propaganda. Because much of National Socialist ideology, as one Bavarian Nazi succinctly put it, was in some sense little more than "applied biology,"[99] it becomes extremely difficult, after 1933, to separate the goals and activities of "professional eugenicists" from the rhetoric and racial policies of Hitler and high-ranking Nazi Party members. For our purposes, however, Nazi "race hygiene" will be defined as the activities of professional eugenicists, the Deutsche Gesellschaft, and the two major eugenics institutes during the twelve-year dictatorship. But no examination of eugenics in the Third Reich can neglect the legacy of the pre-1933 movement, nor can it ignore the connection between race hygiene and such Nazi racial policies as the "euthanasia" program, the extermination of Europe's gypsy population, and the "final solution." Although none of the latter were viewed by Germany's professional eugenicists as belonging to the province of race hygiene, in at least some instances there were both personal and ideological ties between the two.

"Gleichschaltung" and Change

The new political leadership imposed significant changes upon the race hygiene movement. Not long after the triumph of the new order the Deutsche Gesellschaft, like all other organizations in the Reich, was *gleichgeschaltet* (coordinated) and subjected to the "Führer principle." This meant, first of all, that the society was no longer an independent organization. It was placed under a special Reich Commission for National Health Service, which in turn was directly subordinate to the Reich Ministry of Interior. Accordingly, the society was expected to "support the government in the fulfillment of its race hygienic goals." In addition to becoming a de facto government body, the society lost all semblance of democratic control. In November 1933 Rüdin, director of the Kaiser Wilhelm Institute for Psychiatry in Munich, was appointed *Reichkommissar* of the society by Nazi Minister of Interior Wilhelm Frick. He, in turn, was in charge of appointing the business manager as well as the leaders of all the local chapters of the society. Final authority, however, remained in the hands of Frick. The minister of interior could veto all appointments, had to approve any changes in the society's bylaws, and could remove anyone from office at will.[100]

Even before the new statutes were drawn up early in 1934, Rüdin eliminated the word *Eugenik* from the society's official name and reinstated the one used before the compromise of 1931, the Deutsche Gesellschaft für Rassenhygiene.[101]

[99] Robert Proctor, "Pawns or Pioneers? The Role of Doctors in the Origins of Nazi Racial Science" (Ph.D. Diss., Harvard Univ., 1982), p. 37.

[100] "Deutsche Gesellschaft für Rassenhygiene, e.V.," *Arch. Ras. Ges.-Biol.*, 1934, *28*:104–108; and Kröner, "Eugenik in Deutschland" (cit. n. 84), pp. 92–93.

[101] "Notizen," *Arch. Ras. Ges.-Biol.*, 1933, *27*:467.

Unlike the previous name change, this one had more than merely symbolic significance. The Nazi seizure of power eliminated the possibility of a non-Aryanist race hygiene in Germany. Since the Deutsche Gesellschaft was now virtually a government organ, and since race hygiene was central to the new order, there could be little if any deviation from the official line. This meant the end of the "Berlin interpretation" of eugenics—an interpretation that appeared to have won the upper hand at the end of the Weimar period. Two of the most influential relatively nonracist eugenicists, Ostermann and Muckermann (the Social Democrat Grotjahn had died in 1931), were removed from their offices and forced into retirement. After 1937 Muckermann was prohibited from writing on the subject of eugenics.[102] With the removal of Muckermann, the movement lost its best popularizer as well as the director of the Eugenics Division of the Berlin Kaiser Wilhelm Institute for Anthropology, Human Heredity, and Eugenics. These two men were not the only sacrifices of the new regime. Although membership lists for the Deutsche Gesellschaft are not available for the late Weimar and early Nazi years, it can be safely assumed that many of the less prominent critics of the racists left the society of their own accord or were urged to do so by the newly appointed local chapter leaders. It goes without saying that Jewish members, such as the geneticist Richard Goldschmidt, were forced out of the organization; according to the new 1934 statutes, membership in the Deutsche Gesellschaft was restricted to "Germans of Aryan ancestry."[103]

Hand in hand with the elimination of relatively nonracist eugenics and its supporters from the newly "coordinated" Deutsche Gesellschaft came a greater preoccupation with race. In the past not even Aryan sympathizers like Lenz, Ploetz, and Rüdin had made *Aufnorderung* (nordification) a cornerstone of their eugenics policy, nor had they publicly suggested that the preservation and racial purification of the "Aryan" population of the Reich should become a primary focus of their attention. After 1933 race hygiene combined *Rassenpflege* (racial care) and *Erbpflege* (genetic care).[104] The latter component was equivalent to the old, relatively nonracist meritocratic eugenics concerned with the rational management of those mental and physical traits of the population seen as favorable to a more culturally and economically productive Reich. *Rassenpflege* (the management of a population's racial traits) was something new, although Lenz as well as "racial scientists" (racist anthropologists) such as Hans F. K. Günther had earlier suggested that high economic and cultural achievement was a product of certain superior races.

This new blatantly racist (and sometimes explicitly anti-Semitic) line was given clear expression by Germany's race hygienists both in public speeches and in their writings. In an address presented at a special meeting of the Deutsche Gesellschaft in 1934, Rüdin stressed the cultural importance of "the Nordic race in

[102] Lilienthal, "Rassenhygiene im dritten Reich" (cit. n. 2), p. 123; and Hans Ebert, "Hermann Muckermann: Profil eines Theologen, Widerstandskämpfers und Hochschullehrers der Technischen Universität Berlin," *Humanismus und Technik,* 1976, *20*:29–40, on p. 35.

[103] Weindling, "Weimar Eugenics" (cit. n. 2), pp. 309, 315 (quotation); and "Deutsche Gesellschaft für Rassenhygiene, e.V." (cit. n. 100), p. 107.

[104] Otto Freiherr von Verschuer, *Leitfaden der Rassenhygiene* (Leipzig: Georg Thieme, 1941), p. 125; and Arthur Gütt, "Ausmerze und Lebensauslese in ihre Bedeutung für Erbgesundheit und Rassenpflege," in *Erblehre und Rassenhygiene im völkischen Staat,* ed. Ernst Rüdin (Munich: Lehmann, 1934), p. 118.

world history and especially German history" and concluded that as such it "urgently deserves to be preserved and protected." Although he denied that the goal of preserving and protecting Germany's Nordic and closely related stock meant a devaluation of other races, Rüdin rejected out of hand the crossing of "dissimilar races." In his influential *Leitfaden der Rassenhygiene* (Textbook of race hygiene) (1941), Otmar Freiherr von Verschuer (1896–1969), director of the Frankfurt University Institute for Heredity and Race Hygiene, and later director of the Kaiser-Wilhelm Institute for Anthropology, Human Heredity, and Eugenics, discussed the necessity of preserving the "racial peculiarities of the *Volk*" by combating "the penetration of foreign races."[105] Similar statements are found in the writings of other Nazi race hygienists such as Theodor Mollison, Otto Reche, and Martin Staemmler.

Continuities

Yet despite the important changes that the eugenics movement underwent during the Third Reich, there was at least as much continuity as discontinuity. The new preoccupation with race after 1933 in no way lessened the attention devoted to the more traditional concerns of race hygiene (e.g., increasing the birthrate of the "fitter" classes of society, reducing the number of nonproductive elements). Indeed, judging from the plethora of books on the subject, the obsession with reducing the number of the unfit and boosting the ranks of the productive classes through the implementation of a vigorous race hygiene program was far greater than it had been even in the Weimar years. Popular works such as Otto Helmut's *Volk in Gefahr* (Nation in danger), which sold over 26,000 copies, and Friedrich Burgdörfer's *Völker am Abgrund* (Peoples at the abyss) did not focus their attention on "the Jewish menace" but rather used numerous graphs and diagrams to reiterate such longstanding eugenic concerns as the "hereditary defectives' burden on the German people," "the threat of the subhumans [criminals]," and "the decline of the fit, the increase of the unfit." In one diagram entitled "Fertility and Race," Helmut did not compare Aryans and Jews but rather tried to demonstrate the alleged Slavic threat facing Germany—the same fear articulated by Schallmayer and others before and during World War I.[106]

This continuation of earlier themes is found in statements from those outside the movement as well. In an address given in 1933 to the newly instituted Expert Commission for Population and Racial Policy, a committee set up by the Nazi government to deal with various "racial questions," Interior Minister Frick asserted that "in order to raise the number of genetically healthy progeny, we must first lower the money spent on asocial individuals, the unfit, and the hopelessly hereditarily diseased, and we must prevent the procreation of severely hereditarily defective people." In a short article published the same year, the physician Friedrich Maier urged his readership to replace the system of "welfare, which generally served only the weakest and asocial individuals," with one emphasizing

[105] Ernst Rüdin, "Aufgabe und Ziele der Deutschen Gesellschaft für Rassenhygiene," *Arch. Ras. Ges.-Biol.*, 1934, *28*:228–233, on p. 232 (quotation); see also Kröner, "Eugenik in Deutschland" (cit. n. 84), p. 94; and Verschuer, *Leitfaden*, p. 115.

[106] Friedrich Burgdörfer, *Völker am Abgrund* (Munich: Lehmann, 1936); and Otto Helmut, *Volk in Gefahr: Der Geburtenrückgang und seine Folgen für Deutschlands Zukunft* (Munich: Lehmann, 1934), pp. 26, 28, 30, 34 (diagram).

[107] "Ansprache des Herrn Reichministers des Innern Dr. Wilhelm Frick auf der ersten Sitzung des

the "management of the health of those portions of the German nation still racially intact in order both to prevent genetically diseased offspring, and to encourage the hereditarily fit individuals in all segments of the population."[107] Thus, although Schallmayer and other relative nonracists would have viewed the racist and anti-Semitic side of eugenics during the Third Reich as deplorable and "unscientific," they would not have found all parts of Nazi race hygiene objectionable; its logic and many of its aims were too similar to their own.

Of all the various strategies and programs implemented by the Nazis in the interest of improving the racial substrate of the Reich, none reveals the continuity between pre- and post-1933 race hygiene better than the sterilization law. Formally enacted on 14 July 1933, the *Gesetz zur Verhütung erbkranken Nachwuchses* (Law for the prevention of genetically diseased offspring) was based on the 1932 Prussian proposal initiated by Muckermann, Ostermann and others, including the then director of the Berlin-based Kaiser Wilhelm Institute for Biology, Richard Goldschmidt.[108] Unlike the failed Prussian proposal of 1932, however, the Nazi law allowed the mandatory sterilization of those individuals who, in the opinion of an *Erbgesundheitsgericht* (genetics health court), were afflicted with congenital feeblemindedness, schizophrenia, manic depressive insanity, genetic epilepsy, Huntington's chorea, genetic blindness, and genetic deafness. Those suffering from "serious alcoholism" could also be sterilized against their will. The *Gesetz* made no provisions for sterilization based on racial grounds.

Although Ernst Rüdin collaborated in the law's well-publicized interpretative commentary, it is not clear what role, if any, Germany's "professional eugenicists" had in drafting it. The initial impetus for the *Gesetz* came from the director of the Commission of National Health Service, Arthur Gütt.[109] As members of the Expert Committee, Lenz, Ploetz, and Rüdin may merely have enjoyed the function of "rubber-stamping" a proposal originated by the Ministry of Interior. Nonetheless they wholeheartedly approved the new measure. Like other members of the medical community, they had good reason to do so. The statutes called for the establishment of genetic health courts and supreme genetic health courts to adjudicate the *Gesetz,* all of which were presided over by a lawyer and two doctors. The *Gesetz* stipulated that one physician be an expert in the field of heredity and that the second be employed by the state. Moreover, since physicians were required to report all individuals afflicted by any genetic illness mentioned above, they were the ones most often responsible for bringing cases to the attention of the courts. Hence most of Germany's physicians were now afforded ample opportunity to fulfill their obligation as custodians of the nation's health, either directly, through their involvement on the courts, or indirectly, by ensuring that the genetically ill were registered with the courts.[110]

Although initially some 1,700 genetic health courts were envisioned by the Nazis (one in each large city and in each county), probably not more than two to

---

Sachverständigenbeirats für Bevölkerungs- und Rassenpolitik," *Arch. Ras. Ges.-Biol.,* 1933, *27*:412–419, on p. 416; and Friedrich Maier, "Die neue Staatsmedizinische Akademie in München," *ibid.,* 1934, *28*:56–57.

[108] Bock, *Zwangssterilisation* (cit. n. 2), pp. 80–84; Lilienthal, "Rassenhygiene im dritten Reich" (cit. n. 2), p. 124; and Müller-Hill, *Tödliche Wissenschaft* (cit. n. 2), p. 32.

[109] Bock, *Zwangssterilisation,* p. 84

[110] "Gesetz zur Verhütung erbkranken Nachwuchses," *Arch. Ras. Ges.-Biol.,* 1933, *27*:420–423; and Proctor, "Pawns or Pioneers" (cit. n. 99), p. 47.

three hundred were ever established. It thus proved to be impossible to extend the *Gesetz* to cover an even broader group of "defectives"—something that at least some eugenicists desired. Hence, Rüdin's plea that all "burdensome lives, ethically defective and socially unfit psychopaths, and the huge army of confirmed hereditary criminals" come under the surgeon's knife went largely unanswered.[111]

Lenz also believed that the *Gesetz* was too narrow. He spoke of the desirability of sterilizing 1 million feebleminded, 1 million mentally ill, and 170,000 idiots in "the social interest." He at least half-seriously suggested that it would be better if the bottom one third of the entire population did not reproduce.[112] Nonetheless, estimates on the number of people sterilized between 1934 and 1939 range from 200,000–350,000 to 350,000–400,000. All had passed through the genetic courts, and the overwhelming majority of them were sterilized against their will. It is estimated that slightly more than half of all operations were performed on the so-called feebleminded. During the first three years of the *Gesetz,* at least 367 women and 70 men died owing to complications following the procedure. The number of related deaths throughout the six-year period was probably much higher.[113]

During the Nazi period, the research conducted in Germany's academic institutes associated with eugenics was similar to investigations carried out during the Weimar period. In the German Research Institute for Psychiatry in Munich, the heavy emphasis placed on twin studies as a means of investigating the inheritance of mental disorders continued during the Third Reich much as it had during the Weimar years. Formed in the hope that the research undertaken would one day help reduce the enormous financial cost of caring for the army of mental defectives, by 1938 the institute had at least eleven researchers working on material collected from over 9,000 identical and fraternal twins. Although we do not know whether these researchers were also involved in providing genealogies for individuals whose pure "Aryan lineage" was in question, like their counterparts in the Berlin institute, the major task of Rüdin's institute was to provide the hard evidence for the inheritance of pathological mental traits to aid the government's effort to sterilize the "unfit." However, the institute also made a contribution to positive eugenics by studying genealogies of talented individuals; these included shop foremen and their spouses—who were undoubtedly seen as a group of elite workers who rose to low-level managerial positions owing to their genetically endowed abilities.[114] The backgrounds and exact number of researchers in the institute and the precise nature of the investigations carried out between 1933 and 1945 remain unknown.

Far eclipsing the Munich institute in importance, Berlin's Kaiser Wilhelm In-

[111] Proctor, "Pawns or Pioneers," p. 47; Müller-Hill, *Tödliche Wissenschaft* (cit. n. 2), p. 35.

[112] Hans-Ulrich Brändle, "Aufartung und Ausmerze," in *Volk und Gesundheit. Heilen und Vernichten im Nationalsozialismus,* ed. Projektgruppe "Volk und Gesundheit" (Tübingen: Tübinger Vereinigung für Volkskunde, 1982), p. 167n7.

[113] Baader, "Das 'Gesetz zur Verhütung erbkranken Nachwuchses' " (cit. n. 98), p. 865; Müller-Hill, *Tödliche Wissenschaft* (cit. n. 2), pp. 36, 37; Bock, *Zwangssterilisation* (cit. n. 2), pp. 237–238.

[114] Max Planck, ed., *25 Jahre Kaiser-Wilhelm-Gesellschaft zur Förderung der Wissenschaften* (Berlin, 1936), Vol. I, pp. 131–132; and Ernst Rüdin, "20 Jahre menschliche Erbforschung an der Deutschen Forschungsanstalt für Psychiatrie in München, Kaiser-Wilhelm-Institut," *Arch. Ras. Ges.-Biol.,* 1938, *32*:193–203, on p. 195, 198.

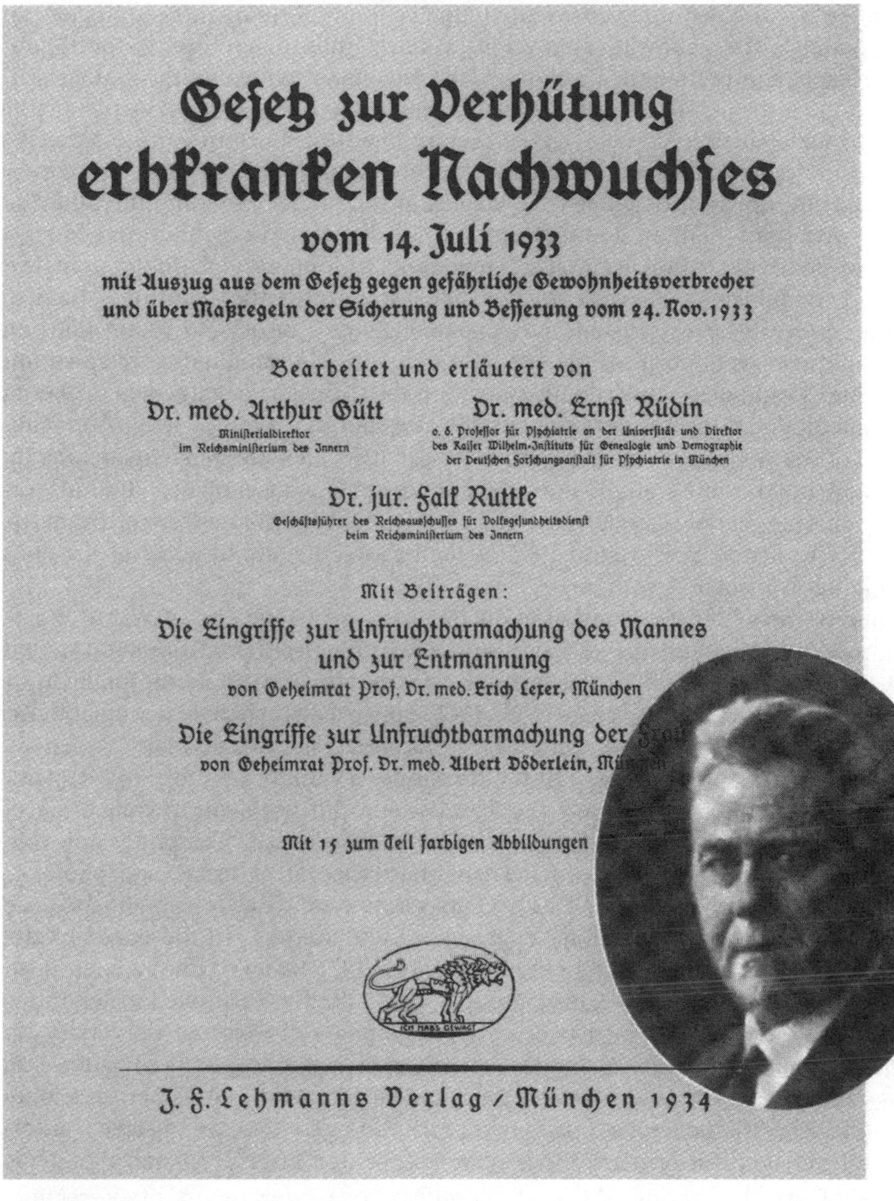

Gesetz zur Verhütung
erbkranken Nachwuchses
vom 14. Juli 1933
mit Auszug aus dem Gesetz gegen gefährliche Gewohnheitsverbrecher
und über Maßregeln der Sicherung und Besserung vom 24. Nov. 1933

Bearbeitet und erläutert von

Dr. med. Arthur Gütt
Ministerialdirektor
im Reichsministerium des Innern

Dr. med. Ernst Rüdin
o. ö. Professor für Psychiatrie an der Universität und Direktor
des Kaiser Wilhelm-Instituts für Genealogie und Demographie
der Deutschen Forschungsanstalt für Psychiatrie in München

Dr. jur. Falk Ruttke
Geschäftsführer des Reichsausschusses für Volksgesundheitsdienst
beim Reichsministerium des Innern

Mit Beiträgen:

Die Eingriffe zur Unfruchtbarmachung des Mannes
und zur Entmannung
von Geheimrat Prof. Dr. med. Erich Lexer, München

Die Eingriffe zur Unfruchtbarmachung der
von Geheimrat Prof. Dr. med. Albert Döderlein,

Mit 15 zum Teil farbigen Abbildungen

ICH HABS GEWAGT

J. F. Lehmanns Verlag / München 1934

Figure 1. *Title page of the published version of the Nazi sterilization law, the* Gesetz zur Verhüting erbkranken Nachwuchses, *issued under the authorship of Arthur Gütt (director of the Commission of National Health Service), Ernst Rüdin, and Falk Ruttke (Munich: J. F. Lehmann, 1934).* Insert: *Ernst Rüdin, from an advertisement for the* Münchener Medizinische Wochenschrift.

stitute for Anthropology, Human Heredity, and Eugenics remained the institutional center of German race hygiene research throughout the Nazi period. It officially opened on 15 September 1927—a date chosen to coincide with Germany's hosting of the Fifth International Congress of Genetics in Berlin. From its very inception the institute represented "the wish of German anthropologists and race hygienists for a central research institute for their disciplines in the Reich." Eugen Fischer, a prominent Freiburg racist eugenicist whose anthropological investigations into the "Reheboth bastards" (mulattos) of Southwest Africa in 1908 launched his academic career, was chosen to head the Berlin institute as well as its anthropological division. In 1933 he was also appointed rector of the University of Berlin, apparently against the wishes of the Nazis. Heading the other two original divisions—human heredity and eugenics—were Verschuer and Muckermann, respectively. Muckermann's connections to influential Catholic industrialists were in no small measure responsible for part of the institute's financial backing. In 1933 he was dismissed despite Fischer's efforts to retain him, and Lenz took over as director of the eugenics division. He remained at his post until 1945 while simultaneously holding a position as professor of eugenics in the faculty of medicine of the University of Berlin. Fischer managed to retain Verschuer as head of the division of human heredity until 1935, despite Nazi suspicions that "he could not be integrated" into the new order because of his "liberal" outlook.[115] In that year Verschuer received a position at the University of Frankfurt; he did not return to Berlin until 1942, when he was chosen director of the entire institute upon Fischer's retirement.

Owing to the willful destruction of documents toward the end of World War II, it is impossible to detail the services and research activities of the institute with any degree of certainty.[116] The surviving documents, as well as the publications of institute workers for the years 1927–1945, suggest that their research activities did not change fundamentally after 1933, although admittedly this evidence is unlikely to tell the entire story. In the divisions of human heredity and eugenics the focus of investigation during the Third Reich did not seem to reflect an obsession with either Aryan supremacy or Jewish inferiority. Verschuer and those who came to work under him studied the inheritance of "normal" morphological and physiological traits as well as the inheritance of disease, intelligence, and behavior. Like Rüdin, Verschuer engaged in twin studies. In the eugenics division, the primary concern both before and after 1933 seems to have been differential birthrates of various social groups. In 1930, for example, Muckermann examined the differential birthrates of 3,947 families of German university professors; six years later Ilse Schmidt, a researcher in the eugenics division, studied the relationship between intelligence and urbanization. Another area under investigation in the eugenics division was radiation genetics. In Fischer's anthropological division the primary focus both before and after 1933 was the genetic analysis of racial crossing. At least according to a later report by Verschuer, virtually every crossing was studied except that between Jews and "Aryans."[117]

While those race hygienists holding research positions seem to have continued with "business as usual" after 1933, their institutional affiliation did obligate

[115] Jürgen Kroll, *Zur Entstehung und Institutionalisierung einer naturwissenschaftlichen und sozialpolitischen Bewegung: Die Entwicklung der Eugenik/Rassenhygiene bis zum Jahre 1933,* sozialwiss. Diss., Tübingen, 1983 (Tübingen: privately printed), p. 161; Müller-Hill, *Tödliche Wissenschaft* (cit. n. 2), p. 78.

[116] Müller-Hill, *Tödliche Wissenschaft,* pp. 24–25.

[117] Otto Freiherr von Verschuer, "Das ehemalige Kaiser-Wilhelm-Institut für Anthropologie, menschliche Erblehre und Eugenik," *Zeitschrift für morphologische Anthropologie,* 1964, *55*:127–174, on pp. 160–161, 156–158, 129–136, 159.

them, willingly or unwillingly, to serve the needs of Nazi racial policy. Both institutes, especially the one in Berlin, were expected to aid the government in its effort to improve the German race. What this meant in practice, as revealed in several memos and reports, is that members of the institutes were called upon to teach eugenics, genetics, and anthropology courses to state-employed physicians and SS doctors; to help carry out the sterilization law by providing *Gutachten* (expert testimony) in cases coming before the genetic health courts; and to compose racial testimonials and genealogies for the ministry of interior after the passage of the Nuremberg Laws. By 1935, for example, over 1,100 physicians had already taken one of the above-mentioned courses; between 50 and 185 doctors participated in a year-long continuation course in "genetic and racial care."[118] The writing of *Gutachten* for the genetic health courts was considered so important that Minister of Interior Frick secured money for a total of five assistants for Fischer, Verschuer, and Rüdin just to help them handle the large case load. Verschuer and Fischer also became members of the Berlin genetic health court and the Berlin supreme genetic health court, respectively. The composition of racial genealogies seems to have been somewhat more unpleasant and time-consuming for the particular eugenicists involved.[119] Nonetheless, insofar as Germany's race hygienists were willing to deliver a verdict on the "racial ancestry" of individuals, they were, at least after 1941, indirectly involved in sending Jews to their death.

Responsibility and Legacy

The eugenicists' willingness to participate in the construction of such racial genealogies raises the question of their connection to other criminal Nazi racial policies. A case in point is the sterilization of the "Rhineland bastards"—the children of German mothers and French African occupation troops stationed in the Rhineland after World War I. Lenz and Rüdin were indeed asked, as members of the Expert Committee for Population and Racial Policy, to give their opinion on what should be done with these children. Interestingly, neither Lenz nor Rüdin was in favor of mandatory sterilization, although their "solutions" to the problem were hardly commendable: Lenz suggested that the children be "exported," while Rüdin opted for their "voluntary" sterilization on pain of deportation. The actual decision to proceed with the forced sterilization of these children was made in 1937 in the Reich Chancellery without further consultation with the eugenicists. Only Fischer and Verschuer were even indirectly involved in this action; both were called upon to write the requisite anthropological testimonials needed to document the children's racial ancestry prior to sterilization.[120] Whether they willingly prepared the genealogies that resulted in the sterilization of 385 "colored" children remains unknown.

German eugenicists also bore at most only indirect responsibility for the "euthanasia action." Officially, about 100,000 so-called useless eaters (mentally ill

[118] Tätigkeitsbericht von Anfang Juli 1933 bis 1. April 1935. In Archiv der Max-Planck-Gesellschaft, West Berlin, folio 2401, doc. no. 49.

[119] *Ibid.*, doc. no. 49b; and Müller Hill, *Tödliche Wissenschaft* (cit. n. 2), p. 39.

[120] Müller-Hill, *Tödliche Wissenschaft,* pp. 34–35; and Reiner Pommerin, *Sterilizierung der Rheinlandbastarde: Das Schicksal einer farbigen deutschen Minderheit 1918–1937* (Düsseldorf: Droste, 1979), pp. 75, 78.

and retarded patients) were exterminated in Germany between 1939 and 1941. However, recent evidence has demonstrated that the killings began much earlier, did not end until the end of the war, and were not limited to German victims: "useless eaters" in Poland and the Soviet Union, many of them Jews, were also exterminated under the program. Since the history of the destruction of "lives not worth living" is well documented elsewhere, it is not necessary to give a full account of it here.[121] Suffice it to say that "euthanasia" was never considered a race hygiene measure by any eugenicist. Only Lenz was in any way involved in a official committee designed to formulate a law permitting "euthanasia"—a law that apparently never saw the light of day since the action remained officially secret.[122] Despite the fact that "euthanasia" was never seen as a eugenics measure, the action was known and at least half-heartedly accepted by most active race hygiene practitioners; it was, after all, the logical outgrowth of the cost-benefit analysis at the heart of race hygiene. Nonetheless, Germany's race hygiene practitioners were neither in charge of the program nor directly involved in sending any individuals to their deaths.

Perhaps the most commonly held assumption about German eugenics and its practitioners is that they are intricately bound to the activities of the death camps, where a large percentage of Europe's Jewish and gypsy populations were exterminated. While there are ideological links between race hygiene and the destruction of unwanted "racial groups," it would be inaccurate to assume that individual German eugenicists or German race hygiene as a whole was directly responsible for the Holocaust.

Although those particular eugenicists most active during the Nazi period were undeniably anti-Semitic, their socially acceptable brand of anti-Semitism was typical of the German conservative academic mandarins as a whole: these were not people who wanted to see Jews gassed. Lenz provides a typical case in point. During the Weimar years he refused to change his allegedly "objective" position regarding Jews just to please Germany's anti-Semitic movement. He bemoaned the fact that so much energy was being converted into such a "useless racket." Not surprisingly, he never seemed to recognize his own anti-Semitic prejudices and hence continued to talk about anti-Semites as if they were a group to which he in no way belonged. During the Nazi period, however, Lenz was willing to support a somewhat more blatant anti-Semitic position, as evidenced in the change in his description of the Jews between the third edition of the first volume of the *Grundriß* (1927) and the fourth edition of the same volume (1936).[123]

However, even after 1933, when it would have been politically expedient, the writings of Lenz and other eugenicists did not emphasize anti-Semitism. Had they been rabid anti-Semites they could have published such views in any number of journals both before and after 1933. Moreover, none of the "professional eugenicists" were involved in any piece of anti-Semitic legislation. Even assuming that many of the eugenicists actually welcomed "early measures" designed to separate and isolate the Jews—an assumption that is by no means firmly established—they had little real influence over any piece of Nazi legisla-

[121] For the best and most exhaustive account of this tragic episode see Ernst Klee, *"Euthanasie" im NS-Staat. Die "Vernichtung lebensunwerten Lebens"* (Frankfurt am Main: Fischer, 1983).

[122] Müller-Hill, *Tödliche Wissenschaft* (cit. n. 2), p. 18.

[123] *Ibid.*, pp. 37–38.

tion, let alone legislation relating to the Jews. The Nuremberg Laws forbidding marriages or extramarital relations between Jews and Aryans were composed without the aid of a single professional race hygienist. Finally, the eugenicists did not take part in the infamous Wannsee Conference of 1942, where plans for the "final solution to the Jewish question" were confirmed by Heydrich and other leading Nazi officials.

Absolving the eugenicists of any *direct* responsibility for the "final solution" is, of course, not to excuse or condone their behavior and actions throughout the Nazi period. Ultimately it was not their anti-Semitism that linked them, however indirectly, to the death camps: in terms of any indirect personal responsibility for the Holocaust, their crimes, like those of large sections of the German population, were largely crimes of omission. By 1933 race hygiene had become an established discipline in Germany, and eugenicists had a vested interest in the continued funding of the field and the institutes to which they belonged. When asked in an interview why Ernst Rüdin wrote an article praising the Nazis, his daughter Edith replied, "He would have sold himself to the devil in order to obtain money for his institute and his research." The only way Germany's eugenicists could preserve their positions and secure financial backing for their work was by cooperating with Nazi officials. This often meant paying lip service to Nazi programs and joining the Party—the latter as evidence of loyalty to the regime. Lenz, Fischer, Verschuer, and Rüdin all became Party members, but only after 1937.[124]

But perhaps more important, they expected that their dream of a meritocratic eugenics-based society would be realized in the Third Reich. A statement made by Lenz in 1931 makes it obvious that he welcomed the National Socialists as the only political party willing to take the "eugenics outlook" seriously.[125] Frustrated by the lack of progress in realizing eugenic ideas during the Weimar years, eugenicists active during the Nazi period expected their plans to be realized under Hitler. Even after it became clear to them that Hitler's ideas of race hygiene were not precisely the same as their own, and even after they realized that they were unlikely to be able to exercise any kind of "positive" or moderating influence on Nazi racial policy, Germany's eugenicists showed few qualms about their positions as scientific legitimizers of the kind of racism that sent millions to their deaths. Throughout the Third Reich they simply continued to insist that *their* understanding of eugenics was the scientific one, while attempting to resist taking a rabid anti-Semitic line whenever possible. They sought to hide behind the cloak of "objective science." Fischer, for example, could not be persuaded to say that *all* Jews were inferior. Science, he undoubtedly felt, would not allow such a statement. Hence, for Fischer, Jews were not necessarily always inferior; they were merely "different." Yet the eugenicists' attempt to preserve their moral and scientific integrity did not prevent them from using material shipped back from the death camps to the Berlin institute to further their own research. If the evidence presented by Benno Müller-Hill is accurate, blood samples and organs extracted from twins and dwarfs were transported from Auschwitz to the Berlin institute so that researchers could advance scientific knowledge.[126] Their

[124] *Ibid.*, pp. 131 (quotation), 79, 125, and 133.
[125] Lenz, "Die Stellung des Nationalsozialismus zur Rassenhygiene" (cit. n. 71), pp. 300–308.
[126] Müller-Hill, *Tödliche Wissenschaft* (cit. n. 2), pp. 78, 73–74.

crime was not so much their specific theories or their "respectable" anti-Semitism, but their willingness to continue with their work as though it were totally unrelated to the bestialities carried out by their masters in the name of race hygiene.

What, then—if anything—is the legacy of pre-1933 eugenics for the extermination programs of the Third Reich? Can one rightfully speak of an ideological connection between the kind of eugenics articulated by relative nonracists such as Schallmayer, Muckermann, and Grotjahn and the atrocities carried out in the name of race hygiene by Nazi officials? Throughout its history, race hygiene was a strategy that aimed at boosting national efficiency through the rational management of population. Whereas before the Third Reich "fitness" had generally been understood in purely meritocratic terms, without emphasizing race, after 1933 race and productivity became the two criteria defining it. It is not difficult to see the usefulness of race hygiene as a means of creating a stronger Nazi *völkisch* state. From the standpoint of efficiency, a racial policy such as the "euthanasia" program—the destruction of "unproductive lives"—is not without its logic, morally perverse though that logic is.

But what about the Holocaust? Although the extermination of millions of European Jews cannot really be viewed as a measure designed to boost national efficiency, the designation of the Jews as an unfit, surplus, and disposable group is not unrelated to the emphasis implicit in German race hygiene regarding "valuable" and "valueless" people. For the eugenicists, human beings were in some sense variables—objects easily managed or manipulated for some abstract "good." In one of humankind's most barbaric acts to date, there is more than a hint of where the desire to be rid of a "valueless" population can lead. Thus, whatever the intentions of even relatively nonracist eugenicists before 1933, the very logic of eugenics—the rational management of population for some "higher end"—was a logic readily amenable to other, far more sinister projects than those envisaged by Schallmayer, Muckermann, and Grotjahn. Hence, in the end, it is the *logic* of eugenics, far more than its racism, that proved the most unfortunate legacy of the German race hygiene movement for the Third Reich.

Wilhelm Schallmeyer and the Logic of German Eugenics

By Sheila Faith Weiss

IN 1917 THE PLANT GENETICIST ERWIN BAUR, a recent convert to the cause of eugenics, wrote a letter to the imperial government in which he attempted to awaken senior health officials in the Reich to the national importance and cost effectiveness of *Rassenhygiene*.[1] Rather than spend large amounts of money on welfare at a time when all of Germany's limited financial resources were needed to defeat the enemy at the front, Baur queried, "Would it not be more expedient to prevent invalidism and hereditary inferiority by means of an energetic race hygiene?" This "prototype of a rational management of human life," as Baur succinctly defined the new science of eugenics, would undoubtedly also prove necessary even after the Great War was over.[2]

Because Baur's letter was designed at least in part to secure badly needed funds for the Berlin chapter of the German Society for Race Hygiene, it is not surprising that he emphasized an argument likely to appeal to a government in serious fiscal trouble. Yet the language employed reveals far more about the nature of German eugenics than it does about Baur's ability to "sell" the new science. It reveals the managerial and technocratic logic underlying German race hygiene. Eugenics in Germany (and probably elsewhere as well) was implicitly viewed by its advocates as a form of rational management or managerial control over the reproductive capacities of various groups and classes. This, of course, implies that race hygiene is a tool. However, since a tool does not define its function, it is necessary to ask why race hygiene was embraced in the first place. My own research into the history of Wilhelmine eugenics suggests that German eugenics can best be understood as a sometimes conscious, oftentimes unconscious strategy to boost national efficiency—national efficiency denoting not only economic productivity but also, in the German context, cultural hegemony. German eugenicists believed that a rational administration of human resources would ensure the necessary level of hereditary fitness thought to be a prerequisite for the long-term survival of Germany, Western Europe, and the superior cultural traditions they supposedly embodied.[3]

An earlier version of this paper was presented at the History of Science Society convention in Philadelphia, October 1982. I am indebted to Gerhard Baader, Michael Hobart, Zachary Schiffman, and my husband, Michael Neufeld, for their suggestions and editorial assistance.

[1] The German term *Rassenhygiene* (race hygiene) had a broader scope than the English word "eugenics." It included not only all attempts aimed at "improving" the hereditary quality of a population but also measures directed toward an absolute increase in population. Despite these differences, I will for convenience employ the two terms interchangeably throughout this essay.

[2] Form letter by Erwin Baur and others for the Berliner Gesellschaft für Rassenhygiene, 18 Dec. 1917, Bundesarchiv Koblenz, R86, 2371, Bd. II, fol. 90.

[3] See Sheila F. Weiss, "Race Hygiene and the Rational Management of National Efficiency: Wil-

This managerial approach to population in the interest of national efficiency, although evident in the works of almost all German eugenicists, was first clearly articulated in the writings of the Bavarian physician Wilhelm Schallmayer (1857–1919). Recognized along with Alfred Ploetz (1860–1940) as one of the cofounders of the German race hygiene movement, Schallmayer was instrumental in shaping the future course of German eugenics. By the time of his death in 1919, he had written two treatises and over forty eugenics-related articles in addition to his major prize-winning textbook, *Heredity and Selection in the Life Process of Nations* (1903).[4] His obituaries clearly reveal Schallmayer's intellectual importance for the development of German race hygiene. Even the German race hygienist Fritz Lenz (1887–1976), a man whose interest in ideologies of Aryan supremacy stood diametrically opposed to Schallmayer's vision of eugenics, was forced to admit that "no one has accomplished more than Schallmayer"; his *Heredity and Selection* remains the "classical masterpiece of German race hygiene" and its author "enjoys a worldwide reputation, especially in England and America."[5]

This article seeks to lay bare the rationale and aims of German race hygiene, at least as they were publicly expressed during the Wilhelmine period, through a close examination of the logic of Schallmayer's eugenics. A careful analysis of Schallmayer's writings reveals the centrality of technocratic and managerial rationality to German eugenics and goes a long way toward unmasking its social and political function. It also challenges the "common sense" assumption, unfortunately still prevalent, that German race hygiene in the pre-Nazi period was preoccupied first and foremost with breeding a better "Nordic race."[6] Although, as will be argued elsewhere, Schallmayer may not have been typical of *some* influential race hygienists in his vehement rejection of all ideologies of Aryan supremacy, his obsession with saving the economically and socially better-situated classes from biological extinction and his desire to limit the number of unproductive types in the interest of national efficiency were a common denominator uniting both racist and nonracist eugenicists within the German movement.[7] Before turning to an analysis of the logic of Schallmayer's eugenics, however, it is first necessary to examine the conjunction of circumstances that collectively account for his adoption of such a strategy.

---

helm Schallmayer and the Origins of German Eugenics, 1890–1920" (Ph.D. Diss., Johns Hopkins Univ., 1983).

[4] Wilhelm Schallmayer, *Vererbung und Auslese im Lebenslauf der Völker* (Jena: Gustav Fischer, 1903). The treatise went through two revised editions (1910 and 1918).

[5] Fritz Lenz, "Wilhelm Schallmayer," *Münchener medizinische Wochenschrift*, 1919, *45*:1294–1296.

[6] There is a substantial body of literature not specifically dealing with the history of German eugenics that has perpetuated this misconception. See, e.g., David Gasman, *The Scientific Origins of National Socialism: Social Darwinism and the German Monist League* (New York: American Elsevier, 1971). Some scholars working on the history of eugenics in Britain and France have also helped, however unwittingly, to bolster the same idea. See Pauline Mazumdar, "The Eugenists and the Residuum: The Problem of the Urban Poor," *Bulletin of the History of Medicine*, 1980, *50*:204–205; Donald Mackenzie, "Eugenics in Britain," *Social Studies of Science*, 1976, *6*:501; and William Schneider, "Toward the Improvement of the Human Race: The History of Eugenics in France," *Journal of Modern History*, 1982, *54*:268.

[7] This matter will be discussed further in Sheila F. Weiss, "The Race Hygiene Movement in Germany," to be published in an anthology edited by Mark Adams of the University of Pennsylvania. The term "nonracist" eugenicists as used in this essay denotes those race hygienists who were opposed to all ideologies of Aryan supremacy. It goes without saying that all eugenicists, insofar as they accepted the racial and cultural superiority of Caucasians as a matter of course, were "racist" by today's standards.

THE SOCIAL, PROFESSIONAL, AND INTELLECTUAL CONTEXTS OF SCHALLMAYER'S EUGENICS

Schallmayer's eugenics was shaped by three main influences: the social problems resulting from Germany's rapid and thoroughgoing industrialization, the general outlook fostered by the professional and intellectual traditions of the German medical community, and finally, the "selectionist" variety of social Darwinism then fashionable among certain German biologists and self-styled social theorists. Taken collectively, these three circumstances explain why Schallmayer should have been concerned about Germany's efficiency in the first place as well as how he sought to improve it.

The first of these circumstances, the social context, provided Schallmayer with a seemingly urgent problem he wished to solve through a program of rational selection. Germany's rapid industrialization and urbanization process not only effected profound changes in the social and economic structure of the German Empire but also precipitated a myriad of serious social tensions and problems that, owing to the country's rigidly authoritarian political foundation, often threatened to upset the very stability of the young nation.[8] The most important of these problems was the rise of a radical labor movement. The growing number of strikes, lockouts and other forms of labor unrest, coupled with the growth of the Marxist Social Democratic Party, provoked fear and anxiety among many middle- and upper-class Germans regarding the seemingly hostile, uncontrollable, and ever-increasing industrial proletariat.[9] In addition to the labor movement, there was a series of other, albeit less serious, social problems that Germany's *Bildungsbürgertum* (educated middle class) viewed as posing a threat to the proper functioning of the state. These included an increase in various types of criminal activity and a rise in prostitution, suicides, alcohol consumption, and alcoholism. The *Bildungsbürgertum* at this point was also concerned about the large numbers of insane and feeble-minded individuals.[10] These so-called mental defectives were singled out by both medical and lay observers as an especially grave social and financial liability for the new Reich.

[8] For a discussion of this issue see Hans-Ulrich Wehler, *Das deutsche Kaiserreich* (Göttingen: Vandenhoeck & Ruprecht, 1977), pp. 60–140; Richard J. Evans, ed., *Society and Politics in Wilhelmine Germany* (New York: Barnes & Noble, 1978), pp. 16–22; and Wolfgang Mock, "Manipulation von oben oder Selbstorganisation an der Basis? Einige neuere Ansätze in der englischen Historiographie zur Geschichte des deutschen Kaiserreichs," *Historische Zeitschrift*, 1981, *232*:358–375.

[9] For a brief discussion of the middle-class fear of the proletariat see Fritz Ringer, *The Decline of German Mandarins: The German Academic Community, 1890–1933* (Cambridge, Mass.: Harvard Univ. Press, 1969), p. 129; Fritz Stern, "The Political Consequences of the Unpolitical German," in *The Failure of Illiberalism: Essays on the Political Culture of Modern Germany* (Chicago: Univ. of Chicago Press, 1975), p. 15; Guenther Roth, *The Social Democrats in Imperial Germany* (1963; New York: Arno Press, 1979), pp. 85–101.

[10] On crime, see Vincent E. McHale and Eric A. Johnson, "Urbanization, Industrialization and Crime in Imperial Germany," *Social Science History*, 1976/77, *1*:212–214; Eduard O. Mönkemöller, "Kriminalität," in *Handwörterbuch der sozialen Hygiene*, ed. Alfred Grotjahn and J. Kaup, Vol. I (Leipzig: F. C. W. Vogel, 1912), pp. 687–688; E. Fuld, "Das rückfällige Verbrechertum," *Deutsche Zeit- und Streit-Fragen*, 1885, *14*:453–484. On prostitution and alcoholism see Richard Evans, "Prostitution, State and Society in Imperial Germany," *Past and Present*, 1976, (70):106–108; James S. Roberts, "Der Alkoholkonsum deutscher Arbeiter im 19. Jahrhundert," *Geschichte und Gesellschaft*, 1980, *6*:226, 232, 237; Alfred Grotjahn, "Alkoholismus," in *Handwörterbuch der sozialen Hygiene*, Vol. I, p. 14. Ludwig Meyer, "Die Zunahme der Geisteskranken," *Deutsche Rundschau*, 1885, p. 83; Alexander von Oettingen, *Die Moralstatistik in ihrer Bedeutung für eine Sozialethik* (3rd ed., Erlangen: Deichert, 1882), p. 671; Arthur von Fircks, *Bevölkerung und Bevölkerungspolitik* (Leipzig: Hirschfeld, 1898), p. 116; Eduard O. Mönkemöller, "Selbstmord," in *Handwörterbuch der sozialen Hygiene*, Vol. II, p. 376; Alfred Grotjahn, "Krankenhauswesen," *ibid.*, p. 643.

The above-mentioned problems were hotly debated by many of Germany's academic social scientists and reform-minded religious leaders under the rubric of the "social question"—a term referring to the social and political consequences of unbridled economic liberalism and the industrialization process.[11] Although those discussing the "social question" embraced different economic and political ideals, all agreed that some kind of *Sozialpolitik* (social policy) was necessary to integrate Germany's proletariat (and asocial subproletariat) into the Reich, thereby preventing the collapse of the state. Having studied economics, sociology, and socialist theory during his early years at the University of Leipzig,[12] Schallmayer, perhaps more than most educated middle-class Germans, was keenly aware of this debate and the social problems underlying it. He became particularly concerned about the increased visibility of "mental defectives" and other nonproductive types and set out to tackle this problem using a novel form of *Sozialpolitik*: race hygiene.

That Schallmayer was inclined to offer a biomedical solution for what were, at least in large measure, social and political problems can be attributed to the second major influence on his eugenic thought: the distinctive social, political, and intellectual traditions of the German medical community—the professional context. Schallmayer became acquainted with these traditions after he decided to give up the social sciences in favor of a more practical career in medicine. He completed a medical degree at the University of Munich in 1883 and then later secured an internship in the university psychiatric clinic. As a physician, Schallmayer not only shared the prejudices and posture of the *Bildungsbürgertum* as a whole but also adopted a well-defined set of assumptions about the hereditary nature of disease and the role of medical professionals in safeguarding the health of the nation.[13]

The medical professionals' perception of themselves as custodians of national health, and hence national wealth and efficiency, had a long if not unbroken history. In Germany it dates back at least as far as the early eighteenth century, when cameralist theorists and administrators enlisted the aid of physicians to help survey and regulate what most mercantile political thinkers agreed to be the greatest source of national prosperity: a large and healthy work force.[14] Later, in

[11] Albert Müssigang, *Die soziale Frage in der historischen Schule der deutschen Nationalökonomie* (Tübingen: J. C. B. Mohr, 1968); Adolf Wagner, *Rede über die soziale Frage* (Berlin: Wiegandt & Grieben, 1872); Gustav Schmoller, "Die soziale Frage und der preussische Staat," in *Quellen zur Geschichte der sozialen Frage in Deutschland*, ed. Ernst Schraepler (2nd ed., Göttingen: Musterschmidt, 1964), Vol. II, pp. 62–66; Friedrich Naumann, "Christlich-Sozial," *ibid.*, pp. 79–84; and Ringer, *Decline of the German Mandarins* (cit. n. 9), pp. 145–147.

[12] Lenz, "Schallmayer" (cit. n. 5), p. 1295; and Max von Gruber, "Wilhelm Schallmayer," *Archiv für Rassen- und Gesellschafts-Biologie*, 1922, *14*:55. Information was also obtained from a 1977 interview I had with Wilhelm Schallmayer's son, Friedrich Schallmayer, in Karlsruhe, West Germany.

[13] For a discussion of the attitudes and prejudices of the *Bildungsbürgertum* see Klaus Vondung, ed., *Das Wilhelminische Bildungsbürgertum: Zur Sozialgeschichte seiner Ideen* (Göttingen: Vandenhoeck & Ruprecht, 1976); and Stern, "The Political Consequences of the Unpolitical German" (cit. n. 9), passim.

[14] George Rosen, "Die Entwicklung der sozialen Medizin," in *Seminar: Medizin, Gesellschaft, Geschichte*, ed. H.-U. Deppe and M. Regus (Frankfurt am Main: Suhrkamp, 1975), p. 78. For an excellent overview of the state's interest in a healthy work force and population see Charles Emil Strangeland, *Pre-Malthusian Doctrines of Population: A Study in the History of Economic Theory* (1904; New York: Augustus M. Kelley, 1966) and Michel Foucault, *The History of Sexuality*, Vol. I: *An Introduction*, trans. R. Hurley (New York: Vintage Books, 1980), pp. 24–26.

Figure 1. *Wilhelm Schallmayer (1857–1919) near the end of his life. Courtesy of Friedrich Schallmayer.*

the mid-nineteenth century, the German physicians' consciousness of their responsibility to the state received further impetus from the so-called health reform movement associated with Rudolf Virchow and Salomon Neumann. And finally, during the third quarter of the nineteenth century, the rise of scientific medicine and hygiene bestowed on academic physicians and the medical profession in general an unprecedented level of social esteem and, indirectly, political importance.[15] At this time many young medical professionals eager to make a contribution to national health turned their attention to bacteriology; others, like Schallmayer, adopted a different line of approach. Their exposure to fields of medicine emphasizing the role of heredity in the etiology of disease (e.g., neurology and psychiatry)[16] led them to question the efficacy of concentrating solely on pathogens. Instead they were convinced that serious disorders such as mental illness, feeble-mindedness, criminality, epilepsy, hysteria, and the tendency to tuberculosis were often inherited and could quite frequently be traced back to a "hereditary diseased constitution."[17] Schallmayer and like-minded physicians argued that the surest way to improve the general level of national health was to upgrade the bodily constitution of all individuals in society—a task to be accomplished by means of an energetic eugenics program.

Schallmayer's interest in the social question and his medical background were prerequisites for his adoption of race hygiene. Equally important, however, was the intellectual context: the "selectionist" variety of "social Darwinism," especially as legitimized by the scientific writings of the Freiburg embryologist August Weismann and popularized by Germany's most outspoken biologist, Ernst Haeckel. Weismann, who came to reject the possibility of an inheritance of acquired characteristics through his work on heredity, gave Darwin's principle of natural selection an even greater role in organic and social evolution than did the author of the *Origin* himself. Haeckel, although technically a Lamarckian, always stressed the importance of Darwin's selection principle; indeed, for Haeckel Darwinism was synonymous with selection.[18] Through his readings first of Haeckel and later of Weismann, Schallmayer came to view natural selection as the sole agent of all organic and social progress. From them and others he also learned, however, that progress was not inevitable. Under certain conditions the "unfit" might prosper, thereby posing a challenge to any further evolutionary development.

[15] For a discussion of the health reform movement see Erwin H. Ackerknecht, "Beiträge zur Geschichte der Medizinalreform von 1848," *Archiv für Geschichte der Medizin*, 1932, *25*:61–109, 112–183; and Rosen, "Die Entwicklung der sozialen Medizin," pp. 99–102. On medicine in the late nineteeth century see George Rosen, *A History of Public Health* (New York: M.D. Publications, 1958), p. 44; Eduard Seidler, "Der politische Standort des Arztes im zweiten Kaiserreich," in *Medizin, Naturwissenschaft, Technik und das zweite Kaiserreich*, ed. Gunter Mann and Rolf Winau (Göttingen: Vandenhoeck & Ruprecht, 1977), pp. 91–92; and Hans-Heinz Eulner, "Hygiene als akademisches Fach," in *Städte-, Wohnungs- und Kleidungshygiene des 19. Jahrhunderts in Deutschland*, ed. Walter Artelt et al. (Stuttgart: Ferdinand Enke, 1969), p. 18.

[16] Erwin H. Ackerknecht, *A Short History of Psychiatry*, trans. S. Wolff (2nd ed., New York/London: Hafner, 1968), p. 82. Although a strong commitment to and reliance on hereditarian explanations of neurological and mental ailments are discernible as early as the 1860s, by the 1880s one can speak of a veritable fetish of heredity in German neurological and psychiatric circles.

[17] A partial list of the available literature discussing the importance of heredity as an etiological factor in insanity is given in the bibliography to K. Grassman, "Kritische Ueberblick über die gegenwärtige Lehre von der Erblichkeit der Psychosen," *Allgemeine Zeitschrift für Psychiatrie*, 1896, *52*:960–1022. With regard to the role assigned to heredity in epilepsy see Oscar Aronson, *Ueber Heredität bei Epilepsie* (Berlin: Wilhelm Axt, 1894). The etiological significance of the hereditary disposition for tuberculosis was also stressed in Germany's leading medical journals at this time; see, e.g., M. Wahl, "Über den gegenwärtigen Stand der Erblichkeitsfrage in der Lehre von der Tuberculose," *Deutsche Medizinische Wochenschrift*, 1885, No. 1, pp. 3–5, No. 3, pp. 36–38, No. 4, pp. 34–36, No. 5, pp. 69–71, No. 6, pp. 88–90. One institutional study available to physicians was "Uber die Vererbung von Geisteskrankheiten nach Beobachtung in preussischen Irrenanstalten," *Jahrbuch für Psychiatrie und Neurologie*, 1879, *1*:65–66.

[18] Ernst Haeckel, *The History of Creation*, trans. and rev. E. Ray Lankester (London: Henry S. King, 1876), Vol. I, p. 120.

Schallmayer's Darwinian perspective and language afforded him novel tools of analysis that enabled him to come to grips with the social question by transforming it into a biological (as well as medical) problem: the asocial individuals created by industrialization became for Schallmayer the biologically and medically "unfit." In the interest of national efficiency they must be effectively managed (i.e., weeded out).

THE LOGIC OF SCHALLMAYER'S EUGENICS: THE RATIONAL MANAGEMENT OF NATIONAL EFFICIENCY

These three contexts—social, professional, and intellectual—are nowhere more beautifully synthesized than in Schallmayer's first treatise, a short work entitled *Concerning the Threatening Physical Degeneration of Civilized Humanity* (1891).[19] As the title indicates, Schallmayer's treatise was concerned with the so-called problem of degeneration. First popularized by the French alienist Benedict Morel, the concept of degeneration had by the 1880s become part of the standard vocabulary of social Darwinists as well as physicians, albeit not without significant differences in meaning. For social Darwinists like Schallmayer, the degeneration of human beings was the inevitable result of the tendency of modern civilization to impede the efficacy of natural selection—the alleged instrument of human evolution and cultural advancement.

In his work Schallmayer outlined many "counter-selective" practices and institutions of modern industrial society. None, however, was believed by him to be more damaging to the long-term fitness of the human race than his own profession: medicine. Echoing concerns about the dangers of modern medicine articulated earlier by both Darwin and Haeckel, Schallmayer accused medical practice of tampering with the "natural" ratio existing between "favored" and "defective" organisms in the struggle for survival. More and more inferior types were being artificially kept alive, thus enabling them to reproduce. The resulting increase in the number of defective individuals, he was quick to point out, "did not lie in the interest of a favorable selection."[20]

Schallmayer's "defective" individuals were none other than the *erblich Belasteten* (defectively constituted)—those people perceived by German physicians to have some kind of socially dangerous or burdensome hereditary illness—individuals such as the feebleminded, criminals, homosexuals, and tuberculars who, in one way or another, upset the harmony of the social body. Not coincidentally, these "defectives," or asocial types, were also becoming increasingly visible because of the effects of industrialization and urbanization. Like the German medical degeneration theorists Paul Möbius and Richard von Krafft-Ebing, Schallmayer was especially concerned about the apparent rise in the number of one class of degenerates: the insane. In his treatise he did not neglect to draw the reader's attention to the consequences of the "superficially humane" care of the mentally defective for the efficiency of the German nation. "The insane," contended Schallmayer, "constitute an enormous burden for the state."[21] Not only

[19] Wilhelm Schallmayer, *Ueber die drohende physische Entartung der Culturvölker* (1891; 2nd ed., Neuwied, 1895).

[20] *Ibid.*, p. 6.

[21] *Ibid.*, p. 13. This and all translations are mine.

were they a financial drain on a country that had better things to do with its revenues, but also their care and maintenance required the attention of a large number of persons who could be more productively employed elsewhere. Even worse, however, was that all this expense did not succeed in reducing the number of these unfortunate victims but instead enlarged it.

Schallmayer summarized his views on the so-called achievements of modern medicine by means of a comparison designed to appeal to his professional middle-class audience:

> Even if medical technology grew to such an extent that malfunctioning human organs could safely be replaced by healthy human, animal, or laboratory-produced ones, the following generations would not be more efficient, rather just the opposite: the more advanced therapeutic medicine becomes, the more succeeding generations will have need of it. Therapeutic medicine affects the improvement of national health in about the same way as poor relief contributes to the improvement of national welfare. Both encourage an increase in the dependent population. . . . [M]edicine, insofar as it aims at treatment rather than prevention, contributes nothing to the the gradual advance of human productivity and human happiness. *It aids the individual but at the expense of the human race.*[22]

Couched in the terminology of efficiency, performance, and productivity, his comparison of the counterselective effect of therapeutic medicine with that of aid to the poor reveals to what extent the medically degenerate and socially dependent were related in Schallmayer's mind. Both groups impeded the overall efficiency of the state; both represented a social and financial liability for the nation. Moreover, there was a large overlap between the two groups, as typified by the "pauper idiot"—individuals who, because of their mental deficiency, would remain part of Germany's *Lumpenproletariat*. Whether the behavior of such degenerates manifested itself as insanity, criminality, indigence, unemployability, or inebriety was of little importance. The defective portion of the population needed to be contained and controlled lest it undermine the fitness of the race and the Reich.

Schallmayer did not suggest that Germany abandon either therapeutic medicine or public hygiene as a means of preventing the onslaught of degeneration; such a tactic, even if it could be implemented, would go against all humanitarian instincts. The solution to the problem, he argued, was the creation of a new branch of hygiene dedicated to the betterment of the hereditary stock of the Reich. It would, he insisted, be far more effective in eliminating disease than the thankless "Sisyphean labor" carried out by therapeutic medicine. Simply put, the goal of this new discipline was to impart a "rational influence upon human selection."[23] Although not mentioned by name in the treatise, this new branch of hygiene would later be termed *Rassenhygiene*.

Schallmayer's insistence that Germany embark on a race hygiene program was intimately linked to his understanding of health. Health for Schallmayer was not merely a private but also an inherently political matter. As such it was imperative that the state have a means of monitoring the level of national vitality. Although he was never quite so blunt, he would certainly have agreed with the statement of his German colleague H. Knieke that "in the never-ending drive between rival

[22] *Ibid.*, p. 7–8 (italics in original).
[23] *Ibid.*, p. 9.

nations for predominance, an increasing measure of government control or activity in the area of health will serve as a weapon in the struggle to preserve and strengthen national power."[24]

Schallmayer's own views on the political significance of health can best be seen in an 1899 article entitled "A Ministry of Medicine." Exasperated by the absence of a national ministry of health in Germany, Schallmayer sought to convince those in authority of the national importance of establishing such a body, directed and staffed of course by physicians.

> Who would have the nerve to maintain that a minister exclusively responsible for the health and strength of the bodily organism for all citizens administers goods (*verwaltet Güter*) less valuable than those of an agricultural, trade, or railroad minister? The state has above all an economic, but also an ethical, and to a certain degree a military interest in improving the physical and mental efficiency of its citizens, as well as alleviating their pain and prolonging their life. First of all, it is the task of government to see to it that disease and accidents as well as all other influences that weaken the physical and mental productivity are prevented; and second of all, that the injuries are healed as quickly and fully and with the least amount of pain as possible. For the same reasons the state should take up the task of insuring that seriously diseased constitutions are not passed on to future generations.[25]

Again, the emphasis is on population as a natural resource that becomes even more valuable when in "good health." This quotation articulates the relationship between medicine, health, and national productivity in Schallmayer's thinking. Rational selection or race hygiene was merely the form of medicine that could most efficiently and effectively boost national productivity. And physicians, being health experts, were the most likely candidates for the task of monitoring the Reich's level of biological vitality upon which national strength and power were based.

SCHALLMAYER'S AWARD-WINNING TREATISE

With the exception of a few book reviews, Schallmayer's first eugenic treatise passed totally unnoticed; in terms of public impact the work was undoubtedly a failure. Indeed Schallmayer's arguments for the necessity of race hygiene as well as his specific eugenic proposals, both of which were merely outlined rather than elaborated on in his 1891 essay, might easily have fallen into oblivion had not an unprecedented event nine years later afforded the author a second opportunity to reach the German public.

It is certainly fair to say that the Krupp competition of 1900 marked a turning point both in Schallmayer's personal career and in the attention paid to eugenics in Germany. Designed at least in part to demonstrate that Darwinian biology did not, as some critics suggested, pose a threat to the political status quo, the competition was secretly sponsored by Friedrich Alfred Krupp (1854–1902), son of the famous munitions baron of Essen, Alfred Krupp.[26] Friedrich Krupp, who in

[24] H. Knieke, "Die Verstaatlichung des Aerztewesens," *Politisch-Anthropologische Revue*, 1903/4, *2*:402–409.

[25] Wilhelm Schallmayer, "Ein Medizinalministerium," *Das neue Jahrhundert*, 1899, *1*:393.

[26] Apparently there is some confusion in the historical literature concerning just which Krupp sponsored the *Preisausschreiben* of 1900. Some historians do not bother to tell their readers whether it was Alfred Krupp or his son, Friedrich, who donated money for the contest. Several scholars,

addition to managing his giant firm indulged himself in the study of aquatic life and oceanography, entrusted the execution of the contest to Germany's most eminent marine biologist, Ernst Haeckel. "Toward the advancement of science and in the interest of the Fatherland," Krupp donated 30,000 marks to be used in a contest to answer the question: "What can we learn from the theory of evolution about internal political development and state legislation?" On 7 March 1903 the prize committee announced that Schallmayer was the winner of the competition.[27]

Schallmayer's award-winning essay, *Heredity and Selection in the Life Process of Nations,* represented, as one judge appraised it, a "hygienic-sociological" approach to the question.[28] His treatise bore witness not only to his "selectionist" outlook but also to his adoption of Weismann's biological theories, especially the "continuity of the germ-plasm." Schallmayer's thirty-eight page discussion of Weismann's views on the transmission of hereditary traits served both to address the requirements of the contest (all contestants had to take a position on the mechanism of heredity) and to legitimize his particular eugenic interpretation of the question.[29] But more important, Schallmayer's work, which in a revised edition was to become the standard German eugenics textbook, sought to prove that the real political lesson to be learned from Darwin's theory was that long-term state power depended on the biological vitality of the nation; any "mismanagement" of the hereditary fitness of its population, such as might result from unenlightened laws and customs, was "bad politics" and would inevitably result in the downfall of the state.[30]

In order to maintain itself in the competitive struggle among nations, Schallmayer asserted, the state must shape its policies such that "the greatest possible increase of power" is guaranteed. How, then, could the state work toward the one "scientific" goal of all political development: national preservation? What, according to Schallmayer, were the necessary "means" to this "end"? To begin with, state power entailed a thoroughgoing rationalization of human society. Political power was no longer simply a function of how quickly a nation could mobilize its armed forces for combat. Indeed, the military might of a state was itself in large measure dependent on the country's level of economic and technical efficiency. As Schallmayer put it, power presupposed both an "ever-increasing division of labor" and a "continuous decrease in the squandering of manpower." Moreover, improved organization would raise the efficiency of individual as well as collective productivity. Only state socialism, Schallmayer believed, could achieve the requisite level of efficiency; the chaos of unbridled laissez-faire capitalism could not.

---

following the incorrect lead of Hedwig Conrad-Martius in *Utopien der Menschenzüchtung: Der Sozialdarwinismus und seine Folgen* (Munich: Köse, 1955), p. 74, wrongly believe the donor to have been Alfred Krupp, who had already died in the 1880s.

[27] Heinrich E. Ziegler, "Einleitung zu dem Sammelwerke Natur und Staat" in *Natur und Staat: Beiträge zur naturwissenschaftlichen Gesellschaftslehre* (Jena: Gustav Fischer, 1903), pp. 1–2.

[28] *Ibid.*, p. 9.

[29] Schallmayer, *Vererbung und Auslese* (cit. n. 4), pp. 32–70. In all subsequent editions of *Vererbung und Auslese* Schallmayer devotes space to a discussion of Mendel's theory of heredity. His discussion of Mendelism, however, was included primarily to legitimize his call for eugenics. Schallmayer's turn to eugenics was independent of his commitment to either Weismann's or Mendel's hereditary theories.

[30] *Ibid.*, pp. 380–381.

But economic rationalization, although necessary for national efficiency, was hardly the entire solution. Economic strength, Schallmayer maintained, presupposed a high level of biological vitality. Since only a form of population management could make possible the level of biological efficiency required to support an economically and culturally productive nation, statesmen needed to formulate national policy with this view in mind:

> Since a large population is an essential element of power and a prerequisite for a higher development of social organization, all efforts in the area of domestic as well as foreign policy must be judged from the standpoint of whether they are likely to strengthen or weaken the ability of the population to *survive* and *procreate*. However, it is not simply a matter of the quantity of population, but also a question of the population's social productivity. . . . [I]n order for the state to hold its own or have supremacy over other peoples, domestic policy must not neglect the hereditary composition of the population. *Future political development will prove more successful the more it secures the effectiveness of a generative selection*—a selection which is the necessary condition not only for all progress but even for the preservation of the status quo.[31]

Thus it was necessary for the state to manage rationally not only its economy but also its population. Here, as so often in his writings, Schallmayer employed an economic analogy: the relationship between state and society, he contended, is the same as that between "the landlord and his tenants."[32] That is, the state must encourage the populace not to squander its labor and procreative power for short-term gain but rather to administer them for the long-term good of society.

Having once established the importance of rationally managing Germany's stock of human resources in the interest of national efficiency, Schallmayer presented his readers with a series of eugenics reforms touching numerous social, political, and economic institutions. Although he did not present his ideas in the form of a clearly articulated list in the first edition of *Heredity and Selection,* his disparate statements on the subject can, for the sake of convenience, be categorized under the headings of negative and positive eugenics.

Included under the rubric of negative eugenics (a term that Schallmayer did not use but that later became current) were all measures—voluntary and involuntary, direct and indirect—that sought to prevent or at least strongly discourage the "unfit" and "defective" from reproducing. Schallmayer proposed the imposition of a tax on those unfit for the military. He was also favorably inclined toward either sterilizing proven hereditary criminals in order to protect society from generations of lawbreakers or permanently placing them in a criminal asylum, an idea advocated at the time by a retired London chief of police.[33]

On the whole, however, Schallmayer proceeded with the utmost caution in the area of negative eugenics. Although he clearly believed that marriage restrictions for the insane, the feebleminded, the chronic alcoholic, the hereditary criminal, the tubercular, and those not fully cured of venereal disease were in the best interest of national efficiency, he refrained from openly supporting state legislation as a means to this end. Until such time as more exact information regarding the laws of heredity were known, and enough detailed genealogies could be

[31] *Ibid.*, pp. 247–248.
[32] *Ibid.*, p. 248.
[33] *Ibid.*, p. 303.

amassed, those interested in the welfare of the nation would have to concentrate their efforts on voluntary measures. In Schallmayer's opinion, much more could be accomplished by instituting a new moral code—a more polite way of suggesting that through propaganda and indoctrination, people would eventually be made to recognize the danger of an unbridled increase in the "unfit."

But "generative ethics" was not only an important element of negative eugenics; it also played a major role in what I have called positive eugenics. Included in this category are all those measures that encourage the "fitter" groups and classes of society to increase their fertility rate—measures that formed, for Schallmayer, the more significant side of his eugenic strategy. The question of course remained, which groups are, biologically speaking, the "fittest"? Schallmayer hoped that the "science of heredity" would eventually settle thc issue in an objective, scientific manner. Until this new biological discipline reached maturity, however, relative "fitness" would be determined by a group's overall contribution to society. "In the meantime," he argued, "it would not be incorrect to view highly socially productive individuals, especially the better educated, as being, on the average, more biologically valuable."[34] Civil servants, officers and teachers were to be encouraged to marry as early as possible. Those who chose to remain single should suffer some sort of financial disadvantage. To encourage high-level civil servants to have larger families, Schallmayer suggested that they be given a bonus for each school-age child.[35] The class bias implicit in his criteria for "fitness" could hardly be more blatant; the biologically "fit" turn out to be individuals from the same socioeconomic group as Schallmayer: the educated middle class.

Schallmayer's equation of fitness with social productivity provided him with a useful scientific defense of his political ideal: meritocracy. It is revealing that the utopian vision depicted in Edward Bellamy's *Looking Backward* so captured the attention and approval of Schallmayer. Bellamy's future socialist society placed a premium on productivity, efficiency, and the proper management of human resources. As Schallmayer put it, such a society is a place where "the effectiveness of natural selection is not curtailed."[36] Whatever reservations he might have had concerning the feasibility of creating such a society, Schallmayer was certain that only a meritocratic society (*Leistungsgesellschaft*), with a meritocracy (*Leistungsaristokratie*) forming the ruling class, would, in the long run, allow both Germany and Western Europe as a whole to escape the fate of ancient Rome.

THE LEGACY OF SCHALLMAYER'S EUGENIC STRATEGY

From 1903 until his death in 1919, Schallmayer remained preoccupied with securing the requisite biological foundations for a meritocratic state. Neither his intellectual opinions nor his politics changed appreciably during this time. This is particularly true of his attitude toward Aryan race theories; throughout his career he continued to dismiss publicly the pretensions of Germany's Aryan enthusiasts, especially those of the racist school of *Sozialanthropologie*, as both un-

[34] *Ibid.*, p. 338.
[35] *Ibid.*, p. 338–339.
[36] Schallmayer, *Physische Entartung* (cit. n. 19), p. 35.

scientific and detrimental to the cause of eugenics.[37] Schallmayer claimed that the advocates of this school accepted the superiority of the "Nordic race" as an article of faith, as an a priori truth, in the same way that theologians believe the teaching of the church. Any attempt to equate the aims and methodology of eugenics with the racial nonsense of "social anthropology"—to link race hygiene and racism—would, he reiterated, "guide the eugenics movement in a direction that leads nowhere or nowhere good."[38]

Although prewar international tensions and the realities of prolonged combat during World War I contributed to his growing obsession with "population quantity," even the war did not undermine his essentially cosmopolitan outlook—at least with regard to Western Europe. Schallmayer wanted to boost the national efficiency of Germany through eugenics, but his primary aim was not to give his country a competitive edge over its Western neighbors. His race hygiene program, like Wilhelmine eugenics in general, had a Western European orientation; insofar as it was nationalist or racist at all, it was aimed at saving his own country and others from the "yellow peril" and the "Russian bear." Indeed one can say that Schallmayer saw his eugenics as a last-ditch effort to rescue both Germany and Western Europe from a kind of biological degeneration that, in his opinion, would inevitably lead to the decline of Western political and cultural hegemony.

Schallmayer's eugenics, as should be evident, presupposed the idea that *power* was essentially a problem in the rational management of *population*. Behind his intention to apply "human reason to human selection" lay a technocratic conception of population as a natural resource that, in the interest of national efficiency and state power, should become subject to some form of rational control. Through a policy of rational selection the "degenerate," the asocial, the unproductive would be gradually eliminated, and the "fit," the socially responsible, the productive—a synonym for the educated middle class—would slowly become a larger and larger proportion of the population. Only such a rational administration of population, Schallmayer asserted, could prevent the nation of Goethe, Schiller, and Kant from falling prey to proletarian mediocrity.

As I have shown elsewhere, the technocratic logic and class biases at the foundation of Schallmayer's eugenics shaped the German race hygiene program during both the Empire and the Weimar Republic.[39] Indeed this emphasis on rationally controlling population in the interest of national efficiency became even more pronounced during the troubled Weimar years than it had been during Schallmayer's lifetime. To what degree does the view of eugenics presented in this paper hold true for Nazi race hygiene?

[37] The meaning of *Sozialanthropologie* or "social anthropology" as used in this text is radically different from its current meaning. Today the term is used to denote the subdiscipline of anthropology that "aims at understanding and explaining the diversity of human behavior by a comparative study of social relations and processes over as wide a range of societies as possible"; Raymond Firth, "Social Anthropology," *International Encyclopedia of the Social Sciences*, Vol. I–II (New York: Macmillan; Free Press, 1968), pp. 320–324. During the late nineteenth and early twentieth century in Germany the term was used to describe a noxious brew of Aryan racism, social Darwinism, and anthropometry that masqueraded as a new scientific discipline. Its intellectual roots go back to the writings of the French diplomat, publicist, and aristocrat Comte Arthur de Gobineau, and it remained, as a movement, largely separate from the early German eugenics movement. For a detailed discussion of Schallmayer's rejection of *Sozialanthropologie* see Weiss, "Race Hygiene" (cit. n. 3), pp. 179–204.

[38] Schallmayer, *Vererbung und Auslese*, 2nd ed. (1910) (cit. n. 4), p. 374.

[39] Weiss, "Race Hygiene Movement" (cit. n. 7).

Yet despite a strong beginning, the Broca school ultimately failed to sustain momentum and by the turn of the century entered a period of marked decline. This decline reflected a larger movement within anthropology away from the racialist preoccupations of the nineteenth century, with which Broca and his cohort were closely identified, but it was also rooted in the structural defects of their organizational base—defects variously attributable to their own tactical errors and to peculiarities in the structure of French scholarly life. Indeed, the Broca school's inability to create a firm basis for anthropological teaching, research, and training in France demonstrates the limited potential of the organizational forms and resources at the disposal of French scholars entering new fields.

These forms were of three principal types: the learned society, which in nineteenth-century France was the most popular organizational framework for nontraditional scholarship; programs within the official structure of higher education; and private institutions for research and training. As we shall see, the limitations of the learned society as a vehicle for perpetuating the discipline, especially in providing training to students, caused leaders in anthropology to seek institutionalization within state-supported institutions of higher learning. In this regard, however, anthropology suffered the fate of many emergent disciplines in France; it was greeted with suspicion and given only ad hoc, provisional support. In consequence, anthropologists of the late nineteenth century devoted considerable resources to establishing private institutions—a school, library, laboratory, and museum—for transmitting their increasingly specialized knowledge and skills. These institutions in turn, however, suffered from many weaknesses. Their funding was precarious, and they lacked clear lines of authority and responsibility. Most important, they were peripheral to the formal degree structure of French higher education and hence offered training of dubious value to students preparing for academic or scientific careers. Accordingly, anthropology failed in the nineteenth century to achieve unequivocal academic standing; with the decline of the Broca school around 1900, it fragmented into a number of small research groups lacking momentum, strong leadership, and common purpose.

THE LEARNED SOCIETY

In the nineteenth century, the *société savante* was a ubiquitous form of organization in French intellectual, commercial, and social life. Bourgeois in character and devoted to specialized pursuits, learned societies replaced the aristocratic academies of science and letters that had dominated the intellectual landscape of the eighteenth century. In scholarship the learned society was essential to the organization and advancement of certain fields, especially those like anthropology that were just beginning to achieve formal expression. The

(Berkeley/Los Angeles: Univ. California Press, 1979); William B. Cohen, *The French Encounter with Africans: White Response to Blacks, 1530–1880* (Bloomington: Indiana Univ. Press, 1980), Ch. 8; Michael Hammond, "Anthropology as a Weapon of Social Combat in Late Nineteenth-Century France," *Journal of the History of the Behavioral Sciences,* 1980, *16*:118–132; and Joy Harvey, "Evolutionism Transformed: Positivists and Materialists in the *Société d'Anthropologie de Paris* from Second Empire to Third Republic" in *The Wider Domain of Evolutionary Thought,* ed. D. Oldroyd and I. Langham (Dordrecht: Reidel, 1983), pp. 289–310.

first body of any kind devoted exclusively to the field was a learned society founded late in the eighteenth century—the Société des Observateurs de l'Homme—and over succeeding decades three major societies and a number of smaller ones were established to promote anthropological inquiry.[2]

In a recent essay Robert Fox has provided a much-needed general perspective on learned societies, whose "proliferation" he rightly calls "one of the most startling, and neglected, cultural phenomena of nineteenth-century France." Fox describes the learned society as a center for pursuing "bourgeois enthusiasms"—natural history, "antiquities," and, late in the century, archeology. Ethnology, which in the early nineteenth century generally meant the joint study of the history and physical character of the races, also counted for a time among these enthusiasms. It blended the antiquarian's interest in local populations with the naturalist's penchant for classifying, numbering, and tabulating. When the Société Ethnologique de Paris was founded in 1839 by the physiologist W. F. Edwards, it attracted a small elite of professionally trained academics and a large majority of "gentleman-scholars." At the outset the society proclaimed broad intellectual and social goals, and in subsequent activities it provided serious scholars a forum for work in a little-recognized field while satisfying for other members a more casual intellectual curiosity. Indeed, the society never drew a rigorous distinction between scholarly and public objectives, and it readily passed from ethnology to political controversy, a tendency that contributed to the society's demise amid the upheavals of 1848.[3]

Despite the failure of the Société Ethnologique, the learned society continued to be the favored vehicle for promoting anthropology. In 1859 the Société d'Anthropologie and the Société d'Ethnographie were launched within months of one another. The anthropological society, founded by Broca and a small group of colleagues who split off from the Société de Biologie, rapidly developed into the major anthropological organization in nineteenth-century France. Within a decade the society had almost three hundred members, a widely circulated journal, and a notorious reputation.[4] While the society attracted "enthusiasts"—particularly in the early years, when anthropology generated heated controversies over Darwin's theory of species change, the archaeologist Jacques Boucher de Perthes's proof for "fossil man," and Broca's publications on hybridity—a core of members worked exclusively in anthropology, and a great many others followed developments closely and eagerly participated in the society's activities.

Also founded in 1859, the Société d'Ethnographie was established by a circle of linguists led by the orientalist Léon de Rosny.[5] The society's avowed purpose was to promote a "comprehensive" science of anthropology, one that, while

[2] Robert Fox, "The *Savant* Confronts His Peers: Scientific Societies in France, 1815–1914," in *The Organization of Science and Technology in France, 1808–1914*, ed. Fox and George Weisz (New York: Cambridge Univ. Press, 1980), pp. 240–282, esp. pp. 245–247. On the Société des Observateurs de l'Homme, see Jean Copans and Jean Jamin, eds., *Aux origines de l'anthropologie française: Les mémoires de la Société des Observateurs de l'Homme en l'an VIII* (Paris: Le Sycomore, 1978); and Benjamin Kilborne, "Anthropological Thought in the Wake of the French Revolution: The 'Société des Observateurs de l'Homme,'" *European Journal of Sociology*, 1982, *23*:73–91.

[3] Fox, "Scientific Societies," p. 244; for a detailed discussion of the Société Ethnologique de Paris, see Elizabeth A. Williams, "The Science of Man: Anthropological Thought and Institutions in Nineteenth-Century France" (Ph.D. diss., Indiana Univ., 1983), Ch. 2.

[4] On Broca and the founding of the Société d'Anthropologie, see Schiller, *Paul Broca*, pp. 130–135.

[5] Williams, "The Science of Man," Ch. 4.

alert to problems of race, did not undervalue man's "moral" and intellectual capacities. Accordingly, its founders proclaimed their hostility to the materialist, anatomically based anthropology promoted by the Société d'Anthropologie. In ensuing years the objective of combating materialism proved to be of central importance to this society, which achieved more success as an ideological standard-bearer than as an organization devoted to anthropological science. Indeed, the society's leaders assiduously cultivated the wealthy, the influential, and the high-born, whose participation they thought essential to the accomplishment of ethnography's ultimate social and ideological purposes. This concern was reflected in the membership: in any given year, only a small proportion of members were professional academics; most were political, social, or business notables.[6]

Over the course of the nineteenth century, learned societies took diverse forms and enjoyed varying degrees of prestige, depending on their location, emphasis, and leadership. The so-called national disciplinary societies, usually based in Paris, tended to pursue specific scientific goals under the direction of prominent professors. Some of these societies achieved high standards and contributed significantly to the development and diffusion of scientific knowledge.[7] In so doing, they filled a real gap in scholarly organization in nineteenth-century France. The public system of higher education was slow to extend support to new branches of science and scholarship, much less to encourage innovation or controversy. Indeed, the uncongenial welcome given new disciplines has often been adduced as one of the more serious obstacles to scientific progress in late nineteenth-century France. And for a good part of the century French higher education suffered not only from official conservatism but from serious practical deficiencies as well. Until late in the century, research facilities were few, and little provision was made for the advanced training of students in specialized subjects.[8]

Learned societies, however, were ill equipped to compensate fully for the of-

[6] See membership lists in the *Actes de la Société d'Ethnographie* (1859–1884) and the *Bulletins de la Société d'Ethnographie* (1887–1900).

[7] Fox, "Scientific Societies," pp. 265–279. As Fox has pointed out (*ibid.*, pp. 241–244, n. 6; pp. 265–269), the dearth of historical material on learned societies is surprising in view of their great importance in the social, cultural, and intellectual life of nineteenth-century France. Their mixed "amateur" and "professional" status is doubtless one reason they have been neglected. The concept of "professionalization" is of limited use, however, in analyzing the history of a fledgling discipline like anthropology. If professionalization is taken to mean the efforts of a corporate association to achieve official standing, control training and credentialing, and define standards of conduct, anthropologists were plainly in no position to professionalize. Their work was directed at attracting participants rather than excluding charlatans, stimulating general interest rather than protecting the interests of an established corps. For a helpful guide to the literature on professionalization, see Nathan Reingold, "Definitions and Speculations: The Professionalization of Science in America in the Nineteenth Century," in *The Pursuit of Knowledge in the Early American Republic*, ed. Alexandra Oleson and Sanborn C. Brown (Baltimore: Johns Hopkins Univ. Press, 1976), pp. 33–69, at n. 3, pp. 64–65.

[8] The defects of the French educational system are the subject of a large literature; see Joseph Ben-David and Abraham Złoczower, "Universities and Academic Systems in Modern Societies," *Eur. J. Sociol.*, 1962, *3*:45–84; Theodore Zeldin, "Higher Education in France, 1848–1940," *Journal of Contemporary History*, 1967, *3*:53–80; and on consequences for the development of the sciences (the "declinist" debate), see the fine discussion by Robert Fox and George Weisz, "The Institutional Basis of French Science in the Nineteenth Century," in *Organization of Science* (cit. n. 2), pp. 1–29. For a critique of the declinist position, see Harry W. Paul, "The Issue of Decline in Nineteenth-Century French Science," *French Historical Studies*, 1972, *7*:416–450. No one has yet attempted to compare developments in the natural and social sciences; on the latter, the key work is Clark, *Prophets and Patrons* (cit. n. 1).

ficial neglect of emerging fields like anthropology. Many of them succumbed to practical difficulties such as poor management, lack of public support, or underfinancing. Of the three major anthropological societies founded in the nineteenth century, only the Société d'Anthropologie enjoyed real success. The earliest of them, the ill-fated Société Ethnologique de Paris, was widely praised by French anthropologists late in the nineteenth century for the "brilliant" work it had sponsored, but even these sympathetic observers recognized the society's many faults.[9] Some of the society's problems were common to learned societies of the period; others were peculiar to the ethnologists (their close identification with abolitionism, for example). One problem all societies faced was securing basic resources—a meeting place, funds for publishing, and a staff to send out announcements, manage publications, and keep scholarly and financial records. Like other societies, the Société Ethnologique obtained money to meet these needs from three sources: membership fees, private donations, and public subsidies. Nevertheless, the budget of the Société Ethnologique was very modest: in 1846 total receipts came to 2,925 francs and expenditures to 1,702 francs; in 1848, the only other year for which financial information is available, total income was 3,316 francs, total outlay 2,704 francs.[10]

While the society's financial resources were undoubtedly limited, effective leadership could have put surplus funds to good use. With several exceptions, however, the officers did not make the society's success a central goal of their activities. Gustave d'Eichtal, secretary from 1841 to 1848, was the most solicitous of the society's fortunes. His principal concern seems to have been the small size of the organization, and accordingly he recommended a membership drive in the mid 1840s. Based on the assumption that greater participation by men in public life was vital to the society's prosperity (an assumption later made by the Société d'Ethnographie as well), the drive was held and attracted some seventy new members, almost all of them representatives of commerce, government, and industry. But this change of character in membership coincided with, and undoubtedly encouraged, the society's rather abrupt abandonment of ethnological questions for extended argument over the abolition of slavery. The society became closely identified in the public mind with the ideological struggles of 1848, and it survived the revolution only as a small club for like-minded gentlemen.[11]

The great importance of individual leadership in learned societies is immediately apparent in the history of the Société d'Anthropologie, which was more successful than any other in overcoming the official neglect of the subject, and whose prosperity was in large part the work of Broca. For two decades Broca worked tirelessly as the society's permanent secretary—editing publications, corresponding with anthropologists worldwide, and securing funds and space for the society's laboratory, museum, and school. During his lifetime the society

[9] Léonce Manouvrier, "La Société d'Anthropologie de Paris, depuis sa fondation, 1859–1909," *Bulletins et Mémoires de la Société d'Anthropologie de Paris*, 11th ser., 1909, *10*:305–328, on p. 309.

[10] Archives of the Société Ethnologique de Paris, Procès-verbaux, 1846–1848, Musée de l'Homme, Paris.

[11] On the drive see *ibid.*, 1846; and *Bulletins de la Société ethnologique de Paris*, 1847, *2*:45. The society's formal dissolution is described in a report made by Armand de Quatrefages to the central committee of the Société d'Anthropologie; see Archives of the Société d'Anthropologie, Procès-verbaux (hereafter Soc. Anthr. Proc.-verb.), Musée de l'Homme, Paris, 9 May 1872.

country and from abroad.[18] The Broca school believed firmly in a direct link between the physical and "moral" capacities of the races, especially between cranial characteristics and intellectual ability, and this contention generated sharp controversy among social scientists in and outside France by the turn of the century. Despite the efforts of its most capable spokesmen, there was an increasingly widespread sense that the French school had failed to prove its case.

In sum, it is unclear whether theoretical or organizational woes were chiefly responsible for the decline of the Société d'Anthropologie and its affiliate institutions in the early twentieth century. On balance, however, it seems highly unlikely that in isolation the theoretical challenge to French anthropological doctrine would have had such a disastrous effect. Had French anthropologists in 1900 enjoyed the stronger institutions supporting anthropologists elsewhere—by that date, chairs of anthropology had been established in universities in England, Germany, Italy, and the United States—they would in all likelihood have weathered this crisis and in time developed a new theoretical consensus.

While the Société d'Anthropologie followed a course of steady expansion and then decline, its antimaterialist rival, the Société d'Ethnographie, remained a modest affair from its founding through the end of the century. The society, poorly organized from the outset, suffered continually from defects in leadership and management; its publications followed an uncertain course; its membership was heavily weighted toward political and social notables rather than serious scholars; and its practical affairs were frequently in disarray. In an intellectual atmosphere that questioned generalization from the elusive facts of social and cultural life, these organizational faults made yet more difficult the ethnographers' already strenuous task of demonstrating the value of inquiry into the social and cultural life of "savages." Ultimately, the society fell victim to the charge that ethnography was mere dilettantism.[19]

Despite their many flaws, all of the anthropological societies enjoyed some success in meeting their original objectives. The Société Ethnologique, though short-lived, brought ethnology into scholarly awareness and served as a forum for discussion and publication on ethnological questions. Perhaps most important, it set a precedent for the founding of organizations devoted to the furtherance of anthropology.[20] And the Société d'Ethnographie, by its very existence, challenged the materialist view of human nature advanced by the Société d'Anthropologie. Like many other conservative groups that led a shadow life under the Third Republic, the society steadfastly upheld principles antagonistic to mainstream ideology. Later, amid the antiliberal and antipositivist revival that swept France at century's end, the society claimed credit for defending not only

[18] See, e.g., Alfred Binet, "Recherches sur la technique de la mensuration de la tête vivante," *L'année psychologique,* 1901, *7*:314–369; and Celestin Bouglé, "Anthropologie et démocratie," *Revue de métaphysique et de morale,* 1897, *5*:443–461. On Franz Boas's "critique of formalism in physical anthropology," see George W. Stocking, Jr., ed., *The Shaping of American Anthropology, 1883–1911: A Franz Boas Reader* (New York: Basic Books, 1974), Pts. VI and VII. Elizabeth Fee describes attacks made on the statistical methods used by craniologists in "Nineteenth-Century Craniology: The Study of the Female Skull," *Bulletin of the History of Medicine,* 1979, *53*:415–433.

[19] After reviewing some publications of the ethnographical society in 1875, the anthropologist A. Issaurat remarked that they were sprinkled with "oddities" of "no very great interest for the Société d'Anthropologie de Paris," *Bull. Mem. Soc. Anthropol.,* 2d ser., 1875, *10*:186–92.

[20] Manouvrier, "La Société d'Anthropologie de Paris" (cit. n. 9), p. 309.

humanistic scholarship but also religion and the social order in difficult times.[21]

Of the three societies reviewed, however, only the Société d'Anthropologie successfully championed the new discipline of anthropology by sponsoring well-attended scholarly discussions and by operating a vigorous publishing program. Still, Broca and the other leaders of the society recognized early on that by itself the society could not perform the function most important to the perpetuation of anthropology: the instruction and training of students. In this perception they grasped the most significant shortcoming of the learned society as a form of scholarly organization. The learned society took for granted that its members had access to outside resources—specimen collections, libraries, document repositories—that were crucial to scholarship. A sophisticated physical plant first became pressing in the natural sciences, but as the nineteenth century wore on, scholars in every discipline required ever more elaborate facilities. Learned societies seldom could meet these needs and therefore late in the century became peripheral to the business of scholarship, which, increasingly, was carried on in universities with formal programs of instruction.

ANTHROPOLOGY AND OFFICIAL HIGHER EDUCATION

Recognizing the central importance of formal instruction and training, anthropologists accordingly sought to forge links with the university system, and they enjoyed some success.[22] In 1867, for example, the Faculté de Médecine granted Broca's request for laboratory space for the Société d'Anthropologie, and the society subsequently opened the world's first private school of anthropology in quarters above the laboratory. Anthropologists also gained some support from the Ecole Pratique des Hautes Etudes, the Collège de France, and the Ecole des Langues Orientales. Moreover, the Muséum d'Histoire Naturelle maintained the only formal chair of anthropology established in France in the nineteenth century.

The Société d'Anthropologic was dominated by medical doctors, and from the time of its founding the Ecole de Médecine had been an important, if informal, base of operations for the anthropologists. The Faculté de Médecine's support lent to anthropology some of the faculty's own prestige and paved the way for the laboratory's affiliation with the Ecole Pratique des Hautes Etudes (discussed below). Moreover, when the Ecole d'Anthropologie and a museum were established on faculty property in 1876, the whole complex of institutions came to enjoy a quasi-official existence that not only conferred academic respectability but also led eventually to regular subsidies from the Minister for Public Instruction, the prefect of the Seine, and the city of Paris.[23]

All the same, this link between anthropology and the Faculté de Médecine was tenuous. Since the arrangement was informal, the laboratory, school, and

[21] In his last work Léon de Rosny identified the goals of ethnography with those of the nationalist movement; he now argued that universal fraternity, long proclaimed as ethnography's ultimate purpose, would follow only in the wake of "national regeneration"; see Rosny, *Traité d'ethnographie théorique et descriptive,* 2 vols. (Paris: E. Leroux, 1900, 1902), pp. 33–34.

[22] On the structure and organization of the French university, see Louis Liard, *L'enseignement supérieur en France, 1789–1893,* 2 vols. (Paris: Armand Colin, 1888, 1894); Emile Durkheim et al., *La vie universitaire à Paris* (Paris: Presses Universitaires de France, 1918); and the sources cited in n. 8.

[23] Topinard, *A la mémoire de Broca* (cit. n. 14), p. 9.

museum collection was greatly expanded and served as the resource for studies such as *Crania ethnica,* a widely praised work on skull types that Quatrefages prepared in collaboration with his assistant E.-T. Hamy.[31] Most significant, under Quatrefages a distinct museum doctrine of anthropology emerged. Quatrefages was Broca's chief rival for leadership among French anthropologists, and he opposed Broca's views on the most controversial questions in the field. He adroitly defended the monogenist theory of the origin of the races. He was a learned and persistent critic of Darwinism and indigenous French theories of species transformation. And although he defended the fossil evidence of the great antiquity of man, his outline of human natural history was in many particulars compatible with the biblical account of creation.[32]

As a highly competent scholar, a museum professor, and an aristocrat, Quatrefages was an august personage, and he was frequently asked to pronounce on anthropological questions. His five-volume *Histoire de l'homme* developed from a lecture series sponsored by the Empress Eugenie for the edification of the working class.[33] In later years, Quatrefages published six other syntheses of anthropology and wrote frequently for widely read journals, both scholarly and popular.

Still, despite his position and very considerable abilities, Quatrefages did not influence the development of French anthropology to nearly the same extent as did Broca, a fact that is explained in large part by the power each exerted over anthropological institutions. While Broca's work undoubtedly had greater theoretical scope and intellectual force than his rival's, it was Broca's control of the Société d'Anthropologie, the laboratory, and the school that assured his preeminence. Curious as it may seem, these private institutions constituted a stronger base than Quatrefages's highly prestigious chair at the museum, even though through much of Quatrefages's career museum professors were the best paid, most lavishly housed, and most generously supported scholars in Paris.[34] Despite the many advantages of their position, museum scholars of the later nineteenth century did not easily establish what Terry N. Clark has called a research "cluster"—a research group consisting of a senior investigator assisted by protégés. Indeed, despite all its grandeur, the museum remained outside the educational establishment in a key respect: it lacked the power to confer degrees. This power was the core of French higher education; in all academic fields the sequence of degrees from *baccalauréat* to *doctorat* was the essential prerequisite to success. Accordingly, any institution that did not grant degrees was in a fundamental sense peripheral to the system.[35]

The museum had no officially enrolled students, granted no degrees, and offered courses of study that reflected the interests of individual chairholders rather than larger curricular objectives. Courses, professors, and subjects moved

[31] Vallois, "L'évolution de la Chaire d'Ethnologue du Muséum national d'Histoire naturelle," *Bulletin du Muséum National d'Histoire Naturelle,* 2d ser., 1944, *16*:43–44; and Armand de Quatrefages and E.-T. Hamy, *Crania ethnica: Les crânes des races humaines,* 2 vols. (Paris: J. B. Baillière, 1882).

[32] See esp. Armand de Quatrefages, *Rapport sur les progrès de l'anthropologie* (Paris: Imprimerie Nationale, 1867), and *Charles Darwin et ses précurseurs français* (Paris: G. Baillière, 1870).

[33] Armand de Quatrefages, *Histoire de l'homme,* 5 vols. (Paris: Hachette, 1867–1868).

[34] Limoges, "Muséum d'Histoire naturelle" (cit. n. 30), p. 215.

[35] Clark, *Prophets and Patrons,* pp. 9–12, 21–22, 67–82.

in and out of favor with both the museum authorities and the *auditeurs,* who were free to come or go as they chose. If these characteristics of the museum were relatively unimportant early in the century, they assumed ever greater significance in later decades, as lines of scientific inquiry solidified and requirements for scholarly careers became increasingly standardized. Late in the century, the museum found it difficult not only to draw students but even to justify before the funding hierarchy the great expenses of the institution and its professoriat.[36]

Whatever its limitations, the museum did provide a stable site for anthropological instruction through the end of the nineteenth century. In contrast, other supporting institutions—the Collège de France, the Ecole des Langues Orientales, and the Ecole Pratique des Hautes Etudes—sponsored offerings in anthropology and ethnography only on occasion. Of these institutions, the Collège de France was the least receptive to these new fields. In 1870 Rosny gave a course there on the goals and meaning of ethnography, but over the years a number of teaching proposals submitted by ethnographers, prehistorians, and archaeologists were rejected.[37] And while several of Rosny's colleagues from the Société d'Ethnographie held chairs in Eastern languages at the Collège de France, their instruction seldom encompassed ethnographic subject matter.[38] Similarly, when in 1899 the anthropologist Léonce Manouvrier sought appointment to the school's newly created chair in modern philosophy, his argument that "all modern philosophies rest ultimately on an anthropology" apparently failed to convince the authorities that the latter subject was of broad interest and application.[39]

Until late in the century, ethnographers secured virtually their sole institutional support from the Ecole des Langues Orientales. A number of the leading figures of the Société d'Ethnographie—including Rosny, his teacher J.-H. Garcin de Tassy, and his fellow orientalist Philippe-Edouard Foucaux—taught there at one point or another in their careers. Although their teaching was primarily linguistic, they also offered instruction in history, "traditions," and commercial and political geography.[40] Unfortunately for the ethnographers, the school not only had few students and lacked degree-conferring power but also was in several other respects ill suited to instruction in the broadly conceived ethnography

[36] Limoges, "Muséum d'Histoire naturelle," pp. 233–240. The uncertain fate of those studying anthropology at the museum appears from a letter the anthropologist René Verneau wrote E.-T. Hamy in 1889. On a field expedition in the Canary Islands, Verneau wrote of having to "renounce anthropology after having given fourteen years to it and having sacrificed everything" to his studies. He was considering taking up a medical practice in the Touraine, a choice between "the certitude of starving here and the possibility of vegetating there." Ironically, Verneau was one of the few anthropology students of his generation to obtain stable positions, the first when he succeeded Hamy in the museum chair. Verneau to Hamy, 11 Sept. 1889, Correspondence de E.-T. Hamy (1888–1889), MS 2257, Muséum National d'Histoire Naturelle, Paris.

[37] Archives Nationales (hereafter AN), F^{17} 13554, Paris.

[38] *Ibid.;* for lists of chairs and professors at the Collège de France in the nineteenth and early twentieth centuries, see A. Lefranc et al., *Le Collège de France (1530–1930): Livre jubilaire composé à l'occasion de son quatrième centenaire* (Paris: Presses Universitaires de France, 1932), pp. 15–23.

[39] AN F^{17} 13554. The chair went to Gabriel Tarde and later to Henri Bergson; see Lefranc et al., *Livre jubilaire,* p. 18.

[40] André Mirambel, "Orientalisme d'hier et d'aujourd'hui," *Revue de l'Ecole nationale des Langues orientales,* 1964, *1*:1–29; AN F^{17} 4064–4065; and *La Grande Encyclopédie,* 31 vols. (Paris: H. Lamirault, 1886–1902), Vol. XV, pp. 383–386, s.v. "Ecole."

Rosny and his fellows endorsed. Since the Ecole des Langues Orientales was first and foremost a language school, teaching ethnography necessarily remained peripheral to the school's chief purpose. Moreover, the school's curriculum generally included only the languages and peoples of "civilizations" thought to be of historical, diplomatic, or commercial interest; peoples traditionally classified as savage or barbarian were excluded. Only very late in the century did the school begin teaching some of the languages of black Africa—Mande in 1898 and Wolof in 1899.[41]

It would be misleading to imply that ethnographers chafed at this limitation on their subject matter: the leading scholars were classicists or orientalists, and for the most part they either passively accepted or actively promoted the standard division between civilized and noncivilized peoples.[42] At the same time, the institutional setting for ethnography helped define the content and boundaries of the field itself, so that ethnographers who were interested in the cultural and social life of the "uncivilized" seldom received institutional support or gained recognition for work in this vein.[43]

Both ethnography and anthropology benefited from the founding in 1868 of the Ecole Pratique des Hautes Etudes. Since the French university system was hostile to innovation, French educators, scientists, and scholars had begun clamoring for reform in the 1850s, and the EPHE was the most significant early response to the reform movement. It was established by Victor Duruy, the most creative and energetic Minister for Public Instruction in the later nineteenth century. The EPHE was not a new school but rather a federation of existing research and instruction programs that had failed to find homes within the faculty system. At its inception, the EPHE was divided into four sections for the mathematical, physicochemical, natural, and historical-philological sciences. Broca's laboratory was precisely the kind of institution the EPHE was intended to encourage. In 1872 the laboratory was brought under the natural sciences section of the EPHE; from that date Broca reported to the Minister for Public Instruction on the progress of research at the lab, the distribution of responsibility to his assistants, and the use made of official funds. Although the lab's mixed public and private status had serious drawbacks—especially the ad hoc nature of any instruction offered there—the EPHE's sponsorship represented a measure of official support, financial assistance, and public recognition that anthropologists were happy to receive.[44]

[41] Mirambel, "Orientalisme," p. 14.

[42] Rosny, *Traité d'ethnographie théorique* (cit. n. 21), p. 156.

[43] Disdain for inquiry into the life and practices of savages was common in French scholarly circles. In 1874 Rosny commented that for years Americanist studies had been "persecuted, denied, and deliberately ignored by official science"; see *Actes Soc. Ethnog.*, 1874–1877, p. 10. A new setting for ethnographical work was created in 1880, when the Musée d'Ethnographie de Paris officially opened. Like other anthropological institutions, the museum suffered from financial and organizational troubles; it served, nonetheless, as the nucleus for the modern Musée de l'Homme. On its history, see E.-T. Hamy, "Les origines du Musée d'Ethnographie," *Revue d'Ethnographie*, 1889, *8*:305–417; and Williams, "The Science of Man," Ch. 4.

[44] Liard, *L'enseignement supérieur* (cit. n. 22), pp. 293–295; Zeldin, "Higher Education" (cit. n. 8), pp. 77–78; Durkheim et al., *La vie universitaire* (cit. n. 22), pp. 182–193; [Broca] *Rapports sur l'Ecole Pratique des Hautes Etudes. Science naturelles: Laboratoire d'Anthropologie* (1872–1873; 1873–1874; 1875–1876), Archives of the Société d'Anthropologie, Laboratoire d'Anthropologie Biologique, Juvisy (hereafter Arch. Soc. Anthr., Juvisy). I am indebted to Denise Férembach for access to these materials.

Ethnography succeeded in gaining support from the EPHE only much later. In 1886 a section for religious sciences was added, and at that point Rosny was appointed adjunct director of the entire institution.[45] Under the auspices of the EPHE, ethnographers could expand the historical and cultural content of their courses, shifting away from the predominantly linguistic focus of their presentations at the Ecole des Langues Orientales.[46] In this setting, culturally oriented "area studies"—including for the first time *américanisme* and *africanisme*—began to enjoy some organizational stability around the end of the century. But again the limitations of the institution itself contributed to ethnography's weak position in the scholarly world. The fifth section of the EPHE drew few students since jobs were scarce in this area and those that did exist went to students who had trained at the Faculté des Lettres and held the appropriate degrees. Although it was not impossible for an EPHE student to succeed in his chosen field—Marcel Mauss, who founded the immensely influential Institut d'Ethnologie in 1925, began his career at the EPHE—such cases were few.[47]

Given this poor record of public institutional support, it is not surprising that anthropologists and ethnographers sank their resources into private institutions of their own making. Virtually from the outset, organizers of both the Société d'Anthropologie and the Société d'Ethnographie recognized that in order to penetrate the official university establishment, they first had to build institutions that would attract the attention and participation of scholars and students.

PRIVATE INSTITUTIONS FOR RESEARCH AND TRAINING

With the definitive establishment of the Third Republic in the mid 1870s, an extended period of educational reform opened. Legislation in 1875 made possible the establishment of *écoles libres*—private institutions of higher education—and the founders of both Catholic universities and a great many anticlerical institutions took advantage of this opening.[48] The Ecole d'Anthropologie was for a generation one of the most successful of these schools. Like the others, it was privately organized, funded by individual contributions as well as by a small grant from the government, and staffed by selected members of a parent learned society.[49]

Broca first outlined his plan for the school to the central committee of the Société d'Anthropologie in June 1875. He announced that the Faculté de Médecine had agreed to provide additional space for the school and that the Seine prefect and the Paris municipal council had promised subsidies for installation and operating costs. An exchange at this early policy meeting illustrates the peculiar status of the school. When one committee member suggested that the

[45] *La Grande Encyclopédie*, s.v. "Ecole" (cit. n. 40).

[46] The limitations set on instruction at the Ecole des Langues Orientales are illustrated by the fate of a proposal made by the ethnographer Edouard Madier de Montjou in 1874 for a broadly conceived course encompassing the geography, customs, legislation, commerce, and contemporary history of the Far East. The proposal was criticized by bureaucrats within the Ministry for Public Instruction for being excessively broad and more appropriate to a "commercial" than a linguistic school; see AN F[17] 4064–4065.

[47] Clark, *Prophets and Patrons* (cit. n. 1), pp. 44–51.

[48] AN F[17] 1340.

[49] The prefect of the department of the Seine gave a 6,000-franc subsidy to the school on its opening; Soc. Anthr. Proc.-verb. 10 June 1875.

municipal council should help decide on appointments to the faculty, Broca replied that the prefect and the council had stressed their desire that the school's administration be left to the founders.[50] The authorities who governed Paris in these uncertain years of the Third Republic were reluctant to be identified closely with the anticlerical, intellectually radical leaders of the Société d'Anthropologie, but in succeeding years, as the notoriety of anthropology diminished, the official attitude toward the school changed. Within several years of its founding, the Minister for Public Instruction agreed to a regular annual subsidy for the school and began to list it in the yearbook of educational establishments falling directly under ministry authority. As Topinard later pointed out, this step amounted to "semi-official recognition."[51]

The Ecole d'Anthropologie was the linchpin of the complex of institutions that came to be known as the Institut Anthropologique. The other constituent parts were the society itself, Broca's laboratory, the library and museum. When it officially opened in 1876, the Ecole d'Anthropologie employed four instructors. By the 1889–1890 academic year there were chairs in prehistoric anthropology, comparative embryology and "anthropogeny," ethnography and linguistics, zoological anthropology, general anthropology, medical geography, physical anthropology, and the "history of civilizations," as well as three supplementary courses in ethnography, "histological anthropology," and anatomical demonstrations.[52] All courses were offered free to the general public without entrance requirements of any kind. Teaching at the school was coordinated with work at the laboratory, which employed a director, assistant director, three to four assistants, and miscellaneous helpers. In annual reports to the administrators of the EPHE, Broca and successive directors described the many projects undertaken in this mixed laboratory and instructional setting: the preparation of skull and bone molds, the collection and cataloguing of pathological specimens from hospitals, the dissection and anatomical description of primates, the design and fabrication of new instruments, the collection and study of fossils, and the recording of thousands of measurements taken on human subjects and specimens.[53] The work of French anthropologists in these areas was highly regarded, and each year the laboratory drew scores of visitors from around the world and French scholars in both the natural and social sciences.[54]

Until the 1890s the various institutions constituting the Institut Anthropologique worked in harmony and enjoyed considerable success. At the same time, they were dependent on private monies that might cease to flow in response to loss of public interest and on public subsidies that might be reduced with little warning. And, indeed, both of these eventualities materialized in the late 1890s, when the budgets of both school and laboratory began to decline.[55] The laboratory continued to function as late as the 1920s, but by then it could afford to maintain only one full-time employee. Later it was joined administratively to the

[50] *Ibid.*, 24 June 1875.

[51] Topinard, *A la mémoire de Broca* (cit. n. 14), p. 10.

[52] AN F[17] 13140.

[53] [Broca] *Rapports sur l'Ecole Pratique des Hautes Etudes* (cit. n. 44).

[54] "Inscriptions du Laboratoire," Arch. Soc. Anthr., Juvisy.

[55] From the late 1890s, complaints that the school was forced to operate on a shoestring budget began to mount; see the comments in *Revue de l'Ecole d'Anthropologie*, 1901, *11*:31, and in *Revue Anthropologique*, 1914, *24*:vii.

anthropology laboratory of Toulouse (also under the control of the Ecole Pratique des Hautes Etudes) and from that point ceased even to have its own director.[56] The school followed a somewhat happier course. Although after the turn of the century its early reputation as an innovative center for diversified anthropological research began to decline (in 1913 Mauss wrote offhandedly that the school's courses were "popular" rather than scholarly), it did continue to enjoy some celebrity for the work of its prehistorians and archaeologists.[57] The Ecole d'Anthropologie continued to operate in the interwar years at a modest level, but it survived in name only after being evicted in 1940 from its quarters in the Faculté de Médecine.[58]

In the later years, drawing students to the Institut Anthropologique became increasingly problematic. If in its heyday anthropology had been an immensely popular subject, by the 1890s it had lost favor to the enthusiasms of the day, particularly sociology. In fact, the school's directors had anticipated this shift in popularity and installed their own professor of sociology in 1885. The first chairholder, Charles Letourneau, was a Latin Quarter favorite, but his successor, Georges Papillault, had little success among students.[59] Like the other *écoles libres,* the Ecole d'Anthropologie attracted students as long as popular instructors were on hand and the subjects taught were in vogue, but since it offered no degrees or career preparation, students were lost as easily as they were gained. So long as the whole complex of institutions prospered, students were always in evidence. But decline, once it set in, reinforced decline. In 1909 Manouvrier lamented the school's lack of students seeking career training and observed that if anthropology were to have a future in France it must find a place within the official establishments of higher education.[60]

CONCLUSION

It has frequently been argued that the peculiar organization of research, scientific activity, and higher education in nineteenth-century France hindered scientific development, but the debate has been pursued largely in reference to the physical sciences. As this study shows, the emergent social science of anthropology faced formidable obstacles to successful institutionalization. Both anthropologists and ethnographers responded by establishing private institutions, but only those founded by the Broca school ever enjoyed any vitality and eventually these too foundered. The end result was severe organizational fragmentation: in the years just prior to World War I there were some ten independent, often highly idiosyncratic, research clusters, in some cases headed by volunteers and in all cases making do with meager facilities and resources.[61] It is small

[56] This administrative change was made in 1937; see Vallois, "Le Laboratoire Broca" (cit. n. 14), p. 4.

[57] Marcel Mauss, "L'ethnographie en France et à l'étranger," *Revue de Paris,* Oct. 1913, pp. 537–560, 815–837, on p. 821; on the work of French prehistorians and archaeologists, see Glyn Daniel, *A Hundred and Fifty Years of Archeology* (Cambridge, Mass.: Harvard Univ. Press, 1976), pp. 59–77, 93–150.

[58] Henri V. Vallois, "La Société d'Anthropologie de Paris," *Bull. Mem. Soc. Anthropol.,* 11th ser., 1960, *1*:293–312, on p. 300, n. 1.

[59] Clark, *Prophets and Patrons* (cit. n. 1), p. 120.

[60] Léonce Manouvrier, "Le classement universitaire de l'anthropologie," *Rev. Ecole Anthropol.,* 1907, *17*:75–96, 109–119; and *ibid.,* 1910, *20*:391–401, on *17*:81.

[61] See Williams, "The Science of Man," Ch. 6.

wonder that in these years French anthropology failed to attract first-rate scholars or that its reputation abroad suffered a serious setback.

It was only in the 1920s that this situation began to change, as Mauss and his protégés at the Institut d'Ethnologie (founded in 1925) overcame organizational obstacles and produced some brilliant work in "ethnology," a reformulation of anthropology along lines suggested by Durkheimian sociology. Despite this bright spot, however, many problems that had beset nineteenth-century anthropology—inadequate financial support for research and especially the lack of degree status for anthropological training—continued. Branches of the discipline not encompassed by Maussian ethnology (physical anthropology, the study of material culture) languished. And even with the success of the Maussian group, anthropology remained a marginal pursuit within French scholarly life—ambiguous in self-definition, confined to ad hoc institutions, and peopled by individuals of diverse and sometimes dubious training—until after World War II. (The creation of a chair in social anthropology at the Collège de France for Lévi-Strauss in 1959 was the clearest sign of changes under way.)[62] While it might be argued that the discipline's very marginality accounts for some of the most intriguing features of its development in modern France—the connections between ethnologists and surrealist aesthetes in the 1920s and 1930s, for example[63]—it remains true that anthropology's failure to achieve definitive institutional status in France in the nineteenth century seriously limited the scope and range of the discipline for the first half of the twentieth.

[62] A *certificat* in ethnology was created in 1927, but no *licence* until after 1945; see Clark, *Prophets and Patrons,* 214, 236. For discussions of continuing organizational and financial difficulties in the later period, see Veronique Campion-Vincent, "Organisation de l'ethnologie en France en 1969," *L'Homme* 1970, *10*:106–24; and *L'Anthropologie en France: Situation actuelle et avenir* (Colloques Internationaux du CNRS, No. 573, Paris 18–22 April) (Paris: Centre National de la Recherche Scientifique, 1979), esp. pp. 381–492.

[63] On this link, see James Clifford, "On Ethnographic Surrealism," *Comparative Studies in Society and History,* 1981, *23*:539–564.

The Eugenics Record Office at Cold Spring Harbor, 1910–1940
An Essay in Institutional History

By Garland E. Allen

In 1883 THE BRITISH NATURALIST AND MATHEMATICIAN Francis Galton (1822–1911) first introduced the term *eugenics* to the vocabulary of science. According to Galton's lofty formulation, eugenics was "the study of the agencies under social control that may improve or impair the racial qualities of future generations, either physically or mentally." By 1911 the chief American advocate of eugenics, Charles B. Davenport (1866–1944), had put it more bluntly; to him, eugenics was no less than "the science of the improvement of the human race by better breeding."[1]

Conceived as a scientifically grounded reform movement in an age of social, political, and economic turbulence, eugenics looked to hereditary factors for the sources of such a vast array of human behavioral problems as alcoholism, feeblemindedness, rebelliousness—even criminality. Eugenicists also thought they had found the causes of many fundamental social problems in measurable hereditary defects. Eugenics as a social movement developed throughout most of the countries of Western Europe, but it enjoyed a particularly robust life in the United States. After 1900 the movement became, in the eyes of its American advocates, a major breakthrough in the application of rational, scientific methods to the problems of a complex urban and industrial society.

EUGENICS AND GENETICS

Although Galton coined the term *eugenics* in 1883, by 1900 neither he nor his followers had been able to establish a serious eugenics movement in England. Both Galton and his disciple Karl Pearson (1857–1936) lacked a firm and workable theory of heredity. Their views, which were based on biometry, the statistical analysis of biological traits measured for large samples, encountered great difficulty when applied to individual families or lines of descent. With the rediscovery of Mendel's laws of heredity in 1900, however, the study of heredity in general and eugenics in particular found fertile ground, particularly in the

This research was supported by the National Science Foundation and the Charles Warren Center for Studies in American History, Harvard University. I am grateful for the comments of Mark Adams, Randy Bird, Donald Fleming, Daniel J. Kevles, Kenneth Ludmerer, Jon Roberts, Barbara Rosenkrantz, and Stephen Thurnstrom.

[1] Francis Galton, *Inquiries into Human Faculty and Its Development* (New York: Dutton, 2nd ed., n.d.): p. 17n.; quoted also on the frontispiece of the *Journal of Heredity*. Charles B. Davenport, *Heredity in Relation to Eugenics* (New York: Henry Holt, 1911), p. 1.

United States. By 1910 most American biologists, except for a stalwart few, agreed that Mendel's theory could be applied to all sexually reproducing forms. The enthusiasm with which biologists—in the United States in particular—began to endorse the Mendelian scheme cannot be overemphasized. Here, for the first time, was what seemed to be a generalized, predictive, and experimentally verifiable concept of heredity that applied to *all* living forms, including human beings. Indeed, in the period 1900–1910 geneticists had concluded that several human traits follow a strictly Mendelian pattern of inheritance: red-green color blindness, the A-B-O blood groups, polydactyly (presence of short, stubby digits on the hands and feet), and several metabolic diseases or inborn errors of metabolism. A revolution in genetics had taken hold.[2]

The application of Mendelian theory to human beings armed eugenicists with a powerful analytical tool. Using pedigree analyses as the data from which possible Mendelian patterns of inheritance could be deduced, eugenicists in the United States began to study a wide variety of physical, mental, and moral traits in humans. Although American eugenicists did not adhere to the view, so common in England, that Mendelism and biometry were mutually exclusive, in practice most emphasized the Mendelian scheme. One of these early American supporters of Mendelism, and a champion of experimental biology, was Charles Benedict Davenport, under whose direction the Station for the Experimental Study of Evolution and the Eugenics Record Office were established at Cold Spring Harbor.

The establishment of the Eugenics Record Office (ERO) in 1910 at Cold Spring Harbor, Long Island (New York), was central to the development of eugenics in the United States. Associated with the larger Station for the Experimental Study of Evolution (SEE), the ERO provided both the appearance of sound scientific credentials and the reality of an institutional base from which eugenics work throughout the country, and even in Western Europe, could be coordinated. The ERO became a meeting place for eugenicists, a repository for eugenics records, a clearinghouse for eugenics information and propaganda, a platform from which popular eugenic campaigns could be launched, and a home for several eugenical publications. Moreover, the ERO was headed by two of the country's best-known eugenicists: C. B. Davenport, as director of both the SEE and the ERO, and Harry Hamilton Laughlin (1880–1943), as his deputy at the SEE and as superintendent of the ERO itself. Thus the ERO became a nerve center for the eugenics movement as a whole. When it closed its doors on 31 December 1939, it was clear that the movement as such no longer existed.

The ERO, whose life spans virtually the entire history of eugenics in the United States, provides an illuminating focus for historical study of the movement. Study of the ERO's activities also exposes the modern investigator to a representative cross-section of the work and concerns of eugenicists throughout the world. Moreover, because its financial needs brought the ERO into direct

[2] Much has been written in recent years about the history of Mendelian theory in the early decades of the century. Among the best general sources are L. C. Dunn, *A Short History of Genetics* (New York: McGraw-Hill, 1965); and E. A. Carlson, *The Gene: A Critical History* (Philadelphia: Saunders, 1966). For more detailed analyses of the first decades of genetics, see Garland E. Allen, *Thomas Hunt Morgan: The Man and His Science* (Princeton: Princeton Univ. Press, 1978); and E. A. Carlson, *Genes, Radiation and Society: The Life and Work of H. J. Muller* (Ithaca, N.Y.: Cornell Univ. Press, 1981).

contact with some of the individual philanthropists as well as the larger philanthropic foundations that were emerging in the first decades of this century, this study also provides historical perspective on the initiation and control of funding for scientific work during that period. In many ways, then, the ERO is a microcosm of the larger social macrocosm that was the American eugenics movement. It also provides a focus for exploring the relationship between the development of eugenics and the changing social, economic, and political life in the United States between 1900 and 1940.

To put the present study in perspective, however, I should emphasize that several other groups also played an important role in the development of the American eugenics movement—groups such as the American Breeders' Association (whose Eugenics and Immigration Committees were the first eugenics organizations in the country), the American Eugenics Society, the Eugenics Research Association, the Galton Society, the Institute of Family Relations, and the Race Betterment Foundation. The ERO, however, was the only major eugenics institution with a building, research facilities, and a paid staff. Although unique in having its own institutional base, it nevertheless could not have done as much without the existence of those other organizations. Another point to keep in mind is that the style and particular focus of the ERO's work was not typical of all aspects of the American eugenics movement. Although the ERO did provide a considerable amount of ideological direction, the American eugenics movement was not monolithic or highly organized. Many eugenicists would have preferred that the movement have more of a unified character, but this proved difficult to accomplish. Eugenicists came from all walks of life, though most were professional middle class or upper class. Often individualistic and independent, they tended to focus on their own projects and were generally not amenable to highly coordinated efforts. Although the ERO tried to provide nationwide coordination, in the long run there was little centralized organization or control. Despite the efforts of Charles Davenport and his staff, the ERO was probably far more effective as a clearinghouse and data repository than as an organizational force.

DAVENPORT AND THE FOUNDING OF THE SEE

Charles Benedict Davenport, who was to spearhead the American eugenics movement, was born in Brooklyn, of New England ancestry. He received an engineering degree from Brooklyn Polytechnic Institute in 1887 and an A.B. from Harvard College in 1889. He immediately enrolled in Harvard graduate school and received his Ph.D. in 1892, writing a thesis on morphology under E. L. Mark (1847–1936). Davenport served as an instructor at Harvard until 1899, when he accepted an assistant professorship at the University of Chicago. There he remained until 1904, when he persuaded the Carnegie Institution of Washington to fund the Station for the Experimental Study of Evolution, with himself as director, at Cold Spring Harbor. Davenport remained director of the SEE, and of the Eugenics Record Office, from its founding in 1910 until his retirement in 1934. During this time he built both institutions into major research laboratories for the study of heredity and evolution—the SEE for the study of plants and nonhuman animals, the ERO for the study of human beings. A rigid and humorless man, Davenport was nonetheless well respected within the scientific community, both as a geneticist and as a statesman of science. He was a member of the National Academy of Sciences and the National Research Council, as well as secretary of the Sixth International Congress of Genetics (Ithaca, New York, 1932).[3]

Davenport's engineering background prepared him well to move from classical descriptive morphology into the quantitative and experimental study of heredity and evolution. Far more familiar with mathematics than most biologists of his era, he was among the first in the United States to appreciate the biometrical work of Galton and Pearson. Indeed, at Pearson's request he served as the American representative on the editorial board of the British biometrical journal *Biometrika,* of which Pearson was editor. Yet he was equally prepared to accept the experimental approach of the Mendelian theory. Beginning in the academic year 1892/93, Davenport taught a course entitled "Experimental Morphology" at Harvard (and later at Chicago), and he published a book by the same title in 1897 (revised, 1899). (Two of Davenport's students in that class were to become future leaders of both Mendelian genetics and eugenics: W. E. Castle, a longtime Harvard professor, and Herbert Spencer Jennings, for many years a protozoologist at Johns Hopkins.) Imbued with the rising tide of experimentalism that was so prominent in biology at the time, coupled with his own strong inclination to quantitative studies, Davenport was immediately receptive to the reports of Mendel's work by Carl Correns and Hugo De Vries in 1900.[4] In 1901 Davenport himself published one of the first papers on Mendelism in the United States.[5] He saw no dichotomy between Mendel's laws and biometrical thinking, though he realized early on that Mendel's notion of particulate, and therefore discontinuous, inheritance was not compatible with Galton's theories of continuous inheritance and regression.[6]

Proposal for a Research Facility

During his stay at the University of Chicago two factors stimulated Davenport to seek funds for establishing an independent research laboratory. One was his own research, which focused at that time on large animals such as poultry and mice (as compared, for example, to insects) and thus required expanded facilities for care and breeding. For a while there was talk at Chicago of acquiring an experimental farm, but by 1902 Davenport was convinced that nothing would come of it and began looking for other alternatives. Coincidentally, the future of the summer school of the Brooklyn Institute of Arts and Sciences, held at a small summer marine laboratory at Cold Spring Harbor, was in doubt. Daven-

[3] The standard biography of Davenport is Oscar Riddle, "Charles Benedict Davenport," *Biographical Memoirs of the National Academy of Sciences,* 1946, *25*:75–110. This sketch contains a complete bibliography.

[4] Charles Rosenberg, "Charles Benedict Davenport and the Beginnings of Human Genetics," *Bulletin of the History of Medicine,* 1961, *35*:266–276; see also A. H. Sturtevant, "The Early Mendelians," *Proceedings of the American Philosophical Society,* 1965, *109*(4):199–204.

[5] C. B. Davenport, "Mendel's Law of Dichotomy in Hybrids," *Biological Bulletin,* 1901, *2*:307–310.

[6] Pearson eventually asked Davenport to leave the editorial board of *Biometrika* because of a dispute between the two men over the interpretation of Wilhelm Johannsen's pure-line experiments. This was a rift in their personal and professional relationship that Davenport always regretted.

port, who had taught at the summer school since 1892, recognized Cold Spring Harbor as an ideal spot for the type of research station he envisaged. There would be room to expand animal care facilities, open space for experimental garden plots, facilities for housing a staff of caretakers and scientists, and plenty of marine organisms available for study. Never one to hesitate when an opportunity for funding, however remote, presented itself, in January 1902 Davenport approached the newly founded Carnegie Institution of Washington, established by the personal bequest of Andrew Carnegie.[7]

Davenport sent his proposal to the Carnegie Institution's secretary, Charles Walcott, through an influential Chicago banker who agreed to act as an intermediary. The laboratory that Davenport proposed was to be for "the analytic and experimental study of the causes of specific differentiation—of race change."[8] Convinced that the Darwinian theory of natural selection was hypothetical because it had not been demonstrated experimentally (that is, no new species had ever been produced by artificial selection, no matter how long or how rigorously selection was carried out), Davenport aimed to recast classical selection experiments in terms of the new Mendelian scheme. Intimately connected with this recasting was the problem of variation. On what types of variations (large, discontinuous or small, continuous) did selection act to produce new species? Did new variants breed true or, as Galton claimed, always regress toward the mean? Were Mendelian traits important to animal and plant adaptation, or were they, as some workers claimed, mostly trivial (such as the number of bristles on a fly's abdomen), in no way affecting an organism's fitness? Moreover, as Davenport was quick to recognize, such questions had an importance that extended beyond theoretical issues of evolution. A more thorough understanding of heredity, variation, and selection had enormous implications for agricultural breeding, an issue that was not lost on the Carnegie Institution's board, or on Andrew Carnegie himself. The board defined its purpose (in part) as sustaining "objects of broad scope that may lead to the discovery and utilization of new forces for the benefit of man." Indeed, just a few years later (1905) the Carnegie Institution was to make a substantial and ongoing commitment ($10,000 a year) to the work of Luther Burbank, specifically as an example of the application of scientific principles to practical problems.[9]

Davenport's initial proposal of 1902 was turned down by the Carnegie Institution of Washington, partly because the Board of Directors was engaged at that time in considerable debate over whether the CIW should fund research organizations or only individual researchers. By 1904, however, the board's Executive Committee had accommodated both views and determined to fund institutions as well as individuals provided that the researchers in the former worked cooperatively and in an organized manner. The CIW concluded that it could serve researchers best by helping them to *organize* their joint efforts: "in the field of research the function of the Institution is organization; to substitute organized for unorganized effort; to unite scattered individuals working independently, where

[7] Riddle, "Charles Davenport," pp. 80–81; see also C. B. Davenport, "Biological Experiment Station for Studying Evolution," *Yearbook of the Carnegie Institution of Washington,* 1902, *1*:280.

[8] *Ibid.*

[9] Minutes of the Executive Committee, 3 Oct. 1902 and 12 Dec. 1905, Record Book, pp. 57 and 468–475, Carnegie Institution of Washington (CIW) Archives, Washington, D.C. I am indebted to Barry Mehler for gathering data and copies of material from these archives.

it appears that such combination of effort will produce the best results; and to prevent needless duplication of work."[10] In this context, Davenport's second application was received more favorably, and on 12 December 1903 he was awarded a grant of $34,250, with fixed annual appropriations "to continue indefinitely, or for a long time." The "Station for the Experimental Study of Evolution" (SEE) was the name adopted for the facility at Cold Spring Harbor, and it was incorporated as the "Department of Experimental Biology of the Carnegie Institution of Washington," with the express purpose of studying "hereditary evolution, more particularly by experimental methods."[11] Edmund Beecher Wilson (1856–1939), a cytologist and chairman of the Zoology Department at Columbia University, was appointed as scientific adviser to Davenport in his work as director of the new research station.

No one could have agreed more than Davenport with the principles outlined by the Carnegie Executive Committee. He had always supported the notion of cooperation in research; more important, however, was his belief that for cooperation to occur an organizational base had to be developed. In his presidential address to the American Society of Naturalists given on 29 December 1907, Davenport emphasized that one of the features differentiating modern from ancient or medieval scientific work was its cooperative nature and thus its organization into societies, institutions, and multidisciplinary or international projects. However, he noted that there remained within the scientific community, especially among biologists, a strain of individualism that militated against cooperative programs and thus hampered research. Davenport reminded his fellow naturalists that the great natural history voyages of the nineteenth century, such as the *Challenger* expedition, were monuments to cooperative efforts; they would not have succeeded had individuals insisted on staking out their private research domains. Looking to the field of astronomy, he cited another example of cooperative effort whereby, beginning in 1887, eighteen observatories organized to produce a comprehensive photographic atlas of the heavens. Davenport urged that naturalists "should do well to adopt principles which have worked successfully in other fields of activity. In the modern commercial world one of the most important principles is cooperation."[12] The Station for the Experimental Study of Evolution was, in Davenport's mind, a perfect example of the spirit of cooperative research that could be fostered by successful organization.

Research at the SEE

The SEE developed into, and remained, a prestigious research institution. Today it is the Department of Genetics of the Carnegie Institution of Washington, with James D. Watson as its director. In the early decades of the century, highly qualified young investigators came to the station for varying periods to work on specific problems relating to heredity and evolution.[13] Davenport himself re-

[10] Minutes of the Executive Committee, 3 Oct. 1902, Record Book, p. 56, CIW Archives.

[11] *Ibid.*

[12] C. B. Davenport, "Cooperation in Research," *Science,* 8 Mar. 1907, *25*(636):361–366.

[13] Among those who figured most prominently were George Harrison Shull (1904–1915), Roswell H. Johnson (1905–1908), A. F. Blakeslee (1915–1942), Ross A. Gortner (1909–1914), J. Arthur Harris (1907–1924), F. E. Lutz (1904–1909), and Oscar Riddle (1914–1945). In addition, a number of Associates—senior investigators who came to the SEE to give seminars, participate in research, and in general to keep the staff in touch with the latest developments—were appointed annually. Among

mained in complete administrative control. It was his kingdom. He administered it scrupulously, autocratically, and sometimes dictatorially, until his retirement in 1934 at the age of sixty-eight. The Carnegie Institution had invested not merely in a facility and a program for research but in one man and his vision of a new direction in biology.

Davenport's vision for the SEE was to bring together three areas of interrelated study: heredity, evolution, and cytology. Researchers were to employ experimental, quantitative, and, where feasible, mathematical methods. They would study heredity through carefully planned breeding experiments, the keeping of detailed, quantitative records of offspring of all crosses, and the analysis of the data by both biometrical and Mendelian means. They would examine evolution through the quantitative study of variation in natural populations (following the methods of Galton and Pearson), as, for example, in Davenport's own work on populations of crabs in the waters around Cold Spring Harbor. They would also pursue selection experiments of the sort that Wilhelm Johannsen had initiated in Denmark (1899–1902) on pure lines of the bean *Phaseolus* and that W. E. Castle was to conduct some years later (1907–1914) on the piebald or "hooded" rat. The central issues of selection were, of course, the degree to which the results of selection can be maintained in a line after selection is relaxed and the possibility of creating new species by many generations of selection in a given direction. Researchers would bring in cytology as an adjunct to their studies, particularly heredity. The microscopic study of chromosomes as they relate to observed genetic differences was to become an important and novel part of Davenport's program; it was this aspect of his research that was picked up and developed so fully by the Morgan group at Columbia after 1910, using the common fruit fly *Drosophila*.

During the first years of the operation of the SEE, Davenport not only served as administrator but also carried out research on his own, studying heredity in poultry, mice, and horses. In this work he employed both biometrical and Mendelian analyses. At the same time he began to apply Mendelian analyses to human traits. With his wife, Gertrude Davenport, he wrote a paper on heredity and hair form in humans and several papers on the inheritance of skin color and other physical traits.[14] In 1910 he published the results of a lengthy study in which he explained for the first time the graded series of skin colors in black-white matings in terms of a polygenic inheritance—that is, several sets of genes interacting to produce what came to be called "quantitative inheritance."[15] At the same time he also applied the newly developed Mendelian concept of multiple alleles to the inheritance of human eye color.[16] Although not highly innovative, Davenport's work was solid, and it earned him the respect of the rapidly growing community of Mendelian geneticists in the United States and abroad.

---

the most prominent in this group were H. E. Crampton and E. B. Wilson of Columbia University, D. T. MacDougal of the New York Botanical Garden, W. E. Castle and E. L. Mark of Harvard, and W. J. Moenkhaus of Indiana University.

[14] Gertrude C. Davenport and Charles B. Davenport, "Heredity of Hair Form in Man," *American Naturalist*, 1908, *42*:341–349; C. B. Davenport, "Heredity of Some Human Physical Characteristics," *Proceedings of the Society for Experimental Biology and Medicine*, 1908, *5*:101–102.

[15] C. B. Davenport, "Heredity of Skin Pigmentation in Man," *American Naturalist*, 1910, *44*:642–672.

[16] Rosenberg, "Davenport" (cit. n. 4), p. 268.

organization, one devoted exclusively to eugenics investigation and education, would be desirable, and he naturally thought of locating any laboratory for the study of human heredity and eugenics in Cold Spring Harbor. As Davenport originally envisioned it, a eugenics institute would be administratively under his control but with the day-to-day supervision of research and operating details given over to a superintendent. Thus Davenport, while overseeing major organizational plans, still could devote most of his time to his research, which by 1910 had become almost wholly concerned with human genetics and eugenics. It was clear that he needed both additional facilities and personnel to get on with the growing work in human heredity, "its outlook so vast that . . . the Director . . . cannot cope with it alone."[23]

Funding for a Eugenics Institute

Davenport's first step was to secure funding, without which nothing else could proceed. Ever the philanthropic entrepreneur, Davenport took advantage of two circumstances that led him directly to the doorstep of Mary Williamson Harriman. The first was the death of her husband, railroad magnate Edward Henry Harriman (b. 1848), in September 1909. Between 1880 and his death, Harriman had amassed a fortune, principally through his control of the Union Pacific, Southern Pacific, and Illinois Central railroads. Harriman's estate, estimated at approximately $70 million on his death, was left exclusively to his widow. Mrs. Harriman managed the estate for the next twenty-five years, turning over portions of it to her sons Averell and Roland as they reached majority and as her judgment allowed. In dealing with this fortune, Mary Harriman developed the principle of "efficient" giving—that is, philanthropy devoted to providing individuals with the opportunity to become more efficient members of society. Like her husband, she gave money to conservation groups (the Harrimans were both strong supporters of their friend John Muir), to hospitals, to the arts, and especially to charity organizations devoted to self-help for the poor. A cardinal principle in her philanthropy was to encourage cooperation and scientific planning in every aspect of society—from good government and urban landscaping to the care of the insane. She opposed the tendency toward individualism and competitiveness that she saw in early twentieth-century life, even though competitiveness had won her husband's fortune. From John Muir and C. Hart Merriam (director of the United States Biological Survey), she and her husband gained an insight into the use of scientific principles to plan a more rational and orderly society—according to an order that existed so clearly in nature if human beings would only learn from it.[24]

Mary Harriman did not accept the foundation concept in philanthropy. She wanted to be in close touch with all the projects to which she gave money. She would not, in fact, give to any project with which she did not feel complete sympathy. Moreover, she particularly disliked the direction in which John D.

[23] Davenport's annual report, *Yearbook of the Carnegie Institution of Washington*, 1910, 9:85.

[24] For more details than one could possibly care to know, the two-volume George Kennan biography, *E. H. Harriman* (Boston: Houghton-Mifflin, 1922), is adulatory but complete. A more manageable source is a short biography and appreciation of Mary Williamson Harriman: Persia Campbell, *Mary Williamson Harriman* (New York: Columbia Univ. Press, 1960), with an introduction by Grayson Kirk. For the data summarized here, see *ibid.*, pp. 12–66, esp. 17–18.

Rockefeller, Jr., was taking the Rockefeller Foundation after 1910. Following an interview with Rockefeller on 9 March 1911, she wrote that for the first time she "saw the Rockefeller mask and heard their formulas." Indeed, she was later to complain when the Rockefeller Foundation engineered a takeover of the New York Bureau of Municipal Research Training School, which she had supported with the provision that the program would be altered according to guidelines set by the General Education Board. At a hearing of the U.S. Commission on Industrial Relations on the Rockefeller move, Mrs. Harriman stated: "Nothing has ever made me realize as does this what a grasp money has on this country."[25] Her style of philanthropy was of an older, more personalized sort, less national in scope than that of the rising foundations. Their aims were the same—social control—but the scale and the methods were quite different.

Within a few months after her husband's death, Mrs. Harriman received more than six thousand appeals for donations to many causes, the requests totaling over $247 million. One of those appeals came from Charles B. Davenport. For propriety's sake, Davenport held off initiating a move until February 1910, but, then again, he had a special connection that gave him an edge over others. Davenport had taught Mrs. Harriman's daughter Mary in the summer of 1906 at the Biological Laboratory School of the Brooklyn Institute at Cold Spring Harbor, and he found it very convenient to renew an old acquaintance.[26] His efforts were not misdirected; Mrs. Harriman was attracted to his project of studying hereditary social traits with a view toward solving social problems.

After several interviews and discussions, Davenport came away with an enthusiastic promise of support for what came to be known as the Eugenics Record Office, to be located at Cold Spring Harbor on a site next to the SEE. The site amounted to almost seventy-five acres and included a huge old mansion that had once been the country home of a wealthy New Yorker. Mrs. Harriman initially agreed to fund the complete operating expenses of the eugenics office for at least five years. This commitment included building a concrete, fireproof vault for storing eugenics records collected in the field and a main laboratory-office complex. The two building operations cost over $121,000. During the seven years that Mrs. Harriman was the major donor, she contributed an additional $246,000 in operating costs, including salaries, equipment, office furniture, and indexing facilities. Between 1910 and 1918, the so-called Harriman period in the history of the ERO, the total cost of all operations came to a little over $440,000.[27] During that time the relationship between Mrs. Harriman and Davenport, cordial from the beginning, developed into an almost daily ritual of communication. The correspondence between them, beginning in July 1910, records the extent to which Davenport presented his ideas, large and small, to her, explained his decisions, sought her advice, and submitted every major decision for her approval. As Davenport wrote on her death in 1932:

[25] Entries from M. W. Harriman's diary, "following an interview . . . on 9 March, 1911"; quoted in Campbell, *Mary Harriman,* pp. 24, 27.

[26] See Frances Hassencahl, "Harry H. Laughlin, 'Expert Eugenics Agent' for the House Committee on Immigration and Naturalization" (Ph.D. diss., Case Western Reserve Univ., 1969).

[27] See Harry H. Laughlin, "Notes on the History of the Eugenics Record Office, Cold Spring Harbor, Long Island, New York," mimeographed report compiled from official records of the ERO, Dec. 1939, pp. 5–6, Harry H. Laughlin Papers, Northeast Missouri State University (NMSU), Kirksville, Mo.

> For us at the Eugenics Office [*sic*] the things that counted most were her understanding of the needs of the work at a time when it was ridiculed by many and disesteemed by many others. As she often said the fact that she was brought up among well bred race horses helped her to appreciate the importance of a project to study heredity and good breeding in man. Though she could turn a deaf ear to many appeals to the emotions, she had a lively sympathy for those things of whose lasting value she felt sure.[28]

In 1917 the Carnegie Institution of Washington agreed to take over responsibility for the annual operating expenses and future expansion of the ERO. At that time Mrs. Harriman transferred the ERO in its entirety to the CIW, with an additional endowment of $300,000, thus giving the ERO a financial independence that virtually none of the other departments of the Carnegie Institution enjoyed. The years from 1918 until the ERO was closed on 31 December 1939 are known as the Carnegie period. During that period the CIW spent approximately $25,000 per year in operating expenses. The Harriman period was one of expansion and growth; the Carnegie period, one of stabilization and eventual decline.

Laughlin's Niche

With funds and space secured, Davenport turned to the search for a manager and planner for the ERO. The position of "superintendent," as it was called, required a person of scientific background, preferably someone who understood the principles and problems of heredity and had experience in practical breeding. It also required someone totally devoted to the eugenics cause, someone who could raise money among the wealthy, carry out educational programs, and promote a far-reaching vision of how eugenics could help to remake society. Many people have compared the advocates of eugenics to religious zealots, a comparison no doubt fostered by Francis Galton's references to the "religion of eugenics." In one sense Davenport was a preacher, and he was seeking someone of similar energy, devotion, and vision as his superintendent. This he found in the person of Harry Hamilton Laughlin (1880–1943), who was then teaching in the agriculture department of the State Normal School in Kirksville, Missouri.[29] Laughlin had first come to Davenport's attention in February 1907, when the young man had written to ask some questions about breeding chickens.[30] Noting Laughlin's interest in heredity, Davenport invited him to attend the Brooklyn Institute's summer course at Cold Spring Harbor in 1908. With their common interests in agricultural breeding and in heredity, Davenport and Laughlin hit it off well from the beginning. Both were highly energetic and serious about their work, utterly humorless and rigid in their approach to life, and totally dedicated to the cause of social reform through eugenics. For Laughlin, born in Oskaloosa,

[28] Draft of a one-page eulogy, "Mrs. Harriman," in file, "Mrs. E. H. Harriman," Davenport Papers, American Philosophical Society (APS), Philadelphia.

[29] Hassencahl's full-length study of the life and work of Harry Laughlin, which unfortunately has never been published, focuses particularly on Laughlin's lobbying activities. It contains a wealth of additional information on his other work, the ERO, and the Nazi *Rassenhygiene* movement. For a discussion of Laughlin's work as surveyed from his papers in Kirksville, see also Randy Bird and Garland Allen, "The J.H.B. Archive Report: The Papers of Harry Hamilton Laughlin," *Journal of the History of Biology,* Fall 1981, *14*(2):339–353.

[30] Laughlin to Davenport, 25 Feb. 1907, Davenport Papers, APS.

Iowa, the chance to study at an East Coast marine laboratory with a figure as well known as Davenport was the experience of a lifetime. Of that first summer, he wrote to Davenport: "I consider the six weeks spent under your instruction to be the most profitable six weeks that I ever spent."[31] Although not formally trained in biology or heredity, Laughlin was a quick learner, and his energy and enthusiasm for projects, usually on a grand scale, were boundless.

Although Laughlin wanted to return to Cold Spring Harbor for the summer course in 1908, his teaching duties made it impossible to be absent from Kirksville for another six-week period. Correspondence between Laughlin and Davenport continued regularly, however, during the next several years, concerned with topics such as filling out Mendelian information cards on students at Kirksville, winglessness in chickens, inheritance of redheadedness, and other genetic matters. Laughlin was particularly attentive in distributing all sorts of information cards on human traits to his students and in making sure the cards were completely and thoughtfully filled out.

Laughlin's thoroughness and energy impressed Davenport, and the possibility of a meeting suddenly arose when, in December 1908, Davenport wrote to Laughlin that he would be journeying to Columbia, Missouri, the first week in January to attend the sixth annual meeting of the American Breeders' Association (6–8 January). Laughlin was ecstatic and immediately invited Davenport and his wife to visit Kirksville prior to the meeting.[32] Laughlin also hoped to attend the sessions himself, since he was now teaching "Nature Study and Agriculture," but he was not sure if the president of the Normal School would allow him to leave. The Davenports did visit the Laughlins in Kirksville, and Laughlin was able to attend the meeting in Columbia after all. Thus the two had the opportunity to discuss many facets of breeding. In Kirksville Davenport was induced to give two public lectures that aroused "great interest in the subject of heredity." For Laughlin, Davenport's visit was of special value because it gave a boost to his ongoing attempts to organize a scientifically based agriculture department. "It takes money to run a department like the one I want," he wrote. "In two or three years I will be able to show—I hope—an agricultural department worthy of the name."[33]

Little did Laughlin know that his plans would not materialize, but only because bigger things were in store for him. Davenport subsequently invited Laughlin to attend the 1910 summer course at Cold Spring Harbor, which included lectures and field trips related to eugenics. Then, in mid July, Davenport approached Laughlin about resigning from Kirksville and taking the job as superintendent. As Davenport wrote to Mrs. Harriman: "I was surprised to see how receptive he was of the idea. He said there would be no financial advantage but that, above all, he desired to go into this work. He made no conditions, even as to the length of appointment. I am more than ever satisfied that he is the man for us."[34] Laughlin accepted, returned with his wife Pansy to Missouri to straighten out their business affairs, and moved to the east in mid September 1910.

[31] Laughlin to Davenport, 30 Mar. 1908, Davenport Papers, APS.
[32] Laughlin to Davenport, 15 Dec. 1908, Davenport Papers, APS.
[33] Laughlin to Davenport, 30 Jan. 1909, Davenport Papers, APS.
[34] Davenport to Mrs. Harriman, 1 Oct. 1910, Davenport Papers, APS.

GOALS AND PROGRAMS OF THE ERO

Laughlin set about organizing matters at Cold Spring Harbor as soon as he arrived. At first, because of a shortage of buildings on the new property, the ERO administrative quarters were located on the ground floor of the large home that had been the center of the former estate. The Laughlins lived on the part of the ground floor not occupied by the offices and on the second floor. Several record clerks, a groundskeeper, and two assistants lived on the third. A fireproof vault for eugenics records was added to the east side of the main house in 1911. The Eugenics Record Office opened its doors on 1 October 1910. Although Mrs. Harriman could not be present for the official opening, Davenport wrote her that it was "a red letter day."[35]

The Eugenics Record Office was organized with two general purposes: to carry out research on human heredity, especially the inheritance of social traits; and to educate laypersons about the importance of eugenic research and the implications of eugenic findings for public policy. The work of the ERO was to be strictly scientific, growing out of the experimental and biometrical studies of Davenport and the Station for the Experimental Study of Evolution.[36] To give the organization scientific credibility, Davenport set up a Board of Scientific Directors, consisting of, in addition to himself, Alexander Graham Bell, chairman; Lewellys F. Barker (professor of medicine, Johns Hopkins Medical School); William H. Welch, vice-chairman (dean, Johns Hopkins Medical School); Irving Fisher (professor of economics, Yale University); and E. E. Southard (a brilliant young psychiatrist at the Boston Psychopathic Hospital). Board members were required to attend meetings (they would be asked to resign if they missed more than two consecutively), which indicated that Davenport wanted the scientific advisers to be more than figureheads. Since minutes of meetings of the advisory board are not available, it is difficult to know how often these meetings were held or how seriously the advisers took their jobs. At any rate, Davenport did manage to assemble a prestigious group of advisers, including the dean of American medicine and medical reform (Welch) and one of the foremost inventors in the United States (Bell).[37]

In his first report, in 1913, Laughlin listed a number of the specific functions that the ERO was intended to perform. The following descriptions of these purposes give an indication of the scope of activities that Laughlin and Davenport envisaged.[38]

To serve eugenical interests as a repository and clearinghouse. First and foremost, the ERO was to become a data bank for information on human hereditary traits. This function was clearly one of research and was an extension of work already carried out through the Eugenics Committee of the ABA. The data would ultimately serve as the basis for analyzing the inheritance patterns of a wide variety of traits. As a clearinghouse and information repository, the ERO could also supply individuals with data about their family history if their families

[35] *Ibid.*

[36] Harry H. Laughlin, "The Eugenics Record Office at the End of Twenty-Seven Months Work," *Report of the Eugenics Record Office,* June 1913, No. 1, p. 1.

[37] Bell was interested in eugenics because of hereditary deafness in his own family and because he had always been fascinated with the breeding of sheep and other large domesticated animals.

[38] Laughlin, "Eugenics Record Office" (cit. n. 36), pp. 2–21.

had participated in any of the studies. A newsletter, *Eugenical News,* contained short, nontechnical articles and items of information about eugenics research throughout the country.

To build up an analytical index of traits in American families. All data coming in to the ERO, from whatever source, were to be carefully indexed in accordance with a complex classification system known as *The Trait Book,* which Davenport had devised in 1910. *The Trait Book* listed all the human physical, physiological, and mental traits imaginable (and some that are hard to imagine)—rowdyism, moral imbecility, train-wrecking, and ability to play chess, to name but a few. It classified every trait by a numbering scheme akin to the Dewey Decimal System. The condition of harelip, for example, is classified as 623, where 6 indicates a condition of the nutritive system; 2, the mouth portion of the nutritive system; and 3, the specific mouth feature of harelip. Similarly, chess-playing ability is number 4598, where 4 signifies a mental trait; 5, general mental ability; 9, special game-playing ability; and 8, the specific game, chess. The ERO stored its information on such conditions in folders filed either by family name or by the caseworker who collected the information. This information was then indexed on 3 × 5 cards and cross-referenced in three ways: by family name, by number (for the trait), and by geographic locality. Thus an investigator could search out, for example, all the cases of harelip by going to the card drawer for the number 623, or all the references to a particular family by checking for its surname. Each card in the drawers provided reference to the appropriate file folder or folders containing all the detailed information. By 1 January 1918, the ERO had accumulated 537,625 cards; there were nearly twice that many by the time the office closed in 1939. The information that was filed and catalogued at the ERO was organized into five main categories of traits: physical traits (e.g., stature, weight, eye and hair color, deformities), physiological traits (e.g., biochemical deficiencies, color blindness, diabetes), mental traits (e.g., intelligence, feeblemindedness, insanity, manic depression), personality traits (e.g., liveliness, morbundity, lack of foresight, rebelliousness, trustworthiness, irritability, missile throwing, popularity, radicalness, conservativeness, nomadism), and social traits (e.g., criminality, prostitution, inherited scholarship, alcoholism, patriotism, "traitorousness"). These groupings were not meant to be mutually exclusive since, for example, a personality trait could have more than one social manifestation. It was nonetheless the hope of Davenport, Laughlin, and others that, through such a detailed breakdown of traits into categories and subcategories, researchers could easily identify and follow the same traits through a wide variety of family lines.

To study the forces controlling and hereditary consequences of marriage-matings, differential fecundity, and survival migration. Today these studies, which include a considerable amount of sociological as well as biological information, would fall roughly under the heading of demography. From the start eugenicists were particularly concerned about the "differential fertility" issue—that is, about which groups in society were showing the higher and the lower birthrates.

To investigate the manner of inheritance of specific human traits. These studies were mainly straight-line applications of Mendelian principles to analyzing human genetic data. Thus eugenicists were interested in determining not only whether a trait was inherited but also whether it was dominant or recessive, whether it was sex-linked, the degree to which its expression might be influenced by environment, whether it was expressed early in life or was of late onset, and so forth. Investigations in this category involved constructing pedigree charts from raw data on families and deducing from the data what the pattern of heredity might be. (The obvious difficulties facing the eugenicist, especially in 1910–1920, in collecting enough reliable data to draw such conclusions will be discussed in a later section of this paper.) In the analysis of inheritance patterns, ERO workers were advised and sometimes aided by members from the appropriate committee of the American Breeders' Association—for example, the Committee on Heredity of the Feebleminded, the Committee on the Heredity of Epilepsy, the Committee on Heredity of Deafmutism, the Committee on Heredity of Eye Defects, and the Committee on Heredity of Criminality.

To advise concerning the eugenical fitness of proposed marriages. Prospective marriage partners could visit or write to the ERO for what today might be called "genetic counseling." Drawing on as much of the individuals' family histories as possible, in conjunction with other data already in the files, ERO workers would discuss with the couple the probabilities of their children inheriting this or that trait and emphasize the importance of good mate selection in marriage. As Laughlin wrote:

> It is one of the cherished beliefs of the students of eugenics that when painstaking research has determined the manner of the inheritance of traits so that, upon examination of one's somatic traits and pedigree, something concerning his or her hereditary potentialities can be determined, social customs will make such hereditary potentialities marriage assets, valued along with—if not above—money, position and charming personal qualities. This belief is based not upon desire alone, but upon a few actual visits and letters from intelligent persons that come with increasing frequency to the Eugenics Record Office, asking for instructions for making a study of the eugenical fitness of a contemplated marriage.[39]

Laughlin noted that as of 22 January 1913 there were seventy-seven such requests on file at the ERO.

To train fieldworkers to gather data of eugenical import. The most reliable data on heredity could be collected, Laughlin noted, by fieldworkers who were trained to gather information in hospitals and asylums as well as in individual homes. Each summer the ERO ran a short training course for fieldworkers, including lectures by Laughlin, Davenport, and occasional guests on endocrinology, Mendelian heredity, Darwinian theory, elementary statistical methods, and eugenic legislation. Students also became familiar with various mental tests (Binet, Yerkes-Bridges, army Alpha and Beta tests) and learned how to administer and interpret them. They memorized classifications of insanity, criminality, epilepsy, and skin and hair color and methods of anthropometrical measurement, with particular emphasis on cranial capacity. The course also involved field trips to nearby hospitals and institutions for mental defectives in New York—Kings Park Hospital for the Insane, Letchworth Village for the Feebleminded—and the receiving stations for immigrants at Ellis Island. To conclude the summer's training program each student produced a research project that involved col-

[39] Laughlin, "Eugenics Record Office" (cit. n. 36), pp. 10–11.

Figure 1. *Summer trainees in the fieldworker course at the Eugenics Record Office, about 1919. Harry Laughlin is in the front row, center (seated, with hat in hand). The building in back is the original house in which all the ERO work was done. It remained the major facility for all ERO business, as well as the Laughlin residence, during the thirty year history of the ERO. Courtesy of the Harry H. Laughlin Papers, Northeast Missouri State University, Kirksville, Missouri.*

lecting and analyzing eugenical data. The summer also had its lighter side, with clambakes, picnics, and boat trips. By 1917 the ERO had trained approximately 156 fieldworkers, 131 women and 25 men, among them 8 Ph.D.s and 7 M.D.s (see Fig. 1).

Those who completed the training program took up positions in various institutions. A few were retained as paid fieldworkers by the ERO. The majority were attached to state mental hospitals, insane asylums, or almshouses, with their salaries either paid wholly by those institutions or, more frequently, shared between the institution and the ERO. The fieldworkers' jobs involved taking family histories of patients within the institution to determine to what degree their conditions were hereditary. These linear studies, as they were called, would then be filed in large folders at the ERO, where they provided the basis for studies on the inheritance of mental deficiency, insanity, Huntington's chorea, and the like. Laughlin's records show that in the first three years (1910–1913) thirty-two fieldworkers amassed 7,639 pages of family case histories (text) and 800 pages of pedigree charts and averaged forty-six interviews per month.[40] The training program was carried out most extensively between 1910 and 1917; thereafter it tapered off somewhat but remained in operation until 1926. During the first seven years, funds for the training program came from the personal bequests of John D. Rockefeller, Jr., amounting to a total of $21,650.[41] From

[40] *Ibid.*, foldout chart opposite p. 19.

[41] Harry H. Laughlin, "Notes on the History of the Eugenics Record Office," mimeographed report (Cold Spring Harbor, 1934), p. 5; original in the Laughlin Papers, NMSU.

then on, for the duration of the program, funds came from the Carnegie Institution as part of the ERO's regular budget.

To encourage new centers for eugenics research and education. Laughlin in particular conceived of the ERO as encouraging the formation of new groups and prompting existing organizations to take up eugenic studies within the context of their established programs. For example, he was quite active in getting the YMCA to take part in eugenical work (making available data on vital statistics of members as well as propagandizing eugenics ideals). He urged women's clubs to get involved and asked the director of the United States Census to include eugenics questions in the 1920 and subsequent censuses. He encouraged colleges to hold programs on eugenics, show eugenics films, teach eugenics courses, and take surveys of their student populations.

To publish the results of research and to aid in the dissemination of eugenic truths. A final specific function of the ERO was education. To Laughlin this included everything from showing films to publishing the results of research on human heredity, monographs on the status of relevant legislation, and analyses of public attitudes toward eugenic ideas. The ERO itself published a list of eugenics monographs, written by such investigators as Henry H. Goddard, Davenport, and Laughlin himself (a number of monographs came from his pen).[42]

DATA GATHERING AND GENETIC ANALYSIS

Because eugenics claimed from the outset to be an objective and scientifically based program, to understand its general history and social impact it is important to see what type of research eugenicists pursued. While it is clearly beyond the scope of this study to examine these projects in depth, a few examples of work carried out at the ERO under the auspices of Davenport and Laughlin will show the style and flavor of eugenicists' scientific work. While the research interests and methods of analysis employed by Davenport and Laughlin are not necessarily representative of eugenics as a whole, they are nonetheless indicative of much of the work going on in the United States between 1910 and 1935.

The raw data from both individual family questionnaires and fieldworker studies collected at the ERO during the years 1910–1939, as well as the index cards cross-referencing them, are now housed in the basement of the Dight Institute of Human Genetics at the University of Minnesota in Minneapolis.[43] The vast bulk of the data (some ten filing cabinets) consists of individual questionnaires; the rest (some eight cabinets) consists of fieldworker studies of individual families. It is a testimony to the energy and dedication of the field and office workers that in the course of less than thirty years they accumulated, indexed, and cross-referenced such a monumental amount of material.

A quick perusal of the data collected by fieldworkers indicates that, despite Davenport's and Laughlin's emphasis on rigorous, quantitative methodology,

[42] Laughlin, "Eugenics Record Office" (cit. n. 36), pp. 21–22.

[43] When Milislav Demerec, director of the SEE, wanted to clear out the old ERO building at Cold Spring Harbor in 1946, he put out a call to various organizations and individuals to see who would take the case studies, index cards, and back issues of *Eugenical News*. The only acceptance came from Sheldon Reed, director of the Dight Institute. I am grateful to Professor Reed for having preserved the material at that time and for his hospitality and guidance when I inspected the records in 1981.

most of the data collected were of a subjective, impressionistic nature. One example will illustrate this point. The fieldworker Anna Wendt Finlayson carried out a study of the Dack family, descendants of two Irish immigrants in western Pennsylvania. She did no mental testing, and the data consist solely of "community reactions," a euphemism for "common gossip." The interviewer talked with family members, neighbors, and local physicians. The write-ups on two of the individuals, James Dack and William Dack, read as follows:

> *James Dack* (II6) was commonly known as "Rotten Jimmy," the epithet was given because of the diseased condition of his legs, which were covered with chronic ulcers, although the term is said to have been equally applicable to his moral nature. He was a thief and general good-for-nothing, but neither shrewd nor cunning. His conversation quickly revealed his childlike mind.
>
> *William Dack* (I2) was born in Ireland and came to the United States about 1815. He settled near a little town in the northern part of the soft coal district of Pennsylvania, which we will designate Bushville, and raised his children (9) in that vicinity. William died almost fifty years ago, but he is remembered by a few of the oldest settlers of the locality as a peculiar, silly old fellow who drank a good deal, stole sheep and household valuables from his neighbors, and did not seem to be very intelligent. He was married twice, his first wife died in Ireland and we know nothing of her. She bore him one child. . . . William's second wife was (I3) Mary Murphy. . . . An old resident of Bushville, now deceased, once stated to a woman who was interviewed by the writer that William and Mary were first cousins.[44]

Because there is no way to verify such information, it is of no value as objective data. Yet on the basis of that "evidence" the researchers drew up a pedigree chart indicating the presence of hereditary feeblemindedness in the Dack family.

Slightly different problems are associated with the data processed from questionnaires sent out to families. In these cases the individual subjects recorded the data about themselves and their family members. These data are subject to the errors introduced when many different observers are involved in measuring the same quantity throughout a population. No two observers measure even the same item in the same way. The problem is obviously compounded when many different observers measure many different quantities. Even the data on height of individual family members (one item on the questionnaire) appear to be guesses, not actual measurements, for they often relate to relatives who are either geographically distant or deceased. In the collection of data known as the "Record of Family Traits," much of the information is secondhand, and none of it is quantitative. As Sheldon Reed, Director Emeritus of the Dight Institute in Minneapolis, has stated, most of the data collected by the ERO are worthless from a genetic point of view.[45]

Even if the raw data collected by ERO fieldworkers and others were considered reliable, their application in determining patterns of heredity was fraught with difficulties. The major method of analysis, of course, has always been the pedigree chart, but this involves two types of problems. First, many families have only a small number of children, statistically speaking, and thus the appearance, or especially the nonappearance, of a trait often says nothing about

[44] Anna Wendt Finlayson, "The Dack Family: A Study in Hereditary Lack of Emotional Control," *Bulletin of the Eugenics Record Office,* 1916, No. 15, pp. 6–7.
[45] Sheldon Reed, personal interview, 30 Oct. 1981, Minneapolis, Minn.

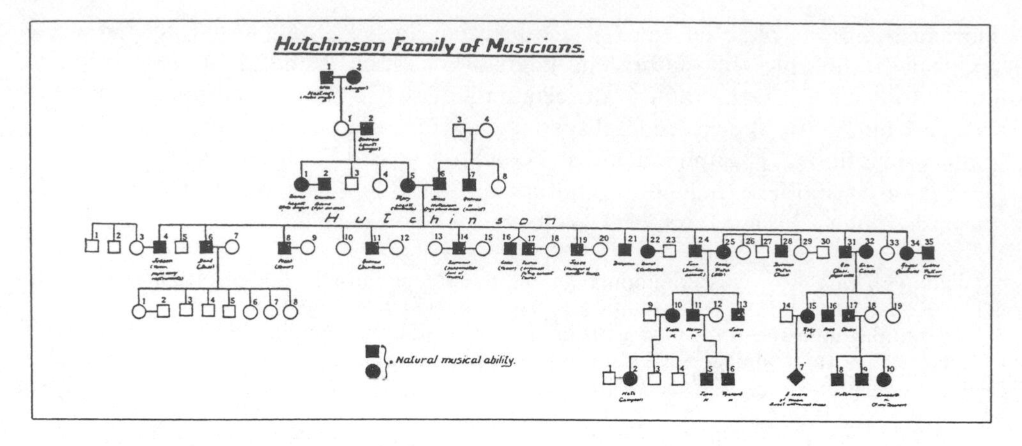

Figure 2. Pedigree chart for "Natural Musical Ability" in the Hutchinson family, as constructed by Davenport in 1915. Solid figures indicate individuals who display "natural musical ability." This sort of chart, which does not distinguish between learned and inherited (genetic) causes for a trait, led critics to note that eugenicists failed to understand the difference between genetics and genealogy. From C. B. Davenport and H. H. Laughlin, "How to Make a Eugenical Family Study," Bulletin of the Eugenics Record Office, 1915, No. 13, p. 27 (2nd ed. 1919).

its actual mode of inheritance—for example, whether the trait is dominant, recessive, or sex-linked. Moreover, pedigree charts are often woefully incomplete—that is, many family members are not included, and thus what might look like a dominant trait (because it appears frequently) appears so only because data on other family members are missing. Second, and probably most critical, pedigree charts provide no way to separate genetically determined from environmentally determined phenotypes. The fact that musical ability, for example, appears repeatedly in the Hutchinson family pedigree (see Fig. 2) says nothing about the actual inheritance of that trait in the genetic, as compared to the social, sense. The more a trait involves social, behavioral, or personality features, the less possible it is to separate genetic from environmental influences. Since eugenicists were far more interested in mental and personality traits than in clinical conditions, their pedigree charts were prone to such misinterpretation.

As an example of the simplistic generalizations in which eugenicists indulged, consider Davenport's study of the inheritance of *thalassophilia* ("love of the sea" or "sea-lust"). In 1919 Davenport published a book-length study, under the auspices of the Carnegie Institution of Washington, entitled *Naval Officers: Their Heredity and Development*. It was a study of why naval careers seemed to run in families. Davenport's explanation was genetic; in fact, he attributed this tendency to a single Mendelian gene! Here is how Davenport reasoned. Nomadism, the impulse to wander, was obviously hereditary because such racial groups as Comanches, Gypsies, and Huns were all nomadic. Searching individual family pedigrees, Davenport found recurrent examples of nomadism in the families of traveling salesmen, railroad workers, tramps, vagabonds, and boys who played hookey from school. Since the trait of nomadism showed up mostly in men, he concluded that it must be sex-linked and recessive, passing from mothers to half of their sons. Thalassophilia, a version of nomadism, is thus also genetically determined:

> Thus we see that thalassophilia acts like a recessive, so that, when the determiner for it (or the absence of a determiner for dislike) is in each germ-cell the resulting male child will have love of the sea. Sometimes a father who shows no liking for the sea . . . may carry a determiner for sea-lust recessive. It is theoretically probable that some mothers are heterozygous for love of the sea, so that when married to a thalassophilic man half of their children will show sea-lust and half will not.[46]

Davenport's method of argument was by analogy, not by direct evidence. Thus, he drew an analogy between thalassophilia and the inheritance of comb size in fowl: "It is possible . . . that the irresistible appeal of the sea is a trait that is a sort of secondary sex character in males in certain races, just as a rose comb is a male characteristic in some races of poultry."[47] By 1919 the inheritance pattern for rose comb was a well-established Mendelian trait. By making the comparison between human beings and poultry, Davenport assumed that superficial similarity in patterns of inheritance between two quite different species implied similarity in genetic causality. More important, he virtually discounted the effect of environmental factors in molding human behavioral traits.

Davenport's genetic determinism led to the obvious view that the source of a social problem was not environment but "bad genes." He urged philanthropists to donate their funds to eugenics, and not to charity, which would only perpetuate hereditary degeneracy. Accordingly, in a report to the Committee on Eugenics of the American Breeders' Association in 1909, Davenport insisted: "Vastly more effective than the million dollars to 'charity' would be ten million to eugenics. He who, by such a gift, should redeem mankind from vice, imbecility and suffering would be the world's wisest philanthropist."[48] In public and in private, Davenport belittled social reform. He apparently was fond of telling the parable of a man who found a bitter gourd and watered and tended it carefully to produce a delicious vegetable. That man was, Davenport claimed, like the trustee of a rehabilitation hospital for the insane. Poverty and lack of social or economic success were *de facto* the phenotypic expressions of genotypic inferiority. In 1912 he advised the National Conference of Charities and Corrections that social reform was futile since "the only way to secure innate capacity is by breeding it."[49] To Davenport, the comparison between breeding humans and breeding strains of domesticated animals or plants was self-evident.

TAKING EUGENICS TO THE PUBLIC ARENA

Since one of the expressed purposes of the ERO was education and the dissemination of "eugenical truths,"[50] it is not surprising to find that Laughlin (in particular) devoted considerable energy to publicity endeavors. One vehicle was the ERO's publication, *Eugenical News,* whose first volume was issued in 1916

[46] C. B. Davenport, *Naval Officers: Their Heredity and Development* (Washington, D.C.: Carnegie Institution of Washington, 1919), p. 29.

[47] *Ibid.,* p. 28.

[48] C. B. Davenport, "Report of the Committee on Eugenics," *Rep. Amer. Breeders' Assoc.,* 1909, *6*:94.

[49] See Mark Haller, *Eugenics: Hereditarian Attitudes in American Thought* (New Brunswick, N.J.: Rutgers Univ. Press, 1963), p. 65.

[50] See Laughlin, "Eugenics Record Office" (cit. n. 36), pp. 19–20, which lists among the ERO's purposes No. 9, "to encourage new centers of eugenics research and education" (p. 19), and No. 10, "to publish the results of researches and to aid in the dissemination of eugenical truths" (p. 20).

with Davenport and Laughlin as editors. *Eugenical News* contained short, popular articles reporting on eugenics research, the menace of the feebleminded, differential fertility, the evils of race-crossing, and the like, as well as reviews of books on eugenics. The editorial board of the *News* remained substantially the same from 1916 through 1939, the only changes being the addition of Roswell H. Johnson for 1920–1929 and Morris Steggerda for 1932–1939. The tone of the *News* as a whole was overtly propagandistic, quite often with few facts and little or no presentation of data.

In addition to *Eugenical News,* the ERO helped to launch and guide through publication popular and semipopular works of other eugenicists who were not directly connected to the institute. Laughlin, for example, was a close personal friend of Madison Grant, a wealthy New York lawyer, conservationist, member of several public commissions, and author of one of the most racist, pro-Nordic tracts written during the period 1910–1920, *The Passing of the Great Race.* Laughlin met regularly with Grant in New York to discuss matters concerning the several eugenics organizations of which they both were members: the American Eugenics Society, the Eugenics Research Association, and later the Pioneer Fund. Grant regularly donated money to these organizations, as well as to specific ERO projects. Laughlin supported Grant in a variety of ways. When Grant was about to publish his second book, *Conquest of a Continent,* in 1932 (it was actually published in 1933), Laughlin went over the manuscript carefully and helped him to avoid some of the most blatant racial slurs.[51] Furthermore, Laughlin bid hard to encourage Yale to award Grant an honorary degree (Grant was a Yale alumnus, class of 1887). To Laughlin, presentation of an honorary degree by a prestigious university to one of the country's foremost eugenicists would provide a big shot in the arm for the movement in general and for the ERO, with which Grant was closely associated, in particular.

Through the ERO, Laughlin also organized a series of research and propaganda efforts, including a nationwide study of racial origins of inventiveness; a study of the hereditary lineage of aviators; a survey of the human resources of Connecticut, in which ancestry was studied in complete detail for the entire population of a small Connecticut town; a study of alien crime, organized in conjunction with Judge Harry Olson of the Municipal Court of the City of Chicago; and the distribution of defectives in state institutions by type of defect and by national and racial origins. He was also in close contact with Charles M. Goethe, a wealthy lumberman from Portland, Oregon, who gave considerable financial support to eugenics projects and was a great publicizer of eugenic ideals (Goethe also left his estate to the Dight Institute of Human Genetics in Minneapolis).[52] Laughlin supported and encouraged Goethe's plan to establish a "clinic on human heredity," a kind of eugenic counseling and birth control clinic that, despite all the effort, never materialized. The list could go on and on, but

[51] See Madison Grant, *The Passing of the Great Race* (New York: Scribners, 1916); and Laughlin to Grant, 10 Nov. 1932, Laughlin Papers, NMSU. Laughlin told Grant he should strike from the manuscript the statement that "if the remainder of the Jews could be prevented from coming to the United States. . . ." As Laughlin remarked, "This has a tinge of 'Damn Jew' about it. It would, I believe constitute a more forceful statement if it were pointed out that the United States has already one out of five of the world's Jews" (p. 2). Laughlin did not disagree with Grant in substance, only in form.

[52] Sheldon Reed, personal communication, 9 Nov. 1981.

the point is this: using the ERO as an operational base, Laughlin developed and kept up a lively network of associations that served to gain financial and moral support for eugenics in general and the work of the ERO in particular. Furthermore, through his activities, Laughlin gave considerable organization and coordination to far-flung and conceptually diverse eugenics projects in the United States and, somewhat later, throughout the Western Hemisphere.

Laughlin also helped to popularize eugenics through his ERO association. He loved exhibits. His correspondence is filled with plan after plan for exhibits at state fairs, genetics meetings, teachers' conferences, and the like. For example, in preparation for the Third International Congress of Eugenics at the American Museum of Natural History in New York in 1932, Laughlin sent out over one hundred letters asking for donations to mount a huge eugenics exhibit in one of the museum's largest halls. It was an ambitious exhibit, for which he finally raised sufficient funds. Laughlin used ERO secretarial and research help in preparing many of his projects, including exhibits. Without this sort of institutional support it would have been difficult, if not impossible, to carry out so many projects and integrate the activities of so many people.

Laughlin also used his institutional base at the ERO as a platform for political activity on behalf of eugenics. The two most notable examples are his research and testimony before the House Committee on Immigration and Naturalization and his effective lobbying for the passage of eugenical sterilization laws in various states. In 1924 the Johnson Act (also called the "Immigration Restriction Act") passed both houses of Congress, and by 1935 some thirty states had passed sterilization laws. Neither of these results can be attributed to Laughlin alone, but he was instrumental in both—perhaps more directly visible in his House testimonies than elsewhere. Laughlin brought forth reams of biological data to prove the genetic inferiority of southern European, central European, and Jewish people. His congressional testimony received wide press coverage, and a transcript was reprinted as part of the *Congressional Record*.

Laughlin's invitation to become the congressional "expert witness" came from Representative Albert Johnson, a rabidly anti-immigrant, antiradical, and anti-Communist journalist and editor from Washington State who had entered Congress in 1912 on a restrictionist platform. Laughlin, long interested in the immigration issue, had made the initial contact with Johnson and, along with Madison Grant, had established a close personal and professional relationship with him. One consequence was that in 1924 Johnson, who was not then even a member, was elected to the presidency of the Eugenics Research Association.[53] As "eugenics expert," Laughlin received congressional franking privileges, and he used them to assemble vast amounts of data about the institutionalized alien and native stock. The Carnegie Institution of Washington in turn officially allowed Laughlin to use his secretarial staff at the ERO to help compile data and figures for the congressional testimony. Later the CIW would regret encouraging Laughlin in this overtly political role, but in the early and mid 1920s the directors had no objection.

[53] See Hassencahl, "Harry H. Laughlin" (cit. n. 26), pp. 206–208. Grant gave moral support to the committee but so far as we know did not appear before it in person; see *ibid.*, pp. 283, 293–300; and Kenneth Ludmerer, *Genetics and American Society: A Historical Appraisal* (Baltimore: Johns Hopkins Press, 1972), pp. 112–113.

The story of Laughlin's work as eugenics expert to the House Committee on Immigration and of his arguments in his major congressional hearings has been told in detail elsewhere.[54] What is striking in these testimonies is the strong racist and antiethnic feeling to which Laughlin, bolstered by charts and graphs, gave vent. Laughlin was already voicing distinctly anti-immigrant sentiment immediately after World War I; like Madison Grant, he now called for a "purification" of the good Nordic stock of the United States to free it from contamination by the "degenerate" sectors of Europe (according to Laughlin, eastern and southern Europe). Laughlin was particularly anti-Semitic, arguing that with respect to immigration "high-grade Jews are welcome, and low-grade Jews must be excluded." "Racially," he argued, "the country will be liberal if it confines all future immigration to the white race, then, within the white race, if it sets up differential numerical quotas which will admit immigrants in accordance not with external demand but on the basis of American-desired influence of such racial elements on the future seed-stock of America."[55] Laughlin further distinguished himself by devoting considerable research energy to showing that recent immigrants and "aliens" were responsible for much of the crime committed in the United States between 1890 and 1920.[56]

In discussing the immigration issue, Laughlin was particularly disturbed by the specter of "race-crossing." He reported that a committee from the Eugenics Research Association had studied the matter and had failed to find a single case in history of two races living side by side and maintaining racial purity. Race mixtures, Laughlin said, are poor mixtures, referring for corroboration to a study on race-crossing in Jamaica in which Davenport was then engaged. Like W. E. Castle, Edward M. East, and other geneticists at the time who had agricultural interests, Laughlin compared human racial crossing with mongrelization in the animal world. The progeny of a cross between a racehorse and a draft horse, Castle once wrote, "will be useless as race horses and they will not make good draft horses. . . . For similar reasons, wide racial crosses among men seem on the whole undesirable."[57] Like Grant, Laughlin felt that immigrants from southern and eastern Europe, especially Jews, were racially so different from, and genetically so inferior to, the current American population that any racial mixture would be deleterious. Even after the phenomenon of "hybrid vigor" was known to be widespread, eugenicists conveniently explained it away by arguing that only a few of the offspring of any hybridization would really show increased vigor. The rest would be decidedly inferior.[58] Using statistics

[54] Ludmerer, *Genetics and American Society,* pp. 87–119; Hassencahl, "Harry H. Laughlin" pp. 161–312; and Garland E. Allen, "The Role of Experts in the Origin and Closure of Scientific Controversies: The Case of the American Eugenics Movement, 1910–1940," in *Scientific Controversies: Studies in the Resolution and Closure of Disputes Concerning Science and Technology,* ed. A. L. Caplan and H. T. Engelhard (New York/Cambridge: Cambridge Univ. Press, forthcoming).

[55] Harry H. Laughlin, *Report of the Special Commission on Immigration and the Alien Insane* (submitted as a study of immigration control to the Chamber of Commerce of the State of New York, 16 Apr. 1934), pp. 17, 18.

[56] See National Commission on Law Observance and Enforcement, *Report on Crime and the Foreign Born* (Washington, D.C.: U.S. Government Printing Office, 1931).

[57] C. B. Davenport, "Race Crossing in Jamaica," *Scientific Monthly,* 1982, *27*:225–238. This was a summary of Davenport's lengthier study, carried out with Morris Steggerda, *Race Crossing in Jamaica* (Washington, D.C.: Carnegie Institution of Washington, 1929); and W. E. Castle, *Genetics and Eugenics* (Cambridge, Mass.: Harvard Univ. Press, 1916), p. 233.

[58] See E. M. East and Donald F. Jones, *Inbreeding and Outbreeding* (Philadelphia: Lippincott,

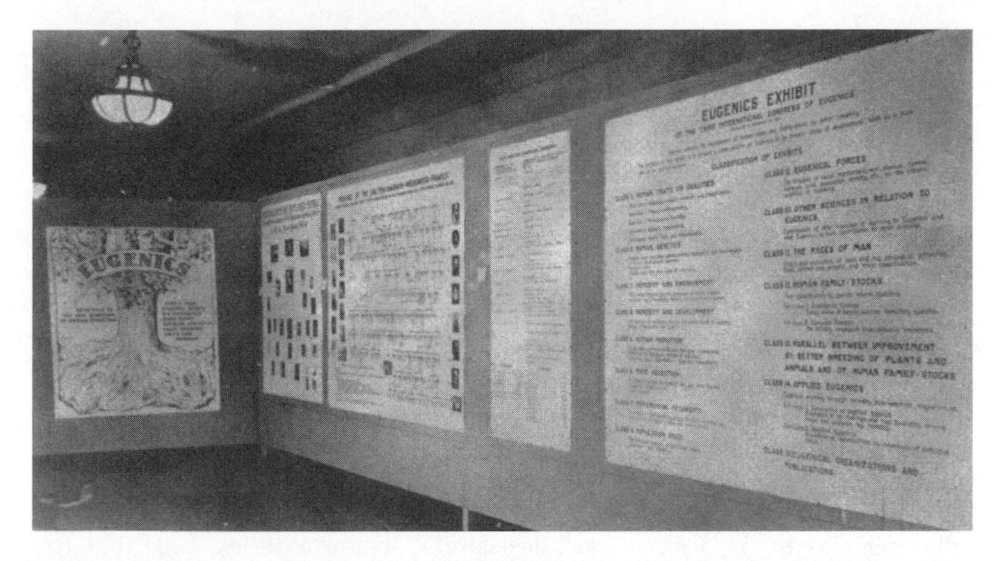

Figure 3. Exhibit introducing the concept of eugenics, as displayed in the American Museum of Natural History during the 1932 International Congress of Eugenics in New York. The large poster at the far end depicts eugenics as a tree whose roots draw upon many areas of study: genetics, anthropology, religion, statistics, physiology, and the like. The museum's services were made available by its president, Henry Fairfield Osborn, noted paleontologist and avid eugenicist. Courtesy of the Laughlin Papers.

and data buttressed by analogies from agricultural breeding, Laughlin managed to provide a "scientific" rationalization in Congress for passage of a highly selective immigration restriction law. The effect of Laughlin's testimony, both on committee members and on the public (through newspaper accounts), was enormous.[59] The groups who were most restricted (Jews, Mediterraneans—particularly Italians—and people from Central Europe) were also the ones Laughlin claimed were the most biologically inferior.

With the immigration debates, the "old-style" eugenics movement hit its zenith. When the Johnson Act was passed in early 1924, Laughlin, Grant, and other eugenicists were euphoric.[60] Laughlin made good use of his position as superintendent of the ERO—not only in terms of the actual services his staff was able to render in preparing for the immigration testimony but also in terms of the prestige afforded by his title and by his association with the Carnegie Institution of Washington. Laughlin immediately aspired to even greater triumphs—advocating a Pan-American eugenics society, trying to convince the U.S. Census Bureau to use the 1930 Census to obtain eugenical data, drawing up model sterilization laws for all the forty-eight states, and presenting a plan

1919). It is ironic that one of the coformulators of the notion of hybrid vigor, E. M. East, was also one of the eugenicists who tried to argue away the analogy to human racial crossing. In the final chapter of his book with Jones, East claims that because some human races are decidedly inferior to others, hybridization between races is not of general value unless the two races are equivalent in genetic endowment. East's argument is somewhat more complex because he admits that some hybridization can on occasion be a stimulus to further variability and thus to favorable new combinations of traits (see pp. 244 ff.).

[59] Hassencahl, "Harry H. Laughlin" (cit. n. 26), pp. 282–283.

[60] Ludmerer, *Genetics and American Society* (cit. n. 53), p. 106.

to have American consulates in foreign countries perform eugenical tests on prospective immigrants before they left their native countries. None of these plans bore fruit. The eugenics movement began to take a new turn, losing some of the groundswell of support it had previously enjoyed from biologists, the wealthy elite, and the general public. The fifteen-year period from 1925 to 1940 saw the decline of old-style eugenics, and that change is reflected in the fortunes of the Eugenics Record Office.

THE DECLINE OF OLD-STYLE EUGENICS

A number of factors contributed to the decline of old-style eugenics between 1925 and 1940: increasing criticism from geneticists and anthropologists; public rancor at the blatantly racist and anti-Semitic statements of Laughlin and Grant, who persisted in making political campaigns out of eugenics; the rise of Nazi race-hygiene, with its explicitly American—and most notably Cold Spring Harbor—connection; and, finally, changing social and economic forces that made eugenics of the ERO variety less useful to the wealthy elites that had previously funded it.

An important though much-belated force in undermining the continued efforts of old-style eugenics was the increasing loss of support for eugenical research among practicing geneticists. Among those who publicly or privately attacked the claims of eugenicists after 1915 were T. H. Morgan, Herbert Spencer Jennings, Raymond Pearl, H. J. Muller, and Sewall Wright. By 1925, even W. E. Castle began to question claims for eugenics. Academic geneticists began to come to the fore over the exaggerated claims about genetic differences between races and ethnic groups that emerged as a result of the immigration debates. Because of the publicity surrounding the debates on and final passage of the Johnson Act (1921–1924), biological arguments became prominent in the public media, and some biologists felt compelled to speak up. Moreover, to many geneticists, the arguments of eugenicists—particularly those of Grant and Laughlin—were totally out of touch with advances in the field of genetics.[61]

These criticisms found expression within the institution of the ERO itself. Partly in response to growing academic skepticism regarding eugenics, the Carnegie Institution of Washington, under its president John C. Merriam, felt called upon to invite a visiting committee to examine the work being carried out at the ERO and to evaluate the office's usefulness and future potential for genetic work.[62] When the first visiting committee convened on 19 February 1929 at Cold Spring Harbor, however, its members included, along with Laughlin and Davenport, men who were largely eugenics sympathizers: A. K. Kidder, the chairman and an archaeologist and associate curator of archaeology at the CIW; the psychometricians Carl C. Brigham of Princeton and Edward L. Thorndike of Columbia; and Clark Wissler, an anthropologist from the American Museum

[61] An example of such attacks is found in Raymond Pearl, "The Biology of Superiority," *American Mercury*, 1927, *2*:257–266. Pearl's views are discussed in detail in Garland E. Allen, "Old Wine in New Bottles: From Eugenics to Population Control in the Work of Raymond Pearl," In *Eugenics: Comparative Studies*, ed. Lyndsay A. Farrall (Dordrecht: Reidel, forthcoming). A shorter version is already published as "From Eugenics to Population Control: The Work of Raymond Pearl," *Science for the People*, July 1980, *4*:22–28.

[62] Hassencahl, "Harry H. Laughlin" (cit. n. 26), p. 330.

of Natural History; only Leslie C. Dunn, a mammalian geneticist from Columbia, was seriously skeptical of eugenics. After a day-long meeting and personal inspection of the index cards and the fieldworkers' folders, the committee drew up its report. Despite the basically sympathetic nature of the committee, there were three major points of criticism. Not enough effort had been put into developing quantitative and precise techniques for assessing individual traits. As a result, the majority of the scientific records depended on the subjective assessments of the individual fieldworkers and were thus of practically no use to other investigators. Finally, the usefulness of the accumulated material would have to be tested on a few selected problems in human genetics to see what basis it formed for actual scientific study.[63]

Matters rested here for some five years. In the meantime, Laughlin continued to proselytize for immigration restriction,[64] spoke repeatedly to nativist groups like the Daughters of the American Revolution, and published strongly adulatory articles about Nazi race-hygiene in the *Eugenical News*. Although Davenport had on more than one occasion warned Laughlin to be more cautious in his public statements and to involve himself less in politically inflammatory issues, the latter seems not to have taken the advice.[65] Probably as a result of this fact, and because of Davenport's retirement early in 1934, Merriam convened another visiting committee in 1935. The new committee's composition was significantly different from that of its earlier counterpart. Absent were all the strong pro-eugenicists (Davenport, Laughlin, Brigham, and Thorndike); the only continuing member was L. C. Dunn, whose anti-eugenics stance was well known by this time.[66] Among the others were Adolph H. Schulz of the Johns Hopkins Medical School, Hobart Redfield of the University of Chicago, and Ernest A. Hooten, an anthropologist at Harvard. (The latter did maintain a somewhat pro-eugenics stance but did not agree with the simplistic Mendelian formulations of Davenport and Laughlin.)

The new visiting committee met on 16–17 June 1935 at Cold Spring Harbor. To try to smooth over the situation, Mrs. Laughlin, for whom cooking was not a delight, prepared meals and offered generous hospitality. The committee did not mince words, however. They found the ERO's total collection of records "unsatisfactory for the study of human genetics" and concluded that the indexing of nearly a million cards, covering over 35,000 case histories, had consumed more time, energy, space, and money than was justified:

> The records, upon which so much effort and money has been expended, have to date been extremely little used, to judge by the number of publications based upon them. Thus the Office appears to be accumulating large amounts of material, and devoting a disproportionately great amount of time and money to a futile system for indexing it, without certainty, or even good probability, that it will ever be of value.[67]

[63] *Ibid.*, p. 331.

[64] See Laughlin, *Report of the Special Commission* (cit. n. 55).

[65] Davenport clearly, though gently, rebuked Laughlin in 1928 after the latter made public statements about the "menace" of Mexican immigration just when CIW president John C. Merriam was in Mexico staying with the U.S. ambassador. See Davenport to Laughlin, 16 Apr. 1928, Laughlin Papers, NMSU.

[66] See Dunn to Merriam, 3 July 1935, Laughlin Papers, NMSU.

[67] "Report of the Advisory Committee of the Eugenics Record Office" (n.d.), pp. 2–3, Laughlin Papers, NMSU.

The committee went on to specify the problems that made the collected records worthless: some traits, such as "personality" or "character," lacked precise definition or quantitative methods of measurements; some traits, such as "sense of humor," "self respect," "loyalty," and "holding a grudge," could seldom be known to anyone outside an individual's close friends and associates, and furthermore, to get an honest recording of such traits is virtually impossible; and even more objectively measurable characteristics, such as hair form, eye color, or degree of tooth decay, become "relatively worthless items of data for genetic study" when recorded by an untrained observer. Ironically, the committee remarked that probably the most reliable records were the individual questionnaires filled out by college students, but these represented such a small fraction of the whole that they hardly constituted a usable resource. "Never again," the committee wrote, "should records be allowed to bank up to such an extent that they cannot be kept currently analyzed."[68] They suggested that all current work of the ERO be discontinued as soon as each project came to completion.

The committee then moved to address the problems of mixing eugenical research with political activity and propaganda. They recommended that the personnel of the ERO discontinue their association with *Eugenical News* and, further, that they "cease from engaging in all forms of propaganda and the urging or sponsoring of programs for social reform or race betterment such as sterilization, birth control, inculcation of race or national consciousness, restriction of immigration, etc."[69]

The entire report, but especially the last insertion, was obviously an indictment of Laughlin. As was his way, Laughlin did not take the criticism lying down. He penned what appear to be the notes for a response, but it is not clear whether the response was ever written up formally. These notes express his frustration that the report was tying his hands for future eugenics research and his continued assertion that "the study of human migration, mate selection, size of family can be pursued objectively by eugenics as a science." The scientist, he argued, is not responsible for the use made of his work. Laughlin denied that his work on immigration and sterilization was propagandistic, and he was resentful that the advisory committee had accused him of sponsoring propaganda and going beyond "scientific evidence."[70]

As if the committee report were not enough, several additional factors contributed to Laughlin's appearance as an embarrassment to the Carnegie Institution of Washington. The first was his involvement with, and enthusiasm for, the German race-hygiene movement. Throughout the 1920s *Eugenical News,* of which Laughlin was chief editor, had continued to run favorable discussions of the German race-hygiene movement, including summaries of articles appearing in the German eugenical journal *Archiv für Rassen- und Gesellschaftsbiologie*.[71]

[68] *Ibid.*, pp. 3–4.

[69] *Ibid.*, p. 6.

[70] Laughlin, "Notes" (n.d.), found in Laughlin Papers, NMSU; summarized from Hassencahl, "Harry A. Laughlin" (cit. n. 26), p. 336.

[71] See *Eugenical News* as follows: "Hitler and Race Hygiene," Mar.–Apr. 1932. *2*:60; "The German Population and Race Politics," Mar.–Apr. 1934, *2*:33; "Sterilization in Germany," *ibid.*, p. 38; "Eugenics in Germany," *ibid.*, p. 40; "The Mother of Nations," *ibid.*, p. 45; "New German Etymology for Eugenics," Sept.–Oct. 1934, *19*(5):125; "Jewish Physicians," *ibid.*, p. 126; "A Letter from Dr. Ploetz," *ibid.*, p. 129; and "The Sterilization Law in Germany, Nov.–Dec. 1934, *19*(6):137–140.

Indeed, *Eugenical News* became the major forum for bringing German eugenics to American readers. For example, in 1929 the *Eugenical News* expressed high praise for the involvement of German medicine with eugenical matters. After the Nazis came to power in 1933, the magazine lauded the German government for putting eugenics into practice. One unsigned article looked to the German sterilization law as a "model law," stating that "from a legal point of view nothing more could be desired." The writer is undoubtedly Laughlin, since the term "model law" as applied to legalized sterilization is a term he coined. Furthermore, in 1929 Laughlin had been invited by the editor to prepare a paper for the *Archiv* on legalized sterilization in America.[72] Davenport, too, had been interested in the German movement in the 1920s and had at one point instructed Laughlin to send complete sets of reprints and information about eugenical sterilization in the United States to Eugene Fischer, Director of the Kaiser Wilhelm Institut für Anthropologie, Menschliche, Erblehre and Eugenik in Berlin-Dahlem.[73] However, the *Eugenical News*'s avid support of Nazi race-hygiene became a cause for alarm and concern among some geneticists and eugenicists.[74] It became clear that Laughlin and the ERO were the main propagandists for the German eugenical cause, a factor that further contributed to his disfavor among Carnegie Institution officials.

To make matters worse for the Carnegie, in the middle of May 1936 Laughlin was notified that he was to be awarded an honorary Doctor of Medicine degree at the approaching celebration of the 550th anniversary of Heidelberg University.[75] Since the Nazis had gained control of all the German universities by 1935, university recognition at that time was equivalent to official Nazi recognition. Laughlin was invited to attend the ceremonies in Heidelberg, 27–30 June, and receive his degree. Deeply honored, he wrote back:

> I consider the conferring of this high degree upon me not only as a personal honor, but also as evidence of a common understanding of German and American scientists of the nature of eugenics as research in and the practical application of those fundamental biological and social principles which determine the racial endowments and the racial health—physical, mental, and spiritual—of future generations.[76]

Laughlin did not actually journey to Germany to receive his degree—indeed he may have been advised against it by Davenport or other officials of the CIW. He did, however, go as far as downtown Manhattan, to the German consulate,

[72] See Harry H. Laughlin, "Die Entwicklung der gesetzlichen rassenhygienischen Sterilisierung in der Vereiningten Staaten," *Archiv für Rassen- und Gesellschaftsbiologie*, 1929, *21*:253–262. Laughlin drafted a "model sterilization law" that was used as the basis for a number of state sterilization laws in the United States and for the German sterilization laws as well. Although this last point cannot be demonstrated with certainty, "Eugenical Sterilization in Germany" notes: "To one versed in the history of eugenical sterilization in America, the text of the German statute reads almost like the 'American model sterilization law' " (p. 89).

[73] Davenport to Laughlin, 21 Dec. 1920 and 6 Oct. 1927, Davenport Papers, APS.

[74] See, e.g., W. K. Gregory of the American Museum of Natural History (paleontology) to Clarence C. Campbell, 6 May 1935, and Gregory to Raymond Pearl, same date, in which Gregory states his intention to resign from the Galton Society (which by that time had taken over the publication of *Eugenical News* from the ERO) unless the *News* stopped publishing favorable articles on Nazi eugenics; Pearl Papers, APS.

[75] Carl Schneider, Dean of the Faculty of Medicine of Heidelberg University, to H. H. Laughlin, 16 May 1936, Laughlin Papers, NMSU.

[76] Laughlin to Carl Schneider, 11 Aug. 1936, Laughlin Papers, NMSU.

to receive his diploma in November 1936.[77] Announcements of the award in various newspapers brought some added criticism—not, however, from mainstream, old-style eugenicists such as Grant or C. C. Little.

A final matter that sealed Laughlin's fate was related to what appeared to be his deteriorating health. Rumor had it that he suffered increasingly from epileptic attacks, some of which had occurred in public places. On one occasion in 1937 he had a seizure while driving down the main street of Cold Spring Harbor; his car was prevented from plunging directly into the water only by its crashing into a retainer wall. There is an irony in the fact that epilepsy was one of the traits that Laughlin and other eugenicists had wanted to purify out of the population; now he and his career became the victims of that neurological disorder.

In October 1937 a letter from the chairman of the board of the Carnegie Institution, George Streeter, to its president, John C. Merriam, first called the latter's attention to reports about Laughlin's condition. From there matters moved quickly. Sensing a chance to avoid yet another embarrassing situation, as well as to change vastly the direction of the ERO, Merriam, under direction from Streeter and the board, asked Laughlin to have a complete medical checkup and submit the results to the CIW. Laughlin obtained a letter from his own doctor that vaguely indicated his health to be satisfactory. Streeter wanted a fuller report from a doctor appointed by the CIW, but such documentation appears to have been unnecessary. Merriam retired as CIW president at the end of 1938, and his successor, Vannevar Bush, wasted no time in bluntly asking Laughlin to retire. Negotiations strung out for a while, but a settlement was reached and Laughlin left Cold Spring Harbor in January 1940. He retired to Kirksville, Missouri, where he died in January 1943, at the age of sixty-three.[78]

On 31 December 1939 the Eugenics Record Office closed its doors for the last time. The old style eugenics movement had lost its major institutional base. New-style eugenicists, led by Laughlin's longtime friend Frederick Osborn, began the transformation of the movement, under Rockefeller auspices, into a series of international population control experiments. Although he was skeptical of the birth control movement during the 1920s and 1930s, Laughlin might well have been pleased by the direction in which Osborn took eugenical concepts after World War II. Population control became the international version of Laughlin's eugenical sterilization principle—this time on a much vaster scale.

THE ERO IN AMERICAN SOCIETY

What conclusions can we draw regarding the founding and development of the Eugenics Record Office? How does it relate to the development of the American eugenics movement as a whole, and to broader currents of American social history in the late nineteenth and early twentieth centuries? Was the ERO the product of the whims of a few zealous biologists and utopian philanthropists? Or was it, as I will argue, the product of a whole complex of social forces converging during the period 1890–1920? I would like to suggest that the founding of the ERO represents several trends in American social history that were set

[77] H. Borchers to H. H. Laughlin, 25 Nov. 1936, Laughlin Papers, NMSU.

[78] Streeter to Merriam, 26 Oct. 1937; Merriam to Laughlin, 31 Dec. 1938; Bush to Laughlin, 4 Jan. 1938; CIW Archives.

in motion by the rapid growth of industrial capitalism in the latter part of the nineteenth century. It was founded and developed by those with the means—both the financial capital, on the one hand, and the scientific knowledge, on the other—to set about combating what they perceived to be a disintegration of the fabric of modern society. Eugenicists and their supporters attributed this disintegration to lack of social planning and to inefficient management of the human germ plasm. Eugenics was but one of many attempts to apply the concepts of scientific management and rational planning to a society that was experiencing severe growth pains as it developed from a rural agrarian to an urban industrial economy.

Mirror of the Progressive Era

The ERO brought together a number of strands of American social history during the late nineteenth and early twentieth centuries. Among these were the spirit of reform that grew out of evangelical notions of social responsibility, as exemplified by Andrew Carnegie's own "gospel of wealth"; the broad economic changes that accompanied widespread industrialization, particularly the shift from laissez-faire to planned capitalism; the cult of efficiency as developed in the industrial, and later the social, spheres; the rise of a new sector of the middle class—the professional managers—including scientific experts, engineers, social workers, sociologists, and foundation directors; and, lastly, concern with the problem of social control, of using scientific (which meant also sociological) knowledge to achieve greater social and economic stability. While it is impossible to discuss all of these aspects of eugenics history in detail, I would like to emphasize here the way in which the ERO mirrored the notions of reform, scientific planning, efficiency, and social control that were harbored by its wealthy benefactors and their middle-class professional advisers.

The last decades of the nineteenth and first decades of the twentieth century in the United States were times of considerable upheaval in economic, class, and demographic patterns. Robert Wiebe has characterized the period as one of a "search for order," that is, for stability, integration, and social control in a society that was changing and fragmenting in a variety of directions. The various responses to this fragmentation have been frequently collected under the term *progressivism,* and the period is commonly called the "Progressive Era." Many historians, however, have found difficulty with the designation "progressive," not only for its denotative meaning (as the opposite of conservative) but also for its suggestion that there was a unified or focused social movement at that time. Despite these reservations about the term, distinct concerns and points of view did come to the fore during those years. James Weinstein has emphasized the important shift that wealthy elites made from laissez-faire to managed or planned capitalism during the period 1890–1920.[79] Not only had industrialization created new economic problems of its own, but it had also magnified many of

[79] See Robert Wiebe, *The Search for Order* (New York: Hill & Wang, 1967), one of the best overall summaries of the period; and James Weinstein, *The Corporate Ideal in the Liberal State, 1900–1918* (Boston: Beacon Press, 1968), a close second. Others with the same basic message, though focused on different aspects of the period or with different interpretations, include Paul A. Carter, *The Spiritual Crisis of the Gilded Age* (DeKalb: Northern Illinois Univ. Press, 1971); and Graham Adams, *The Age of Industrial Violence* (New York: Columbia Univ. Press, 1966).

the problems already encountered in earlier stages of the development of American capitalism: vast price fluctuations, inflation, small business failures (accompanied by the growth of monopolies), and depressions (there had been several depressions between 1870 and 1900, particularly serious ones in 1873 and 1893). In response to these recurrent problems, radical trade union organizing, from outside as well as from within the movement, had increased dramatically from the 1880s on. The Haymarket Affair in Chicago in 1886, the Homestead Steel Strike of 1892, and the Pullman Strike of 1894 had already terrorized the upper classes before the turn of the century. After 1900 militancy continued with the San Francisco General Strike in 1900, the Lawrence, Massachusetts, textile strike of 1912, the Ludlow Massacre in Colorado in 1915, strikes in Patterson and Bayonne, New Jersey, also in 1915, and the Seattle General Strike in 1919, to name only a few of the more prominent outbreaks in the United States alone. Labor agitation and the instability it brought became issues of major concern among the wealthy elites. Various sectors of the wealthy business class had gradually been won by these realities to the concept of a more planned economy—and that meant using scientific and technical knowledge where possible to organize and regulate economic activities. This was the era that saw passage of the Sherman Antitrust Act (1890) and the beginnings of federal regulatory activities: the creation of the Interstate Commerce Commission in 1887, further strengthened in 1906; the enactment of the Pure Food and Drug Act in 1906; and the creation of the Federal Trade Commission in 1914.

A cornerstone of the era of regulation was the notion of efficiency, an ideology that spilled over into a number of areas, fron conservation to urban politics, criminology, education, and medicine. In the marketplace as well as the workplace, a concern, almost a fanaticism, about efficient use of time, space, and energy engulfed American business leaders. Efficiency experts such as the engineer Frederick Winslow Taylor (1856–1915) created a science of efficient business operation. Efficiency came to mean analysis of the input and output of factory machinery and human activities alike, the breaking down of complex tasks into discrete components under the control of planners, and, finally, accentuation of the process of division of labor, including such innovations as the assembly line. Applied to the social sphere, efficiency meant correcting problems at their source, not in the aftermath of damage already done. Prevention became a central organizing concept within the efficiency movement. Efficiency also involved knowledge and use of scientific principles, and it became a commonplace to talk of certain kinds of reform as the scientific solution to social problems.

The application of rational planning in general, and of concepts of efficiency in particular, required the active participation of scientifically trained experts, professionals whose job it was to bring technical concepts and knowledge to bear on problem solving. The professional expert became an indispensable agent for the modernizing of American business in the early twentieth century.[80] Experts served two different functions: as advisers who provided technical infor-

[80] See, e.g., S. Haber, *Efficiency and Uplifts: Scientific Management in the Progressive Era, 1890–1920* (Cambridge, Mass.: Harvard Univ. Press, 1964); and Alfred D. Chandler, *The Visible Hand: The Managerial Revolution in American Business* (Cambridge, Mass.: Harvard Univ. Press, 1977); see also M. S. Larson, *The Rise of Professionalism* (Berkeley: Univ. California Press, 1967).

mation or problem-solving skills, and as managers who oversaw the day-to-day operation of a business, a government agency, a medical facility, or an educational institution. The professional expert thus became part of the new managerial class, those trained individuals whose specialized knowledge was increasingly needed in business and in society and who thus played a vital role in the shift toward planned or managed capitalism. The professional experts even found their way into American philanthropy as the managers of the new foundations (for example, Frederick Gates of the Rockefeller Foundation and R. S. Woodward of the Carnegie Institution of Washington). Although these experts were solidly middle-class professionals, they carved a niche for themselves in American society that allied them with the needs and aims of the wealthy elites who employed them.

Several writers have emphasized the importance that business leaders and their professional managers attached to the social and natural sciences for their evolving concepts of economic and social control.[81] Economic as well as social instability, especially surrounding the labor unrest of the period 1880–1920, had become cause for enormous concern among the wealthy elites. (Andrew Carnegie was even prompted to devote an essay from his *Gospel of Wealth* to the topic of labor violence and anarchy.[82]) The topic of controlled social change informed much of the advice that Columbia University economist Wesley Clair Mitchell gave to the Rockefeller Foundation in the teens, and it was the basis of his own program as director of the National Bureau of Economic Research (founded 1919–1920).[83] Social control meant more than merely not upsetting the apple cart or than not altering the *status quo*. Conceptually, it was also linked with the notion of efficiency, equating control and order with more effective production. Social control implied integration of the fragments of society into a coordinated, interacting whole. That the wealthy elites and their professional advisers also saw social control as a way of preserving their own values and hegemony is not a negligible factor. However, it is most important to see the interrelationships between the idea of social control and the ideology of efficient planning in order to understand the role that eugenics was expected to play in the new social order.

In the eyes of many, the necessity of social control was underscored by the threat, both real and imagined, of labor unrest caused by foreign (alien) radicals. With the changing patterns of immigration (increased numbers from eastern Europe, the Mediterranean countries, and the Balkans) and the concentration of new immigrants in slum areas of the larger cities, the fear of alien radicalism

[81] See, e.g., Herman Schwendinger and Julia R. Schwendinger, *The Sociologists of the Chair: A Radical Analysis of the Formative Years of North American Sociology, 1883–1922* (New York: Basic Books, 1974); and Clarence Karier, "Testing for Order and Control in the Corporate Liberal State," *Educational Theory,* 1972, *22*:154–180.

[82] Andrew Carnegie, *The Gospel of Wealth* (Cambridge, Mass.: Harvard Univ. Press, 1962); Essay VI, "Results of the Labor Struggle," was originally published in the *Forum* in Aug. 1886 and is in large part a strong attack on labor violence and agitation as a means of achieving social change.

[83] "Reform by agitation or class struggle is a jerky way of moving forward. Are we not intelligent enough to devise a steadier and more certain method of progress?" W. C. Mitchell, *The Backward Art of Spending Money* (New York, 1950), p. 5. Mitchell was recommended to the Rockefeller as the most capable man to found a social science research facility; see David M. Grossman, "Professors and Public Service, 1885–1925: A Chapter in the Professionalization of the Social Sciences" (Ph.D. diss., Washington Univ., 1973), esp. Ch. 7, "Organizing Research."

bordered at times on hysteria.[84] Immigrants as a group were accused of spawning every sort of social ill, from criminality to rebelliousness, alcoholism, prostitution, socialism, Bolshevism, and trade unionism.[85] They became a national scapegoat for many of the economic and social problems experienced by urban industrialized society. Nativist feeling was brought into the movement for social control by portraying the radical, alien worker as innately (biologically) inferior—not only to old American stocks but also to earlier generations of immigrants—and thus incapable of adjusting to complex industrial society.

It is not difficult to see how the organization and ideology of the ERO in many ways fit most, if not all, of the general characteristics of the Progressive Era.[86] First and foremost, eugenics was a reform movement aimed at correcting existing social problems—that is, the problem of the "defective classes." It was also a movement based on the concept of rational, scientific planning in the cause of national efficiency. Thus Davenport, writing in 1910 about the "socially defective" portion of the population, reflects the aims of eugenics in general:

> This three or four per cent of our population is a fearful drag on our civilization. Shall we as an intelligent people, proud of our control of nature in other respects, do nothing but vote more taxes or be satisfied with the great gifts and bequests that philanthropists have made for the support of the delinquent, defective classes? Shall we not rather take the steps that scientific study dictates as necessary to dry up the springs that feed the torrent of defective and degenerate protoplasm?

Davenport's booklet, a distinct call for rational social control, begins with a primer of Mendelian genetics and the explicit assertion that "were our knowledge of heredity more precisely formulated there is little doubt that many certainly unfit matings would be prevented."[87]

Davenport, Laughlin, and other eugenicists were loud in claiming to apply knowledge of biology and statistics to social problems. As Laughlin put it succinctly in 1913:

> In justice to the new science, it must be said for most of the traits *specifically* and *extensively* studied, the student of eugenics can confidently predict the nature of the offspring of two parents of known ancestry. To the degree that this prediction can be made, eugenics is justified in calling itself a science, for here, as in all other cases, predictability is the criterion of the understanding of nature.[88]

Conceiving of themselves as scientists with expert knowledge, and of the ERO as a "scientific institution," Davenport and Laughlin sought a role in social planning along with those who were emerging as social planners in industry and

[84] See Barbara Solomon, *Ancestors and Immigrants* (Chicago: Univ. Chicago Press, 1972).

[85] Charles Leinenweber, "The Class and Ethnic Bases of New York City Socialism, 1904–1915," *Labor History,* 1981, *22*:31–56.

[86] Among those noting the close ideological affinity between eugenics and the loose fabric of progressive thought are Donald Pickens, *Eugenics and the Progressives* (Nashville, Tenn.: Vanderbilt Univ. Press, 1968); and (more satisfactory, though more limited in scope) Rudolf Vecoli, "Sterilization: A Progressive Measure?" *Wisconsin Magazine of History,* 1960, *43*:190–202; and Michael Freeden, "Eugenics and Progressive Thought: A Study in Ideological Affinity," *Historical Journal,* 1979, *22*(3):645–671.

[87] C. B. Davenport, *Eugenics: The Science of Human Improvement by Better Breeding* (New York: Henry Holt, 1910), pp. 31–32, 4.

[88] Laughlin, "Eugenics Record Office" (cit. n. 36), p. 1.

government. It was time to get the layperson, the amateur, out of social planning. They would have agreed strongly with eugenicist Raymond Pearl, who wrote in 1912 (long before his later denunciation of old-style eugenics): "Hitherto everybody except the scientist has had a chance at directing the course of human evolution. In the eugenics movement an earnest attempt is being made to show that science is the only safe guide in respect to the most fundamental of social problems."[89]

The ERO was involved in scientific management, or planning, at two levels: planning of actual marriages, for which the thousands of index cards and other genealogical information stored at the ERO could be used; and social legislation and education through large-scale programs aimed at reducing the number of "defectives" actually brought into the society. The latter was by far the more important aspect of the ERO's work, though Laughlin never tired of emphasizing its marriage-counseling role as well.

The efficiency of eugenical planning was a central feature of the ERO's *raison d'être*. Laughlin emphasized this point in one of its periodicals in 1914:

> Eugenics, which Davenport defines as "the improvement of the human race by better breeding," is one of these agencies of social betterment, which in its practical application would greatly promote human welfare, but which if neglected would cause racial, and consequently social, degeneration. Eugenics, then, is the warp in the fabric of national efficiency and perpetuity. As an art it is as old as mankind; as a science it is just now taking definite form.[90]

Science applied to social degeneration, as previously applied to medicine, dictated seeking the causes of a malady, not merely treating the symptoms. The older solutions for social misfits constituted charity, and Davenport, in particular, was absolutely clear on the misspent efforts that charity represented. Science dictated drying up the springs that "feed the torrent of defective and degenerate protoplasm." A curious indication of the prevalence of the cult of efficiency in ERO circles is a newspaper clipping found in Laughlin's papers in Kirksville, Missouri. The clipping comes from the *Birmingham Mail* (Birmingham, England) and is dated 1930. The column-wide headline, in three parts, reads:

RATIONALISING MANKIND

Big Business Methods
in Evolution

Eugenic Reform

The article reports on a British eugenics meeting devoted to applying the methods of business to human evolution. The folder containing the clipping was labeled in Laughlin's own hand. There are no notes or comments on the substance of the article, but it would be difficult to imagine that he disapproved of

[89] Raymond Pearl, "First International Eugenics Congress," *Science,* 1912, *36*:395–396.

[90] H. H. Laughlin, "Report of the Committee to Study and to Report on the Best Practical Means of Cutting Off the Defective Germ-Plasm in the American Population," *Bull. Eugen. Rec. Off.*, Feb. 1914, No. 10, p. 100.

the comparison. Significantly, the clipping demonstrates that at least some eugenicists at the time appear to have seen a clear relationship between their own work and the broader currents of Progressive solutions to economic and social problems as a whole.

Many of these very problems converged on the issue of immigration, particularly on the nature of the newer immigrants and the problems they faced in adapting to American life. For eugenicists, the source of the problem, as we have seen from Laughlin's testimony before the House Committee on Immigration and Naturalization, was the alleged genetic inferiority of the new immigrants from Southern and Eastern Europe, notably those of Jewish extraction. Immigrants became criminals, alcoholics, or radicals because they were biologically incapable of competing with the older, established Anglo-Saxon and Nordic stock. They took to antisocial ways as a means of dealing with their frustrations and incapabilities. In his own way, Laughlin made his feelings on this point quite clear. In *Europe as an Emigrant-Exporting Continent and the United States as an Immigrant-Receiving Nation* (1924), Laughlin praised those American consuls abroad who weeded out as many of the unfit as possible from those who applied for visas. Among those "unfit" that he actually mentioned were "white slavers, anarchists, or Bolsheviks."[91] Restriction of immigration became the efficient solution to stop propagation of defective immigrants. The immigration problem was not merely a faddish cause tacked on to eugenical programs; its presence as a strong focus of attention and research by the ERO reflects the importance that the problem assumed in "the search for order" that pervaded late nineteenth- and early twentieth-century American society.

Philanthropy for Social Control

The concerns of Progressives and eugenicists were the same concerns to which the wealthy and the coterie of managers and advisers who clustered around them were also responding. Both Mrs. Harriman and the Carnegie philanthropies were concerned with what the former called "efficient giving." This meant, among other things, using philanthropic dollars to get at the cause of a problem and treat it at its source. It also meant, somewhat more indirectly, using philanthropic dollars to bring organization and order into research itself through emphasis on cooperative efforts, creation of research institutions, and the like. Foundations in particular (as opposed to individual donors) played an extremely important role during the Progressive Era in translating the concerns of the wealthy elites into concrete, scientifically grounded research projects, or into actual social planning. Wesley Clair Mitchell, a Columbia University economist and an adviser to the Rockefeller Foundation, explicitly related the use of science to achieving social control in a memo to the foundation trustees in 1914: "Just as science affords the chief means of improving the practice of medicine, so science affords the chief means of improving the practice of social regulation."[92]

[91] H. H. Laughlin, *Europe as an Emigrant-Exporting Continent and the United States as an Emigrant-Receiving Nation,* Report of the U.S. Congress House Committee on Immigration and Naturalization (Washington, D.C.: U.S. Government Printing Office, 1924), p. 1235.

[92] W. C. Mitchell to Trustees of the Rockefeller Foundation, Jan. 1914, Rockefeller Foundation Archives, Draft Report 22. A useful study of the ideological and organizational debates within the

Table 1. ERO funding in the "Harriman period," 1910–1917

Donation of original 74.85 acres (next to the SEE at Cold Spring Harbor) and buildings	$ 80,680
Construction of new office building (1915)	15,000
Contribution for annual operating expenses (Mrs. E. H. Harriman) (total for 7-year year period)	246,000
Additional endowment at time of transfer to CIW (Mrs. Harriman)	300,000
Total	$641,680

SOURCE: Records in the Harry H. Laughlin Papers, Northeast Missouri State University (NMSU), Kirksville, Mo., and in the Charles B. Davenport Papers, American Philosophical Society (APS), Philadelphia.

Efficiency, rational planning, and expert knowledge thus were all concerns common to both the professional, middle-class eugenicists (such as Davenport and Laughlin) and to the wealthy elites and their advisers. It was a convenient marriage of like minds and concerns. It is not surprising, then, that the wealthy contributed significantly to the funding base and continuing support of the ERO. From the fairly complete financial records of the ERO—available both in the Laughlin Papers in Kirksville, Missouri, and in the Cold Spring Harbor Series of the Davenport Papers at the American Philosophical Society—it is possible to reconstruct something of the endowment as well as the annual budget (operating expenses) of the ERO for its entire history. The records show that there were two distinct funding periods: first, the Harriman period (1910–1917), which provided endowment, seven years of operating expenses, and property—in short, the necessary start (see Table 1); and second, the Carnegie period (1918–1939), which provided twenty years of operating expenses but nothing in the way of additional endowment (see Table 2). In addition to these two major sources of funding, other benefactors, large and small, provided sums for specific new projects or, occasionally, to rescue an ongoing project from disaster (see Table 3).

Endowment for the ERO came exclusively from the Harriman family. There was the bequest of the original property and houses at Cold Spring Harbor, valued at $80,680 at the time of purchase; the new office built in 1915 and valued at $15,000; and the additional $300,000 donation given when the ERO was transferred to the Carnegie Institution in 1917. The total endowment, including properties, thus amounted to approximately $395,000. Operating expenses from both periods (from 1910 through 1939) amounted to $720,000 ($246,000 plus $474,000). Add to this the major individual gifts, and the operating budget for a thirty-year period tops $820,000. Combined operating funds and endowment thus gave the ERO an overall financial base of $1,217,308 for the period 1910–1939. The Harriman funding was crucial to getting the ERO on its feet, while the Carnegie support enabled the institution to consolidate its efforts and pursue a long-range course of planning and development.

A significant problem in trying to understand the sources and motives of

Rockefeller Foundation during this period can be found in Grossman, "Professors and Public Service" (cit. n. 83).

Table 2. Expenditure for eugenics by the Carnegie Institution of Washington, 1918–1935

Year	Total CIW budget for Dept. of Genetics (SEE and ERO) ($)	Sum from total budget expended for ERO[a] ($)	% of total budget expended for ERO
1918	84,790.00	25,000.00	29.5
1919	95,910.00	26,836.00	28.0
1920	109,129.00	30,785.76	28.2
1921	125,974.03		
1922	125,205.00		
1923	121,290.00		
1924	124,055.00		
1925	129,125.00		
1926	125,960.00		
1927	131,510.00	283,429.36	15.0
1928	133,727.67		
1929	139,380.00		
1930	143,666.67		
1931	146,384.99		
1932	158,460.00		
1933	143,550.00		
1934	141,242.00		
1935	135,780.00	22,903.15	16.8
1936	144,135.00	25,256.78	17.5
1937	138,980.01	20,180.00	14.5
1938	145,745.00	20,943.37	14.4
1939	143,220.00	18,680.00	13.0
Total	2,887,219.37	474,014.69	16.4

SOURCE: Harry H. Laughlin, "Notes on the History of the Eugenics Record Office, Cold Spring Harbor, Long Island, New York," mimeographed report compiled from official records of the ERO, Dec. 1939, p. 5; from Laughlin Papers, NMSU.

[a] Figures for 1918–1920 are exact; total for 1921–1934 is an estimate, equal to 15% of the total Department of Genetics budget; figures for 1935–1939 are exact, from the files of the chief clerk of the ERO.

funding for the ERO is that of finding a meaningful basis on which the scale of funding can be judged. A given donation—or even the collective budget of the ERO—may sound like a trifling amount or a fortune, depending on one's basis for comparison during the same period. How, then, do we determine what are useful and valid comparisons? For example, should donations to eugenics be compared to donations to genetics as a whole? Or to physics? Or to the American Museum of Natural History? Or to the Rockefeller Institute for Medical Research? Is it valid to compare grants to an institution such as the ERO with aggregate individual grants for all genetics research? Or is it valid only to compare one institution with another? There are no easy answers to these methodological questions. I have therefore approached the comparative issue ex-

Table 3. Additional contributions to eugenics work at the ERO, 1910–1930

Rockefeller family contributions	
John D. Rockefeller, Jr., for training program for ERO fieldworkers (1910–1917)	$ 21,650
Individual gifts	
Walter J. Salmon, for study of inheritance of racing capacity (Lexington, Kentucky)	75,726
Bleeker van Wagenen, to underwrite publications such as *Eugenical News*	1,737
Mrs. Lucy W. James, for a field lecturer on eugenics	2,500
Total	$101,613
Total funds for operations from Tables 1, 2, and 3	$821,628[a]

SOURCE: Records in the Laughlin Papers, NMSU, and the Davenport Papers, APS.
[a] Figure from Table 2 rounded.

peditiously, largely by making only the comparisons for which some data are reasonably available. However, if several such comparisons are made, some idea about the scale of funding for eugenics in its time period emerges.

First, within the Carnegie's biological funding alone, Table 2 shows that the ERO's operating budget for the period 1918–1939 averaged approximately 16 percent of that for the Station for Experimental Evolution. Note also that over the years the annual percentage tended to drop, starting from a value of 29 percent in 1918 and dropping to 13 percent by 1939. It is clear from the start, then, that the ERO's budget rarely exceeded 25 percent of the budget for genetics and evolution as a whole at Cold Spring Harbor. Nonetheless, over the years the Carnegie did pump over half a million dollars into the ERO for direct operating expenses.

For comparisons outside of biology, some figures from Daniel J. Kevles's *The Physicists* are helpful. In 1906 two Princeton alumni gave, as initial expenses, $200,000 to equip a laboratory for the study of electron emission; ten years later the total annual operating expenses for the Princeton physics department came to only $1,600. In 1916 the entire operating budget for the Ogden School of Science at the University of Chicago was $14,531, while for all the Carnegie years the ERO's operating budget averaged approximately $20,000 annually. By contrast, a government agency such as the National Bureau of Standards had a budget of $350,000 in 1910 and $700,000 in 1915. In 1919, when the California Institute of Technology was founded, George Ellery Hale convinced one philanthropist to pledge $4 million toward endowment; at the same time Robert A. Millikan had operating expenses of approximately $100,000 per year for the Division of Physics. Finally, the endowment of the Carnegie Institution of Washington itself amounted to $10 million at its founding in 1902, a sum equal to the endowment of Harvard and greater than the endowment for research in all other American universities combined.[93] What these comparative figures suggest is

[93] Daniel J. Kevles, *The Physicists* (New York: Knopf, 1978). For some more generalized trends and figures, see Spencer Weart, "The Physics Business in America, 1919–1940: A Statistical Reconnaissance," in *The Sciences in the American Context: New Perspectives,* ed. Nathan Reingold (Washington, D.C.: Smithsonian Institution Press, 1979), pp. 295–358.

that, on the scale of research money available in the early twentieth century, the ERO fell in roughly the median range for total operating expenses and in the lower bracket for endowment. Leaving out the funding for Caltech, the ERO fared better than average at the time.

Thus, it would seem that the funding afforded to eugenics during the years 1910–1940 was not insignificant. It was also not the product of a fringe of eccentric philanthropists. Indeed, eugenics received its funds from mainstream philanthropic sources of the day. This evidence implies that, at least for a period, the ERO enjoyed considerable respect and offered significant hopes among the wealthy, philanthropic elites for improved scientific management and control of social policy.

CONCLUSION

In an age that saw the decline of laissez-faire economic, political, and social philosophy and the concomitant rise of theories of scientific planning and social control, eugenics emerged as an efficient panacea for a variety of social ills. While eugenic solutions were not so prominent as to be a cornerstone of Progressive thinking, they were very much in the mainstream of the Progressive Era. Eugenics was the scientific management of human evolution, and as such it brought human society and culture into line with biological realities hitherto ignored or too easily dismissed. Eugenics was the perfect biological theory of society in an era that was rapidly accepting notions of scientific management and control (1900–1930), just as social Darwinism had been the appropriate theory for an earlier generation committed to classical economic and social policies of laissez-faire (1870–1900). I do not suggest that as economic and social theories changed biological theories changed in direct and conscious response. However, it would be folly to ignore such a patent parallel shift in views of both economics and society on the one hand and biological models of society on the other. Although it is difficult to demonstrate directly, I suggest that the primary and driving force for the initial founding and continued support of the ERO came from the new economic environment of planned capitalism, designed to insure more effective economic and social control. Funding from such major philanthropic sources as the Harriman family and the Rockefeller and Carnegie foundations formed a part of the new concern of the business community with planning for order and stability. The funding of research in the medical, biological, and social sciences was part of the new ("progressive") view of approaching problems rationally and seeking long-range solutions. The Harriman, Carnegie, and Rockefeller philanthropies had their own styles, advisers, and agendas. But they were all generated by large capitalist enterprises in the latter part of the nineteenth century and were subject to the same problems of economic and social stability—the same perceived fragmenting of society. The attempt to bring order into various spheres of economic and social life, including scientific research and the planning of human evolution itself, was a common response to a common set of problems.

"It May Be Truth, But It Is Not Evidence": Paul du Chaillu and the Legitimation of Evidence in the Field Sciences

By Stuart McCook

IN THE SPRING OF 1861, gorillas had captivated the public imagination in England. The recent publication of Darwin's *On the Origin of Species* had fueled the ongoing debate about the place of humans in nature. A key piece of evidence in the debate was the recently discovered gorilla. In the pages of learned journals and in the more popular *Athenaeum,* Richard Owen and Thomas Henry Huxley debated the significance of gorilla skeletons recently brought to England. They based their arguments about the uniqueness of man on the physiology of the gorilla, which was anatomically closer to man than was any other animal. Owen argued that the human brain had a lobe, the hippocampus minor, that distinguished it from all other living organisms. Huxley countered by arguing that the hippocampus minor was not peculiar to the human brain. The gorilla skulls over which Huxley and Owen debated had been brought to England with great fanfare by a little-known explorer named Paul du Chaillu (see Fig. 1).[1]

From 1856 to 1859, Du Chaillu had traveled through western equatorial Africa, studying the lands and people and collecting flora and fauna. He brought a choice selection of his collections to England early in 1861, where he was quickly made a celebrity. He described his adventures to packed halls at the Royal Geographical Society and at the Royal Institution in London, with a row of stuffed gorillas on the stage and gorilla skulls on hand beside the lectern.[2] The evening after du Chaillu

This article is based on my master's thesis, "The Monkey Business: Paul Du Chaillu and Nonscientist Roles Within the Scientific Community" (Rensselaer Polytechnic Institute, 1990). Prof. Sydney Ross generously provided me with the manuscript that introduced me to the world of Paul du Chaillu. I would also like to thank P. Thomas Carroll, Deborah J. Coon, Gerald L. Geison, Robert E. Kohler, Henrika Kuklick, Sal Restivo, two anonymous reviewers, and my classmates at Rensselaer Polytechnic Institute and Princeton University for their comments and guidance.

[1] For two perspectives of the scientific world of Victorian London, see Adrian Desmond, *Archetypes and Ancestors: Paleontology in Victorian London, 1850–1875* (Chicago: University of Chicago Press, 1982), particularly pp. 74–76; and Nicholaas Rupke, *Richard Owen: Victorian Naturalist* (New Haven: Yale University Press, 1994). Owen and Huxley refer to du Chaillu's gorilla skeletons in, among other places, "The Gorilla and the Negro," *Athenaeum,* 23 March 1861, no. 1743:395–396; and T. H. Huxley, "Man and the Apes," *Athenaeum* 13 April 1861, no. 1746:498.

[2] Lynn Barber, *The Heyday of Natural History, 1820–1870* (New York: Doubleday & Co., 1980), p. 276.

Figure 1. *"Paul du Chaillu in his african costume," from* Stories of the Gorilla Country *(New York: Harper and Brothers, 1897), xii.*

lectured at the Royal Institution, Owen presented a paper there on the distinction between the Negro and the gorilla, based on skulls given to him by du Chaillu.

Du Chaillu's honeymoon with the British natural history community was short-lived. *Explorations and Adventures in Equatorial Africa,* du Chaillu's narrative of his travels in Africa, was published in May 1861. Inconsistencies and ambiguities in the book provoked a politely fierce debate that temporarily eclipsed the debate over humanity's place in nature. The argument over du Chaillu's reliability was one of the few instances when scientists in Victorian England publicly discussed issues of what constituted good scientific practice and who made a good scientist, rather than what constituted good scientific theory.

Although Paul du Chaillu is deservedly not considered one of the great figures in Victorian science, his case brings to light some issues of interest to historians of science. In recent years historians of science have begun to look at people who were essential to the production of scientific knowledge, but who remained hidden from public view. Steven Shapin has written about the "invisible technicians" who worked on Robert Boyle's experiments.[3] The Victorian era had its own share of invisible technicians, most of whom normally remained out of the public eye. Invisibility was somewhat harder to establish in the field sciences, however, since the field collectors could not easily be written out of the story of how natural history specimens were obtained. Still, the scientific communities in England and the United States developed strategies of accommodating field collectors that ensured that they remained relatively invisible. The controversy over du Chaillu threw a spotlight onto a process that normally remained behind the scenes.

Historians of science have also recently begun to look at the production of knowledge in the field sciences and to explore how it differs from knowledge production in the laboratory. The controversy over du Chaillu illustrates the fragility of the process of legitimation of scientific knowledge in natural history during the Victorian era. Field observations in natural history depended very heavily on the perceived credibility of the person who made them. Observations without clear provenance had greatly diminished value. As with art objects, the provenance of natural history objects depended partly on concrete records and partly on the credibility of the person who offered it. Credibility was, in turn, strongly tied to the social standing of the observer. Du Chaillu's class, educational background, and race quickly became key issues in the debate over the scientific worth of the *Explorations.* The strategies that du Chaillu's detractors used to tear him down and those that his supporters used to defend him give a detailed picture of the process by which objects collected in the field became scientific evidence.

THE MAKING OF A FIELD EXPERT, 1831–1861

Paul du Chaillu's voyage to equatorial Africa took place during the heyday of scientific exploration, which began with the voyages of Cook in the late eighteenth century. Scientists, bureaucrats, and businessmen at the European and American centers of power sought information about the unknown regions of Asia, Africa, and America. This knowledge was part of the establishment of European political, economic, and intellectual hegemony over the region.[4] Metropolitan governments, museums, botanical gardens, zoos, and scientific societies eagerly sought natural history objects and observations from travelers to these unknown regions.

Du Chaillu was probably born around 1831. His father, Charles-Alexis Duchaillu, was a French trader who had worked on the Île Bourbon (Réunion) before moving to the Gabon in the mid-1840s.[5] Little is known about his mother, who certain

[3] Steven Shapin, "The Invisible Technician," *American Scientist,* 1989, *77:*554–563.

[4] See Michael Adas, *Machines as the Measure of Men: Science, Technology, and Ideologies of Western Dominance* (Ithaca, N.Y.: Cornell Univ. Press, 1989), 133–165; William H. Goetzmann, *New Lands, New Men: America and the Second Great Age of Discovery* (New York: Penguin, 1987).

[5] The best historical reconstruction and analysis of du Chaillu's life appears in H. H. Bucher's "Canonization by Repetition: Paul du Chaillu in Historiography," *Revue Française d'Histoire d'Outre-Mer,* 1979, *66:*15–32. Through painstaking research, Bucher reconstructs du Chaillu's early years, which du Chaillu deliberately obscured. Michel Vaucaire's *Paul du Chaillu: Gorilla Hunter. Being the Life and Extraordinary Adventures of Paul du Chaillu* (New York: Harper & Bros., 1930) is an uncritically celebratory biography of du Chaillu. Unfortunately, it is vague about du Chaillu's life before he came to America. Other reliable biographical fragments are Helen Everston Smith,

historians and contemporaries of du Chaillu have suspected was black or of mixed race. Others, however, conclude that she was Italian, apparently because of the Italian ring of "Belloni," du Chaillu's middle name. Du Chaillu apparently spent some of his teenage years in Paris and was almost certainly there during the February Revolution of 1848. The seventeen-year-old Paul first arrived in the Gabon late in 1848, where he enrolled at the American Protestant mission school.[6] During du Chaillu's formative years, both Europeans and Americans had extensive commercial interests in Africa. By the mid-1850s a well-established trading network had been formed between Europeans and Africans along the African coast, reaching far inland. In the late 1840s and early 1850s du Chaillu made several trips to the interior on behalf of his father's company. On these voyages, he learned about the culture, flora, and fauna of equatorial Africa.

Du Chaillu had become fascinated with the United States during his time at the American mission school. In 1852 the chief missionary, John Wilson, arranged for du Chaillu to teach French at a girl's school in Carmel, New York. When he arrived in the United States, he found that his background gave him an expertise valued by American scientists. Western commercial, political, and religious involvement in Africa had spawned a great deal of popular curiosity about Africa. Travel narratives such as those of Stanley, Humboldt, and Burton were devoured by a public eager to find out about an exotic world. As a European who had lived in Africa for many years, du Chaillu became a popular dinner guest and speaker. He wrote articles on Africa for the *New York Tribune* and lectured at learned societies. His articles attracted the attention of some members of the Academy of Natural Sciences of Philadelphia, to whom he donated a collection of natural history specimens that he had brought with him from Africa.[7] Duly impressed, several members of the Academy of Natural Sciences of Philadelphia raised money for du Chaillu to make a collecting expedition to Africa. Natural history specimens from Africa had been of great interest in the United States. Ten years earlier, a missionary named Savage had published the first description of the gorilla in the *Journal of the Boston Society of Natural History,* gaining worldwide attention for the society. Du Chaillu's sponsors in Philadelphia hoped for a similar coup.

Du Chaillu returned to the Gabon, on the west coast of Africa, in October 1855. His purpose was "to spend some years in the exploration of a region of territory lying between lat. 2° north and 2° south, and stretching back from the coast to the mountain range called the *Sierra del Crystal,* and beyond as far as I should be able to penetrate."[8] He returned to the United States in 1859, having spent several years exploring the region and mapping the outlets of the Ogobay, Fernand Vaz, and Mexias rivers. He had also collected plants and animals, sending periodic shipments and descriptions to his sponsors in Philadelphia. Many of these were duly described in the *Proceedings* of the Academy. John Cassin, vice president of the Academy, de-

"Reminiscences of Paul Belloni du Chaillu," *The Independent,* 1903, *55:*1,146–1,148; and Edward Clodd, *Memories* (New York: G. P. Putnam's Sons, 1916), pp. 71–75. The entries in the *Dictionary of American Biography* and *Who Was Who in American History—Science and Technology* are unreliable.

[6] Bucher, "Canonization," pp. 16–18; Smith, "Reminiscences," p. 1,147.

[7] Bucher, "Canonization," p. 18.

[8] Paul Du Chaillu, *Explorations and Adventures in Equatorial Africa* (New York: Harper & Brothers, 1861), p. 25.

scribed over three hundred bird specimens, all of which had been sent to him by du Chaillu.[9] By du Chaillu's own account, he had travelled some 8,000 miles on foot, "unaccompanied by any other white men." He claimed to have found sixty new species of birds and twenty new species of quadrupeds, as part of the total of three thousand animals and birds he collected on the voyage.[10] Although the details of these claims were later questioned, it was clear that he had gone to Africa and made substantial collections.

Upon du Chaillu's return late in 1859, his relations with the Academy of Natural Sciences of Philadelphia began to sour. Du Chaillu had apparently been under the impression that the Academy itself, rather than individual members, had formally sponsored his voyage.[11] When he presented the bill for his work and for the specimens sent to the Academy and displayed in its hall, however, the Academy denied any obligation to pay him. The secretary of the Academy regretted "that Mr. Du Chaillu should have erroneously supposed that he was making explorations and collections in western Africa under the patronage and at the cost of the Acad[em]y, when in fact he had merely the favor of many of its members as joint purchasers of parts of his collections." The cash-strapped Academy could simply have been trying to get out of having to pay his substantial bill of $866.50. "The Academy of Philadelphia owes me money still," wrote an aggrieved du Chaillu over a year after his return, "the amount I paid to the vessels which took me home is more than the whole amount of money I have received or will ever receive for the objects they have bought."[12] His negotiations with the Academy lasted well over a year before he finally gave up in frustration.

Du Chaillu spent much of 1860 touring the United States with his collections of exotic stuffed animals, giving popular lectures. He also began work on the narrative of his voyage to Africa, which was to be published by Harper's publishing company. Du Chaillu continued his negotiations with Richard Owen, the superintendent of the British Museum of Natural History, for the sale of parts of his collection, especially

[9] John Cassin, "Description of New Species of Birds from Western Africa, in the collection of the Academy of Natural Sciences of Philadelphia," *Proceedings of the Academy of Natural Sciences of Philadelphia,* April 1855, *7:*324–328; "Descriptions of new species of African Birds, in the Museum of the Academy of Natural Sciences of Philadelphia, collected by Mr. P. B. du Chaillu, in Equatorial Africa," *Proc. Acad. Nat. Sci. Phil.,* August 1856, *8:*156–159; "Catalogue of Birds Collected at Cape Lopez, Western Africa, by Mr. P. B. du Chaillu, in 1856, with notes and descriptions of new species," *Proc. Acad. Nat. Sci. Phil.,* December 1856, *8:*316–322; "Catalogue of Birds collected on the Rivers Camma and Ogobai, Western Africa, by Mr. P. B. du Chaillu, with notes and descriptions of new species," *Proc. Acad. Nat. Sci. Phil.,* January 1859, *11:*30–55; April 1859, *11:*133–144; June 1859, *11:*172–176.

[10] Du Chaillu, *Explorations,* p. x.

[11] Members of the Academy also gave that impression, although the wording was ambiguous. An account written in 1857, while du Chaillu was still collecting in Africa, describes his original arrangement with the Academy as follows: "At a meeting of the Academy of Natural Sciences of Philadelphia, held October 16, 1855 [,] 'Mr. Cassin announced that M. du Chaillu was about to return to Western Africa, for the purpose, exclusively, of geographical exploration, and the collection of objects of Natural History. Arrangements have been made to secure, for the cabinet of this Society, the collections of Birds especially, and also of some other objects.'" In Alfred Maury, *Indigenous Races of the Earth* (Philadelphia: J. B. Lippincott & Co., 1857), p. 324, fn. 243. See also Bucher, "Canonization," p. 23.

[12] The secretary of the Academy to Paul du Chaillu, 5 March 1860, MSS Coll. #464, Library of the Academy of Natural Sciences of Philadelphia; Paul du Chaillu to W. J. Vaux, 31 January 1860. MSS Coll. #567, Library of the Academy of Natural Sciences of Philadelphia; Paul du Chaillu to Richard Owen, 1861. MSS 323A, Smithsonian Institution Libraries.

the gorillas. Du Chaillu had first made the offer to Owen in 1859, while he was still in Africa. Jeffries Wyman, the prominent Harvard anatomist who had coauthored the first scientific paper on the gorilla, reviewed du Chaillu's collections while du Chaillu was lecturing in Boston. Du Chaillu presented a paper on the gorilla to the Boston Society of Natural History, of which Wyman was president. The Society elected du Chaillu a corresponding member of the Society. Although he remained a corresponding member until his death in 1903, he never presented another paper there. Du Chaillu's participation in the Society was largely symbolic.[13]

Prior to 1861, du Chaillu would have been considered the ideal field collector. He had specialized knowledge about Africa that was not widely shared. He avidly collected objects of natural history that were greatly valued by metropolitan scientific communities. Tensions with the Academy of Natural Sciences of Philadelphia notwithstanding, du Chaillu enjoyed good relations with prominent scientists and scientific societies on both sides of the Atlantic. He encouraged scientists to study and describe his collections and offered to sell them some of the most scientifically significant objects. In his public lectures he emphasized the more dramatic aspects of his travels—encounters with cannibal tribes and savage animals—leaving the scientific analysis of his discoveries to the natural history community. When he attempted to do a scientific analysis of his own, however, his comfortable relationship with the natural history communities quickly fell apart.

KNOWLEDGE CLAIMS AND SOCIAL STANDING IN THE FIELD SCIENCES

In the nineteenth century the only people who had the authority to produce scientific knowledge were the self-selecting "gentlemen of science."[14] The gentlemanly specialists who studied geology or natural history were caught between the traditional disdain for manual labor on the one hand and the Baconian ideal of empirical research on the other. Geologists, for example, saw fieldwork as a rite of initiation and passage into the community. Like geology, most of natural history was necessarily tied to specific places. One had to travel to obtain samples. But scientists in Victorian England and the United States had strong ties to their metropolitan communities and so were often unable or unwilling to be away from the metropolis for the long periods that collecting expeditions required. John Cassin, an ornithologist and vice president of the Academy of Natural Sciences of Philadelphia, cautioned that

[13] Richard Owen, "On Some Objects of Natural History from the Collection of M. Du Chaillu," *Reports of the British Association for the Advancement of Science,* 1861:155. In the Jeffries Wyman papers at Harvard Medical School is a series of letters from du Chaillu to Wyman discussing the identification and measurement of du Chaillu's primate skeletons, dated beginning October 1859 and continuing into 1860. The letters also discuss the sale of the primates to the British museum and du Chaillu's work on the manuscript of his book. A letter from Richard Owen to Wyman dated November 1861 thanks Wyman for his letter of introduction for du Chaillu. Owen had been "well satisfied with [du Chaillu's] conduct of business." Jeffries Wyman Papers, Countway Library of Medicine, Harvard Medical School. Paul du Chaillu, "Descriptions of the Habits and Distribution of the Gorilla and other Anthropoid Apes," *Proceedings of the Boston Society of Natural History,* 1860, *7:*276–277.

[14] For discussions of the role of social standing in the English scientific community, see Jack Morrell and Arnold Thackray, *Gentlemen of Science: Early Years of the British Association for the Advancement of Science* (Oxford: Clarendon Press, 1981), Pts. 1 and 3; and Martin Rudwick, *The Great Devonian Controversy: The Shaping of Scientific Knowledge Among Gentlemanly Specialists* (Chicago: Univ. Chicago Press, 1985), Chs. 2 and 16; for an earlier period, Steven Shapin, "The House of Experiment in Seventeenth-Century England," *Isis,* September 1988, *79:*373–404.

> [The delights of the wild] are temporary, and only to be as a teacher,—we must return ever to the social life as the ark of safety, bringing, we may hope, the olive branch of peace with knowledge. For all that I have said, or that anyone else has said, our greatest and truest interests are in society. There only we acquire true cultivation and elevation. Science, Literature, Art, the greatest civilizers, there only flourish. Betake thyself not to the wilderness, or for a period only, and never longer than forty days,—never!—if there is any help for it.[15]

Gentlemen of science interested in African natural history had to rely on amateur and professional collectors to gather data for them. These collectors were from a diverse range of class backgrounds. A very few were from the same social class as the scientists they served. For example, Henry A. Ward set up a business of collecting and selling natural history specimens in Rochester.[16] Ward was a professor at the University of Rochester whose primary occupation was the collection and sale of natural history specimens to other naturalists.

The natural history collectors in west Africa were from an unusually wide variety of geographical and social backgrounds. The native Africans who sold gorilla skeletons to whites on the African coast were socially and culturally the furthest removed from the social world of the metropolitan scientific communities. Although the scientists were quite happy to get hold of the skeletons, they only gave credence to native accounts of gorilla behavior if there were no accounts from white witnesses. European travelers, missionaries, traders, and consular officers became a valuable source of natural history objects and observations in the nineteenth century. In 1819, an explorer saw what was later described as a gorilla, although he did not recognize it as such. The American missionary Savage obtained the first gorilla skeletons in the late 1830s from his fellow missionary John Wilson, who was later du Chaillu's teacher in the Gabon. Ship's officers from the French and British naval vessels that patrolled the coast supplemented their incomes by selling exotic plants and animals to museums and collectors in Europe and America. Other gorilla specimens had made it to European museums, but it is not clear how they were obtained.[17]

The transformation of an object collected in the field to an object that appeared in a scientific paper was a long and often tenuous process of intellectual legitimation. Ideally, the collector collected the "natural" object in the field, making careful observations of the circumstances. Then the object and the description of its manner of collection were passed back to the metropolis through as few hands as possible, since each transfer of the object and observations implied a possible degradation of the information. Collectors such as du Chaillu raised problems of authority for the scientists back in the metropolis. The missionaries, ship's officers, and professional collectors were generally not practicing scientists, so their discoveries were treated as tentative knowledge until the collections had been processed and described by a

[15] John Cassin, *Illustrations of the Birds of California, Texas, Oregon, and British and Russian America* (Philadelphia: J. B. Lippincott & Co., 1856), p. 292.

[16] Sally Gregory Kohlstedt, "Henry A. Ward: The Merchant Naturalist and American Museum Development," *Journal of the Society for the Bibliography of Natural History,* 1980, *9:*647–661.

[17] For a brief overview of the history of the discovery of the gorilla, see Robert Yerkes and Ada Yerkes, *The Great Apes: A Study of Anthropoid Life* (New Haven: Yale University Press, 1926), Ch. 4, "The Emergence of the Gorilla." See also Isidore Geoffroy Saint-Hilaire, "Description des Mammifères Nouveaux ou Imparfaitement Connus, Famille des Singes," *Archives du Museum D'Histoire Naturelle,* 1858–1861, *10:*1–21.

metropolitan scientist.[18] Confirmation by a scientist was essential for an object of natural history to become scientific knowledge.

Data from the jungles of Africa, however, was quite valuable to scientists in Europe and America. Describing a significant new plant or animal could greatly improve a scientist's standing within the community. In practice, then, the metropolitan scientists bent the norms for the collection of field data from Africa and other inaccessible regions. Scientists legitimated the findings of the collectors by symbolically incorporating the collectors into the scientific community. The collector would first be made a corresponding member of a scientific society. Then, with the "assistance" of a regular member, the collector presented a paper describing his findings to the society. The collector thereby became a provisional scientist, superficially vested with all the credibility of a professional.

This strategy of making collectors provisional members of scientific societies could yield great benefits. When the missionary Savage obtained the first gorilla skeleton, he was elected a corresponding member of the Boston Society of Natural History. In 1848, he prepared a paper describing his findings with the help of the Harvard anatomist Jeffries Wyman. The paper was duly published in the Society's journal and the Society basked in the glory of a significant new discovery. Savage never published another scientific paper, and seems not to have participated much in any subsequent activities of the Society.[19] The Academy of Natural Sciences of Philadelphia, the Boston Society of Natural History, and the British societies all treated du Chaillu much the same way a decade later. This pattern was repeated with most other collectors; their participation in the scientific community was symbolic, honorary, and quite limited.

This strategy also carried risks. The process of legitimating knowledge brought from far afield involved many unknowns. Scientific societies took a considerable amount on trust when they elected collectors as members. Both socially and geographically, collectors were often remote from the centers of metropolitan life. Metropolitan scientists often knew little about the social and educational backgrounds and the field practices of the collectors who supplied them. Nor were the collectors subject to the same social and intellectual standards and control that the scientific establishment could impose upon its regular members.[20] Censure from the scientific community meant little to people who were only marginal members of the community to begin with. Most of the time, neither the collector nor the specimen was subject either to intense scrutiny or controversy, and everything worked out well. On occasion, however, the whole fragile structure of legitimation collapsed, as it did when Richard Owen and Sir Roderick Impey Murchison tried to vest du Chaillu with the authority of a scientist in the British scientific community.

[18] Rudwick, *The Great Devonian Controversy,* pp. 424–425.

[19] How Savage obtained the skeleton is described in "Notice of the External Characters and Habits of the Troglodytes Gorilla, A New Species of Orang from the Gaboon River," *Boston Journal of Natural History,* 1847, *5:*117–119, reprinted in *Science in America: Historical Selections,* ed. John Burnham (New York: Holt, Reinhart, and Winston, 1971).

[20] See, for example, Brian Wynne, "Between Orthodoxy and Oblivion: The Normalization of Deviance in Science," in *On the Margins of Science: The Social Construction of Rejected Knowledge,* ed. Roy Wallis. Sociological Review Monograph 27. (Keele, U.K.: University of Keele, 1979), pp. 67–84. Wynne describes how the British physics community of the 1920s, 1930s, and 1940s accommodated the work of the renegade Nobel laureate C. G. Bakala to avoid a controversy that could seriously damage the community.

Figure 2. *"Death of my hunter," from* Explorations and Adventures in Equatorial Africa *(New York, Harper and Brothers, 1861), 342. The illustrations in the* Explorations *reflect the tensions between popular narrative and scientific reporting that also characterize the text. Here, an illustrator at Harper's depicted the death of one of du Chaillu's hunters. Du Chaillu did not see the attack himself, but only described its aftermath.*

DU CHAILLU IN ENGLAND: THE BEGINNING OF THE "GORILLA WARS"

Du Chaillu's *Explorations and Adventures in Equatorial Africa* enjoyed immediate popular success when it appeared in early May of 1861. It was snapped up by scientists and nonscientists alike. The *Explorations* was an impressive work over five hundred pages long, but it contained internal tensions. It was supposed to be simultaneously a personal travel narrative and a scientific analysis. The account described new tribes, including the cannibal Fans; new geological features, such as an equatorial mountain range; and new animals, including, as du Chaillu claimed, over "twenty new species of quadrupeds."[21]

The structure of the book reflected the difficulty of maintaining a stimulating narrative alongside accurate scientific description. Some passages were written in lurid prose to attract general readers (see Fig. 2). One of the most famous of these passages described du Chaillu's dramatic encounter with a gorilla:

> [B]efore us stood an immense male gorilla. . . . Nearly six feet high, with immense body, huge chest, and great muscular arms, with fiercely-glaring large deep gray eyes, and a hellish expression of face, which seemed to me like some nightmare vision: thus stood before us the king of the African forests. . . . His eyes began to flash fiercer fire as we stood motionless on the defensive . . . [and] now truly he reminded me of nothing but some hellish dream creature—a being of that hideous order, half man half beast. . . .

[21] Du Chaillu, *Explorations,* p. x.

> Here, as he began another of his roars and beating his breast in rage, we fired, and killed him.[22]

The colorful, dramatic narrative of the book is interrupted two thirds of the way through for a series of analytic chapters on topics such as on the seasons and fevers of Equatorial Africa, tribal politics, the great apes, and the gorilla. These chapters were written in a sober, scientific style in which du Chaillu began behaving as if he were a scientist. For example:

> All the features of the gorilla, especially in the male, are exaggerated; the head is longer and narrower; the brain is backward; the cranial crests are of immense size; the jaws are very prominent, and show great power; the canines are very large. The proper cavity of the brain is marked by immense occipital ridges. But the remainder of the skeleton of the gorilla comes much nearer to man than that of any other ape. And, after a careful examination of the osteological facts which have been mentioned; after having observed the live gorilla and studied carefully its mode of progression, I came to the conclusion that the gorilla is the nearest akin to man of all the anthropoid apes.[23]

Du Chaillu's use of the scientific narrative style was an attempt to reinforce the scientific validity of his findings. The vivid narrative resumed clumsily after this dry interlude.[24]

Early reviews of the *Explorations* criticized du Chaillu's writing style. One of the first published reviews of the book noted that

> [f]or a solid book of serious travel, the style is rather airy, and the matter sometimes contradictory, but this latter may arise from the opposite accounts the writer was likely to hear of the same circumstances. To students of Natural History, the volume will, deservedly, be attractive; but we are not disposed to yield full credence to Mr. Du Chaillu's assertion that he is the first white man who has described the gorilla, or the Nest-building ape.[25]

This was a polite foreshadowing of the controversy to come.

The most persistent critic of du Chaillu was John Edward Gray. Gray was the keeper of the zoological collection at the British Museum and vice president of the Royal Zoological Society. As keeper, Gray had taken a small collection and built it into one of the finest in Europe. He had published well over a thousand scientific papers in his lifetime. Gray had first examined du Chaillu's specimens at the quarters of the Royal Geographical Society in April.[26] His criticisms of du Chaillu were perfectly legitimate from a scientific standpoint, but the vigor with which he undertook his task suggested more than simple intellectual dissatisfaction. Richard Owen had been appointed superintendent of the museum some years before, to the displeasure

[22] *Ibid.,* pp. 98–101.

[23] *Ibid.,* p. 417.

[24] The tensions between the narrative and scientific modes of discourse are not unique to du Chaillu's work. Mary Louise Pratt describes the history of the interaction of the two modes in "Fieldwork in Common Places," *Writing Culture: The Poetics and Politics of Ethnography,* eds. James Clifford and George E. Marcus (Berkeley: Univ. California Press, 1986), especially p. 35.

[25] Unsigned review of *Explorations and Adventures in Equatorial Africa, Athenaeum,* 11 May 1861, no. 1750: 623.

[26] The key accounts of du Chaillu's arrival in London are Vaucaire, *Gorilla Hunter,* pp. 131–137, and John Edward Gray, "Zoological Notes on Perusing M. du Chaillu's 'Adventures in Equatorial Africa,'" *Annals of Natural History,* 1861, *7:*463–470; 1861, *8:*60–65.

of Gray and some of the other section heads. They resented Owen's interference in what they considered to be their personal fiefdoms. The problems Gray found with the *Explorations* gave him the perfect chance to attack Owen vicariously by attacking his protegé du Chaillu.[27]

Gray focused on questioning du Chaillu's competence in the practice of science. In a series of letters to the editor of the *Athenaeum* spanning several months, Gray attacked the credibility of the book and du Chaillu's authority as a witness. Dissatisfied with the *Athenaeum*'s review of the *Explorations,* Gray lashed out in a letter to the editor in the following week's issue: "I have examined the collection of mammalia with care," he contended, "and there is not a specimen among them that indicates that the collector had traversed any new region." The gorilla was not new to European science, despite popular assumptions to the contrary: "we have been receiving specimens of them for the last fifteen years . . . until almost every museum in Europe is provided with specimens." The book, he continued, was "full of improbable stories," and some of the illustrations in the book were "copied from figures prepared in this country to represent other kinds, and without acknowledgement."[28]

Gray was unimpressed by Du Chaillu's attempts at naming and identifying new species: "We are overburdened with useless synonyma, and Natural History may be converted into a romance rather than a science by traveller's tales, if they are not exposed at the time." Gray quibbled with almost every new species that du Chaillu had named, saying that most of them had been named elsewhere. There is, in fact, a great deal of ambiguity in identifying new species. "Every man thinks his own geese to be swans," wrote one scientist in the *Natural History Review,* "and the error of describing old species as new, is one of such ordinary occurrence, that we fear there is scarcely a living Naturalist, who could wash his hands, and say that he was innocent of the offense."[29] It appears that Gray mostly resented having an amateur name new species, since naming was normally the province of professional scientists.

Gray commented that the popularity of du Chaillu's lectures and the success of the *Explorations* "become important when the writer is put forward as an authority on subjects upon which I believe he is not qualified to speak; and it is only against the work being considered as what I conceive he himself never intended it to be when it was first compiled (that is, a regular and veracious work of travels and natural history) that I have ever objected or desire to object." In a detailed critique of the zoology of the *Explorations,* Gray noted that du Chaillu had included reptiles under the mammalia. He also questioned du Chaillu's ability as a collector, specifically his ability to preserve animal skins. Gray described the skins of du Chaillu's gorilla specimens as rotten and fragile because they had been inadequately stuffed. Du Chaillu's method of stuffing the gorilla skins led Gray to conclude that "they could not have been prepared very far inland; for no one would have adopted such an inconvenient and dangerous practice if the skins had to be carried for many miles on the backs of men; and even near the coast it would have been better if the skins

[27] "Gray, John Edward," *Dictionary of National Biography* (hereafter, *DNB*) 23, pp. 9–10; *DNB* 42, pp. 435–444; A. E. Gunther, *The Founders of Science at the British Museum, 1753–1900* (Suffolk, England: Halesworth Press, 1980), pp. 122–124; Rupke, *Richard Owen,* p. 319.

[28] John Edward Gray, "The New Traveller's Tales," *Athenaeum,* 18 May 1861, no. 1751: 662–663.

[29] *Ibid.,* p. 662; "The Fauna of Equatorial Africa," *Natural History Review* July 1861, *1:*290.

had been spread out flat and allowed to dry on both sides, and then had been packed in a small space, as is usual with good collections."[30]

In subsequent letters to the *Athenaeum* through the summer of 1861, Gray backed up his original charges with even more detailed evidence. Gray showed, for example, that the frontispiece of the *Explorations* was plagiarized from an article on the gorilla that Isidore Geoffroy Saint Hilaire had published some three years previously in the *Archives du Museum.*[31] The version in du Chaillu's book added a strategically placed branch to cover that part of the gorilla's anatomy that would presumably offend Victorian sensibilities. He also showed that many other illustrations in the book were copied from other sources. Gray appears to have been angry at du Chaillu for overstepping his bounds as a collector.

The debate began to center as much on du Chaillu himself as it did on his discoveries. Neither Gray nor the *Athenaeum* made outright slurs against du Chaillu's character. The editors of the *Athenaeum* edited one letter that they felt had too much personal invective, and Gray always denied that he was carrying out a personal war ("Gorilla war," Gray punned) against du Chaillu. In fact, he said that "it is no fault of M. Du Chaillu that, to meet a supposed want of one or more of our scientific Societies, he was seized upon and put forth as a scientific traveller and zoologist." Even in private, Gray maintained this charitable tone, while also making more serious accusations about du Chaillu than he dared make in print: "I have always regarded Du Chaillu 'as more sinned against than sinning.' He only wanted strength of mind to tell the truth that the book was not written by him."[32] Publicly and privately, Gray's tone, while not personally hostile, was still one of paternalistic sympathy. Du Chaillu's credibility among scientists was damaged by repeated accusations that a ghost-writer at Harper's had embellished his accounts to produce a more profitable book.

One way du Chaillu's detractors discounted his claims was to exclude him from the community of scientists. "Traveler's tales" was a term of contempt that cropped up several times during the debate, concerning the works of several people. Du Chaillu used it in describing the work of previous explorers, and then Gray used it to describe du Chaillu, also calling him "an uneducated collector of animal skins for sale and an exhibitor of them in Broadway, New York." Thus, Du Chaillu's success with the public did little to help him: the true gentlemanly specialist sought recognition from his peers, who were qualified to judge the quality of his work, rather than from the public, who were not.[33] The *Athenaeum*'s use of "the New Traveller's Tales" as the heading for the section of letters about du Chaillu and the *Explorations* suggests what the editors thought about the issue.

Gray was not the only person to discover problems with du Chaillu's book. An anonymous writer to the *Athenaeum* pointed out that the implicit chronology of the voyages given in the book was impossible. Du Chaillu had given two different and

[30] John Edward Gray, "Zoological Notes on perusing M. Du Chaillu's *Adventures in Equatorial Africa*," *Ann. Nat. Hist.* 1861, 8:60–61; *ibid.*, pp. 64–65.

[31] John Edward Gray, "The New Traveller's Tales," *Athenaeum,* 25 May 1861, p. 695; Bucher, "Canonization," p. 20, n. 32.

[32] Gray, "Zoological Notes," *Ann. Nat. Hist.*, 1861, *8*:60; John Edward Gray, MS note in his personal copy of T. H. Huxley's *Evidence as to Man's Place in Nature* (London: Williams and Norgate, 1863). Collection of Sydney Ross, Troy, N.Y.

[33] Gray, "Zoological Notes," (cit. n. 30), p. 60; Susan Faye Cannon, *Science in Culture: The Early Victorian Period* (New York: Science History Publications, 1978), p. 150.

mutually incompatible accounts of what he had been doing in the early months of 1858.[34] With the entire chronology of the narrative thrown into question, all the individual parts of the narrative also became suspect.

Another of du Chaillu's detractors was an American trader named Walker, who had known du Chaillu in the Gabon. Walker contradicted many of du Chaillu's most sensational descriptions. Du Chaillu claimed that he had traveled alone and that he spoke the language of the natives. Walker asserted that du Chaillu had been accompanied by a Frenchman and had an interpreter. Du Chaillu wrote about being stranded on the coast for four months and being miraculously rescued by a trading ship. Walker countered that du Chaillu spent most of the time comfortably at an American mission and was only on the coast for two weeks. Du Chaillu had, in fact, arranged in advance to meet the ship. With these and other criticisms, Walker felt that he had "sufficiently shown that M. Du Chaillu has been guilty of many incorrect statements—in fact, his work contains almost as many errors and inconsistencies as there are paragraphs. It is, moreover, teeming with vanity; and, taking it as a whole, it is hard to say whether the author, in his attempt to impose upon and in fact humbug the scientific world, displays most mendacity or ignorance."[35]

The evidence provided by Walker and Gray seemed pretty damning, but there were other perspectives on the validity of du Chaillu's work. The reviewer in the *Natural History Review* of July 1861 argued that du Chaillu was neither particularly mendacious nor ignorant. Rather, the reviewer attributed the defects of the narrative to be the result of "imperfectly kept notes, assisted by the efforts of a rather vivid imagination and a not very perfect memory." He attributed the energy of the attacks against du Chaillu to jealousy: "We fear, owing to the somewhat over-zealous way in which he has been taken up and made a 'lion' of, that M. Du Chaillu has provoked such severe criticisms upon his performances; such, indeed, as, in our opinion, ought not to have been put forward, until the most positive and satisfactory evidence of the untruth of his statements had been obtained."[36] This review makes explicit one of du Chaillu's key sins in the eyes of many British scientists: his public acclaim.

Science reviewers, naturally enamored of classification, identified du Chaillu's precise place in the constellation of scientific explorers. The reviewer for the *Natural History Review* argued that du Chaillu could not be a "scientific naturalist" because "he . . . who 'feels the breath of a serpent against his face'; and who 'turns turtles' in fresh water lakes and then classifies them among the Mammalia in his list of newly described species, is no doubt a vigorous voyager and a lively narrator, but wants the knowledge and the sobriety of a man of science." Nor was du Chaillu even a "scientific traveller," for "he took no observations, either astronomical, barometrical, meteorological, or thermometrical." The reviewer found evidence of du Chaillu's honesty in these very mistakes and omissions—"that such errors as we have last described, are of the very kind that any one, intentionally deceiving, would most surely avoid." In spite of all du Chaillu's "not inconsiderable" shortcomings, then, the reviewer found du Chaillu "to be an energetic and active explorer."[37] Such was the subtle hierarchy between the different categories of traveler.

[34] Gray, "The New Traveller's Tales," (cit. n. 31), p. 695.

[35] R. B. Walker, "M. Du Chaillu and his book," *Athenaeum,* 21 September 1861, no. 1769: 374; Bucher, "Canonization," pp. 20–21.

[36] "The Fauna of Equatorial Africa," *Nat. Hist. Rev.,* July 1861, *1:*291.

[37] *Ibid.,* pp. 288–289.

Behind the polite facade of gentlemanly scientific debate, some gentlemen of science found other reasons to doubt du Chaillu. Charles Waterton, another pillar of the British scientific community, wrote "[Du Chaillu's] adventures with the gorilla are most formidable and false. . . . I suspect strongly that the traveller has been nothing but a trader on the western coast of Africa, possibly engaged in kidnapping negroes."[38] This was a harsh judgment coming from a man whose own travel narrative had been widely doubted and ridiculed just a few decades earlier. In October 1861 George Ord, a member of the Academy of Natural Sciences of Philadelphia, told Waterton that members of the Academy thought that du Chaillu was of mixed race. "If it be a fact that he is a mongrel, or a *mustee,* as the mixed race are termed in the West Indies, then we may account for his wondrous narratives; for I have observed that it is a characteristic of the negro race, and their admixtures, to be affected to habits of romance." Even du Chaillu's *friends* suspected he was of mixed parentage. Du Chaillu's "diminutive stature, his negroid face, and his swarthy complexion," wrote one of his friends, "made him look somewhat akin to our simian relatives."[39] It is not surprising that racism, which was such a strong part of the social fabric of the day, should also play an important role in the assessment of scientific claims. Members of disparaged races were also disparaged as scientific witnesses.

Du Chaillu's authority as a witness was undermined for several different reasons. The process by which his authority was undermined illustrates the original bases of that authority. Authority derives partly from the more obvious elements one would expect: logical consistency, originality, accuracy, and honesty in reporting observations. But it also derives from more subtle factors, such as respecting the intellectual territory of other scientists by not making claims in their areas of expertise, being of the right race, and not being too blatantly commercial and publicity-seeking. Du Chaillu had clearly faked parts of his book or allowed them to be faked, and so it was perfectly natural for the scientists to question the validity of his data. The criticisms of du Chaillu, however, went beyond the criticism warranted by his transgressions. Some of du Chaillu's claims for having discovered new species of animals would probably have ordinarily fallen within the bounds of normal scientific uncertainty. But because he was suspect as a witness, his ambiguous claims were dismissed out of hand.

SAVING THE PHENOMENA: MURCHISON AND OWEN COME TO THE RESCUE

Historians and sociologists of science have noted that there is normally a circular relationship between the observer and what is observed. Credible observations lend

[38] Charles Waterton, *Essays on Natural History* (1871), p. 582; cited in Barber, *Heyday* (cit. n. 2), p. 277.

[39] George Ord to Charles Waterton. 20 October 1861. Ord-Waterton correspondence, Library of the American Philosophical Society, Philadelphia. Du Chaillu's life had many similarities with that of the famous artist and naturalist John James Audubon. Both were born illegitimately of mothers of (at the time) unknown race in French colonies. Both men distorted their personal histories, with a tendency to dramatize their lives, and both functioned on the margins of the scientific establishment. Finally, both were subjects of the frequently vitriolic letters between George Ord and Charles Waterton. For a further discussion of Audubon, see Alice Ford, *John James Audubon: A Biography* (New York: Abbeville Press, 1988); Edward Clodd, *Memories* (New York: G. P. Putnam's Sons, 1916), p. 72.

authority to the observer; a credible observer lends authority to the observations.[40] Du Chaillu's detractors questioned both the observer and the observations. Du Chaillu had powerful critics, but he also had powerful allies. Sir Roderick Impey Murchison, president of the Royal Geographical Society, and Richard Owen, the superintendent of the British Museum of Natural History, were his two most prominent defenders in England. Murchison and Owen, both of whom had much of their own credibility invested in du Chaillu, fought hard to reestablish du Chaillu's reputation for competence as an observer. Murchison tried to reassert du Chaillu's good character to support the evidence, whereas Owen used the evidence to try to deduce du Chaillu's good character.

The annual meeting of the Royal Geographical Society took place shortly after Gray had first publicly criticized the *Explorations*. Murchison took pains to defend du Chaillu in his presidential address to the Society. He was also defending the reputation of the Royal Geographical Society, where many of du Chaillu's discoveries had been displayed. Murchison had a personal stake in establishing the credibility of du Chaillu's narrative since some of du Chaillu's geographic observations confirmed one of his pet theories about the existence of a chain of mountains in Central Africa.[41] Murchison affirmed du Chaillu's good character and his reliability as a witness. In describing du Chaillu's accounts of the gorilla, Murchison cited the authority of Owen to establish du Chaillu's credibility. Describing du Chaillu's accounts of the countryside, Murchison said that "strikingly attractive and wonderful as are his descriptions, they carry in themselves an impress of substantial truthfulness." Du Chaillu had "brought to us what most will admit to be unanswerable evidences of his fidelity of observation—evidences which the Council of this Society has considerately allowed him to exhibit in our own apartments in Whitehall place." Yet Murchison also distanced himself from du Chaillu, reminding his audience that du Chaillu "never claimed to be a man of science."[42]

The editors of the *Athenaeum* quickly refuted Murchison's defence of du Chaillu. The editors acknowledged the importance of Murchison's endorsement of du Chaillu but noted that "Sir Roderick's favourable opinions will not reconcile the singular discrepancies of dates in M. du Chaillu's book. No amount of friendly oratory will make it possible for a man to have been in two distant places at the same time.[43] Prominent scientists could use their social standing to reinforce weak or controversial evidence and observations, but social standing could not overcome logical inconsistencies.

Owen's turn to defend du Chaillu came at the September 1861 meeting of the British Association for the Advancement of Science (BAAS) in Manchester. Owen

[40] This is the "experimenter's regress," discussed in Harry M. Collins, *Changing Order: Replication and Induction in Scientific Practice* (London: Sage, 1985), Ch. 4. Steven Shapin has formulated an observational version of this concept in "The House of Experiment," (cit. n. 18), p. 376, n. 6.

[41] P. B. du Chaillu, *Explorations et Aventures dans l'Afrique Équatoriale* (Paris: Michel Lévy Frères, Libraires, Éditeurs, 1863), pp. vi–vii. Murchison was no stranger to controversy himself. His role in controversies in Victorian geology is documented in Martin J. Rudwick, *The Great Devonian Controversy* (cit. n. 18); James A. Secord, *Controversy in Victorian Geology: The Cambrian-Silurian Dispute* (Princeton: Princeton Univ. Press, 1986); and David Oldroyd, *The Highlands Controversy: Constructing Geological Knowledge Through Fieldwork in Nineteenth-Century Britain* (Chicago: University of Chicago Press, 1990).

[42] Sir Roderick Impey Murchison, "Anniversary Address," *Journal of the Royal Geographical Society*, 1861, *31*:clxxxii–clxxxiii.

[43] "Our Weekly Gossip," *Athenaeum*, 1 June 1861, no. 1753: 729.

emphasized du Chaillu's natural history collections, which could be independently verified, and avoided discussion of du Chaillu's personal observations, which could not. Owen presented a paper discussing the specimens that du Chaillu had sold to the British Museum, but only peripherally mentioned any of du Chaillu's claims that could not be independently confirmed. In this way, many of du Chaillu's claims became Owen's while du Chaillu was eased out of the picture. Du Chaillu was present at that BAAS meeting and at the discussion of the paper, but appears to have had little to say.[44]

Owen lent credence to du Chaillu's narrative by focusing on du Chaillu's natural history collections. In one of the most dramatic passages of his book, du Chaillu described how he shot an attacking gorilla in the chest. Owen's paper confirmed this account, saying that the wounds suggested that the gorilla had indeed been shot in the chest. Gray subsequently published a letter in the *Athenaeum* saying the wounds were consistent with the gorilla having been shot in the back. Gray invited anyone who was interested to look at the wounds and decide for themselves (see Fig. 3). "I might cite many names of high authority in corroboration of what I have here advanced," he wrote, "but I am not disposed to appeal to any authority, however great, where the facts are open to all. On these, and these only, I shall rest my case."[45]

The "facts" did not speak as clearly as Gray claimed. His own authority was important in establishing them as facts. Nor was he entirely successful in establishing them as such. Owen, who was defending himself as well as du Chaillu, rose to the challenge and argued that the facts did not speak for themselves. Owen cited the authority of a well-known hunter who argued that the same "facts" could be consistent with the gorilla having been shot from the front. But since the facts did not speak for themselves, and neither man had enough authority to defeat the other decisively, the debate over the bullet holes was simply dropped. The evidence remained suspect, neither proved nor disproved.

Du Chaillu's text was clearly tainted, but the task of sorting out the fraudulent from the merely exotic was beyond the capability of anyone in London. Du Chaillu had a geographic advantage over his detractors: he was the only person involved in the debate who had actually been to Africa. Although some evidence suggested that a few of du Chaillu's trips were faked, nobody in London could decisively prove them so. Since so few other travelers had been in the region, no other accounts existed by which du Chaillu's could be judged. Because of du Chaillu's geographic advantage, attempts in London either to support him or to discredit him had to focus on his character instead. T. H. Huxley, who had been watching the debate but not participating in it actively, remarked that "[S]o long as [du Chaillu's] narrative remains in its present state of unexplained and apparently inexplicable confusion, it has no claim to original authority respecting any subject whatsoever. It may be truth, but it is not evidence."[46] Huxley recognized the potential value of du Chaillu's observations, arguing at the same time that du Chaillu's credibility was not sufficient for the observations to be categorized as evidence.

[44] "Section D.—Zoology and Botany," *Athenaeum,* 14 September 1861, no. 1768: 347–349. Du Chaillu's presence was noted, but no comments of his were recorded.

[45] "Prof. Gray to the President of Section D," *Athenaeum,* 21 September 1861, no. 1769: 373.

[46] Thomas Henry Huxley, *Evidence as to Man's Place in Nature* (London: Williams and Norgate, 1863), reprinted in Huxley's *Man's Place in Nature and Other Anthropological Essays* (New York: D. Appleton & Co., 1898), pp. 71–72.

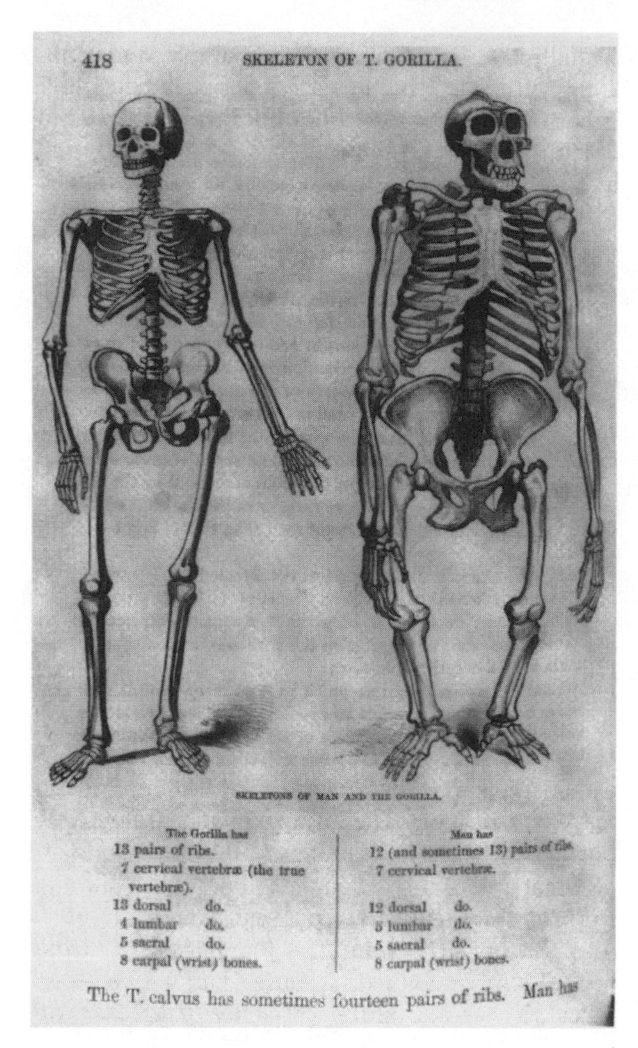

Figure 3. *"Skeletons of Man and the Gorilla," from* Explorations and Adventures in Equatorial Africa *(New York: Harper and Brothers, 1861). John Edward Gray argued that this drawing was copied from a photograph of a gorilla skeleton. These photographs were sold as postcards at the British museum for a few pence. Even these simple drawings were also misrepresentations, since "the skeleton of the Gorilla is made stronger, broader, and more powerful, while that of the Man is proportionally diminished, so as to give anything but an accurate comparison." John Edward Gray, "Zoological Notes on Perusing M. Du Chaillu's* Adventures in Equatorial Africa," Annals of Natural History, *1861,* 7:*465.*

Du Chaillu's silence throughout the debates over the *Explorations* was a clear indication of his marginal status in the scientific community. Apart from one letter he wrote to the editor of the *Athenaeum,* his defense within the scientific community was carried out by others, gentlemen who were members of the scientific community. Even his supporters moved him to the periphery of the arguments, discussing du Chaillu only in the third person. The facts did not "speak for themselves" and du Chaillu did not speak for the facts. They always needed an interpreter of standing from within the community.

While the *Athenaeum* had been highly critical of du Chaillu, the editors saw a conclusion to the controversy only when Winwood Reade went to Africa in 1861 specifically to check out du Chaillu's claims. Reade was a twenty-three-year-old "traveller, novelist, and controversialist" who had studied at Oxford, although he did not take a degree. When he heard du Chaillu's stories, he raised money and went to the Gabon. For five months, Reade followed the path of du Chaillu's travels and interviewed many of the native porters and others with whom du Chaillu had worked. He confirmed many of du Chaillu's accounts about the geography and tribes of Western Africa, but rejected almost all of du Chaillu's accounts of the gorilla.[47] When Reade returned to England in 1862, he presented his findings to the Royal Geographical Society and published articles in several magazines. In an observation that echoed the commentary on du Chaillu, Murchison said that "the Paper showed that Mr. Winwood Reade was an enterprising traveller, who had visited many parts of Africa, but not, as he modestly said, as a scientific man."[48]

Reade's final assessment of du Chaillu and the *Explorations* is a curious mixture of criticism and sympathy. Reade concluded that the *Explorations* was a mix of truth and fiction, "prepared by a gentleman well known in the New York literary world" using du Chaillu's notes. Reade also found that du Chaillu's "labours as a naturalist have been very remarkable. And a fellow labourer, though a humble one, may be permitted to regret that, actuated by a foolish vanity or by ill-advice, he should have attempted to add artificial flowers to a wreath of laurels which he had fairly and hardly earned.[49] Although Reade never names this mysterious figure in New York who helped du Chaillu write the book, his claim is plausible. Accounts of du Chaillu when he first arrived in the United States say that his English was not very good. Certainly it is very awkward in his early correspondence with Jeffries Wyman, during the time that du Chaillu was supposedly working on the book. The level of language used in the book is far superior to the language in his personal correspondence, suggesting that he had at least some strong editorial help. Quite possibly, the editorial process went from simply cleaning up du Chaillu's language to spicing up the facts.

Reade's letter was published in the *Athenaeum* in November of 1862, almost a year and a half after the controversy had broken. The editors of the *Athenaeum,* ignoring many of the debates that had taken place in the pages of their own journal, said that "[p]eople refused to believe that a book full of amusing contradictions and absurdities was true; and for this refusal they have now received from an *English witness* at the Gaboon a further and conclusive warrant."[50] Although people from the Gabon gave convincing evidence of some fraud on du Chaillu's part within seven months of the beginning of the debate, it was not until a year later that Reade, an English witness and a gentlemanly emissary of the scientific community, provided an assessment credible enough that the debate could publicly be declared finished. Even though Reade was not "a scientific man," as Murchison pointed out, his Englishness and his rectitude made him a better witness than du Chaillu. Certainly Reade's own accounts of Africa were never given the same kind of critical scrutiny as du Chaillu's. Du Chaillu's observations had been superseded by those of a more credible witness.

Despite the judgments of the *Athenaeum* and John Edward Gray, much of du Chaillu's work did not fade from the public or scientific eye. Many of du Chaillu's observations fell legitimately within the bounds of scientific uncertainty. Reade's

[47] "Reade, William Winwood," *DNB,* 47, pp. 361–362.

[48] W. Winwood Reade, "Travels in western Africa," *Journal of the Royal Geographical Society,* 1863, *7:*106–107.

[49] Winwood Reade, "News From the Gorilla Country," *Athenaeum,* 22 November 1862, no. 1850: 662–223. Nobody has yet found the manuscript for the *Explorations* or correspondence relating to the publication of the work that gives any details about how it was put together.

[50] "Our Weekly Gossip," *Athenaeum,* 22 November 1862, no. 1830: 665 (emphasis in original).

explorations, for example, had also *confirmed* many of du Chaillu's observations. For the next sixty years most people who did field work on the gorilla took the *Explorations* as their starting point. Even if du Chaillu's claims were not proved, they did give subsequent explorers to the region an important list of observations to be confirmed, modified, or refuted. Du Chaillu had unintentionally helped to define the research agenda for subsequent natural historians, explorers, and anthropologists who visited the Gabon.

In 1863, du Chaillu returned to the same region of Africa to prove some of his more controversial points. He spent another several years there and got another book out of it. This second book, *A Journey to Ashango-Land,* was popular but not as controversial as the *Explorations.* The *Explorations* were also a motivating force behind Richard F. Burton's expedition to the area a few years later. "Travelling with M. Paul B. du Chaillu's "First Expedition" in my hand," Burton wrote, "I jealously looked into his every statement, and his numerous friends will be pleased to see how many of his assertions are confirmed by my experience." Burton devoted a chapter of the book to the gorilla. A French surveying party had by then largely confirmed many of du Chaillu's geographical discoveries. Nonetheless, many of the facts concerning the habits of the gorilla were still disputed as late as 1876.[51]

Later assessments of du Chaillu varied. An 1881 account of du Chaillu's first book notes that some of the facts that du Chaillu had discovered on his first trip had "excited unreasonable and unreasoning hostility in England."[52] This assessment was not universally shared. In their pioneering study on primate life, the famous primatologists Robert and Ada Yerkes found du Chaillu to be "grossly careless in exposition, and perhaps also handling of the gorilla exhibits. At any rate his publications were so far repudiated by the scientific world that even today [1928] they are considered first rate examples of 'nature faking'. . . . However charitable his inclination, one must admit that his showing is bad."[53] The explorer Carl Akeley argued that du Chaillu had grossly misrepresented the behavior of the gorilla. "If you read the tale as Du Chaillu wrote it, it gives an impression that the gorilla is a terrible animal. If you read merely what the gorilla did, you will see that he did nothing that a domestic dog might not have done under the same circumstances. . . . All I want to point out is that the gorilla should be judged by what he does, not by how the people who hunt him feel."[54] Whatever the final assessment, it is significant that for more than fifty years after the *Explorations* were written, people who worked with primates often felt obliged to address du Chaillu and his discoveries. This is a telling indication of the extent of his impact.

Paul du Chaillu was far from ruined by this scientific controversy. While this study has focused on the reception of du Chaillu and the *Explorations* by the scientific

[51] Richard F. Burton, *Two Trips to Gorilla-Land and the Cataracts of the Congo,* 2 vols. (London: Sampson Low, Marston, Low, and Searle, 1876; reprint edition, New York: Johnson Reprint Corporation, 1967), p. vii; Paul du Chaillu, *A Journey to Ashango-land: And Further Penetration into Equatorial Africa* (London: J. Murray, 1867).

[52] Charles H. Jones, *Livingstone's and Stanley's Travels in Africa, also; the Adventures of Mungo Parke, Clapperton, DuChaillu, and other Famous Explorers in the Land of the Palm and the Gorilla* (New York: Hurst & Co., 1881), p. 220.

[53] Robert M. Yerkes and Ada W. Yerkes, *The Great Apes: A Study of Anthropoid Life* (New Haven: Yale University Press, 1929; reprint edition, New York: Johnson Reprint Corporation, 1970), p. 34.

[54] Carl E. Akeley, *In Brightest Africa* (Garden City, N.Y.: Garden City Publishing, 1923), pp. 237–239.

community, scientists were never a major part of his audience. Months after most scientists had washed their hands of him, he could still pack lecture halls and earn several hundred pounds in a single evening lecturing to the public. His second voyage to Africa was financed by the proceeds from the animals he sold to the British Museum. He also wrote a highly successful series of children's stories loosely based on his two voyages to Africa. Du Chaillu supported himself from the proceeds of these books and by giving public lectures about his experiences in Africa. With the proceeds he spent several years in Scandinavia, resulting in the two-volume ethnography entitled *Land of the Midnight Sun,* which earned him even more money and got him invited to more lectures. He was working on a similar ethnographic study of Russia when he died in 1903.[55]

CONCLUSION

The field sciences in Victorian England and the United States depended heavily on trust. Out of necessity, metropolitan scientists accepted reports and collections from people who were on the social margins of the scientific world but who had privileged access to the field. Their elaborate strategies for incorporating collectors and their reports and collections into the scientific sphere generally worked well. Both the explorers and the societies benefited from the exchange. Where the field reports fit with what was previously known, they were accepted without a murmur of dissent and indeed sometimes with celebration. The controversy over the *Explorations and Adventures in Equatorial Africa* made the fundamental weakness of this system clear to all. It illustrated a crucial difference between the laboratory sciences and the field sciences. In the laboratory, all experiments are at least in principle independent of time and space. Doubts can be resolved by repeating the experiment. In contrast, the field sciences depend heavily both on time and on space. Replicability, particularly in the nineteenth-century field sciences, was difficult, if not impossible. Accordingly, the field scientists depended much more heavily on authority than did the laboratory sciences.

Authority could be vested in objects and texts as well as people. The substantial natural history collections offered proof of du Chaillu's accomplishments in Africa. The authority of other natural history objects was, however, subject to different interpretations. The bullet holes in the gorilla skin were not sufficient to prove du Chaillu's account one way or another. The *Explorations* was criticized for its "fantastic" accounts, but equatorial Africa was also a place where Europeans expected to encounter the fantastic, so little that du Chaillu said was inherently implausible. The chronological contradictions and basic scientific mistakes in the narrative effectively undermined its independent authority. Given that neither the authority of the evidence nor that of the text was sufficient to establish truth, attention turned to du Chaillu himself. Just as they dissected the mammals and insects in their collections,

[55] Paul du Chaillu, *The Land of the Midnight Sun: Summer and Winter Journeys Through Sweden, Norway, Lapland, and Northern Finland* (New York: Harper & Brothers, 1882). Some of the more popular of du Chaillu's children's books include *Lost in the Jungle: narrated for young people* (New York: Harper & Brothers, 1870); *Stories of the Gorilla Country: narrated for young people* (New York: Harper & Brothers, 1881); *Wild Life under the Equator* (London: Sampson Low, Son & Marson, 1869); *The Country of the Dwarfs* (New York: Harper & Bros., 1899); and *My Apingi Kingdom: with life in the great Sahara, and sketches of the chase of the ostrich, hyena, &c.* (New York: Harper & Brothers, 1871).

scientists in Victorian London minutely dissected Du Chaillu's credibility into its component parts—personal background, education, race, social class—and found it wanting. Nor were Owen and Murchison's authority sufficient to reestablish the credibility either of du Chaillu or of his narrative.

Since appeals to authority did not resolve the question, attention shifted back to the field. Although it was impossible to replicate du Chaillu's observations precisely, it was at least possible to follow in his footsteps. Even so, no crucial experiment could decisively prove du Chaillu wrong. If subsequent travellers did not see the things that du Chaillu claimed to have seen, such observations did not, in and of themselves, refute du Chaillu's claims. Observing docile gorillas did not prove that gorillas never attacked. Instead of being decisively refuted, the authority of du Chaillu's observations was gradually eroded over the next fifty years. During these years, scientists began to resolve the problem of field observations either by going into the field themselves or by sending trained observers to the field. Instead of moving parts of the field back to the metropolis, scientists increasingly moved parts of the metropolis out to the field. By doing so, they got around the weakest link in the field sciences, their traditional dependence on untrained collectors. Replicability was still impossible, but at least scientists could bring their knowledge and authority to field observations with no mediations. Anybody could discover truth, but as far as scientists were concerned, only scientifically trained people could turn that truth into evidence.

Francis La Flesche (courtesy National Anthropological Archives, Smithsonian Institution).

Francis La Flesche: The American Indian as Anthropologist

By Joan Mark

IN 1893, AS SHE WAS READING the final proofs for her pioneering monograph *A Study of Omaha Indian Music,* Alice Fletcher wrote to F. W. Putnam, the Director of the Peabody Museum of American Archaeology and Ethnology, which was to publish the study: "I find that Francis has a great deal of feeling concerning the recognition of his share in the work involved in this monograph. He wants his name to appear on the title page . . . 'aided by Francis La Flesche.'" Fletcher went on to justify La Flesche's "feeling" to Putnam:

> He has spent several hundred dollars and given much time and labor, moreover he first had the idea of writing out this music. He had several songs taken down by Zimmelgave (or some such name), a professional musician, but the work was a failure. After that I took it up with him, having made a few attempts alone and failed. I shall be glad to have his name added . . . and hope you will be willing.[1]

Putnam readily agreed, and the change was made. The issue cannot rest there where they left it, however, for Fletcher's rare admission—that the idea of doing the work had originally been La Flesche's—and the latter's insistence that his contribution be recognized challenge us to look more closely at Alice Fletcher and Francis La Flesche and in general at the relation between anthropologists and the American Indians who have worked with them.

Conscientious anthropologists in the American tradition have repeatedly stressed the importance of the "informant" or native American collaborator to their work. Lewis Henry Morgan dedicated his first book to Ely Parker. Franz Boas acknowledged his indebtedness to George Hunt, listing Hunt as co-author of two collections of Kwakiutl tales, and that tradition continues. Periodically there have been more general expressions of professional indebtedness to native collaborators, as recently in *In the Company of Man* and *American Indian Intellectuals*.[2]

Some of these informants have eventually been recognized and employed as

I am grateful to Margot Liberty, whose two recent studies of Francis La Flesche stimulated this one.

[1]Alice Fletcher to F. W. Putnam, 13 April 1893, Peabody Museum Papers, Harvard University Archives, Cambridge, Mass. By permission of the Harvard University Archives.

[2]Lewis Henry Morgan, *League of the Ho-dé-no-sau-nee, or Iroquois* (Rochester, N.Y.: Sage, 1851); Franz Boas and George Hunt, *Kwaikutl Texts* and *Kwaikutl Texts: Second Series* (Memoirs of the American Museum of Natural History, 5, 14) (Leiden: Brill; New York: G. E. Stechert, 1902, 1906); Joseph Casagrande, ed., *In the Company of Man: Twenty Portraits by Anthropologists* (New York: Harper, 1960); Margot Liberty, ed., *American Indian Intellectuals* (Proceedings of the American Ethnological Society, 1976) (St. Paul, Minn.: West Publishing Co., 1978).

anthropologists in their own right. Francis La Flesche, who was appointed Ethnologist in the Bureau of American Ethnology in 1910, was one of the first of these. But that recognition came almost thirty years after he had begun to work with Alice Fletcher. Why did it take so long for him to be accorded professional status? The answer to this question indicates something of the process of professionalization in the early years of anthropology and hints at the difficulties that lay in the path of those American Indians who wanted to become anthropologists.

FRANCIS LA FLESCHE

Francis La Flesche was born on the Omaha Reservation in Nebraska in 1857. His mother was an Omaha woman, Tainne. His father, the prosperous trader Joseph La Flesche, was half Ponca Indian and half French. Although Joseph La Flesche had once been prominent in Omaha tribal affairs, during the years in which Francis was growing up he moved increasingly to the periphery of Omaha life, for he had become convinced that the Omahas must prepare themselves to live in a white man's world. In 1866 he gave up his position as a tribal chief, and he and his first wife Mary became leaders of the small group of Christians among the Omahas. The missionaries talked to Joseph La Flesche about his three wives, and he sent home the youngest one, who had no children. But Tainne he was reluctant to part with. Not until after she had given birth in 1873 to a third child, their son Carey, was she quietly removed from the family circle, although still provided for.[3]

Francis La Flesche therefore grew up in a doubly marginal position. His family was part white and increasingly moving toward white ways, a situation typical of many anthropological informants. But Francis was also marginal in his own family, for he was the eldest child of the second wife, the Omaha wife, whose place in the family group was becoming increasingly precarious. This situation encouraged his intense interest in the "salvage anthropology" of his day, the recording of customs and beliefs that were vanishing. Joseph La Flesche wanted Francis to learn the new ways of white civilization, and Francis did. But emotionally he was drawn to the old ways that his mother represented. Out of this complex situation was born not only his life's work, but also his relation to Alice Fletcher. In working with her and in allowing himself to be adopted by her, he took a new mother, one who represented both sides of the conflict he felt. Alice Fletcher was in many ways the epitome of white Victorian civilization toward which his father urged him. Yet her activities in studying and recording the Omaha way of life constituted, unconsciously if not consciously for him, a vindication of his Omaha mother.

Francis La Flesche studied at the mission school on the reservation until it was closed in 1869. Then he kept up his English by reading the Bible aloud to his father at the latter's insistence. As a young child he had participated in Omaha ceremonies, but in his early teens his father refused for him the tribal honor of piercing of the ears, fearing that this would hinder him as he moved outside the reservation. We know little of his life in his late teens and early twenties other than

[3]Norma Kidd Green, *Iron Eye's Family: The Children of Joseph La Flesche* (Lincoln, Neb.: Johnsen Publishing Co., 1969), p. 45; see also Margot Liberty, "Francis La Flesche: The Osage Odyssey," in Liberty, *American Indian Intellectuals,* pp. 44–59; and Liberty, "Native American Informants: The Contribution of Francis La Flesche," in John V. Murra, ed., *American Anthropology: The Early Years* (Proceedings of the American Ethnological Society, 1974) (St. Paul, Minn.: West Publishing Co., 1976), pp. 99–110.

that several times he took a major role in the annual buffalo hunts, probably against the wishes of Joseph La Flesche, who no longer participated. The question was soon academic, because by 1876 the buffalo were gone.

By a happy historical accident, however, we do know what Francis La Flesche was thinking and doing in the years just before Alice Fletcher arrived on the reservation. In 1878 John Wesley Powell, anticipating the founding of his Bureau of Ethnology, sent the linguist and former missionary James Owen Dorsey to collect information among the Omahas. Dorsey spent two years on the reservation, where Joseph La Flesche was one of his best informants and Francis La Flesche acted both as an interpreter and as a source of information. Of Francis, Dorsey wrote: "He has a fair knowledge of English, writes a good hand, and is devoted to reading. He is the only Omaha who can write his native dialect." Because Dorsey was primarily a linguist and was interested in making a record of the Omaha language as spoken by different individuals, he worked the material over very little and did not worry about reconciling conflicting opinions. The result is invaluable because it reveals differences in attitude among the Omahas.[4]

Joseph La Flesche is matter of fact about the old ways and frequently skeptical, as of claims that worship of thunder can stop the rain, and he often says that he does not know or cannot verify what someone has told Dorsey because he has never attended such ceremonies or feasts. Francis La Flesche, on the other hand, appears in Dorsey's transcriptions as young and wide-eyed, and as a much more up-to-date participant in Omaha tribal rites. Francis is especially interested in the sacred objects and in accidents or sudden deaths that followed violations of ritual procedure, and with good reason, for one of these involved the death of his young wife, Alice Mitchell, in 1878. She, her brother, and her father had all died suddenly after her father, the keeper of the sacred peace pipes, had yielded to pressure from the Omaha chiefs to take the pipes out of their coverings without first performing the mandatory preliminary rituals, which had been forgotten. Francis was also keenly interested in the ceremony of the anointing of the sacred pole, again in part for a personal reason. Many years before Joseph La Flesche had had to have his leg amputated, the result of an infected wound, and old men in the tribe had said it was punishment for his opposition to the ritual of the pole.[5]

When Dorsey asked Francis about the songs connected with the ceremony of the sacred pole, Francis told him it would be very difficult to get them, for it was considered sacrilegious to sing them apart from the ceremony and only those in charge of the ceremony knew them. But, Francis acknowledged, "I myself would

[4]James Owen Dorsey, *The Ȼegiha Language: Myths, Stories, and Letters* (U.S. Geographical and Geological Survey of Rocky Mountain Region, Contributions to North American Ethnology, 6) (Washington, D.C.: GPO, 1890), p. 2. Edward Sapir based an entire essay on Dorsey's frequent admission, "Two Crows denies this," i.e., denies what someone else had told Dorsey. Sapir argued that a cultural group ought to be considered a collection of individuals, not a monolithic whole. Edward Sapir, "Why Cultural Anthropology Needs the Psychiatrist," *Psychiatry*, 1938, *1*:7–12, rpt. in David G. Mandelbaum, ed., *Selected Writings of Edward Sapir in Language, Culture and Personality* (Berkeley/Los Angeles: Univ. California Press, 1958), pp. 569–577.

[5]J. Owen Dorsey, "Omaha Sociology," in *Annual Report of the Bureau of Ethnology, 3, 1881–1882* (Washington, D.C.: GPO, 1884), pp. 211–368, pp. 224, 235. Francis's mention of his wife's death is corroborated in a letter Joseph La Flesche wrote to his brother in September 1878, in which he noted without comment, "Frank took a wife last summer, but she is dead" (Dorsey, *Ȼegiha Language*, p. 488). The event is shrouded in confusion, however, because Francis La Flesche was granted a divorce from Alice Mitchell in 1884, according to local records. Either Alice Mitchell did not actually die a physical death but was considered "dead" by the family for some unexplained reason, or two different women were at various times called Alice Mitchell in court records and local newspapers. See Green, *Iron Eye's Family*, pp. 177–178.

like to know it all."[6] It seems likely that Francis La Flesche got the idea of collecting songs in their ceremonial context through his work with Dorsey.

In 1879, while Dorsey was still on the Omaha reservation, Francis left to travel east with his half sister Susette, the Ponca chief Standing Bear, and the Omaha journalist Thomas H. Tibbles on the famous journey that was to set off a wave of interest in Indian welfare among reformers. Helen Hunt Jackson heard them speak of the forced removal of the Poncas from their reservation in Nebraska and was inspired to write *A Century of Dishonor* (1881), her indictment of the nation's betrayal of Indian peoples. The trip also led to major changes in Francis La Flesche's life. His abilities came to the attention of Senator Kirkwood of Iowa, soon to be Secretary of the Interior, who got him a job as clerk in the Bureau of Indian Affairs in Washington, and in Boston the travelers met Alice Fletcher, a woman in her early forties just beginning her career in anthropology.

ALICE FLETCHER

Alice Cunningham Fletcher was born in 1838 in Cuba, where her parents had gone in a vain attempt to restore her father's health. Her father, Thomas Gilman Fletcher, a graduate of Dartmouth College and a lawyer in New York City, died when Alice was twenty months old. Her mother, Lucia Adeline Jenks, a "well-educated lady of Boston,"[7] eventually married again, a strait-laced man named Gardiner.

Alice Fletcher attended private girls schools, traveled in Europe, and worked for a time as a governess. In 1873 she helped to found the Association for the Advancement of Women, one of the first national women's organizations and modeled both in name and format on the American Association for the Advancement of Science. The group was led by very prominent women. Julia Ward Howe was the near-perennial president, and other officers included Mary Putnam Jacobi, the nation's leading woman physician, Maria Mitchell, the astronomer, Mary A. Livermore, a well-known journalist and public lecturer, and Frances B. Willard, the temperance leader. They sponsored an annual Women's Congress, held each year in a different city, at which papers were read on a wide variety of topics, from the serious—"Scientific Study and Work for Women" and "More Desirable Clothing for Women"—to the sentimental—"The Value of Simplicity in Childhood." After the papers there was discussion open only to members, and it was soon generally agreed that the discussions were the most interesting features of the sessions. One newspaper reported: "The ladies take the floor readily, and speak with remarkable aptness, pith, and beauty, without ranting or staginess, in gentle but distinct tones, with a felicity of expression, and directness rarely found, except in the best speakers." Alice Fletcher is mentioned in this account as one of those who is "always ready for the debate." For several years she organized the annual meetings, which were so popular that in Boston in 1877 one thousand women attended the sessions during the day and three thousand people came at night, when guests and husbands were allowed.[8]

[6]J. Owen Dorsey, "Omaha Sociology," p. 291.

[7]Edward H. Fletcher, *The Descendants of Robert Fletcher of Concord, Mass.* (Rand, Avery & Co., 1881), p. 50; Alice C. Fletcher, "Biographical Sketch," quoting p. 1, MS in Peabody Museum Papers.

[8]*Report of the Association for the Advancement of Women, Fifth Congress*, 1877 (Boston: Gunn, Bliss & Co., 1879); *Report of the Association for the Advancement of Women, Seventh Congress, 1879* (n.p., n.d.), quoting p. 2.

Meanwhile Fletcher had also begun a career in public speaking, and she was soon being described as a "lady of high literary attainments and a successful popular lecturer."[9] What is striking about these descriptions is that the emphasis is on the form, not the content, of her activities. Fletcher was well educated by the genteel standards of the day, earnest of purpose, and formidably skilled at organizing and public speaking. What she lacked, in common with many other similarly situated persons, was a cause to which to give these talents. She was to find a cause, not in the women's rights movement, which she soon abandoned, but in the science of anthropology and in her attempts to help the Indians.

Fletcher's career in anthropology began when she went to the Peabody Museum in Cambridge, seeking information for her series of "Lectures on Ancient America." At the Peabody she met the young and energetic curator, F. W. Putnam, and he encouraged her not just to popularize the activities of others but to take up the work herself.[10]

Fletcher seized the opportunity. She was assigned a seat at one of the work tables in the museum, and there for some weeks she examined artifacts and began to absorb Putnam's vision of anthropology, his scientific method, and his hopes for his museum. Above all Putnam stressed the need to get the facts. A student of Agassiz, he had broken with Agassiz to embrace Darwinism, but he was skeptical of the theory of social evolution of Lewis Henry Morgan. He wanted anthropology, which he thought of as the natural history of mankind, to be based on a solid foundation of empirical data—on the bodily forms, artifacts, languages, customs, and beliefs of various peoples, in particular, of the early inhabitants of the American continents. Putnam was interested in materials from Europe and elsewhere for comparative purposes, but he clearly thought that American anthropology ought to begin at home. He envisioned his museum as a great research center for anthropology, a place that would not just collect objects but would study them and publish the results of such study. Fletcher's loyalty to Putnam was to be steadfast for twenty-five years. Sitting there in the museum she set herself two goals: to make a contribution to anthropology and to help "the good professor."

Fletcher brought visitors to the museum, young women from her lecture audiences and wealthy friends whom she hoped to interest in its activities. In 1881 she brought Susette La Flesche and Thomas Tibbles, now husband and wife, who were on their second speaking tour in the East. Within a few months she was on her way to Nebraska to travel with them and experience Indian life at first hand.

Fletcher went west in 1881, intending to study the life of Indian women, but she found the topic difficult because it was closely related to sociological theorizing and because she did not know what was and what was not already known about Indian family life.[11] She soon turned to politics, where she was more at ease. The Omahas feared that they might be moved to Indian Territory as the Poncas had been. That winter Fletcher went to Washington, where she spoke in homes and churches and lobbied in Congress for a bill providing for individual allotments of reservation land to the Omahas. In Washington working for the Omahas, she

[9]E. Fletcher, *Descendants,* p. 50.

[10]In encouraging not only Fletcher but other women, including Cordelia Studley, Zelia Nuttall, and Erminnie Smith, to do work in anthropology, Putnam began the tradition (continued by Franz Boas) which has made anthropology one of the professional fields in the United States in which women have always been prominent.

[11]Fletcher to Putnam, 22 Mar. 1881; Fletcher to Lucien Carr, 3 Aug. 1881; Peabody Museum Papers.

conferred occasionally with young Francis La Flesche, and that summer when she returned to the West she had a new project, presumably at his suggestion: the study of Indian ceremonies. This was a much better topic. It would satisfy Putnam's desire for facts and could at the same time be an important original contribution, for there were very few full and accurate accounts of such ceremonies.

THE COLLABORATION OF FLETCHER AND LA FLESCHE

Fletcher spent much of the summer of 1882 among various bands of Sioux in Dakota Territory, investigating the Sun Dance, Ghost Lodge, and Elk Mystery ceremonies. That fall she went to Washington to work up her notes with Francis La Flesche. Her expenses were paid by William Thaw, the Pittsburgh railroad and steel magnate who had become interested in Fletcher's work while she was his wife's house guest. "I mean to get the music of the various dances from Frank, and write out all I can that is available," she told Putnam. Shortly thereafter she sent a progress report: "The Sec'y of the Interior is holding Red Cloud in the city that I may use him in my work and I am busy day and night. Frank is rendering very valuable assistance." She would have five Indian ceremonies ready for publication within a month, she promised, adding: "The Sun dance will have the ritual music, as well as illustrations. The music I can get from Red Cloud. thro Frank. I am negotiating for a person to write it out." In May she sent a description of the Omaha calumet or pipe dance, the final part of the study which was to be her first significant publication. She wrote: "Frank has gone over and over it and been very patient. It is all here but two ritual songs. These I could not get for Frank's mother died a week ago and he will not even hum a song. The ms. is entirely unique, and important."[12]

Meanwhile Fletcher had been appointed Special Agent of the Office of Indian Affairs to make land allotments to the Omahas, in accord with the Act of Congress of 7 August 1882. Francis La Flesche was detailed to go with her as interpreter. Hardly had they begun the Omaha allotments when Fletcher fell ill of inflammatory rheumatism. For five months she lay in bed at the Agency, and ever after she was crippled and walked with a limp. Francis La Flesche saw that she was well cared for and, as she began to recover, helped her continue the allotting work from her room. Their close association led to much gossip on the reservation and elsewhere, gossip which continued for years and which Francis attributed, probably rightly, to their political enemies. These included Thomas and Susette Tibbles, who wanted citizenship and legal rights for the Omahas rather than the complete program of acculturation (land allotments, Eastern schooling, frame house building, and communal grazing schemes) that Fletcher brought. "I wouldn't mind if they smeared me all over with mud, their nasty stories, but I am *very* sorry that they should talk about Miss F.," Francis wrote his half-sister Rosalie.[13]

Through the 1880s both Alice Fletcher and Francis La Flesche remained in the employ of the Office of Indian Affairs, he continuing as a clerk and she in a series of appointments as special agent. She wrote a report on Indian education and civilization. After the passage of the Dawes Act in 1887, she was sent to make

[12]Fletcher to Putnam, 20 Jan., 1 Feb., 1 May 1883, Peabody Museum Papers.

[13]Francis La Flesche to Rosalie Farley, 16 Dec. 1886, La Flesche Family Papers, Nebraska State Historical Society, Omaha, Nebr.

land allotments to the Winnebagos and the Nez Perces. Meanwhile they continued their ethnographic studies of traditional Omaha life. The Omahas, or at least the minority "citizens' party" led by Joseph La Flesche, were so grateful to Fletcher for helping them get legal titles to their land that they arranged to have their sacred objects, the sacred pole, sacred pack, and sacred shell, given jointly to Fletcher and La Flesche for transfer to the Peabody Museum for safekeeping.

The terms of the transfer show that the Omahas perceived Fletcher and La Flesche as working together as colleagues, as equals. But in fact they were not equal. La Flesche had the knowledge and contacts in the tribe and a desire to preserve the traditions of his people, but Fletcher had professional and social contacts with anthropologists, politicians, and potential patrons of anthropology, as well as strong professional aspirations. She needed Francis La Flesche, and it seems scarcely to have occurred to her that he could and perhaps would like to be something more than her assistant. Every year she hurried to meetings of the AAAS, to present papers herself and to learn what others were doing. In 1884 she was somewhat taken aback at James Dorsey's suggestion that Francis should join the association and that he, not she, should give a paper on the sacred pipes.

She agreed, although reluctantly and wrote rather petulantly to Putnam: "Frank wants to write on these pipes, and all things considered it is not best for me to object. Altho. I feel sure he cannot go into the matter as deeply as it ought to be done from lack of objective knowledge and general culture. Still I will help him. . . ." At the same time she was worried about what she would present if La Flesche spoke on the pipes. What topic would be suitable? "The truth is my dear Professor I want you to tell me just what to do." She continued: "Let me tell you my friend, it is a hard thing to work with Indians, to mix with them, to keep patience, pluck and a steady head. If I were to yield to the sentiments of my education I should wish I were dead. I get so weary and lonely, and feel so strange, but that is only when I am weak. I shan't give up. . . ." At other times she worried that Frank would be stolen from her: "They are after Frank in Washington. Mr. D. [Dorsey] tried to get a promise that F. would work for the Bureau, but I can hold him and he is valuable." And later: "You can't fancy how many plans are on foot to get him away from me and turn his work off from the Museum. He is to be a member of the Anthropological Society and he is also offered other things. He holds so far. This is all between us."[14]

In the summer of 1888 Francis La Flesche experienced once again what might be construed as the power of the Omaha sacred objects. His father had persuaded Smoked Yellow, the old man who had been the keeper of the Omaha sacred pole, to reveal to Francis La Flesche and Alice Fletcher its legend, but the old man was full of misgivings and did not begin to speak until Joseph La Flesche offered to accept for himself any penalty that might follow the revealing of these sacred traditions. The interview lasted for three days. Very shortly thereafter a fatal disease struck Joseph La Flesche, and within two weeks he was dead.[15]

Alice Fletcher wrote Putnam of this "great grief and disaster:"

> Francis' Father lies dead. He died Monday and F. is en route, too late to see his Father alive. Joseph La Flesche was to spend the winter in Washington and F. and I were to

[14] Fletcher to Putnam, 28 July, 11 Sept. 1884, 11 Dec. 1885, Peabody Museum Papers.

[15] Alice C. Fletcher and Francis La Flesche, *The Omaha Tribe* (Annual Report of the Bureau of American Ethnology, 27) (Washington, D.C.: GPO, 1911), p. 224.

> get all we needed from him. How this sore calamity will effect [sic] our work I cant tell. Poor F. will be heart broken. His Father was his idol. and I fear the boy will hardly have courage to go on. but he is noble and manly and his father's great interest in this work will be a help.

Francis La Flesche himself wrote matter-of-factly to Putnam several months later to report that they had secured "the sacred pole, idol or whatever name it can be given." He was not sure they would be able to carry out his father's plan of having the entire ceremony performed the following summer so that they could photograph it and take down all the songs, for "the people are yet in the shades of superstition and it will be hard to make them believe that my father's death was in no way the result of the taking away of the pole. . . . But still there may be some way of getting a few of the songs, at least." At the bottom of this letter, a note in Putnam's handwriting says, "This is an Omaha Indian!"[16]

LA FLESCHE'S BID FOR RECOGNITION

A crisis in the relation between Francis La Flesche and Alice Fletcher came in 1890, when Mrs. Mary Copley Thaw created the lifetime Thaw Fellowship for Alice Fletcher at the Peabody Museum, in honor of her late husband. The purpose of the fellowship was to enable Fletcher to write up her Indian studies unhampered by the need to earn a living. Francis La Flesche must have wondered why such support was forthcoming for her and not for him, but his protests came later and very covertly. One was his sudden expression of feeling about being listed on the title page of the music monograph. The other was his decision in 1891 to go to law school in the evenings. This particular move seems related to Fletcher's sponsorship at this time of the career of a Nez Perce Indian leader, James Reuben, whom she encouraged to enroll in law school. La Flesche received an LL.B. from National University in 1892 and an LL.M. the following year.[17]

Fletcher was noncommittal about La Flesche's decision. She seems never to have recognized the source of his discontent. Increasingly she regarded him not only as her assistant but as an adopted son, and she signed a written statement to that effect in 1891. In adopting La Flesche, Fletcher gave expression to her warm feelings for him but also and perhaps unconsciously to her sense of her relation to Indians, whom she tended to regard and occasionally to speak of as "children." Mary Copley Thaw helped Fletcher buy a home, and she chose to live not in Cambridge but in Washington, so that she could be with Francis and they could continue their work.

Almost throughout the 1890s Fletcher continued to talk about "my" work with the Omahas. She began to receive professional honors, beginning in 1895 when she was named vice-president (presiding officer) of Section H of the AAAS, the main gathering place for anthropologists before the founding of the American Anthropological Association in 1902. She was president of the Anthropological Society of Washington in 1903 and of the American Folk-Lore Society in 1905.

Despite a full time job and his law studies, Francis continued to help her with their ethnographic work. In 1891 he took John C. Fillmore, a music expert whom

[16]Fletcher to Putnam, 26 Sept. 1888; Francis La Flesche to Putnam, 3 Dec. 1888; Peabody Museum Papers.

[17]Green, *Iron Eye's Family*, p. 184.

Fletcher had hired to help them, out to the reservation to verify their transcriptions of Omaha songs. In 1892 he took the artist who was doing the illustrations for Fletcher's four articles for *Century Magazine,* and he supervised the final products to insure their accuracy. In 1895 he took a graphophone to the reservation and returned with eighty records of songs and ritual material. That winter his Ponca uncle Francis, for whom he had been named, spent ten days with them in Washington, and they recorded and transcribed several Ponca ceremonies to compare with the corresponding Omaha ceremonies. Fletcher reported to Putnam:

> Francis has worked night and day, and will have to work night and day for some time to come, with me. It would have been absolutely impossible for anyone to have gained this material without him. His uncle's regard for him has opened many a secret. Francis said jokingly that he thought he was doing enough for the Museum to win him a place among its workers. He certainly does labor assiduously and loyally.[18]

Did she really think he was only joking? It seems likely that she simply preferred not to think about it.

Around this time Francis was finally the recipient of a project of his own, when Edouard Seler in Germany wrote asking him to collect Omaha objects for the Berlin Museum. He worked hard and enthusiastically on this, and Fletcher, almost perversely unwilling to be excluded from anything, wrote to Putnam: "Francis is going to make an illustrated catalogue for his Berlin collection; would it not be well for me to make one for the [Peabody] Museum of the Omaha collection there?"[19]

The long overdue change in their working relationship finally began in 1898. The precipitating event was an invitation they received to make a presentation on "Aboriginal Music" at a National Congress of Musicians, to be held in Omaha, Nebraska in July 1898, in conjunction with the Trans-Mississippi Exposition. The invitation was addressed to La Flesche, but Fletcher immediately began to plan the event, deciding that Francis should go to the Omaha reservation, gather up a group of singers, and bring them to Omaha, where they would perform to illustrate papers to be given by John C. Fillmore and herself.[20]

Exactly what happened next is unclear, but somehow La Flesche made her understand that he wanted more recognition for their work. It seems unlikely that there was a major outburst, for that would have been very uncharacteristic of La Flesche, and furthermore, six months later Fletcher was still not exactly sure what had happened.[21] Francis La Flesche's friends frequently spoke of his "instinctive kindliness."[22] He was genuinely devoted to Fletcher and would not have wanted to

[18]Fletcher to Putnam, 22 Jan. 1896, also 25 Sept. 1895, Putnam Papers, Harvard University Archives.

[19]Fletcher to Putnam, 29 Sept. 1897, Peabody Museum Papers. There is no evidence that she did so.

[20]Fletcher to Putnam, 5 Apr., 26 June 1898, Putnam Papers.

[21]On 27 July 1898 Fletcher wrote Miss Mead, Putnam's secretary, praising the paper Francis had written for the Congress. Then she added, rather cryptically: "(Please say privately to Prof. Putnam not to hint to anyone or in any way the personal matter I spoke of last spring—it seems to have been but a passing shower. Science has greater charms. I would not have my little confidence betrayed, for it would do harm. You will do this, please but say nothing of it to me in a letter. You will understand the delicacy of my position. I spoke of it hastily, because I was so disappointed on account of the work, but I did wrong and was too anxious, it seems)" (Peabody Museum Papers).

[22]Hartley B. Alexander, "Francis La Flesche," *American Anthropologist,* 1933, *35*:328–331, p. 330.

Putnam at the Peabody Museum or John Wesley Powell at the Bureau of American Ethnology) who would then supervise his or her work and provide jobs, institutional support, research funds, and publishing outlets as needed. Francis La Flesche had none of these things, except through Alice Fletcher. His case illustrates the danger in having as a mentor someone on the middle level in the professional hierarchy, for instead of promoting him she used his work to promote herself. By the time Powell learned of his work and tried to attract him to the Bureau of American Ethnography, La Flesche's loyalty to Alice Fletcher evidently did not allow him to abandon her and the rival institution, the Peabody Museum, for which she worked.

In addition, La Flesche, like other American Indians, had to struggle with powerful deterrents in his own personal and cultural situation. These included the fear of supernatural sanctions against the revealing of sacred knowledge, for which there was much seeming evidence, and the gossip and hostility of his own people who were jealous of his position or feared the consequences of what he was doing. The awkwardly phrased letter to Putnam speaks to this issue: La Flesche was trying on the posture of detachment he considered appropriate for Putnam, but it was not quite what he himself believed. La Flesche would also have been uncomfortably aware that the white people whose institutions were to be repositories for the Omaha heritage might very well laugh at the sacred things they were being given.[28] Alice Fletcher did not laugh, and he did put his trust in her.

Finally, Francis La Flesche had difficulty being fully accepted as a professional anthropologist because as a member of another culture, his concerns and his categories of thinking were not always congruent with those of Western investigators. Here the best example is Fletcher and La Flesche's study of Omaha Indian music.

Francis La Flesche's original interest, which he passed on to Alice Fletcher, had been in preserving Omaha songs as part of the rituals and ceremonies in which they were embedded. Together they transcribed the melody lines, texts, and rhythms of eight songs belonging to the Omaha calumet or peace pipe ceremony and included these with a full account of the ceremony and its cultural context. Some of the songs were assigned to particular keys, such as E major or D flat major, but Fletcher noted: "The Indian musical scale having different subdivisions from our own, the original songs lose a little of their native tone by being forced into our conventional scale."[29] This was an insight she clearly owed to Francis La Flesche, for he was a good singer and knowledgeable both in the traditions of his tribe and in the conventions of Western harmonic music, thanks to his years in the mission boarding school.

When, however, in 1890 a rival enterprise, the Hemenway Southwestern Expedition under the patronage of Mrs. Mary Hemenway, began to study Zuni Indian music with the help of Dr. Benjamin Ives Gilman, lecturer at Harvard on the psychology of music, Fletcher feared that her work with Francis La Flesche on Omaha music (by now a collection of several hundred songs with their cultural context) might be overshadowed, and she decided to hire her own expert. She persuaded John C. Fillmore, a pianist and founder of the Milwaukee School of

[28]Fletcher had to contend with Indian awareness that whites laugh at Indian ways; see Alice C. Fletcher, "Indian Ceremonies," *Sixteenth Annual Report of the Peabody Museum*, 1884, pp. 260–333, p. 280 (n.).

[29]*Ibid.*, p. 308 (n.).

Music, to work with her in Washington and also sent him to the Omaha reservation with La Flesche so that he could verify her transcriptions. For the next seven years, until his death in 1898, Fillmore was the leading authority in the United States on the technical aspects of Indian music. He and Gilman were soon embroiled in a controversy over the nature of Indian music. Briefly, Fillmore, who transcribed the Indian songs in four-part harmonies on a Western diatonic scale, argued that the music was based on a "natural," latent feeling for the harmonic relations between notes, in fact, for the same harmonies that are more overtly expressed in Western classical music. Gilman, on the other hand, used a harmonium and other instruments to record the exact tone that was being sung, accurate to within one twelfth of the interval between two notes on a distonic scale, and he concluded that the Indians did not yet have fixed scales. Eager to assign Indian music to a rank on an evolutionary ladder, he ascribed to it a primitive, random character.[30]

From a contemporary perspective they were both wrong, Francis La Flesche's original insight, that Indian music uses fixed scales but scales which are different from those used in Western classical music, being more nearly correct. This theoretical controversy, based on one component of Indian music, scale, drew attention away not only from La Flesche's more accurate perceptions but also from his concern to study the intimate connection between music and all other aspects of Indian life.

That La Flesche's case was not unique is illustrated by the career of a near contemporary of his, the part Tuscarora Iroquois anthropologist John Napoleon Brinton Hewitt (1859–1937).[31] Hewitt's career in anthropology also began when he served as assistant to an ethnographer, in his case Mrs. Erminnie A. Smith of Jersey City, New Jersey. Mrs. Smith's scientific credentials were superior to Alice Fletcher's (she had studied at the Freiberg School of Mines as well as at Heidelberg and Strasburg while accompanying her four sons during their European education), but she had had no training in anthropology. With Hewitt's help she published "Myths of the Iroquois," compared various Iroquois languages, and compiled a grammar and dictionary of the Tuscarora language. Hewitt worked with her for six years, supporting himself during part of that time with a job with a railroad company. He might have had a career very like Francis La Flesche's except that Erminnie Smith died suddenly in 1886 at the age of forty-nine. Almost immediately thereafter the Bureau of Ethnology hired Hewitt to carry on her work, and he remained in the employ of the Bureau for the rest of his life.[32]

[30]John C. Fillmore, "A Structural Analysis of the Peculiarities of the Music," in Alice C. Fletcher aided by Francis La Flesche, *A Study of Omaha Indian Music* (Archaeological and Ethnological Papers of the Peabody Museum of American Archaeology and Ethnology, 1[5]) (Cambridge, Mass., 1893); Fillmore, "A Study of Indian Music," *Century Magazine*, 1894, *47*:616–623; Fillmore, "What Do Indians Mean to Do When They Sing, and How Far Do They Succeed?" *Journal of American Folk-lore*, 1895, *8*:138–142; Fillmore, "The Scientific Importance of the Folk-Music of Our Aborigines," *Land of Sunshine*, 1897, *7*:22–25; Fillmore, "The Harmonic Structure of Indian Music," *Amer. Anthropol.*, 1899, *1*:297–318; Fillmore, "The Zuni Music as Translated by Mr. Benjamin Ives Gilman," *Music*, 1893, *5*:39–46; Benjamin Ives Gilman, "Zuni Melodies," *Journal of American Archaeology and Ethnology*, 1891, *1*:63–91; Gilman, "Hopi Songs," *J. Amer. Archaeol. Ethnol.*, 1908, *5*.

[31]John R. Swanton, "John Napoleon Brinton Hewitt," *Amer. Anthropol.*, 1938, *40*:286–290. Hewitt described his work with Smith in a letter published in *In Memoriam: Mrs. Erminnie Adele Platt Smith, 1837–1886* (Boston: Lee & Shepard, 1890), pp. 47–49.

[32]It is significant that both La Flesche and Hewitt worked with woman ethnographers, and in part this contributed to their dilemma. Alice Fletcher and Erminnie Smith were remarkable women,

American Indians of a slightly later generation had an easier time beginning careers in anthropology, as their educational opportunities increased and as the profession of anthropology expanded. Both Arthur C. Parker and William Jones came into contact with F. W. Putnam while they were still young students (Jones as a freshman at Harvard, Parker as one of "Putnam's boys" at the American Museum of Natural History), and Putnam encouraged them to take up professional careers in anthropology, which they did.[33] But American Indians and members of other much-studied groups are still more likely to meet anthropologists eager to make use of their knowledge than anthropologists able to help them up the ladder of a professional career.[34] That Francis La Flesche's case is but an exaggeration of a common situation only adds to its poignancy. It also serves to remind us how fundamental and creative (and not yet adequately recognized) have been the contributions of native Americans to anthropology.

---

talented, energetic, appreciative, and as generous as they were able to be in their situations. But as women they were on the periphery of their profession, outside the positions of power. They could not offer professional positions to their associates, for they were still trying to get them for themselves.

[33]See Hazel W. Hertzberg, "Nationality, Anthropology, and Pan-Indianism in the Life of Arthur C. Parker (Seneca)," *Proceedings of the American Philosophical Society,* 1979, *123*:47–72; Henry Milner Rideout, *William Jones: Indian, Cowboy, American Scholar, and Anthropologist in the Field* (New York: Frederick A. Stokes, 1912). Jones's career came to an abrupt and tragic end. After receiving a Ph.D. from Columbia in 1904, he wanted to return to the Sauk and Fox Reservation to do research among his own people, but funds were available only for overseas research. He was murdered while in the field in the Philippines in 1910.

[34]Once again fiction leads the way in revealing us to ourselves. In *Tent of Miracles* (*Tenda Dos Milagres,* 1969) the Brazilian novelist Jorge Amado brilliantly imagines the situation of a native anthropologist. I am grateful to James Ito-Adler for this reference.

in Algeria, France's first and most important colony, older works have looked at how medicine eradicated endemic disease and improved sanitary conditions; they have also examined the social significance of such developments for the colonizer or the colonized, including the resistance of the latter to the scientific blandishments of the former. It is now considered a truism that colonial health policies were seldom impartial, geared as they were to preparing insalubrious areas for colonial settlement and otherwise promoting colonial rule. Recently, attention has shifted to the medical discourses in and on Algeria and to the emergence of medicine as a factor in the empowerment of the colonizers.[2]

One aspect of the role of medical agencies in Algeria, hitherto overlooked, is their contribution to the creation of the hierarchical paradigms of culture and ethnicity essential to the development of a French social, cultural, and political domain. Algeria was a multiethnic society, comprising diverse groups of peoples, of which the two largest, the Arabs and the Kabyles (Berbers), provoked the most interest.[3] The medical corps accompanying the expeditionary force was present primarily to minister to the troops, but as the period of conquest extended into pacification and then into occupation, the indigenous population was inevitably drawn into the French medical sphere. The first concern, from a medical point of view, was the transmission and eradication of disease. This provoked an interest in the various diseases present among the local population. Specific research instructions to follow up this interest were in fact given to physicians going to Algeria. The question then arose as to whether ethnicity or race could be disease specific. Was the local population immune to certain diseases? Or, more important, did their lifestyle, culture, and morality influence the transmission of diseases and epidemics? Customs and mores thus became relevant subjects of inquiry.[4] There was also the question of medical methodology

Mackenzie, ed., *Imperialism and the Natural World* (Manchester: Manchester Univ. Press, 1990); Lewis Pyenson, *Civilizing Mission: Exact Sciences and French Overseas Expansion* (Baltimore: Johns Hopkins Univ. Press, 1993); Paolo Palladino and Michael Worboys, "Science and Imperialism," *Isis*, 1993, *84*:91–102 (for a discussion of Pyenson's work); and Michael A. Osborne, *Nature, the Exotic, and the Science of French Colonialism* (Bloomington: Indiana Univ. Press, 1994). On the role of medical agencies see Roy Macleod and Milton Lewis, eds., *Disease, Medicine, and Empire* (London: Routledge, 1988); David Arnold, ed., *Imperial Medicine and Indigenous Societies* (Manchester: Manchester Univ. Press, 1988); Nancy E. Gallagher, *Medicine and Power in Tunisia* (Cambridge: Cambridge Univ. Press, 1983); Paul Rabinow, *French Modern* (Cambridge, Mass.: MIT Press, 1989); and Teresa Meade and Mark Walker, eds., *Science, Medicine, and Cultural Imperialism* (New York: St. Martins, 1991).

[2] Among older works that look at the history of medical agencies in Algeria see J.-B. Vincent, "Les médecins de l'Algérie au temps de la conquête" (Ph.D. diss., Univ. Algiers, 1914); and Jean Tremsal, *Un siècle de médecine coloniale française en Algérie* (Tunis, 1929). Akin to these are Léon Lapeyssonnie, *La médecine coloniale: Mythes et réalités* (Paris: Seghers, 1988); and Yvonne Turin, *Affrontements culturels dans l'Algérie coloniale: Écoles, médecine, religion, 1830–1880* (Paris: Maspero, 1971) (Turin examines the areas of contact and conflict in the medical "dialogue" between the French and their Algerian subjects and analyzes the confrontation as a clash between modernity and tradition). Works that stress the "partiality" of colonial health policies include Ann F. La Berge, "The Conquest of Algeria and the Discourse on Public Health in France: The Context of Colonial Medicine in Algeria," in *Mundialización de la ciencia y cultura nacional*, ed. Antonio Lafuente *et al.* (Madrid: Doce Calles, 1991), pp. 539–544; Anne Marcovich, "French Colonial Medicine and Colonial Rule: Algeria and Indochina," in *Disease, Medicine, and Empire*, ed. Macleod and Lewis, pp. 103–117; and Radhika Ramasubban, "Imperial Health in British India, 1857–1900," *ibid.*, pp. 38–60.

[3] The Kabyles, from the region of what is now known as Kabylia, were the largest of four groups of Berbers. The others were the Chaouia of the Aurès Mountains in southeastern Algeria, the Mozabites of the Mzab in the northern Saharan region, and the Tuareg of the central Sahara. The Kabyles resided in the three mountain ranges that today make up Greater and Lesser Kabylia. The remaining ethnic groups were Turks, Kouloughlis (descendants of Turks and North African women), Andalusians (descendants of the Moors exiled from Spain), blacks (mainly soldiers, emancipated slaves, and slaves), Jews, and "infidels" (non-Muslim slaves and renegades, many of whom held high office under Turkish occupation).

[4] For examples of interest in diseases of the local population see Dr. Haspel, "Des maladies du foie en Algérie," *Recueil de Mémoires de Médecine de Chirurgie et de Pharmacie Militaires*, 2nd Ser., 1845, *58*:1–231; and M.

and an interest in Arab medical practices. The publication of an early article attempting to rehabilitate Arabs and their medical achievements suggests that the concept of European superiority was prevalent from the outset. The military physician Auguste Warnier, who was to become a leading figure in the colony, claimed in his doctoral thesis on Arab medicine that in certain areas it was superior to French medicine and that the French would do well to emulate such practices. But these were minority views. By the time the Scientific Commission for the Exploration of Algeria (1840–1842) was under way, the medical persona as a civilizer as well as a healer had been created.[5]

In this essay I examine two aspects of French imperial medicine in Algeria. The first is the development of medical practitioners' ideas about the indigenous population, from their ideological underpinnings in France to their application in the colonial context. As we shall see, medical discourse in the colony was a factor in the formation of both intellectual and popular attitudes in the metropolis toward the indigenous peoples of the colony. The periodization of this process is of special significance, for it took place prior to a time when the superiority of French medicine could be delineated in technological terms and was thus measured according to philosophical, sociological, or moral indexes. Second, I attempt to show that by establishing themselves as a cultural, social, and political presence in Algeria, the members of the medical profession made an important contribution to the development of a French colonial identity and to the creation of a French intellectual space in the colony.

THE MEDICAL PENETRATION OF ALGERIA

The period 1830–1870 was one of military rule in colonial Algeria. (See Figure 1.) Although the territory occupied by France was transformed into three French departments in 1848, conquest and pacification were not completed until 1857, with the fall of the mountainous region of Kabylia. During this period the medical corps in Algeria was essentially a military one. The doctors, surgeons, and pharmacists were army personnel who were conscripted, posted, and remunerated in accordance with army directives. Some, like Eugène Bodichon and Auguste Warnier, eventually set up practices as civilians, but the majority remained within the confines of the military. Ostensibly their purpose in the army was to maintain the health of the soldiers and, thus, the efficiency of the fighting force, but in fact their role extended well beyond this primary objective.

The medical personnel accompanying the army to Algeria represented a quarter of France's peacetime medical corps. Ninety-five pharmacists and 176 surgeons disembarked at Sidi Ferruch, and their number was soon augmented as civilian hospitals, run mainly by military doctors, were set up in successive Algerian towns as they fell to the French. These hospitals—three of them in the city of Algiers alone—were very well kept by contemporary standards and much larger than the number of patients warranted. They were

Hattute, "Des gangrènes spontanées chez les Kabyles," *ibid.*, 3rd Ser., 1868, *21*:518–533. For the instructions see Etienne Serres, "Instruction médicale pour la Commission scientifique d'Afrique (lue à l'Académie des sciences dans sa séance du 26 mars 1838)," *Gazette Médicale de Paris*, 1838, *6*:225–229, on p. 227. For an essay considering a local custom see M. Delange, "La circoncision chez les Arabes," *Rec. Mém. Méd. Chir. Pharm. Milit.*, 1868, *21*:181–185.

[5] J. Amoureux, "Histoire (essai) et littéraire sur la médecine des Arabes," *Gaz. Méd. Paris*, 1831, *2*:207, 214; and Auguste Warnier, *Du traitement des plaies d'armes à feu chez les Arabes bedouins de l'Algérie* (Montpellier: Jean Martel Ainé, 1839). In the last paragraph of its medical instructions to the Scientific Commission (1840–1842) to Algeria, the *Gazette Médicale de Paris* spells out its civilizing role. See Serres, "Instruction médicale," p. 229.

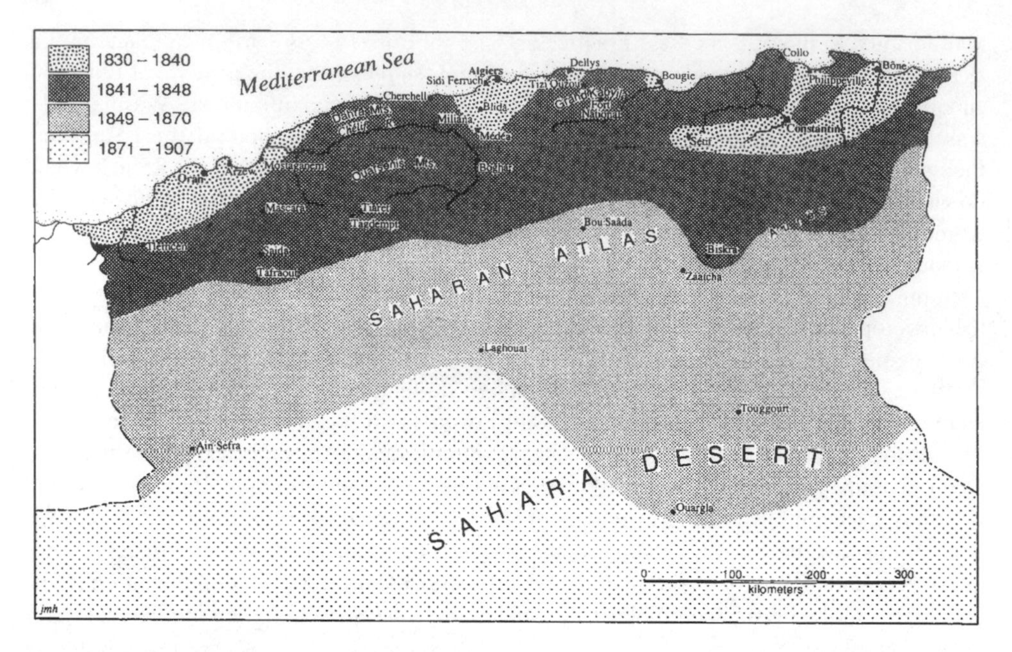

Figure 1. Map showing the stages of the French conquest of Algeria. (From John Ruedy, Modern Algeria *[Bloomington: Indiana University Press, 1992]; used by permission of Indiana University Press.)*

not, therefore, the makeshift arrangements of a temporary occupation but, rather, facilities with long-term potential. By 1840 there were nine such hospitals. As French penetration proceeded and military administrative units, the *cercles* of the Bureaux Arabes, were set up, a military physician was assigned first to the divisional headquarters and then to the *cercles* themselves. By 1867 even remote tribal areas had their own doctors.[6] Who then were these doctors, and what was their role in Algeria? To answer these questions, we must first look at the circumstances and intellectual influences that shaped these physicians' thought.

FORMATIVE INFLUENCES

The Revolutionary and Napoleonic Eras

Although the methodological, philosophical, and ideological factors that shaped the thought and activities of the French medical practitioners in Algeria are well documented, they need to be mentioned briefly here to put the colonial physicians' activity into its political and cultural perspective. The politics of the French Revolution and the ensuing Napoleonic era were primary factors in this development. In the first place, between 1789 and 1795 Revolutionary reforms were instituted that combined French medicine and surgery in a single discipline for the first time and encouraged the emergence of new concepts

[6] On the size of the medical corps in Algeria see Comité d'Histoire du Service de Santé, *Histoire de la médecine aux armées*, 3 vols. (Paris: Lavauzelle, 1984), Vol. 2, p. 133. On the setting up of hospitals see Dr. Pointe, "Relation médicale d'un voyage à Alger," *Gaz. Méd. Paris*, 2nd Ser., 1835, *3*:609–613, 625–629, 641–647, on pp. 642, 643; and Turin, *Affrontements culturels* (cit. n. 2), p. 13. Kenneth Perkins, *Quaids, Captains, and Colons* (New York: Africana, 1981), pp. 132–135, reports the penetration of doctors to remote areas.

and practices, such as social hygiene and statistical science. Second, the Revolution initiated the politicization of French physicians. Medical involvement in French politics continued throughout the nineteenth century, and political activism in the colonies was its logical extension.[7] This activism was to place physicians among the movers—and in some cases the prime movers—of the colony.

The Napoleonic period saw the linking of French medical agencies both to the military and to imperial ventures. During the early decades of the century the majority of the graduates from the Paris medical school went straight into military service. As far as the medical corps was concerned, the Napoleonic battlefields served both as a laboratory and as a political arena. The sociability of the military ensured that medical personnel came into daily contact with influential figures who had distinguished themselves not only on the battlefield but in the political arena as well. Such contact reinforced the tendency to mix medicine with politics. Methodologically, too, the Napoleonic period had relevance for later colonial activities. Surgeons and physicians, who encountered a wide variety of medical situations and dealt with large numbers of cases, established the tradition of comprehensive reports utilizing newly created timesaving systems of numerical analysis.[8] Initially the accumulation of such data was a type of medical shorthand from which to analyze patterns of disease and evaluate diagnostic methods. Later in the century it would encompass not only the physical but also the social and moral ills of society. This system of reports was continued in Algeria. Hygiene also received a boost during the Napoleonic campaigns, when casualty experience demonstrated its value in the prevention of infection and further disease. As a result, large numbers of French military and ex-military men became active hygienists. Casualty hygiene was of course highly relevant to Algeria; moreover, its wider role in maintaining a healthy European population in the face of high morbidity due to unexplained fevers and disease would elevate the perceived value of hygiene still further. Hygiene was also important in the colonial context for its link to the notion of "health through civilization," an aspect of medical ideology in France encouraged by the well-known professor of medicine Fédéric Bérard (1789–1828). As part of the emerging concept of social medicine—whose aim, expressed in the first issue of the *Annales d'Hygiene Publique et de Médecine Legale* (1829), was the perfection of mankind—"health through civilization" became an important part of colonial discourse, although in Algeria, as in France, demographic concerns were implicated in the drive to improve conditions of hygiene.[9]

[7] For the medical revolution carried out during the Revolutionary and Napoleonic periods see Williams, *Physical and Moral*, pp. 67–114. For the significance of social hygiene to the development of French medicine see *ibid.*, pp. 151–166; Jacques Léonard, *La médecine entre les pouvoirs et les savoirs* (Paris: Aubier Montaigne, 1981), pp. 150–159; and Erwin H. Ackerknecht, *Medicine at the Paris Hospital, 1794–1848* (Baltimore: Johns Hopkins Press, 1967) (hereafter cited as **Ackerknecht, *Medicine at the Paris Hospital***), pp. 149–153. For accounts of the emergence of statistical science and its impact on society see Hacking, *Taming of Chance* (cit. n. 1); Porter, *Trust in Numbers* (cit. n. 1); and Matthews, *Quantification and the Quest for Medical Certainty* (cit. n. 1). For nineteenth-century developments see Ackerknecht, *Medicine at the Paris Hospital*, pp. 183–186; Léonard, *Médecine entre les pouvoirs et les savoirs*, pp. 204–221; and Jack Ellis, *The Physician-Legislators of France: Medicine and Politics in the Early Third Republic, 1870–1914* (New York: Cambridge Univ. Press, 1990).

[8] William Coleman, *Death Is a Social Disease* (Madison: Univ. Wisconsin Press, 1982), pp. 15, 23. General Maximilien Foy and Field Marshal Jean de Dieu Soult, for example, were both military and political figures.

[9] On military men as hygienists see Ackerknecht, *Medicine at the Paris Hospital*, p. 149. On the hygiene movement in France and its regional differences see Williams, *Physical and Moral*, pp. 140–166. There is disagreement as to whether circumstances in Algeria influenced the ideas espoused by metropolitan hygienists. La Berge, "Conquest of Algeria" (cit. n. 2), p. 543, argues that they did not. Michael Osborne, on the other hand, argues that they did; see Osborne, "La renaissance d'Hippocrate: L'hygiène et les expéditions scientifiques en

It was during Napoleon's Egyptian campaign that the triangular relationship of military activity, medicine, and imperialism was truly established. In Egypt military activity extended beyond mere conquest to cultural and scientific concerns. The French occupation engendered a period of intense scientific research that subjected the country to exhaustive scrutiny and culminated in the publication of the twenty-three-volume *Description de l'Égypte.* The medical corps, as an essential part of the military presence, was called upon to participate.[10] These activities contributed to establishing the tradition among imperial physicians of looking beyond the immediate demands of their patients to the appraisal not only of the local population but also of the flora, fauna, and topography of the area in which they practiced.

The École de Médecine de Paris and Utopian Philosophy

Institutionally, the École de Médecine de Paris threw the longest shadow over the medical corps in Algeria. It was here, and in the medical schools of the French provinces, for which it was the prototype, that the majority of the physicians and surgeons serving in Algeria received their training. Not only did it hold pride of place as the center of medical innovation in the first half of the nineteenth century; of equal relevance was its link to Saint-Simonianism. Saint-Simon and his Utopian disciples were closely associated with three major state institutions, of which the École de Médecine de Paris was the only nonmilitary one.[11] The value placed by Saint-Simonians on the role of science and individual scientists

Égypte, en Morée et en Algérie," in *L'invention scientifique de la Méditerranée: Égypte, Morée, Algérie,* ed. Marie-Noelle Bourget *et al.* (Paris: École des Hautes Études en Sciences Sociales, 1998), pp. 185–204 (a translation of "Resurrecting Hippocrates: Hygienic Sciences and the French Scientific Expeditions to Egypt, Morea, and Algeria," in *Warm Climate and Western Medicine: The Emergence of Tropical Medicine, 1500–1900,* ed. David Arnold [Amsterdam: Rodopi, 1996], pp. 81–99). The first issue of the *Annales d'Hygiene Publique et de Médecine Legale* is quoted in Léonard, *Médecine entre les pouvoirs et les savoirs* (cit. n. 7), p. 150. Health through civilization was the subject of Bérard's maiden lecture on taking up the chair of hygiene at Montpellier in 1825. For Bérard's activities as a Montpellier "eclectic" see Williams, *Physical and Moral,* pp. 140–151.

[10] Published under the full title *Description de l'Égypte; ou, Recueil des observations et de recherches qui ont été faites en Égypte pendant l'expédition de l'Armée française publié par les ordres de sa Majesté l'Empreur Napoléon le Grand* (Paris: Imprimerie Impériale, 1809–1828), it comprised nine volumes of monographs and fourteen volumes of plates, maps, etc. It was divided into three sections entitled "Antiquités," "État Moderne," and "Histoire Naturelle." For a contribution from a medical military man see M. le baron Larrey, "Notice sur la conformation des différentes races qui habitent en Égypte, suivie de quelques reflexions sur l'embaumement des momies," *ibid.,* Vol. 6. The importance of the Egyptian campaign to the development of Western discourse on the Orient is discussed in Edward W. Said, *Orientalism* (New York: Vintage, 1979); and Said, *Culture and Imperialism* (New York: Vintage, 1993). For a contemporary account of French activities in Egypt see Al-Jabarti, *Chronicle of the French Occupation,* trans. Shmuel Moreh (Princeton, N.J.: Markus Wiener, 1993). For the impact of the Egyptian campaign on imperial activity in Algeria see Patricia Lorcin, *Imperial Identities* (New York: Tauris/St. Martins, 1995) (hereafter cited as **Lorcin, *Imperial Identities***), pp. 102–107. For its role in the scientific imaging of the Mediterranean see Bourget *et al.,* eds., *Invention scientifique.* See also Charles C. Gillispie, "Scientific Aspects of the French Egyptian Expedition, 1798–1801," *Proceedings of the American Philosophical Society,* 1989, *133*:447–474.

[11] In 1808 the École de Médecine de Paris came under the aegis of the newly created university and was known as the Faculté de Médecine de Paris. To avoid confusion I will refer to it here as the École de Médecine. Of the provincial medical schools, Montpellier and Marseilles had the strongest links to Algeria. See Ackerknecht, *Medicine at the Paris Hospital,* p. 147. For an account of the conceptual development of medical science in Paris in the early part of the century see Matthews, *Quantification and the Quest for Medical Certainty* (cit. n. 1), pp. 3–86. The other two institutions with links to Saint-Simonianism were the École Polytechnique and the École d'Application de Metz, both military institutions at the time. For the link between Saint-Simonianism and medicine see Williams, *Physical and Moral,* pp. 213–224. For the impact of Saint-Simonianism and other Utopian philosophies on the École de Médecine de Paris see Gaston Pinet, *Écrivains et penseurs polytechniciens* (Paris: Ollendorff, 1898), pp. 224–228.

in shaping an improved society and on the necessity of a hierarchical order to achieve this end led to the association of the Utopian philosophers with scientific and military establishments.

Utopian philosophy contained concepts that were relevant to the colonies. Saint-Simon, like Hegel, Condorcet, and Courtet de l'Isle (himself a disciple of Saint-Simon), considered civilization to be a ladder up which people progressed in accordance with the degree of social harmony and political or scientific development in their society. Saint-Simonian doctrine advocated the achievement of well-being through economic endeavor, efficient technology, and industrial development. The ideal society was a paternalistic hierarchy in which a natural elite (of scientists, technocrats, and industrialists) guided the masses to improved living standards and higher states of civilization. The postulate implicit in the doctrine—that class defined man's nature and destiny—was picked up by one of Saint-Simon's disciples, Augustin Thierry, who linked it to race. Saint-Simon himself believed in the superiority of Europeans and their civilization and advocated populating the globe with Europeans so as to transform it along European lines. These beliefs were carried to Algeria by the graduates of the institutions to which the Utopian philosophers were linked.[12] But the impact of Saint-Simonianism extended beyond French institutions to intellectuals and laymen throughout Europe, thus incorporating Algeria into a frame of intellectual interest that extended beyond national boundaries. Doctors were enthusiastic disciples, and Saint-Simon himself had numerous close medical friends. In Algeria, three of the colony's leading advocates of Saint-Simonianism were physicians.[13] The interlacing of Utopian philosophy and French medical science, with its moral and social dimensions, was of considerable significance to the activities of the medical personnel in Algeria.

THE MEDICAL CORPS IN ALGERIA IN THE 1830S: CONSTRUCTING THE COLONIAL PARADIGM

We come now to Algeria and the role the medical corps played there. Generally speaking, the "new" French physicians were expected to be "vigilant sentinels of the truth," true protagonists of the *médecine d'observation* or, as the *Gazette Médicale de Paris* put it, "medical journalists." As the "representatives of a scientific democracy," they were to report anomalies and deviations, scrupulously analyze available data, and shape the future of medical science by the judicious marriage of new trends and the best of the past.[14] The medical corps embarking on their assignments in Algeria, therefore, meant both to minister to the needs of the troops and to accumulate large amounts of data with a view to furthering scientific research.

[12] Augustin Thierry, *Histoire de la conquête de l'Angleterre par les Normands* (Paris, 1825); and Claude Henri de Saint-Simon, "De la réorganisation de la société européen," in *Oeuvres de Saint-Simon,* Vol. 1 (Paris: Dentu, 1868), p. 204. On Saint-Simon and his doctrine and influence see *The Doctrine of Saint-Simon: An Exposition,* ed. and trans. G. G. Iggers (New York: Schocken, 1972); Frank E. Manuel, *The New World of Saint-Simon* (Cambridge, Mass.: Harvard Univ. Press, 1956); Manuel, *Prophets of Paris* (Cambridge, Mass.: Harvard Univ. Press, 1962); and Pinet, *Écrivains et penseurs.* For the impact of Saint-Simonianism in Algeria see Marcel Emerit, *Les Saint-Simoniens en Algérie* (Paris: Belles Lettres, 1941); Magali Morsy, ed., *Les Saint-Simoniens et l'Orient: Vers la modernité* (Aix-en-Provence: Édisud, 1989); and Lorcin, *Imperial Identities,* pp. 97–146.

[13] On the wider European impact of Saint-Simonianism see G. G. Iggers, "Introduction," in *Doctrine of Saint-Simon,* ed. and trans. Iggers, p. xxiv. For doctors' interest see Ackerknecht, *Medicine at the Paris Hospital,* pp. 149, 184; and Emerit, *Saints-Simoniens en Algérie,* p. 40. The doctors in question in Algeria were Antonini, Millot, and Warnier.

[14] Introduction to the first edition of the *Gaz. Méd. Paris,* 1830, *1*:1. The concept of *médecine d'observation* was promoted by Georges Cabanis and his contemporaries. See Williams, *Physical and Moral,* pp.80–81; and Ackerknecht, *Medicine at the Paris Hospital,* p. 3.

Figure 2. *The siege of Constantine. (From the archives of the Musée du Service de Santé des Armées au Val-du-Grâce.)*

In the years immediately following the French arrival in Sidi Ferruch, the doctors and surgeons were primarily occupied with treating the sick and wounded and establishing the necessary institutional framework for dealing with them. As the conquest fanned out from Algiers, eastward to Bougie (1833) and Constantine (1837) and westward to Mascara (1835) and Oran (1836), and the hospitals set up in the wake of this progression settled into an established routine, physicians and surgeons could devote more time to matters beyond immediate medical care. (See Figure 2.) By 1837, according to Jean-Pierre Bonnafont, *médecin principal,* who served in Algeria from 1830 to 1842, the physician taking up his post in Algeria was expected to study the geography, climate, and conditions of hygiene prevailing in the area. Assessing the salubrity of a given region was of "vital importance" in newly occupied territory destined to accommodate a new population.[15] In addition to treating the sick and cataloguing endemic diseases, therefore, physicians were expected to use their scientific training to accumulate any data that could be of interest in sizing up the colony. A thorough knowledge thus obtained would permit a decision as to the desirability of long-term colonization. (The debate over colonization, which had started very soon after the 1830 invasion, gathered momentum throughout the first decade of the

[15] Jean-Pierre Bonnafont, "Amélioration de l'état sanitaire des environs d'Alger," *Gaz. Méd. Paris,* 2nd Ser., 1837, *5*:481–489, on p. 481. Bonnafont was a graduate of Montpellier. Titles such as "*médecin principal*" denoted military rank, as follows: *élève sous-aide*/second lieutenant; *médecin sous-aide*/lieutenant; *médecin aide*/captain; *médicin-major*/major; *médecin principal*/lieutenant colonel; *médecin principal inspecteur*/colonel; *inspecteur general*/brigadier general. See Adolphe Armand, *Souvenirs d'un médecin militaire* (Paris: Rozier, 1858), pp. 214–215.

French presence and was finally resolved when Algeria became a departmental extension of France.[16])

The success of the French medical corps in fighting disease was initially modest, particularly with regard to the epidemics that swept through different regions of the colony at regular intervals. As Yvonne Turin has shown, after an initial curiosity and spontaneous contact the indigenous population tended to avoid French physicians. Nonetheless, medical contacts were made, and they colored the way in which the indigenous population was assessed. The first such contacts came during the cholera and smallpox epidemics that swept the coastline in the early 1830s. If these epidemics did not provide an occasion for the triumph of French medicine, they at least offered French physicians an opportunity to examine local customs and hygiene. By contemporary French standards, indigenous conditions during these epidemics were appalling. One doctor, describing his ministrations among smallpox patients in the Dellys area, declared the odor of their homes to be suffocating. The sick, covered with pustules and devoured by flies, lived in hothouses of infection. The situation was so bad that vaccination was out of the question; the only possible treatment was to teach the rudiments of hygiene: fumigation with aromatic herbs and bathing.[17] Experiences such as these reinforced the didactic nature of the relationship between occupier and occupied, which was, of course, the basis of the *mission civilisatrice.* Furthermore, the implicit comparison between poor local hygiene and sanitary French practices confirmed existing notions of the inferiority of indigenous society and culture. These developments were not just occurring at the grassroots level of the medical corps, however.

Among the leading medical personalities to emerge in the first decade of the French presence was Lucien Baudens, the army's foremost surgeon, a professor of medicine, and personal physician to the duke of Nemours (son of Louis-Philippe and commander of one of the battalions in the Constantine expedition). (See Frontispiece.) He was to dominate colonial medicine until the end of the military regime in 1870. Baudens is an example of the "well-rounded" medical man who ably accomplished his task as a physician and his role as a "medical journalist." While serving in Algiers he transformed one of the hospitals there into a military teaching hospital, the first of its kind on the continent, where he taught pathology and therapeutic medicine. According to a contemporary, only the best military physicians taught at this hospital, making it an unsurpassed center of clinical study. Baudens's activities as a surgeon were matched by his efforts as a data gatherer for the colony. He collected geological and botanical samples, which he then sent to the Academy of Sciences, and was a corresponding member of the medical academies of Marseilles, Montpellier, and Lyon, keeping them in touch with medical developments in the colony.[18]

In 1838 Baudens's report of the French expedition to Constantine was published in book

[16] It is not within the scope of this essay to examine this aspect of medical activity in Algeria. Nonetheless, it is a fact that physicians took an active part in the colonization debate. See, e.g., Louis-François Trolliet, *Mémoire sur la nécessité et les avantages de la colonisation d'Alger* (Lyon, 1835); Jean-Christian Boudin, *Médecine politique: De l'acclimatement en Algérie* (Paris, 1848); and Dr. Topin and F. Jacquot, *De la colonisation et de l'acclimatement en Algérie* (Paris, 1849).

[17] Turin, *Affrontements culturels* (cit. n. 2), pp. 87, 318.

[18] On Baudens see *ibid.,* p. 98. For his geological and botanical reports see Lucien Baudens, *Relation de l'expédition de Constantine* (Paris: Bailliere, 1838), p. 22. Baudens's teaching hospital is discussed in Comité d'Histoire du Service de Santé, *Histoire de la médecine aux armées* (cit. n. 6), Vol. 2, p. 157; the contemporary's praise is in Henri Jean François Edmond Pellissier de Reynaud, *Annales algériennes,* 3 vols. (1836–1839; Algiers: Bastide, 1854), Vol. 1, p. 161. The hospital ceased its teaching activities when Baudens left Algiers but resumed them in 1857.

colonial identity was to be established. The instructions called for comparative medical statistics. Diseases, births, marriages (including the incidence of polygamy), women's maladies, and methods of hygiene—to name but a few of the pertinent topics—were to be recorded, analyzed, and compared with their counterparts in France in order to establish the role of civilization in eradicating (or encouraging) certain conditions. Comparison and contrast between the French and the indigenous population was to become a methodological feature in the categorization of colonized peoples. In the medical reports themselves, where statistics were annotated and analyzed, objectivity suffered as physicians went beyond the immediate concerns of their reports and extrapolated about character, intelligence, and morality. Value judgments masquerading as objective observations thus became incorporated into images of the indigenous population.

Although developed expressly for the Scientific Commission, the physicians' instructions were printed in the *Gazette Médicale de Paris*. They were thus diffused to a more general medical public than that of the colonial establishment. Indeed, the physicians and surgeons posted in Algeria seem to have responded to the call in sufficient numbers that the *Recueil de Mémoires de Médecine de Chirurgie et de Pharmacie Militaires* could enthuse in 1842 that "the numerous reports on the topography and diseases of this *decidedly French* territory are such as to excite huge interest. They prove that the lofty ideal of observation is ever alive and a credit to the members of the military medical corps practicing in Algeria, who distinguished themselves both by their scientific research and their devotion to the health of their patients."[23] As we have seen, these reports were not relegated to the realm of medical journals but appeared in less specialized reviews as well. By the end of the first decade of the French presence, therefore, the medical corps in Algeria had not only established itself medically and institutionally; it had also created a pattern of research and data gathering that extended well beyond immediate medical concerns.

The period was also important in that it saw the arrival in the colony of several military physicians who would make their mark politically and culturally or use their colonial experiences to further their scientific reputations in the metropole. Those relevant here were Auguste Warnier (1810–1875), Eugène Bodichon (1810–1885), Auguste Vital (1810–1874), Joanny-Napoléon Périer (1809–1880), and Jean-Christian Boudin (1806–1867). From a political standpoint, the most involved of these men was Warnier. The path he was to take, which would transform him into one of the foremost spokesmen of the colony and lead him to the posts of prefect of Algiers in 1870 and deputy for Algeria in 1871, was already being mapped out in the 1830s.

In 1837 Warnier joined the Service des Affaires Arabes. This was an implicitly political act. The indigenous population—specifically, its relationship to the colonizing power—was central to colonial politics, and involvement in the institutions that shaped this relationship was a stepping stone to colonial empowerment. From 1837 to 1839 Warnier served in Mascara combating cholera. He was in close and constant contact with the tribes in the vicinity, tending their sick. The proximity to the local population afforded by this experience provided him with a sound knowledge of Arab customs and mores and contributed

[23] "Nouveaux documents relatifs à l'histoire médicale de l'Algérie," *Rec. Mém. Méd. Chir. Pharm. Milit.*, 1842, *52*:115 (my emphasis). This particular issue included reports from the regions of Blida, Bône, Constantine, Philippeville, and Oran. Topographical articles were often long, encompassing details on geography, climate, water supplies, inhabitants, mores, etc. See, e.g., M. Cambay, "Topographie physique et médicale du territoire de Tlemcen et compte-rendu des maladies qui ont été traités à l'hopital militaire de cette ville pendant l'année 1842 dans le services des fievreux," *ibid.*, 1844, *57*:1–65. Similar articles appeared regularly in both the *Recueil* and the *Gazette Médicale de Paris* after 1830.

to the creation of his reputation as an expert in local affairs. Such expertise was greatly valued both in the colony and in the metropole, and "experts" in indigenous affairs inevitably found themselves propelled into colonial politics either as advisors or as actual decision makers. Warnier served in Mascara with Captain Eugène Daumas (1803–1871), who went on to become one of the colony's leading military, political, and ethnological lights. Daumas found Warnier congenial and medically talented, and he wrote enthusiastically of his capabilities and growing reputation in the colony to his superiors. Warnier's relationship with Daumas and other well-placed figures in the colonial hierarchy furthered the emergence of his political persona. So too did his ties with the colony's Saint-Simonian movement, which counted prominent colonial personalities among its members and served as a vehicle for establishing and maintaining political contacts and exchanging ideas. At this stage of his career, Warnier was one of the colony's leading Saint-Simonians, and although he eventually left the fold his political activities in the context of the movement and the reputation he acquired as one of the best-informed individuals on Algeria and its peoples would persist until well after his death in 1875.[24] Warnier came into his own politically when the civilian administration took over in 1870, but his activities in the decades of military colonization brought him the prominence necessary for his political empowerment.

THE MEDICAL CORPS IN THE 1840S: ESTABLISHING A FRENCH DOMAIN

By the early 1840s, therefore, two trends had been established that would lead members of the medical corps into essentially nonmedical terrain. The first was the creation of official and personal relationships leading to active involvement in colonial politics and society. The second was a pattern of research, often statistical in nature and motivated by the need for reconnaissance, that extended beyond the medical realm. Both were tentative steps in the construction of a domain in which the political, cultural, and social paradigms and discourse were French.

In the second decade of French involvement in Algeria more territory was conquered and "pacified" and the debate over colonization was apparently settled in favor of the procolonial faction. (See Figure 3.) The two trends that had emerged in the first decade were reinforced: medical men became more involved in the political aspects of colonization and research intensified. The medical profession took up the Scientific Commission's objectives of exploring and understanding every aspect of the colony to facilitate its administration. In a report published in the *Recueil de Mémoires de Médecine de Chirurugie et de Pharmacie Militaires* in 1841, J.-A. Antonini, the principal doctor of the occupying

[24] "Daumas à Auvray, Mascara, le 7 janvier 1838," and "Daumas à Raptal, Mascara, le 11 février 1838," in *Collections de documents inédits sur l'histoire de l'Algérie après 1830: Correspondance du Capitaine Daumas, consul à Mascara, 1837–39*, ed. Georges Yver (Algiers: Jordan, 1912), pp. 61, 105. Daumas arrived in Algeria with Eugène Clauzel and served as consul in Mascara and as an officer in the Bureaux Arabes in a variety of posts before becoming director of Arab affairs in Algeria and later of Algerian affairs at the Ministry of War. He rose to the rank of general and became a senator. He published numerous books on Algeria and its inhabitants; his magnum opus was *La grande Kabylie: Études historiques* (Paris: Hachette, 1847), on which he collaborated with Captain Fabar. He was president of the Société de Géographie de Paris in 1858–1859 and a regular contributor on Algeria to numerous scholarly journals. See Emerit, *Saint-Simoniens en Algérie* (cit. n. 12), for the influence of Saint-Simonianism on colonial activity. References to Warnier as an "expert" on indigenous affairs occur repeatedly in the ethnology of the period—see Lorcin, *Imperial Identities,* pp. 129–130—and were still appearing after independence. See, e.g., Charles-André Julien, "L'insurrection de Kabylie 1870–71," *Preuves,* Dec. 1963, pp. 60–66. For Warnier's role in the Saint-Simonian movement in Algeria see Emerit, *Saint-Simoniens en Algérie.*

Figure 3. *Major Arcelin, a military physician, is borne in triumph by an old sergeant from the Zouaves for having defended and saved a convoy of wounded men. (From the archives of the Musée du Service de Santé des Armées au Val-du-Grâce.)*

forces and an ardent Saint-Simonian as well as a tireless worker, drew attention to the value of the monthly statistical reports and the comprehensive quarterly reports in providing the information necessary for a deeper understanding of the situation in Algeria. Every doctor, he wrote, "will eagerly provide his tribute of daily observations and his reflections on them." While some of these reports were limited to medical data, many included observations on nonmedical topics. The collection of data within the framework of their military obligations inspired many doctors to undertake their own nonmedical research.[25] Subject matter ranged from botany to archaeology; most prevalent were observations on the indigenous population.

It was in the second decade of French occupation, when it became increasingly clear that the French would not relinquish the colony, that attention turned in earnest to the indigenous population that would somehow have to be incorporated into the colonial framework. For various reasons, the Arabs and the Kabyles, the largest Berber group in Algeria, were the focus of this attention. First, they were the only two groups that put up a sustained resistance to French conquest. After Abd-el-Kader, the leader of the Arabs, surrendered in 1847, the French turned their attention to Kabylia and the Kabyles, who were finally vanquished in 1857. For twenty-seven years, therefore, one or the other of the two groups was in conflict with the French, a fact that colored the colonial imagery of them. Moreover, as each successive area was conquered, security became a prime preoccupation. The indigenous population—the only true obstacle to securing the country for French settlement—came under close scrutiny. Underlying all ethnological research was the need to discover which were the most subversive elements of the population and which, if any, were those most likely to cooperate with French rule. There was also the question of Islam. Always a barrier to the implementation of the *mission civilisatrice,* its ideological impenetrability and practical inaccessibility consistently stymied French attempts to come to terms with it, not only as a religion but, especially, as an alien culture. Its cultural strength could be neither circumvented nor overruled; it could be rendered socially and culturally ineffective only if marginalized. Such colonial concerns shaped the viewpoint of the medical corps; but the physicians brought an added dimension to their research, namely, the practice and ideology of their scientific training. For this reason, furthermore, their opinions were considered reliable testimony on indigenous ethnology.[26]

One of the first full-length accounts of the situation in Algeria, *Considérations sur l'Algérie,* written by the physician Eugène Bodichon, was published in 1845. Bodichon, who served in Algiers, was a prolific writer, a moderate republican and one-time Radical candidate for the legislative elections. He was an advocate of civilian rule, for he believed that the replacement of the military regime by a civilian one was the sole route to permanent French occupation. In considering the indigenous population, Bodichon concentrated on the character traits, morality, and religion of both Arabs and Kabyles, discussing them in the light of the French occupation and its civilizing capacities. He believed that intellectual and moral traits, as well as certain diseases, were passed from generation to generation. Tracing Arab characteristics from their pre-Islamic days, Bodichon concluded that the modern-day Arab was, like his ancestors, a pillager and a thief. This trait, an inherently

[25] J.-A. Antonini, "Rapport médicale sur l'Algérie," *Rec. Mém. Méd. Chir. Pharm. Milit.,* 1841, *50*:181–238, on p. 181; and Emerit, *Saint-Simoniens en Algérie,* p. 40. Lucien Baudens collected botanical and geological samples; Dufour, a military physician in the region of Bône from 1834 to 1840, undertook the botanical exploration of parts of Kabylia and, in conjunction with E. Cossou, another doctor, produced the *Càtalogue des plantes phanérogames;* Périer produced several works on ethnology, including "Des races dites berbères et leur éthnologie," which was published in the *Mémoires de la Société d'Anthropologie de Paris,* 2nd Ser., 1873, *1*:1–54; Leclerc turned his attention to archaeology in "Antiquities de la Kabylie," *Revue Africaine,* 1857, *2*:140–145; as did Léopold Buvry, "Voyage dans le Sahara oriental algérien: Description archéologie, histoire naturelle, nature du sol, position, limites, hydrographie," *Revue de l'Orient et des Colonies,* 2nd Ser., 1859, *10*:479–501.

[26] For the imagery of warfare created during this period see Lorcin, *Imperial Identities,* pp. 17–34. Bertherand, Boudin, Baudens, Bodichon, Périer, Vital, and Warnier, to name but a few, were all referred to in anthropological or ethnological works. See, e.g., Paul Broca, "Histoire des travaux de la Société d'Anthropologie," *Mémoires d'Anthropologie,* 1874, *2*; Armand Quatrefages, *Rapport sur le progrès de l'anthropologie* (Paris: Imprimerie Impériale, 1867); and issues of the *Bulletin de la Société d'Anthropologie* and other anthropological journals.

Figure 4. A member of the ambulance brigade carrying out an emergency operation on the battlefield while accompanying a column moving into the interior. (From the archives of the Musée du Service de Santé des Armées au Val-du-Grâce.)

Studies of the indigenous population, then, provided the best opportunities for imposing a French cultural identity, in that social, moral, and even religious comparisons between the two different cultures were the most convenient yardstick by which to measure the disparities in the degree of "civilization" and hence to justify the French presence on Algerian soil as a progressive inevitability. Other intellectual pursuits on the part of the medical corps contributed to this process. Archaeological, historical, legal, geological, and botanical studies provided the means of creating an essentially French intellectual space in a foreign land. In the case of archaeology, studies of Roman ruins became a way to reclaim a European past for North Africa, a past—"interrupted" by the invasion of "barbarians"—that France would take it upon herself to recall and reestablish. Dr. Worms, a military physician serving in Bône, examined Muslim property rights and laws. His research facilitated colonial land acquisition by pointing out legal "loopholes" that the French could use against Algerians whose ignorance of French legal practices prevented them from claiming what was rightfully theirs.[32] Although research of this sort was carried out by nonmedical officers, medical personnel did participate, and at all levels.

Officially, the importance of the medical corps in the colonizing process had been acknowledged from the earliest stages of occupation. (See Figure 4.) In the first place, their ability to heal was seen as a sure way of convincing the indigenous population of the benefits of French rule. Second, physicians and surgeons were considered harbingers of French culture and authority: it was their duty to ensure that their patients experienced the

[32] For a discussion of reclaiming a European past for the city of Constantine see James Malarkey, "The Dramatic Structure of Scientific Discovery in Colonial Algeria: A Critique of the Journal of the 'Société Archéologique de Constantine,' 1853–1876," in *Connaissances du Maghreb* (Paris: CNRS, 1984), pp. 137–160. For a discussion of the importance of Roman antecedents in North Africa as a colonial exemplar see Lorcin, *Imperial Identities,* pp. 21–23. Works by Worms are *De la constitution territoriale des pays musulmans* (1842) and *Recherches sur la propriété territoriale dans les pays musulmans: De la propriété rurale et urbaine en Algérie* (1844). See Julien, "Insurrection de Kabylie" (cit. n. 24), p. 104.

well-being that French civilization could offer. Health and civilization were, therefore, linked from the start. As pacification proceeded and the arena of medical activity enlarged, a third advantage became obvious, namely, social accessibility. In the course of their professional activities doctors had access to the homes of the indigenous population, a privilege whose value neither the physicians themselves nor governing officials underestimated. "We could visit their villages, enter their homes and study their customs and mores in a way no other officer could," wrote one doctor.[33] From an official viewpoint, home visits enabled doctors to describe and situate the different groups constituting the indigenous population, a valuable reconnaissance service that was part of the drive to ensure the security of the new colony. For their own part, doctors recognized that they could use their special access to further their own ambitions. Some merely recorded their observations on the character, morals, and domestic arrangements of the villagers they were treating as asides in their regular medical reports; others assumed the role of ethnologist, spending considerable periods observing one or another of the different indigenous groups.

THE MEDICAL CORPS IN THE 1850S AND 1860S: KNOWLEDGE INSTITUTIONALIZED

By the early 1850s the debate on colonization had shifted: the main concern was not *whether* but *how* to colonize. Should colonization be directed by the military or become exclusively civilian? How should land be distributed? What role should the indigenous population play? To be sure, dissenting voices continued to be heard, but Algeria had been administratively incorporated into France, and its very designation as a part of France suggested that the process of its conversion into a French cultural, social, and political domain would endure. If the first two decades saw the tentative beginning and gradual consolidation of French social and political paradigms and the introduction of a French discourse, the next two decades saw their institutionalization. Politically, an administrative structure was imposed. Culturally and intellectually, societies were created to catalogue and store the data being collected in the name of culture and science or to exhibit the evidence of the French cultural and scientific presence. Socially, a colonial hierarchy started to take shape, one whose lower rungs were occupied by the indigenous population.

Physicians in the colony were involved in all these processes to a greater or lesser degree. (See Figure 5.) In 1847, on the eve of relinquishing his post as governor and leaving Algeria, General Thomas Robert Bugeaud founded a society of sciences, letters, and arts. In the same year, a certain Charles Texier was put in charge of compiling a list of the historical monuments in the colony. By 1851, according to the *Revue Africaine*, the colony's foremost scholarly journal, the scientific impulse, hitherto dormant in Algeria, was truly awakened. The Société Archéologique de la Province de Constantine was founded in 1852 and the Société Historique Algérienne four years later. These were to become the leading scholarly societies of the colony and to serve as models for similar organizations in other parts of Algeria. Physicians were active in all these societies and contributed to their journals. Among the founding members of the Société Historique Algérienne, for example, was Dr. Bertherand. He was also founding president of the Algerian Medical

[33] Jules-René Anselin, *Essai de topographie médicale sur la ville de Bougie et le pays Kabyle limitrophe* (Paris: Rignoux, 1855), p. 63. The physicians' duty to bring the benefits of French culture to the indigenous population was made clear in Serres, "Instruction médicale" (cit. n. 4), p. 229. Recognition of the benefits that accrued from access to local homes appears in Delange, "Circoncision chez les Arabes" (cit. n. 4), p. 181; and P. Ribourt, *Le gouvernment de l'Algérie de 1852 à 1858* (Paris: Panckoucke, 1859), pp. 31–32.

Figure 5. *An Army of Africa surgeon. (From the archives of the Musée du Service de Santé des Armées au Val-du-Grâce.)*

Society, editor-in-chief of its *Gazette Medicale,* and a regular medical columnist for the bilingual *Mobacher,* which targeted the literate sectors of the indigenous population. Bertherand directed the first hospital at Blida in 1840 and served in the Arab Bureau of Algiers from 1848 to 1855, when he published *Médecine et hygiene des Arabes,* a work widely read by young doctors and students in the colony. In it he expounded the belief that French medicine would counteract the pernicious clerical (*sacerdotale*) influence that managed to grip even the liveliest minds.[34]

It was, however, not only in Algeria that these colonial physicians disseminated their

[34] "Introduction," *Rev. Africaine,* 1856, *1*:8. The Académie d'Hippone was founded in 1863, the Société de Géographie et d'Archéologie de la Province d'Oran in 1875. Among the journals to which physicians contributed were the *Revue Africaine* (journal of the Société Historique Algérienne), the *Annuaire de la Société Archéologique de la Province de Constantine,* and the *Bulletin de l'Académie d'Hippone.* Bertherand's view of the clerical influence is reported in Turin, *Affrontements culturels* (cit. n. 2), p. 28.

knowledge and opinions of the indigenous population. From its inception, physicians were the most prominent professional group among the members of the Société d'Anthropologie de Paris, founded in May 1859. Most were graduates of the École de Médecine de Paris. Doctors were also well represented in the short-lived Société Ethnologique (1839–1849). The medical profession thus became closely involved in ethnological and anthropological research, and the fruit of their work was either published or presented in the form of memoirs at the newly formed Paris societies dedicated to such investigations. In her study of the social context of the scientific debates of the Société d'Anthropologie, Joy Harvey indicates that colonial doctors responded well to the opportunity to "contribute to science" by studying the colonized peoples, their diseases and immunities, their physical dimensions, and their skull measurements and by relaying this information back to the society.[35]

A case in point was Joanny-Napoléon Périer, a military physician and member of the Scientific Commission who made a name for himself as an anthropologist. Périer believed that doctors were the professionals best suited for the study of man; at one of the early meetings of the Société d'Anthropologie de Paris, he chastised a colleague for not being thorough enough in his efforts in this direction. Périer's anthropological career, which included a term as president of the Société d'Anthropologie (1866), stemmed directly from his activities in Algeria. He was closely involved in the medical treatment of the indigenous population, especially during the periodic epidemics that afflicted the colony. This experience led him to branch out into local ethnology. From an interest in Algerian ethnology he moved on to Egyptian and, finally, to European ethnology. His concerns included the influences of climate on morality and race and the question of miscegenation. In an obituary address delivered at the Société d'Anthropologie, Paul Broca drew attention to Périer's contribution to the preparatory research necessary for successful colonization and to the importance of his work on the indigenous peoples of Algeria. He noted specifically that Périer's ethnological focus on the Arabs and the Kabyles, two totally different races, had scientifically demonstrated that the latter were more suitable for assimilation than the former. In Périer's words, the Kabyle was "an honorable type, loyal and reliable, an ally with a rare fidelity."[36] Périer's conclusions on the Arabs and Kabyles were not intellectual innovations: by the 1860s the belief that the Kabyles could be more easily assimilated was widely held. Périer's work and similar contemporary research serve as examples of the unscientific ideas that were being presented as truth to scholarly societies in France.

Jean-Christian Boudin was another medical man who served in Algeria and contributed to the emerging colonial discourse on the indigenous population. He too was a president

[35] Joy Dorothy Harvey, "Races Specified, Evolution Transformed: The Social Context of Scientific Debates Originating in the Société d'Anthropologie de Paris, 1859–1902" (Ph.D. diss., Harvard Univ., 1983), p. 127. In 1861, 82 of the 111 national members (full, associate, and corresponding) of the Société d'Anthropologie de Paris were physicians; all 7 of the overseas corresponding members were doctors. Of the 82 national physician members, 59 were either professors at or graduates of the École de Médecine de Paris. See *Bulletin de la Société d'Anthropologie,* 1861, *2*:xii–xvi. For physicians in the Société Ethnologique see *Bull. Soc. Ethnol.,* 1839–1841, *1*:xvi–xviii. For examples of these physicians' work see M. Armand, "Aperçu sur les variétés de races humaines observées de 1842 à 1862 dans les diverses campagnes de l'armée française," *Bull. Soc. Anthropol.,* 1862, *3*:553–558; Périer, "Races dites berbères" (cit. n. 25); M. Bleicher, "Sur l'anthropologie de la province d'Oran," *Bull. Soc. Anthropol.,* 1876, *11*; and Jean-Pierre Bonnafont, *Pérégrinations en Algérie 1830 à 1842* (Paris: Challamel, 1884).

[36] Périer, "Races dites berbères," p. 30. For Périer's criticism of his colleague see his introduction to Anselin's "Essai de topographie médicale sur la ville de Bougie et le pays kabyle limitrophe," *Bull. Soc. Anthropol.*, 1860, *1*:155. For his medical work among the indigenous population see Annie Rey-Goldzeiguer, *Le Royaume Arabe: La politique algérienne de Napoleon III, 1861–1870* (Algiers: SNED, 1977), pp. 450–457. For the obituary address see Paul Broca, *Bull. Soc. Anthropol.*, 3rd Ser., 1880, *3*:402.

By the 1860s, however, race was no longer just a theoretical preoccupation: colonization and race had become entwined issues as the establishment of a French colonial hierarchy increasingly pushed the indigenous population to the lower end of the scale. Race—specifically, the presumed racial inferiority of the native population—was not merely an *ex post facto* justification of colonization, however. It was as much a political tool of the anticolonialists (i.e., those who were against civilian colonization) as of procolonialists. The last decade of military rule, the period of the Royaume Arabe, was characterized by considerable unrest. The arabophile policies of Napoleon III were endorsed by the military but were not popular among the civilian population, which had grown considerably. In two letters published in 1861 and 1865 Napoleon III presented his ideas on French rule in Algeria. Their contents were ambiguous, but, in essence, he proposed the creation of an "Arab kingdom" (under the aegis of the French) where the practices and religion of the indigenous population would be respected. This suggested that the French role would be one of administration and not colonial exploitation. The debate polarized colonial society and was enshrined in a power struggle between the military, perceived by their opponents to be arabophile, and the "land-grabbing," arabophobe civilians.[43] Physicians participated in both sides of the political debate, which became increasingly strident as the decade progressed. The clamor surrounding the struggle has tended to obscure the fact that arabophilia did not imply a belief in equality—or indeed in assimilation.

Some physicians, like Bodichon, had left the military camp and were unequivocally in favor of civilian colonization; others, such as Auguste Vital, a military doctor who had arrived in Algeria in 1836, were of the opposite opinion. Vital fell into the arabophile camp, but his arabophilia, although humanistic, did not extend beyond the parameters of Saint-Simonian racial hierarchies that granted Europeans pride of place. Vital practiced in Constantine and was highly respected throughout Algeria for his intellect, his culture, and his honesty. He was a close friend of the influential Arabist Ismael Urbain and, like him, a Saint-Simonian. Although Vital shared many of Urbain's views, including the belief that colonization of Algeria was undesirable under the prevailing conditions, he did not believe in the possibility of assimilation, whatever the form. Writing to Urbain, he declared: "One colonizes an empty country or one whose people have enough in common with the invader to blend together, but one does not colonize a fairly well populated country whose people are morally the antithesis of the conquering race." Arab morality was the stumbling block: "It is not only their profoundly and actively hostile religion which separates us, but their moral inferiority."[44] Vital believed that Arabs and Europeans were too different morally,

les populations musulmanes du Nord de l'Afrique (Paris: Dupont, 1864); and Ferdinand Quesnoy, *L'Algérie* (Paris: Fume, 1885). Ricque's work was highly derogatory of Islam; Quesnoy, who also wrote a history of the French army in Algeria from 1830 to 1871, dealt with every aspect of Algeria from topography, climate, and history to morality and religion. For the importance of medical discourses and ideas in the population at large during the nineteenth century see "Introduction," in *French Medical Culture*, ed. La Berge and Feingold (cit. n. 1), pp. 1–22; and Léonard, *Médecine entre les pouvoirs et les savoirs* (cit. n. 7), p. 333. For the involvement of the medical profession in politics see Ackerknecht, *Medicine at the Paris Hospital*, pp. 185–186; and Ellis, *Physician-Legislators of France* (cit. n. 7).

[43] See Rey-Goldzeiguer, *Royaume Arabe* (cit. n. 36), for a detailed exposé of the paradoxical nature and the political ramifications of the Royaume Arabe.

[44] Auguste Vital to Ismael Urbain, 27 Sept. 1861, 17 May 1861, in André Nouschi, ed., *Correspondence du Dr. A. Vital avec I. Urbain (1845–1874): Collection de documents inédits et d'études sur l'histoire de l'Algérie*, Vol. 5 (Paris: Larose, 1959), pp. 56, 53; see also p. 43. Urban was an interpreter involved in the administration of Algeria. A fervent arabophile, he became an advisor to Napoleon III. In the 1860s the debate as to whether the colony should remain in the hands of the military administration or be transferred to a civilian administration gathered momentum and became acrimonious. The civilians accused the military of being arabophile to their detriment. See Lorcin, *Imperial Identities*, pp. 13, 77. The most comprehensive work on the 1860s is Rey-Goldzeiguer, *Royaume Arabe*.

culturally, and socially to be able to fraternize to the degree necessary for successful colonization. As a result of his anticolonization stance, Vital disapproved of another politically active medical worker, Auguste Warnier.

By the 1860s Warnier was devoting himself exclusively to colonial politics. Initially a devoted Saint-Simonian, he had left the fold—but not before acquiring the sobriquet "petulant Danton" from a fellow Saint-Simonian. He had also ceased to practice and had retired from the military in 1861. Until his death in 1875 he spent his time promoting civilian colonization both in the political arena and in print. In 1863 a collection of his articles that had appeared in *L'Opinion Nationale* was published under the title *L'Algérie devant le Sénat.* It was widely read in France and Algeria and was rapidly followed by six more publications of a similar nature. Warnier's writings, particularly *L'Algérie devant le Sénat,* became the intellectual framework of civilian colonization. In 1870 the military administration was replaced by a civilian one and the worst insurrection prior to the outbreak of the war of independence occurred in Kabylia. The rebellion was put down, but Kabylia, which had hitherto escaped large-scale land sequestration, now became a victim. Warnier was involved in these developments and took an active part in promoting the settlers' interests with regard to land acquisition. In 1873 a law bearing his name, *la loi Warnier,* made all indigenous land holdings subject to French law. Dubbed the law of the "colons," it eliminated recourse to Islamic law on the part of indigenous landowners and thus removed existing obstacles to land seizure.[45]

CONCLUSION

Two significant points emerge: the importance of the period under review for the general interaction of medicine and imperialism overall, and the significance of the role played by the French medical corps in Algeria. The 1830–1870 period is noteworthy for a number of reasons. The conquest of Algeria and its establishment as a departmental extension of France preceded the Scramble for Africa (ca. 1876–1914) and the extension of European influence throughout the continent. Thus, alone among the increasingly science- and discovery-conscious European nations, France had a convenient site and experimental ground for research. This provided French researchers and scientists in Algeria with opportunities to distinguish themselves intellectually in the eyes of their colleagues in France and hence acquire metropolitan recognition. It also meant that Algeria could, and did, serve as an example for future colonial enterprises, both political and scientific.[46] But these were years of military conquest and rule, which meant that the pool of civilian scientists and researchers was relatively small. Military physicians and engineers were the scientific personnel best trained to carry out research, in both an official and a personal capacity.

[45] Auguste Warnier, *L'Algérie devant le Sénat* (Paris: Dubuisson, 1863); Warnier, *L'Algérie devant l'opinion publique* (Paris: Challamel, 1864); Warnier, *L'Algérie devant l'empreur* (cit. n. 42); Warnier, *Cahiers algériens* (Algiers: Duclaux, 1870); Warnier, *Algérie et les victimes de la guerre* (Algiers: Duclaux, 1871); Warnier (with Duval), *Un programme de politique algérienne* (Paris: Challamel, 1868); and Warnier (with Duval), *Búreaux arabes et colons* (Paris: Challamel, 1869). For details on Warnier's progress in colonial politics see Rey-Goldzeiguer, *Royaume Arabe*, esp. pp. 68–69, 360–361, 663–664; for the influence of his writings see p. 247. Captain de Neveu, serving in the Arab Bureau of Dellys, coined the sobriquet: Edouard Neveu to Urbain, 7 Feb. 1861. Archives Nationales Outre-Mer, Aix en Provence, 31 Mi9, letter no. 53. For the implications of *la loi Warnier* see Charles-Robert Ageron, *Histoire de l'Algérie contemporaine*, Vol. 2 (Paris: Presses Univ. France, 1979), pp. 94–96.

[46] See Edmund Burke III, "The Image of the Moroccan State in French Ethnological Literature," in *Arabs and Berbers*, ed. Ernest Gellner and Charles Micaud (London: Heath, 1972), pp. 175–199; and Lorcin, *Imperial Identities*, pp. 221–225, 227–232.

The military nature of the medical corps in Algeria distinguished it from the European medical presence in other parts of Africa at the time in two ways. In the first place, it was nonmissionary. This meant that off-duty hours were not devoted to religious duties or missionary activity but could be spent in the pursuit of personal interests and ambitions, whether in research or in activities of a cultural or political nature. The potential productivity of this situation was enhanced by the military obligation to produce regular station activity reports. Many of the reports drawn up by the medical corps in Algeria went beyond medical specifics to encompass regional topography, detailed descriptions of the local population, and so forth. Contemporary interest in such subjects was sufficient to warrant their publication in both medical and nonmedical journals. A second distinguishing feature was the considerable size of the medical corps in Algeria. This set it apart from European activity in other parts of Africa, where medical missionary representation was sparse. The sheer numbers of physicians and surgeons in Algeria provided a large potential pool of research power and led to significant output, which in turn created a demand for institutions that could collect and catalogue the data being produced. The demand was eventually met by the founding of scholarly societies in both France and Algeria.

The years 1830–1870 also coincided with a growing European preoccupation with scientific explanations and, above all, with race. They straddled the "prescientific" and the "scientific" eras, that is to say, the period before and after 1850, when European scientism is now acknowledged to have emerged. Scientific and colonial endeavors were therefore inevitably entwined. During this period, furthermore, French medical practices were in the European vanguard owing to the Revolutionary reforms of 1789–1795 and the emergence of concepts and practices such as social hygiene and statistical science. The Napoleonic wars were the occasion for the practical application of these developments. The renown that French medicine acquired in the ensuing period lent weight to the activities of French medical personnel and attracted disciples from all over Europe. Research and publications undertaken by the medical corps in Algeria benefited from this prestige. Moreover, any scholarly ambitions that physicians practicing in Algeria may have had were encouraged by the establishment, as of the 1830s, of professional societies such as the Société Médicale d'Observation and the emergence of numerous publications that could serve as a forum for their activities.[47]

Finally, the years 1830–1870 preceded the era of bacteriology and immunology that not only led to the prevention and treatment of many formerly recalcitrant diseases but also served as a scientific and technological endorsement of European claims to medical superiority. Early nineteenth-century French medicine as practiced in Algeria was not spectacularly successful from either a medical or a colonial point of view. Initially, therefore, French medical superiority could be circumscribed only in hygienic, moral, and social terms, rather than in the scientific or technological ones that later discoveries would make possible. This evaluative framework was extended to examine and assess the indigenous population, its mores, and its lifestyle.

We now come to the role of the medical corps in Algeria. Over and above their therapeutic skills, which maintained and safeguarded the army of occupation and were deemed an essential inducement to the civilization of the indigenous population, there was a so-

[47] See Stocking, *Victorian Anthropology* (cit. n. 39), for whom the symbol that separates the "scientific" from the "prescientific" period is the Crystal Palace exhibition of 1850. On the renown of French medicine after the Napoleonic wars see Ackerknecht, *Medicine at the Paris Hospital*, p. 147; and Léonard, *Médecine entre les pouvoirs et les savoirs* (cit. n. 7), p. 14. See also Williams, *Physical and Moral*, pp. 224–243; Pyenson, *Civilizing Mission* (cit. n. 1), p. 331; and Ackerknecht, *Medicine at the Paris Hospital*, pp. 116–117.

ciological and ideological dimension to the activities of the French medical corps in Algeria that greatly assisted the establishment of a colonial domain. The methodology and philosophical underpinnings of their training encouraged the doctors serving in Algeria to assist in the initial reconnaissance of the colony by observing and recording a wide range of information both on the indigenous population and on the country as a whole. This information was important in assessing the desirability of retaining or relinquishing the colony. Although the decision to remain in Algeria was definitively taken in 1848, many doctors were voicing their approval nearly a decade earlier. Doctors also provided essential information in the debate as to the salubrity of the area for European settlement. Even though they initially confronted high mortality among French inhabitants, their preoccupation with improving the hygiene and living conditions in the area and with alerting the administration to the health hazards inherent in the climate and topography indicated their belief in the inevitability of French rule.

The notion of European superiority that the doctors brought with them to Algeria was a philosophical and cultural concept rather than a technological one. This idea shaped relations with the indigenous population: denigration of those parts of the indigenous cultural heritage that clashed with French interests became a necessary step in the imposition of French civilization. A particular target was Islam, which, as an all-encompassing religion, was a serious obstacle to the secularizing trends of the French administration. The secularization of the French state, begun during the Revolutionary epoch and fully achieved in 1905, cast a shadow over perceptions of Islam throughout the nineteenth century in Algeria. Moreover, the negative image, conceived during this period, of Islam as an essentially fanatical and obscurantist religion informed attitudes in France and persisted into the twentieth century.[48]

The concept of European superiority, and hence supremacy, was buttressed by racial concepts that placed Europeans at the top of the ladder of being. These ideas, which were beginning to emerge at the time of the French conquest, were given a boost in Algeria, where doctors applied their knowledge of contemporary philosophical and scientific trends to the study of the indigenous population. Their extrapolations about local morality and mores contributed to the negative stereotypes and categorizations of the indigenous population that were part of the panoply of colonial identity formation. Furthermore, by presenting their research in academic and intellectual forums—largely for reasons of professional ambition—doctors stimulated the emerging racial debate in France. The involvement of the medical corps in scholarly societies both in Algeria and in France and its contributions to academic and popular journals and reviews enhanced the dissemination of ideas held in the colony.

But the physicians in Algeria did not restrict themselves to intellectual or scholarly pursuits. They actively partook of the politics that shaped the colony's development. Whether they were concerned with the indigenous population, the feasibility of colonization, the political ins and outs of administration, or intellectual forays into history and archaeology, doctors actively collaborated in the theory and practice of transforming North Africa into a French cultural space, establishing a colonial domain in Algeria, and transmitting its image to France.

[48] How the negative image of Islam was formed and evolved in Algeria and how it informed French attitudes both in France and in the colony is fully developed in Lorcin, *Imperial Identities*. Other historical works that document French attitudes to Islam are Norman Daniel, *Islam and the West: The Making of an Image* (Edinburgh: Edinburgh Univ. Press, 1962); Daniel, *Islam, Europe, and Empire* (Edinburgh: Edinburgh Univ. Press, 1966); Christopher Harrison, *France and Islam in Africa, 1860–1960* (Cambridge: Cambridge Univ. Press, 1988); and Carl Brown and Matthew S. Gordon, eds., *Franco-Arab Encounters* (Beirut: American Univ. Beirut Press, 1996). In a much broader context see Said, *Orientalism* (cit. n. 10).

Racism and Medical Science in South Africa's Cape Colony in the Mid- to Late Nineteenth Century

By Harriet Deacon

ABSTRACT

Racism has been a particular focus of the history of Western medicine in colonial South Africa. Much of the research to date has paradoxically interpreted Western medicine as both a handmaiden of colonialism and as a racist gatekeeper to the benefits of Western medical science. This essay suggests that while these conclusions have some validity, the framework in which they have been devised is problematic. Not only is that framework contradictory in nature, it underplays differences within Western medicine, privileges the history of explicit and intentional racial discrimination in medicine, and encourages a separate analysis of racism in law, in the medical profession, and in medical theory and practice. Using the example of the Cape Colony in South Africa, this paper shows how legislation, class, institutional setting, and popular stereotypes could influence the form, timing, and degree of racism in the medical professional, and in medical theory and practice. It also argues for an analytical distinction between 'racist medicine' and 'medical racism.'

INTRODUCTION

IN RECENT YEARS, CONSIDERABLE ATTENTION HAS BEEN PAID TO the interplay between racism, medicine, and empire.[1] This work has concentrated on the history of racial inequality or discrimination in medical professionalization, medical institutions, public policy, and in the incidence, perception, and treatment of specific diseases.[2] It is not surprising that racism has been a key issue in the

[1] Key texts include Sander Gilman, *Difference and Pathology: Stereotypes of Sexuality, Race and Madness* (Ithaca: Cornell Univ. Press, 1985); selected papers in Roy MacLeod and Milton Lewis, eds., *Disease, Medicine and Empire* (London: Routledge, 1988) and David Arnold, ed., *Imperial Medicine and Indigenous Societies* (Manchester: Manchester Univ. Press, 1988); Megan Vaughan, *Curing their Ills: Colonial Power and African Illness* (Cambridge: Cambridge Univ. Press, 1991); Mark Harrison, *Public Health in British India: Anglo-Indian Preventive Medicine, 1859–1914* (Cambridge: Cambridge Univ. Press, 1994); Warwick Anderson and Mark Harrison, "Race and Acclimatization in Colonial Medicine," *Bulletin of the History of Medicine*, 1996, *70:*62–118; selected papers in Dagmar Engels and Shula Marks, eds., *Contesting Colonial Hegemony* (London: British Academic Press, 1994) and in Andrew Cunningham and Bridie Andrews, eds., *Western Medicine as Contested Knowledge* (Manchester: Manchester Univ. Press, 1997).

[2] In African (other than South African) medical history, significant work on race and colonial medicine includes Adell Patton, *Physicians, Colonial Racism and Diaspora in West Africa* (Gainesville:

relatively new field of South African medical history, given the significance of racism in the country's history.[3] However, since the 1980s, revisionist medical history has concentrated mainly on the politics, rather than the ideology, of race in Western medicine—asking whether medicine, like the missionary endeavor, was a handmaiden of colonialism.[4] Historians have been interested in the practitioners of Western medicine and the extent to which they aided colonial domination by initiating and implementing the racist medical policies of colonial governments or industries.[5] A small but growing body of research examines indigenous medical traditions and practitioners in South Africa.[6]

Many South African medical historians have adopted the dual framework of colonial and underdevelopment theory. This means that they have been trapped in a "catch-22," arguing *(a)* that Western medicine was a detrimental agent of colonialism, but also *(b)* that black people were disadvantaged because they did not have equal access to its practitioners and therapies. Racist theory and practice in Western medicine certainly encouraged both *(a)* the treatment of black patients as inferior and different and *(b)* the limitation of black patients' access to its services.[7] As Roy MacLeod has noted, "the political uses of medical knowledge are not unambiguously one-sided; its effects are not simple."[8] While the theoretical framework of

---

Univ. Press of Florida, 1996); Megan Vaughan, "Idioms of Madness: Zomba Lunatic Asylum, Nyasaland, in the Colonial Period," *Journal of Southern African Studies*, 1983, *9*, 2:218–38; Philip Curtin, "Medical Knowledge and Urban Planning in Tropical Africa," *American Historical Review*, 1985, *90*, 3:594–613; John Cell, "Anglo-Indian Medical Theory and the Origins of Segregation in West Africa," *Amer. Hist. Rev.*, 1986, *91*, 2:307–35; Jock McCulloch, *Colonial Psychiatry and the "African Mind"* (Cambridge: Cambridge Univ. Press, 1995); Heather Bell, *Frontiers of Medicine in the Anglo-Egyptian Sudan, 1899–1940* (Oxford: Oxford Univ. Press, 1999).

[3] The first serious historical analysis was Edmund Burrows, *A History of Medicine in South Africa up to the End of the Nineteenth Century* (Cape Town: Balkema, 1958).

[4] One of the texts that set the parameters for this research was Shula Marks and Neil Andersson, "Issues in the Political Economy of Health in Southern Africa," *J. Southern African Stud.*, 1987, *13*, 2:177–86.

[5] On the racialization of the nursing profession in South Africa, see Shula Marks, *Divided Sisterhood: The Nursing Profession and the Making of Apartheid in South Africa* (London: Macmillan, 1994). On the Western medical profession, see Harriet Deacon, "Cape Town and Country Doctors in the Cape Colony During the First Half of the Nineteenth Century," *Social History of Medicine*, 1997, *10*, 1:25–52. On the relationship of Western medicine to the colonial state, see, for example, Elizabeth van Heyningen, "Agents of Empire: The Medical Profession in the Cape Colony, 1880–1910," *Medical History*, 1989, *33*:450–71; Maynard Swanson, "'The Sanitation Syndrome': Bubonic Plague and Urban Native Policy in the Cape Colony, 1900–1909," *Journal of African History*, 1977, *18*, 3:387–410 and "The Asiatic Menace: Creating Segregation in Durban 1870–1900," *International Journal of African Historical Studies*, 1983, *16*, 3:401–21; Randall Packard, *White Plague, Black Labor: Tuberculosis and the Political Economy of Health and Disease in South Africa* (Pietermaritzburg: Univ. of Natal Press and James Currey, 1989); Elaine Katz, *The White Death: Silicosis on the Witwatersrand Gold Mines, 1886–1910* (Johannesburg: Witwatersrand Univ. Press, 1994).

[6] See, for example, Harriet Ngubane, *Body and Mind in Zulu Medicine* (London: Academic Press, 1977); Catherine Burns, "Louisa Mvemve: A Woman's Advice to the Public on the Cure of Various Diseases," *Kronos: Journal of Cape History*, 1996, *23*:108–34; David Gordon, "From Rituals of Rapture to Dependence: The Political Economy of Khoikhoi Narcotic Consumption, c. 1487–1870," *South African Historical Journal*, 1996, *35*:62–88; Harriet Deacon, "Understanding the Cape Doctor Within the Context of a Broader Medical Market," in *The Cape Doctor: A History of the Medical Profession in the Nineteenth-Century Cape Colony*, eds. Harriet Deacon, Elizabeth van Heyningen, and Howard Phillips, forthcoming.

[7] This is an interesting contrast to the feminist critique of Western medicine for treating women as inferior but *over*medicating them. See Elaine Showalter, "Victorian Women and Insanity," *Victorian Studies*, 1980, *23*, 2:157–81.

[8] Roy MacLeod, "Introduction," in MacLeod and Lewis, *Disease, Medicine and Empire* (cit. n. 1), p. 11.

colonialism and underdevelopment can explain why Western medicine in the colonial context was often racist in conception and application, it cannot explain how Western medicine could at the same time be politically, economically, and culturally loaded, and also useful.

In South African medical history as a whole, as perhaps elsewhere, there has been a tendency to see racism occurring mainly where white colonists met black indigenes.[9] This perspective constructs racism as an issue that attained its full expression in the European colonies: racism is the one topic about which colonial historians feel they can write authoritatively. This essentialist view of race has been challenged by recent work on the (re)construction of whiteness in European colonies, and on the importance of understanding racism through filters of gender and class.[10] There are other problems, too: within the history of medicine, as in other branches of South African history, the intellectual history of racism stands strangely neglected in contrast to the history of racial discrimination.[11] Even in charting the history of racial discrimination, we have documented official pronouncements, policies, and legislation at the expense of understanding local practices. Little has been written on the relationship between racist theory and practice, diverse forms of discrimination in different contexts, and the correlations between racism and other types of discrimination.[12] Indeed, in spite of recent interest in the history of racist medical and scientific theories, we are still far from understanding the trajectory of racism in colonial medicine.[13]

In order to understand the variations, trends, and contradictions within colonial Western medicine, we require a more nuanced analysis of its relationship to racism. We must explore differences in the application and elaboration of racist ideologies in various colonial contexts.[14] At the same time, Western medicine cannot be treated as a homogenous entity, as doctors from different European medical traditions practiced in different political, economic, cultural, and institutional situations in the colonies.[15] In order to understand racism in Western medicine, we should combine analyses of racialization within the profession, discrimination in medical practice, and

[9] In this framework, racism is "about black people" in the same way that gender is "about women."

[10] See for example, Ann Laura Stoler, "Sexual Affronts and Racial Frontiers: European Identities and the Cultural Politics of Exclusion in Colonial Southeast Asia," and Lora Wildenthal, "Race, Gender and Citizenship in the German Colonial Empire," in *Tensions of Empire: Colonial Cultures in a Bourgeois World*, eds. Frederick Cooper and Ann Laura Stoler (Berkeley: Univ. of California Press, 1997). See also Anne McClintock, *Imperial Leather: Race, Gender and Sexuality in the Colonial Context* (London: Routledge, 1995).

[11] Saul Dubow, *Scientific Racism in Modern South Africa* (Cambridge: Cambridge Univ. Press, 1995), p. 3.

[12] For a comparison of discrimination in hospitals and prisons, see Harriet Deacon, "Racial Segregation and Medical Discourse in Nineteenth-Century Cape Town," *J. Southern African Stud.*, 1996, *22*:287–308. On the relationship between racial and gender discrimination in medicine, see Sally Swartz, "Colonialism and the Production of Psychiatric Knowledge in the Cape, 1891–1920" (Ph.D. diss., Univ. of Cape Town, 1996).

[13] See, for example, Paul Rich, "Race, Science and the Legitimization of White Supremacy in South Africa, 1902–1940," *Inter. J. African Hist. Stud.*, 1990, *23*:665–86; Dubow, *Scientific Racism* (cit. n. 11); Andrew Bank, "Liberals and their Enemies: Racial Ideology at the Cape of Good Hope, 1820 to 1850" (Ph.D. diss., Univ. of Cambridge, 1995).

[14] Bank, "Liberals and their Enemies" (cit. n. 13), suggests that European settlers elaborated different types of racist ideologies in the eastern and western parts of the Cape Colony.

[15] For example, Deacon, "Cape Town and Country Doctors" (cit. n. 5) argues that Continental and British educational and professional backgrounds produced doctors with different approaches to professionalization at the Cape in the early nineteenth century.

racism in medical theory. This essay explores this idea by considering the dynamics between racism and Western medicine in the Cape Colony during the mid- to late nineteenth century.

The Cape Colony, which now comprises most of the western, northern, and eastern Cape Provinces of South Africa, was under British rule from 1795–1803 and from 1806–1910, when it became part of the Union of South Africa. The inhabitants of the Cape at this time consisted mainly of indigenous Africans, including Nguni speakers in the northern and eastern areas, Khoekhoen (then called "Hottentots"), and San or Bushmen; slaves (mainly from the East Indies) and their descendants; Dutch-Afrikaners descended from Dutch and German settlers; and British settlers. As this essay suggests, professional and institutional factors acted as key influences on the emergence of a highly differentiated, although broadly racist, field of colonial medicine in the Cape Colony. It also suggests that, in understanding racism in colonial medicine, we should differentiate between racist medicine (the institution of discriminatory practices in medicine based on broader social discrimination) and medical racism (the application of racially discriminatory practices in medicine justified on medical grounds).

SCIENCE AND RACISM IN SOUTH AFRICA

The relationship between racism and science is hotly contested terrain. Some have argued that Western science and medicine were never free of racial content. Indeed, Keenan Malik and others have argued that the liberalism of the European Enlightenment, which "introduced a concept of human universality which could transcend perceived differences," made the modern concept of race possible.[16] The fact that (racial) difference was recognized within liberal discourse created a space for the paternalism of "civilization"; the construction of an egalitarian moral economy created the opportunity for exclusions from it.[17] On the other hand, others have implied that racism, when present, was just a temporary and undesirable adjunct to scientific theory.

It is not, of course, an easy matter to decide whether racism in science is best identified by examining explicit theoretical content, or the way in which theories were used or applied. It is hard to separate racist theory from practice because both emerged out of broader, often unarticulated, racist discourses, only some of which were arbitrarily elevated by contemporary science to the status of theories. In understanding the trajectory of racism within science it is, however, useful to differentiate (if only in degree) between "scientific racism," which created scientific justifications for racist practice, and "racist science," which incorporated elements of popular racism in the theory and practice of an otherwise universalist science. (In the same way, medical racism can be differentiated from racist medicine.) This distinction can help us to understand the degree to which notions of race became an essential part of the way in which the scientific enterprise was defined.

In the eighteenth century, a biologically based concept of race gained currency in

[16] Keenan Malik, *The Meaning of Race* (Basingstoke: Macmillan, 1996), p. 42.

[17] David Goldberg, *Racist Culture: Philosophy and the Politics of Meaning* (Oxford: Blackwell, 1993), p. 6.

Europe and became central to the articulation of a bourgeois identity.[18] Although Europe remained the dominant partner in the formulation of racist doctrines, the colonial contribution was significant. Not only did empire provide an important political motivation and ideological touchstone for the elaboration of racial difference as a fixed biological reality, but it was a key testing ground for racist theory and practice, some of which was then assimilated into popular and scientific discourses in Europe. Metropolitan and colonial science had common origins and enjoyed frequent crossfertilization, but they were formulated and practiced in different political and ideological contexts. They were also shaped in diverse ways by the patchy and delayed transfer of ideas and people. Scientific theory and practice, and its racism, thus exhibited different features in different contexts.[19]

The relationship between metropole and colony was, of course, by no means equal. The development of an independent colonial science was thus not just a matter of time, as Basalla's work seems to suggest.[20] Colonial scientists often "functioned as collectors of facts, while metropolitan scientists acted as theorists or gatekeepers of scientific knowledge."[21] The development of independent, national scientific communities was thus difficult, and perhaps impossible, to achieve in European colonies.[22] Like colonial historians on racism, colonial scientists, especially in Africa, felt more confident in espousing theories about racial difference considered specific to the colonies, than in competing with metropolitan scientists' theories considered applicable to both colony and metropole. Even compared with other settler colonies, colonial South Africa was slow to develop scientific institutions and a home-grown scientific identity. The South African Association for the Advancement of Science was not founded until 1903, and the first signs of a confident "South Africanized" (albeit not perhaps independent) science emerged only in the late 1920s.[23]

Even though racist practice was commonplace, scientific racism was slow to develop in South Africa. A kind of professional schizophrenia characterized the Cape medical profession, which in practice was deeply rooted in popular colonial racism but in theoretical orientation remained largely universalist and metropolitan.[24] Doctors oscillated between asserting the differences between black and white patients and their essential sameness. South African discussions of eugenics were less intense and less nuanced or consistent than elsewhere.[25] Rich has suggested that

[18] Ann Laura Stoler and Frederick Cooper, "Between Metropole and Colony: Rethinking a Research Agenda," in Cooper and Stoler, *Tensions of Empire* (cit. n. 10), pp. 2–3.

[19] For a Latin American example, see Nancy Stepan, "Race, Gender and Nation in Argentina: The Influence of Italian Eugenics," *History of European Ideas*, 1992, *15*:749–56, p. 749. On South Africa, see Dubow, *Scientific Racism* (cit. n. 11).

[20] George Basalla, "The Spread of Western Science," *Science*, 1967, *156*:611–22.

[21] Susan Sheets-Pyenson, *Cathedrals of Science: The Development of Colonial Natural History Museums during the Late Nineteenth Century* (Kingston: McGill-Queen's Univ. Press, 1988), p. 15.

[22] For a review of the debates on the development of a national scientific community in Australia, see Jan Todd, *Colonial Technology: Science and the Transfer of Innovation to Australia* (Cambridge: Cambridge Univ. Press, 1995), pp. 7–8. For a general overview of the comparative literature, see Sheets-Pyenson, *Cathedrals of Science* (cit. n. 21), pp. 12–15.

[23] Dubow, *Scientific Racism* (cit. n. 11), pp. 12–13.

[24] Harriet Deacon, "Racial Categories and Psychiatry in Africa: The Asylum on Robben Island in the Nineteenth Century," in *Race, Science and Medicine*, eds. Waltraud Ernst and Bernard Harris (London: Routledge, forthcoming).

[25] Dubow, *Scientific Racism* (cit. n. 11), pp. 16–17.

> Until the rise of anthropological research in the 1920s and 1930s, there was little attempt to back up propositions [in the South African debate on racial segregation] with systematic evidence. Arguments often fell back upon well-worn stereotypes [couched] in a 'scientific' vocabulary.[26]

It was only in the 1920s that scientific racism would fully catch up with discriminatory practice as South Africa sought to present itself as a "human laboratory" for "race relations."[27] Even then, "the newer scientific discourse . . . ended up perpetuating older conceptions of African society inherited from nineteenth-century travelers and missionaries."[28]

Before the 1880s, few Cape doctors styled themselves as "scientists." Andrew Sparrman, a Swedish doctor and scientist who visited South Africa in the late eighteenth century, was disappointed in the standard of Western doctors practicing at the Cape, and suggested that they were more interested in making money than in science or healing.[29] Sparrman, a student of Linnaeus, was unusual in documenting his investigation of racial differences between patients. In 1775, he visited the Caledon Warmbaths where he helped to treat a Madagascan slave from Cape Town who was suffering from an ulcerated leg. His comments on the case indicate he expected to find that racial differences were more than skin deep, but he could not support that assumption through his observations of the black man's wound:

> Being curious to examine a negro's flesh, I had for some time undertaken to look after the sore myself . . . The raw flesh appeared exactly of the same colour with that of an European . . . [As] the ulcer began to heal, [it threw out] fresh fibres in the same manner as ours do, with something whitish on the side of the skin, which otherwise was of a dark colour.[30]

Undaunted by finding similarity where he sought difference, in November of the same year Sparrman treated several farm workers (slaves and Khoisan) and a settler girl for "bilious fever." He treated them all with a strong mixture of tobacco, water, and alcohol. He gave racial explanations for the differences he observed in the patients' symptoms, disease progress, and reactions to his treatment. He explained these differences in cultural rather than biological terms, suggesting that those indigenous Khoekhoen ("Hottentots") who had recently "made too sudden a transition from their strange manner of living" responded less quickly to the medication than settler patients and Khoekhoe servants brought up with the family.[31] At this time in Europe, the idea of race was still linked to culture as well as to biology. By the 1880s, Cape doctors had begun to develop cultural explanations for racial difference in the field of psychiatry, a shift in perspective that would soon characterize other fields of medicine, too.[32]

At least until the end of the century, the British model of "gentlemanliness" was

[26] Rich, "Race, Science and White Supremacy" (cit. n. 13), p. 667.
[27] Dubow, *Scientific Racism* (cit. n. 11), p. 14.
[28] Rich, "Race, Science and White Supremacy" (cit. n. 13), p. 667.
[29] Andrew Sparrman, *A Voyage to the Cape of Good Hope towards the Antarctic Polar Circle and round the world . . . from the year 1772 to 1776*, 2 vols., 2nd ed. (London: Robinson, 1786), p. 48.
[30] *Ibid.*, p. 143.
[31] *Ibid.*, pp. 351–3.
[32] Deacon, "Racial Categories and Psychiatry" (cit. n. 24).

more important in the self-definition of the South African medical profession than was the idea of a colonial or national scientific community. Sparrman found few medical colleagues in the Cape Colony with whom he could discuss his ideas on racial difference. It was not that Cape doctors were antiracist, but that they did not put their racist ideas into scientific terms within a medical discourse subject to experimental proof. Well into the twentieth century, the colony lacked the resources, libraries, and universities needed to encourage the development of a scientific community. Earlier, some doctors had contributed to scientific discourse generated abroad (e.g., the trial of anaesthesia in Grahamstown in 1847), but there was little scientific innovation within the colony itself.[33] The two professional medical organizations established before the 1880s were both short-lived.[34] The *Cape Town Medical Gazette*, which began in 1847, also had a short life, and its editor's calls for a medical society went unheeded.[35] In the 1850s, Collis Browne, a military surgeon quartered at the Cape Town Castle and a keen inventor, complained that military doctors had failed to keep up with the "progress of modern science" in the medical field.[36] It was little better among civilian doctors.

THE RACIALIZED MEDICAL GENTLEMAN

By the beginning of the nineteenth century, when Britain had just taken over the Cape from its Dutch colonizers, immigrant medical practitioners in the Cape, as elsewhere, had begun to aspire to being "medical gentlemen," although there was always a gap between the gentlemanly ideal and the reality of colonial life.[37] The notion of the medical gentleman was racialized, gendered, and class specific, circumscribing a community of middle-class, European-trained doctors who were almost all white men. The Cape medical profession situated itself in the strange half-light between European identity and colonial reality. While the doctors did come into contact with black patients and indigenous ways of healing, they practically ignored the scientific study of indigenous medical pharmacology, although the Western Cape boasts a unique floral kingdom.[38] Cape doctors were more likely to look to their European (mainly British) colleagues and institutions for professional affirmation than to each other. This particularly intense form of "colonial cringe" led to a much greater delay in the establishment of colonial medical schools at the Cape than in India, Canada, and Australia.[39]

The spread of the gentlemanly ideal was associated with medical professionalization. In the eighteenth century there was little legal control over who could practice medicine. Indigenous and slave healers were left to their own devices, and Western

[33] On the Grahamstown trial, see Burrows, *A History of Medicine* (cit. n. 3), p. 170.

[34] *Ibid.*, p. 133.

[35] *Cape Town Medical Gazette*, 1847, *1*, 3:32. See also Burrows, *A History of Medicine* (cit. n. 3), p. 350 and C. Blumberg, "The South African Medical Society and its Library," *Cabo*, 1978, *2*, 4:18–25.

[36] Thomas Lucas, *Camp Life and Sport in South Africa* (1878; reprint, Johannesburg: Africana Book Society, 1975), pp. 33–5.

[37] On medical gentlemen elsewhere, see Penelope Corfield, *Power and the Professions in Britain, 1700–1850* (London: Routledge, 1995); Robert Gidney and Winnifred Millar, *Professional Gentlemen: The Professions in Nineteenth-Century Ontario* (Toronto: Univ. of Toronto Press, 1994).

[38] Deacon, "Understanding the Cape Doctor" (cit. n. 6).

[39] Howard Phillips, "Medical Education," in Deacon, Van Heyningen, and Phillips, *The Cape Doctor* (cit. n. 6).

doctors were not organized into a profession with strict rules governing membership and practice. In 1807, however, the autocratic military government in Cape Town tried to impose price and quality controls on the sale of imported medicines. Western-trained doctors used this as an opportunity to seek government regulation of the whole medical profession. Not surprisingly, this won for the doctors government endorsement and a legal monopoly over professional medical practice at the Cape.[40] Indigenous healers were prohibited from charging for their services, and female midwives had to come under the control of licensed male doctors.[41]

Almost all Cape doctors were European settlers, and many were recent immigrants.[42] Almost no black or female Western doctors were in practice; a handful were licensed late in the century. Although there was no racist legislation, the Cape profession excluded most black and female practitioners in practice by defining itself in masculine, Western terms and by insisting on a European medical training. Most colonial families could not afford to send their sons to study overseas. The European medical schools did not usually admit women, and moreover, few Cape and European families wanted to go to the expense of educating their daughters at a time when women were supposed to become housewives. It was also difficult to overcome official, medical, and settler prejudice against black people becoming Western doctors. Most black communities consulted indigenous healers by preference, and there were few opportunities for black doctors outside missionary hospitals. The overwhelming number of white immigrant medical practitioners was also an important factor. Many more black doctors were trained in the Western tradition in West Africa—"the white man's grave"—where from midcentury on there was an Anglicized black elite whose members sought posts in a government service shunned by white doctors.[43]

By the early nineteenth century, the Cape medical profession was thus highly racialized in terms of its professional identity and racial composition, in spite of the absence of formal legal barriers to the entry of black doctors.

RACIST MEDICINE

Within different institutions and areas of medical specialization, racism was expressed in different ways and the timing of segregation varied. This was only partly due to the interplay between racist medicine and medical racism; disease profiles, patterns of institutionalization, and popular stereotypes also played a role. How doctors viewed medicine was closely connected to their conception of their status and responsibility as settlers and medical gentlemen.[44] They were, however, obliged to make a living by selling medical services. Their relationships with black patients were thus influenced by the unequal relations of colonialism, but also by class relations in the context of their consultations (private, charitable, or public). Moreover,

[40] In the other British colony in South Africa, Natal, indigenous practitioners were allowed a limited practice.

[41] Deacon, "Cape Town and Country Doctors" (cit. n. 5).

[42] *Ibid.*, p. 33; Van Heyningen, "Agents of Empire" (cit. n. 5), p. 452.

[43] Patton, *Physicians, Colonial Racism and Diaspora* (cit. n. 2).

[44] John Harley Warner has suggested that even in fiercely Republican antebellum America, doctors' professional identity was predicated partly on "moral character" or social respectability. John Harley Warner, *The Therapeutic Perspective: Medical Practice, Knowledge and Identity in America, 1820–1885* (Princeton, N.J.: Princeton Univ. Press, 1986), pp. 1, 15–16.

the extent and timing of racial segregation in medical and other institutions was influenced by site-specific functional and professional factors.[45] These circumstances can also help explain variabilities in the theory and practice of racism in colonial medicine.

Given a white male profession whose members aspired to be "gentlemen," as well as the racism of settler society, one might expect both the theory and practice of medicine at the Cape to have been racist. This is broadly true, but we have to be very aware of the ways in which we measure the form and extent of this racism. Many historians have focused on documenting overt or covert racism in medical legislation and the racist implementation of "color-blind" medical legislation in South Africa by the end of the nineteenth century.[46] This focus obscures some key issues, however: *(a)* that private black patients may not have experienced the same degree of racist practice as institutionalized black patients, and *(b)* that circumstances within institutions played a key role in shaping the extent and timing of racial discrimination.[47]

The profile of black patients attended by colonial doctors was by no means homogenous, and neither was their treatment.[48] Black patients went to Western doctors for a few specific ailments; some slaves or workers went at the behest of their employers; and many sought free medical care, food, and lodging in colonial hospitals because they were destitute or abandoned. Some black patients actually chose Western medicine in preference to other types of care. Others were treated under duress: Western doctors were often called in by slave owners to ensure that slaves could continue working. Doctors were also used to testify in favor of an owner accused of maltreating a slave.[49] Since doctors identified with and depended on settlers as clients, they seldom asserted the rights of sick slaves against the demands of their masters. When they did, it caused great friction between doctors and the settler community.[50]

Given the lack of detailed records, our understanding of the size and nature of private practice serving black clients is limited. However small, this group of clients is important. The records of Peter Chiappini, an Edinburgh-trained doctor practicing in Cape Town in the 1840s, suggest that he had a varied clientele, ranging from wealthy white settlers to poor black laundresses. Some of his poor black clients consulted him frequently and paid bills as private patients, although on the lowest

[45] Deacon, "Racial Segregation" (cit. n. 12).

[46] Swanson, "The Sanitation Syndrome" (cit. n. 5).

[47] Some work has begun to consider these issues. On psychiatric theory and practice see, for example, Swartz, "Colonialism and Psychiatric Knowledge" (cit. n. 12); Harriet Deacon, "Madness, Race and Moral Treatment at Robben Island Lunatic Asylum, 1846–1910," *History of Psychiatry*, 1996, *7:*287–97. On institutional contexts, see Deacon, "Racial Segregation" (cit. n. 12).

[48] White patients (particularly Dutch-Afrikaans settlers) also sometimes consulted black medical practitioners, whether midwives or traditional healers. See Harriet Deacon, "Midwives and Medical Men in the Cape Colony before 1860," *J. African Hist.*, 1998, *39:*271–92.

[49] Cape slavery is often described as relatively "mild," but this does not mean that slaves escaped severe (even fatal) beatings, malnourishment, and psychological abuse at the hands of their masters.

[50] For example, Daniel O'Flinn, surgeon at Stellenbosch, suffered financially because of his "kindness and attention" to slaves ("Report on O'Flinn," *Confidential Reports on Civil Servants, 1843–51*, Cape Archives, Cape Town, CO 8551). Occasionally, a slave owner was charged with murder as a result of a doctor's report: see, for example, the case of the contradictory postmortem reports on a dead slave by Drs. Shand and Tardieux in Burrows, *A History of Medicine* (cit. n. 3), pp. 56–7, 87.

scale.[51] Western medical care was probably more expensive than other options, so we might assume that only wealthier black clients could afford it. Yet Chiappini had black clients from the lowest income groups. Chiappini's black clientele may have been introduced to Western medical care in the hospitals, at the weekly free clinics some urban doctors offered, or by former employers and slave owners. Given the history of abuse of female slaves by male settlers, it is particularly interesting that not only were some of Chiappini's private black clients women, but that his specialty for Muslims was in an area of gynecology (removal of the afterbirth).[52] Chiappini probably attracted private black patients because he spoke Dutch-Afrikaans and practiced close to the tenements of Cape Town's burgeoning urban working class. (Afrikaans is a Creole language, derived from Dutch, Malay, and African languages, that was spoken by some settlers of continental European origin, slaves, and Khoekhoen, hence the term Dutch-Afrikaans to designate people who were identified with the white settler community.)

In rural areas with few hospitals and Western doctors, and great distances between farms and settlements, black people consulted Western doctors for specific ailments only. Blacks with eye problems came from afar to consult Dr. John Fitzgerald at the Kingwilliamstown hospital in the eastern Cape, for example.[53] Although hospitals and free clinics were not always available, there were other ways of getting access to Western medicine. Missionaries often freely administered it to potential converts. Apothecaries and traveling salesmen sold medical advice as well as drugs, the most popular of which were patent medicines made in Halle. Similar remedies, marketed as "Dutch medicines," are in widespread use among black South Africans even today.

The willingness of black clients to use Western medicine does not mean that it was free of racial bias. Nor was racism in Western medicine by any means homogenous, since Western doctors came from different traditions of racism (ranging from liberal to openly racist), and since settler stereotypes of Khoekhoe, Muslim, and African patients varied. The extent to which Western doctors could and did discriminate against their patients was also influenced by legislation, existing practices among their peers, and the financial aspect of the doctor-patient relationship. Some black patients paid for consultations themselves (reimbursement was often scaled to income), but most did not. In general, there was greater leeway for systematized discrimination and the application of racist policies in institutions where patients were not paying for medical services.

Given the constraints on private practice, government and missionary hospitals were the main point of contact between black patients and white doctors. These colonial institutions were conducive environments for the development of racist theory and practice, as their custodial overtone lent itself to discrimination. In rural

[51] Unfortunately, we do not yet know what the comparative cost of consulting slave or indigenous practitioners might have been in Cape Town during the 1840s, but we can assume it was fairly low, and even a visit to the local apothecary would probably have been cheaper than Chiappini's lowest rate.

[52] On abuse of women slaves, see Pamela Scully, "Rape, Race and Colonial Culture: The Sexual Politics of Identity in the Nineteenth-Century Cape Colony, South Africa," *Amer. Hist. Rev.*, 1995, *100*, 2:335–59. On Chiappini, see Deacon, "Midwives and Medical Men" (cit. n. 48), p. 287, n. 86.

[53] See, for example, Burrows, *A History of Medicine* (cit. n. 3), p. 182.

areas the town jail was often used as a "hospital" or holding place for the homeless and destitute, and hospitals were often established in old barracks or prisons. Most hospital patients were poor, and many were black. Treatment was generally free or nearly so. Patients often had nowhere else to go; some were forcibly institutionalized. Their interaction with doctors was thus less individualized, encouraging discrimination against black patients.

Hospital care in the nineteenth-century Cape (as elsewhere) was not particularly effective, and was shunned by most middle-class patients.[54] Racial discrimination was almost inevitable when a doctor had to divide his time and resources between white and black: if the doctor did not prioritize white patients, they often protested. Once separate institutions for black and white patients had been established in the 1890s, discrimination was entrenched, but at the budgetary, staffing, and policy level rather than at the discretion of institutional staff.

Despite similarities between hospitals, prisons, and other colonial institutions, and between hospitals of different sorts, they presented different patterns of racist practice.[55] At the Robben Island leper, mental, and chronic sick hospitals, for example, racial segregation emerged at different times. The timing depended partly on the numerical balance between black and white patients and the group's class profile. The social and medical stereotypes of the diseases treated also affected patterns of segregation. But medical racism—racist medical theories of insanity and leprosy—did not materially influence the timing of racial discrimination. These theories did little more than justify differential treatment based on race and the creation of separate asylums for black and white mental patients.[56]

MEDICAL RACISM

Racist medical theory did not always accompany racist medical practice. Racism in medicine could arise from circumstances other than the elaboration of formal medical arguments for racial difference in patient diagnosis and care. Colonial doctors often used general justifications for racial discrimination (social entitlement) rather than specifically formulated medical reasons. At the Cape, racist medical theories were slow to develop because of conditions in separate black institutions, a desire to underplay differences between metropolitan and colonial science, and an unquestioning acceptance of racism in colonial society. In Cape psychiatry, for example, doctors treated black and white mental patients differently from the beginning, but only in the 1880s and 1890s did they advance theoretical arguments to justify this, suggesting that black patients responded better to physical rather than psychological therapy because of their supposedly lower developmental status.[57] Such racist theories emerged from doctors' experiences with large numbers of black patients in insti-

[54] Exceptions were the curative wards of the New Somerset Hospital and middle-class lunatic asylums, such as those in Grahamstown or Valkenberg.

[55] Deacon, "Racial Segregation" (cit. n. 12).

[56] The exception to this rule was the reversal of some forms of racial discrimination in the 1860s at the Robben Island Lunatic Asylum with the application of "moral management" treatment, which assumed that insanity was "everywhere the same" (Deacon, "Racial Categories and Psychiatry" [cit. n. 24]).

[57] Deacon, *ibid.*

Figure 1. *A leprosy patient at Robben Island c. 1900.*

tutional settings where their prejudices were often confirmed. These theories were often simply tacked on to pre-established racist practices.[58]

The case of leprosy provides an important example of the emergence of racist medical theory from a popular stereotype. Portraying the leper as black was central to medical theory and public health policy in the Cape. Of course, most leprosy

[58] A similar pattern can be observed in British psychiatry, which did not use the "scientific" evidence collected to produce insights into the types, causes, and possible results of insanity during the nineteenth century. See Richard Russell, "The Lunacy Profession and its Staff in the Second Half of the Nineteenth Century," in *The Anatomy of Madness III: Essays in the History of Psychiatry*, eds. William Bynum, Roy Porter, and Michael Shepherd (London: Routledge, 1988), p. 298.

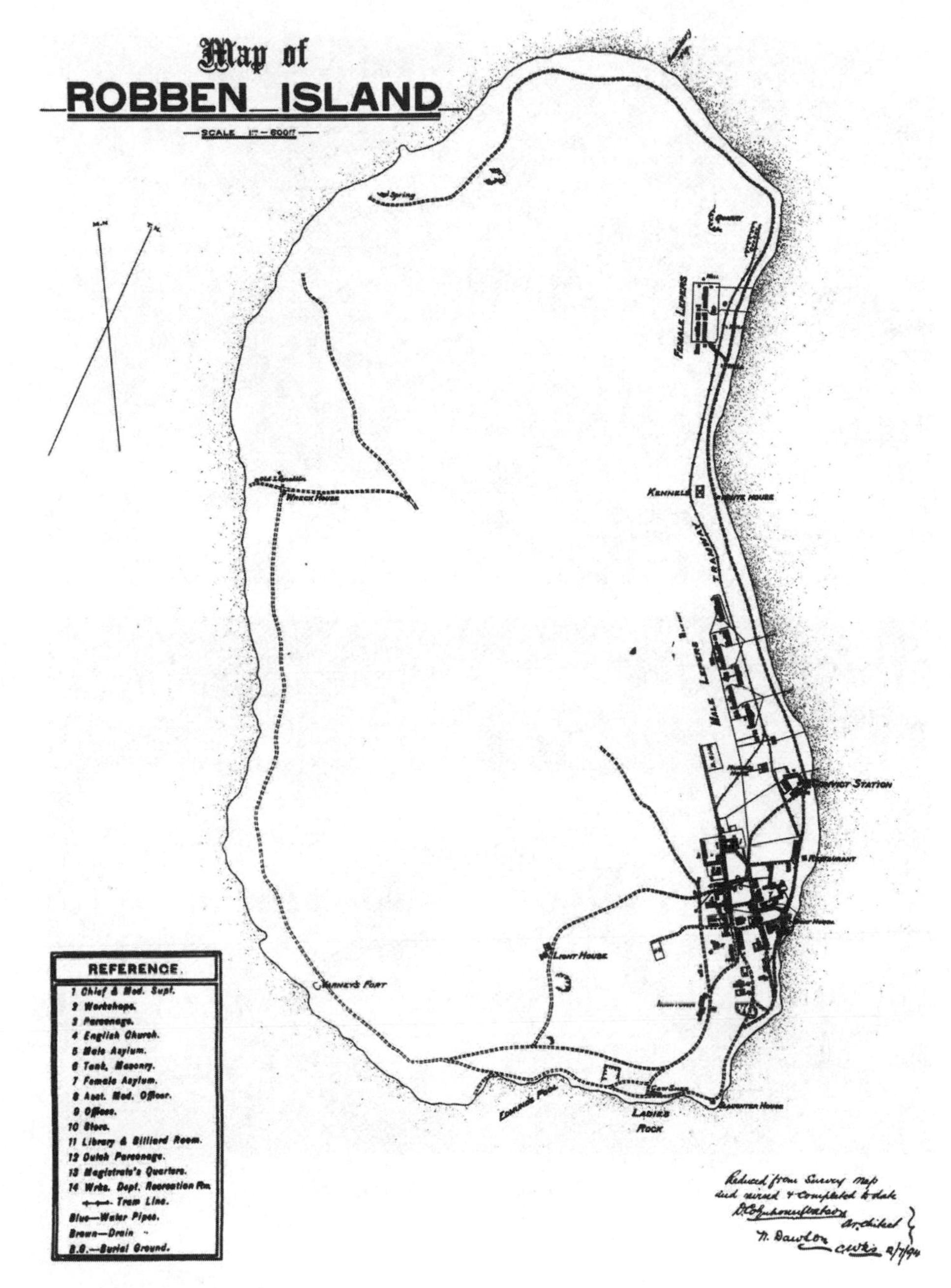

Figure 2. *Robben Island, 1894. The leprosy hospital wards for men and women were separated from the main village area in 1891.*

Figures 3 and 4. *As these pictures show, most leprosy patients at Robben Island were black, while most staff members were white.*

patients in Cape institutions were black. Although it is possible that different populations may have had different levels of generationally acquired immunity to the disease, poverty—more widespread among blacks—may also have influenced its spread and the pattern of institutionalization.[59] But both medical and popular explanations for differential rates of institutionalization for leprosy focused on existing settler stereotypes of black culture, rather than on socioeconomic conditions.

[59] Packard, *White Plague, Black Labour* (cit. n. 5), p. 31.

From the beginning of the nineteenth century, for example, it was popularly believed that Khoi ("half-castes") and later "Malays" (Muslims) and Africans were particularly susceptible to leprosy.[60] The negative stereotype of the Hottentot, which stressed idleness, unpleasant smell, promiscuity, "dirty" habits, and nomadic lifestyle, was translated into a popular aetiology of disease.[61] Once the *lepra* bacillus had been identified as the causal agent during the 1870s, specific black groups ("mixed breeds" and the highly visible small trader, often a Muslim, selling vegetables or retailing other food items), rather than black "habits," were identified as a source of the disease.[62] After 1883, officials, doctors, and colonists often voiced fears about leprosy being transmitted from black farm servants to white farmers or their children.[63] In 1896, Dr. Samuel Impey attributed the historical spread of leprosy in South Africa almost entirely to various black groups, especially "Bushmen."[64] By the 1890s, both medical theory and popular discourse attributed blame for the contraction and spread of leprosy to the supposedly "pathogenic" culture of black people in general.[65]

This led to a particularly harsh approach to the management of the disease. Unlike many other countries that passed public health regulations to control leprosy, the Cape, after almost a decade of concern that the disease was spreading and affecting whites, opted for compulsory segregation of all lepers under the Leprosy Repression Act of 1891. Officials and doctors believed that only a limited number of people would be affected, and that the disease could be stamped out using quarantine techniques like those used for smallpox or plague. But because leprosy was also perceived to be a black disease, it was particularly easy to apply procedures that infringed basic civil liberties. At the time, blacks in general were losing civil rights and being shifted out of the public sphere in the colony.[66]

The racist stereotyping of the leper as black, and therefore the selection of harsh public health legislation, created problems for Cape institutions that had to impose

[60] On Khoi, see Minutes of Evidence, "Report of the Commission of Inquiry into the General Infirmary and Lunatic Asylum on Robben Island," *Cape Parliamentary Papers*, G31-1862, p. 161. Especially Muslims with low morals were suspect; see Joseph Matthews, *Incwadi Yami or Twenty Years' Personal Experience in South Africa* (New York: Rogers and Sherwood, 1887), p. 350.

[61] Minutes of Evidence, "Report of the Commission of Inquiry" (cit. n. 60), p. 161.

[62] Minutes of Evidence, "Report of the Select Committee on the Spread of Leprosy," *Cape Parliamentary Papers*, A23–1883, pp. 11, 33; "Report of the Select Committee Inquiring into the Spread of Leprosy," *Cape Parliamentary Papers*, G3-1889, p. x.; Minutes of Evidence, "Report of the Select Committee . . . 1883," *Cape Parliamentary Papers*, A23-1883, p. 22; Minutes of Evidence, "Report of the Select Committee . . . 1889," *Cape Parliamentary Papers*, G3-1889, p. 4.

[63] "Abstracts of Replies to Interrogations," "Precis of Report of the . . . Leprosy . . . Commission," *Cape Parliamentary Papers*, G4-1895, p. 89; Arnold Simons, Minutes of Evidence, "Report of the Select Committee . . . 1889" (cit. n. 62), p. 8. After 1882, a related contraction scenario was commonly presented for syphilis in which a black nursemaid transmitted the disease first to the child and then, through the child, to the whole family (Aberdeen and Caledon, "Reports of the Civil Commissioners, Resident Magistrates and District Surgeons for 1882," *Cape Parliamentary Papers*, G91-1883; "Reports of the Civil Commissioners, Resident Magistrates and District Surgeons for 1883," *Cape Parliamentary Papers*, G67-1884).

[64] Samuel Impey, *Handbook on Leprosy* (London: Churchill, 1896), chap. 1.

[65] See, for example, "Leprosy Notification," *South African Medical Journal*, July 1898: 63–7, p. 63, and Hutchinson to Black, 6 Feb. 1902, Cape Archives, Cape Town, Health Branch, Letters re Leprosy Commission of 1894–5, CO 7663. Hutchinson's views were echoed in the Indian Medical Service for much of the nineteenth century, where these views had roots in indigenous knowledge (Mark Harrison, pers. comm., 30 Jan. 1994).

[66] Harriet Deacon, "Leprosy and Racism at Robben Island," in *Studies in the History of Cape Town*, ed. Elizabeth van Heyningen (Cape Town: Univ. of Cape Town Press, 1994), vol. 7, pp. 45–83.

these regulations on whites. In practice, black and white leprosy patients were treated differently within institutions, and campaigns were started to highlight the plight of white patients who were forced to stay there. A system of "home segregation" was set up to cater almost exclusively to white patients whose families could support them in a separate dwelling. This was gradually followed by medical rationales for differences between black and white leprosy sufferers. By the 1940s, a visiting leprosy specialist speculated that the reasons for a higher incidence of the more severe lepromatous leprosy in whites were *(a)* that Europeans had less close contact with each other and thus only the more susceptible contracted it in a severe form, and *(b)* that Europeans were at a climatic disadvantage in South Africa, or *(c)* that the disease produced in Europeans more mental depression and therefore more physical weakness.[67] The colonial stereotype of the leper had attained a "scientific" gloss, in which racial or cultural differences were linked to clinical pictures.

CONCLUSION

The introduction to this paper pointed out the contradictions inherent in arguing *(a)* that Western medicine was an agent of colonialism and therefore evil in its effects, and also *(b)* that black people were disadvantaged because they did not have equal access to Western medicine. It suggested that we seek a more nuanced approach to understanding the relationship between racism, medicine, and empire.

To some extent, both *(a)* and *(b)* are true. Given the social profile of the medical profession at the Cape and the close relationship it enjoyed with the colonial state and the settler community, Western medicine inevitably shared many of the aims and assumptions of colonialism. Doctors acted as agents of the colonial state when they represented its authority in courtrooms or enforced its legislation in their practices. They rejected integration with indigenous healing systems, pathologized indigenous cultures, and deferred to European professional standards. Racism in medicine grew out of racism in colonial society. These attitudes were not only reproduced at the individual level, but they played a role in the construction of the colonial medical profession and the definition of the ideal "medical gentleman." Black people had restricted access to Western medicine because it was expensive and because its curative energies and expenditures were mainly targeted at white settlers. Although hospitals admitted large numbers of black patients, they were still underrepresented relative to the general population. Racial discrimination within hospitals meant that black patients received less and lower-quality medical care than did white patients. Racist medical theories helped to entrench and justify such practices.

To describe Western medicine as just an agent of colonialism, however, oversimplifies its form and impact, assumes that colonized peoples were no more than victims, and underestimates differences among practitioners and patients. The colonial state apparatus (and its medical wing) was neither homogenous nor omniscient, and it could not fully predict or control the consequences of Western medical interventions.[68] Western hospitals were used to remove the homeless, insane, or contagious

[67] Ernest Muir, "Report on Leprosy in the Union of South Africa," *Leprosy Review*, 1940, *11*:43–52, p. 43.

[68] Engels and Marks, *Contesting Colonial Hegemony* (cit. n. 1).

from colonial cities and settler farms. Some patients, many of them black, were forced to stay against their will in hospitals such as the leper institution on Robben Island. But other patients, both black and white—and relatives who wished to be rid of them—used illness as an opportunity to obtain free food and lodging. While colonial doctors often discriminated against black patients who were not paying for their own treatment, private practice serving black clients may have encouraged a more equitable approach. The fact that some black patients sought assistance from Western medicine in certain areas (such as surgery or placental delivery) but not others suggests that they were able to shop for what was perceived to be the most effective medical care without necessarily buying into the overall philosophy of any one system.

We cannot assume without comparative evidence that the primary discrimination experienced by black patients, even in the racist environment of the colonial hospitals, had to do with race rather than with gender or type of disease. If some of what Western medicine could offer was useful to black patients, and it probably was, this was not necessarily an unintended consequence of colonialism, nor a kind of false consciousness. Doctors were supposed to cure people, and colonial subjects as well as doctors recognized this as a key element of professional identity. In comparing how doctors treated black and white patients we can see, however, clear evidence of discrimination, the effects of which need to be documented and assessed. We can also note the discrepancy in the proportions of white and black people treated by Western medicine, but it would be foolish to ascribe this solely to restricted access. Some doctors actively sought black clients but were unable to persuade black communities of their worthiness.

Western medicine in the colonies was certainly racist in its professional makeup and in the construction of its theory and practice. Yet this generalization tells us little that is interesting about racism or, indeed, about Western medicine. This essay suggests that, in seeking to understand the trajectory of racism in colonial medicine, we should distinguish between rich and poor black patients and between hospital and private practice. We should contextualize racist practices within the framework of different professional formations, types of disease, stereotypes of disease, and medical theories. We cannot understand the form, timing, intention, and function of racism in colonial medicine without considering these factors together.

Debating Racial Science in Wartime Japan

By Tessa Morris-Suzuki

> One of the greatest enemies of science is pseudo-science. In a scientific age, prejudice and passion seek to clothe themselves in a garb of scientific respectability; and when they cannot find support from true science, they invent a pseudo-science to justify themselves. . . . Nowhere is this lamentable state of affairs more pronounced than in regard to "race." A vast pseudo-science of "racial biology" has been erected which serves to justify political ambitions, economic ends, social grudges, class prejudices.

With these words, Julian Huxley and A. C. Haddon began their influential popular study, *We Europeans: A Survey of "Racial" Problems,* published in 1935.[1] The work (which also included a chapter on the migration of European peoples by the geneticist A. M. Carr-Saunders) both symbolized and encouraged an important turn of the tide in racial theorizing. Since the late nineteenth century, large amounts of time, money, and effort had been devoted to the tasks of defining precise dividing lines between the races and relating racial heritage to various physical, intellectual, and emotional characteristics. Skulls had been measured, I.Q.'s tested, blood types compared, and countless volumes of speculative hypotheses and proposed solutions to the "race problem" had been published.

Among these, interestingly enough, were the earlier writings of Julian Huxley himself. A renowned scholar of evolution (and grandson of Darwin's friend Thomas Henry Huxley), Julian Huxley had begun his career as a firm believer in racially determined differences in intelligence and personality. In a 1924 essay on "the Negro problem," for example, he described African Americans as having infantile minds—"the typical negro servant . . . is wonderful with children, for the reason that she really enjoys doing the things that children do"—and recommended the division of the United States into a racially mixed South and a segregated white North. By the mid 1930s, however, Huxley's views had undergone a radical transformation. Like a number of other scientists, he was increasingly disturbed both by the failure of racial science to define its terms with scholarly rigor and by the sinister uses to which the science of race was being put by Nazi ideology. *We Europeans* represents a response to these problems. Huxley's main coauthor, the anthropologist A. C. Haddon, had always tended to emphasize the importance of environment rather than heredity in explaining differences between human groups, but his early writings,

[1] Julian Huxley, A. C. Haddon, and A. M. Carr-Saunders, *We Europeans: A Survey of "Racial" Problems* (Harmondsworth, Middlesex: Penguin, 1935), p. 7.

hazy. Modern racial science emerged out of simultaneous nineteenth-century developments in a number of disciplines, including anthropology, biology, and human anatomy, and reached the peak of its influence in the first four decades of the twentieth century. Its development was closely associated with the rise of the eugenics movement, whose origins are generally traced to the work of Charles Darwin's cousin Francis Galton (though similar ideas were also propounded by a number of Galton's contemporaries). Scientific racism and eugenics, in other words, evolved both as areas of academic research, pursued by prominent scholars like Karl Pearson at University College London and Charles Davenport at the University of Chicago, and as social movements whose ideologies were propagated by bodies like the German Society for Racial Hygiene (established in 1904), the British Eugenics Education Society (1908), and the U.S. Eugenics Record Office (1910), all of which came to cooperate in cross-border ventures like the International Society for Racial Hygiene.[9]

As Elazar Barkan has emphasized, however, the ideological and social positions of prewar racial theorists were varied, and it is not always easy to divide them simply into "racist" and "antiracist." Because of the widespread popular influence of race as an organizing category of thought, even scholars like Franz Boas, who was famous for his opposition to discriminatory policies and his emphasis on the importance of environment rather than heredity, could at times make statements that sound racist to contemporary ears. Besides, scientific racism was not confined to the political right: the leading British advocate of eugenics, Karl Pearson, was a Fabian socialist who split with the Eugenics Education Society because of his opposition to its conservative stance on feminism and other social issues.[10]

From the Japanese point of view, racial science was part of the baggage of modern Western thought imported from the late nineteenth century onward. But the ready acceptance of this particular genre of thought in Japan can be explained both by preexisting Japanese intellectual traditions and by the political circumstances of late nineteenth- and early twentieth-century Japan. The scientific study of race was new to Japan, but during the seventeenth and eighteenth centuries Japanese scholars had developed something of a passion for encyclopedic listings of "peoples of the world," often accompanied by vivid and dramatic illustrations.

These classificatory lists (which developed simultaneously with an enthusiasm for listing and classifying plants, animals, and mineral resources) drew on Chinese notions of the world as made up of concentric circles of increasing strangeness. Like late medieval European representations of the outside world, they often included the bizarre and grotesque: giants, dwarfs, centaurs, and winged people abound in the remoter regions of the earth. At the same time, though, classifications were not simply based on physical appearance, but also on habits or customs, and this approach encouraged some quite careful ethnographic description of neighboring foreign peoples by explorers like **Mamiya** Rinzō (who traveled through Hokkaido and Sakhalin to Far Eastern Siberia in the first decade of the nineteenth century). It was a relatively small step, therefore, from existing visions of the outside world to the idea

[9] Adrian Desmond and James Moore, *Darwin* (Harmondsworth, Middlesex: Penguin, 1991), pp. 556–557; and Ivan Hannaford, *Race: The History of an Idea in the West* (Baltimore: Johns Hopkins Univ. Press, 1996), p. 332.

[10] See, e.g., Boas's comments on the relative brain sizes of different races, quoted in Barkan, *Retreat of Scientific Racism* (cit. n. 2), p. 86; on Pearson see *ibid.*, pp. 151–153.

(popularized in the 1870s by the famous Westernizer **Fukuzawa** Yukichi) of a world divided into five major "races" (translated into the Japanese neologism *jinshu*) or to the concept of *minzoku*—a term popularized by the nationalist geographer **Shiga** Shigetaka around the beginning of the 1890s and used by a wide variety of writers to mean anything from "ethnic group" to "nation."[11]

From a quite different point of view, Japan's response to racial science was also influenced by traditions that provided ready receptors for imported genetic theories. European genetics, including Lamarckian ideas on the inheritance of acquired characteristics and neo-Darwinian techniques of biometry, were introduced into Japan in the late nineteenth century, and Mendelian theories were imported in the early years of the twentieth century, after the rediscovery of Gregor Mendel's ideas in Europe. Before the mid-nineteenth century, however, Japan already had a strong tradition of selective breeding in agriculture and horticulture. Particularly in the two centuries before the beginning of the Meiji era, the skill of "improving varieties" (*hinshu kairyō*) had been energetically developed in rice farming and above all in silk production. Because silkworms have a short life cycle and are reared in highly controlled conditions, Japanese silk farmers were able to achieve remarkable results in improving the health and yield of their worms through empirical and experimental cross-breeding—results that were studied and emulated by European sericulturists when they first began to have contact with Japan in the middle of the nineteenth century.

The arrival of foreign scientific theories that could explain the mechanisms of selective breeding was therefore greeted with great excitement in Japan. Genetics became a major area of research in newly established scientific institutions like Tokyo Imperial University's College of Agriculture (set up in 1890), and a number of Japanese scientists who had started their careers studying the inheritance of characteristics in rice plants or silkworms went on to speculate about the possible extension of these genetic ideas to human heredity.

At the same time, though, ideas of race were eagerly debated in Japan, not just because there were indigenous pegs on which to hang these borrowed clothes, but also because of the ideological dilemmas of nationalism and imperial expansion in late nineteenth- and early twentieth-century Japan. The Japanese state embarked upon its policy of rapid modernization at a time when ideas of social Darwinism and scientific racism were becoming immensely influential in Western Europe and North America, and the relationship between race and power was from the first an issue of intense concern to the Meiji political elite. During the 1880s a number of Japanese intellectuals proclaimed the necessity of intermarriage with Europeans as a means of "strengthening" the Japanese race, and the prominent politician **Itō** Hirobumi even wrote to ask the advice of Herbert Spencer on the subject. Spencer's reply began by citing recent genetic experiments in sheep that, he claimed, had shown

[11] **Mamiya** Rinzō, *Kita Ezo zusetsu* (An illustrated description of north Ezo) (1855), rpt. in *Hokumon sōsho* (Northern Gate series), Vol. 5, ed. **Ōtomo** Kisaku (Tokyo: Kokusho Kankōkai, 1972); and **Fukuzawa** Yukichi, *Sekai kunizukushi* (A complete account of the countries of the world) (1869), rpt. in *Fukuzawa Yukichi zenshū* (The complete works of Fukuzawa Yukichi), Vol. 2 (Tokyo: Kōmin Tosho, 1926). On the term *minzoku* see **Yoon** Keun Cha, *Minzoku gensō no satetsu* (The failure of the ethnic illusion) (Tokyo: Iwanami Shoten, 1994); and Kevin M. Doak, "Culture, Ethnicity, and the State in Early Twentieth-Century Japan," in *Competing Modernities: Issues in Culture and Democracy, 1900–1930,* ed. Sharon Minichiello (Honolulu: Univ. Hawaii Press, 1998).

that cross-breeding very dissimilar strains produced inferior offspring. He went on to observe that such eugenic concerns lay behind U.S. restrictions on the immigration of Chinese (restrictions with which Spencer clearly sympathized). Asian immigrants, he argued, had to be excluded from U.S. society because, once allowed in, they would either form an enduring separate underclass or interbreed with the white population, and this racial mixing would cause a "deterioration" in the quality of the population. The same decline would occur in Japan, he warned, if Japanese people were encouraged to intermarry with Europeans.[12]

As Japan began its imperial expansion into Asia, the issue of racial and cultural similarities and differences between the Japanese and their Asian neighbors also became a subject of intense controversy. On the one hand, emphasis on similarities between Japanese, Koreans, and Chinese could be used as a justification of Japan's right to "guide" or "lead" the political and economic destiny of neighboring countries and as an ideological prop for assimilationist policies in the colonies. On the other, ideas of racial or ethnic difference were often mobilized, both at the official level and in popular discourse, to justify discrimination against colonial subjects. In both cases, the very fact that the empire extended outward into contiguous regions with which Japan had ancient historical links (rather than being a far-flung seaborne empire like Britain's) meant that the issues of ethnic dividing lines in the Japanese colonies were always particularly ambiguous and contentious.

RACIAL SCIENCE BETWEEN THE WARS

In Japan, as in other countries, debates on race grew out of the confluence of several disciplines. On the one hand, they were encouraged by the development of genetic research, particularly from the beginning of the twentieth century onward. **Sotoyama** Ryūtarō (1867–1918), the agricultural geneticist who introduced Mendelian theory to Japan, was among the first to discuss the scientific application of eugenics to human beings. In an article published in 1910, he applied his expertise in the selective breeding of silkworms to the problem of human inheritance, rehearsing the familiar eugenicist argument that the future health of economically advanced societies was threatened by social systems that permitted weak and "defective" individuals to survive and breed where, in a natural state, they would have been allowed to die.[13] These ideas were later taken up and developed by a number of geneticists and medical scientists, including **Tanaka** Yoshimaro, who similarly moved from the study of silkworms to advocating the importance of human eugenics, and the physician **Nagai** Hisomu, who was to play a central role in the Japanese "racial hygiene" movement.

At the same time, however, the study of race was also being pursued from the point of view of the social sciences, particularly of the newly emerging discipline of anthropology. Anthropological ideas had been introduced to Japanese audiences by the work of foreign academics like the American biologist E. S. Morse, who was employed as a professor at Tokyo Imperial University in the 1870s and 1880s and is also remembered as one of the founders of Japanese archaeological research. The

[12] Quoted in **Koyama** Eizō, *Minzoku to bunka no shomondai* (Problems of ethnicity and culture) (Tokyo: Hata Shoten, 1942), p. 167.

[13] See **Suzuki** Zenji, *Nihon no yūseigaku: sono shisō to undō no kiseki* (Eugenics in Japan: The course of its ideology and activities) (Tokyo: Sankyō Shuppan, 1983), pp. 65–69.

first formal university course in anthropology, however, was introduced in 1892 by **Tsuboi** Shōgorō (1863–1913), the pioneering Japanese anthropologist and archaeologist, who had previously studied in London, where he had attended lectures by Edward Tylor and visited the research laboratory of Thomas Henry Huxley.[14] Like many early Japanese anthropologists, Tsuboi devoted a large part of his career to pursuing the origins of the Japanese people—investigating the relationship between majority Japanese, the indigenous Ainu people of northern Japan, and the various "tribes" mentioned in the earliest Japanese historical writings of the eighth and ninth centuries A.D.

Tsuboi himself developed a hypothesis (later abandoned by most anthropologists) that the ancestors of the Japanese had been the mysterious people referred to in Ainu legends as the "Koropukur." By the early twentieth century, however, many of the leading figures in Japanese anthropology—including Tsuboi's most eminent student, **Torii** Ryūzō—had adopted the view that the Japanese were of mixed racial origins, the product of repeated waves of migration from Siberia, China, Korea, Southeast Asia, and the Pacific.

The Japanese eugenics movement, which emerged from around the time of World War I, brought together elements from genetics, anatomy, and anthropology. An important precursor of the movement was the journal *Humankind* (*Jinsei;* the front page also bore the German title *Der Mensch*), which was published from 1905 under the editorship of the medical scientist **Fujikawa** Yū.[15] Amongst a wide range of academic papers on human biology, evolution, and diversity, the journal carried a number of articles introducing eugenic theories and advocating eugenic policies. Several of the writers who contributed to *Humankind* went on to publish in the journal *Eugenics* (*Yūseigaku;* founded in 1924) and to contribute to the establishment of the Japan Eugenics Association (Nippon Yūseigaku Kyōkai), established in the same year.

Rising interest in eugenics during the 1920s was fueled by issues of international migration. Between 1898 and 1901, almost fifty thousand Japanese had migrated to the U.S. mainland, their growing numbers aggravating the fears (already alluded to by Herbert Spencer decades earlier) that local wages would be undercut by Asian "cheap labor." In 1907 the United States and Japan entered into a "gentlemen's agreement" whereby Japan agreed to restrict the emigration to the United States of what were described as "working-class citizens." This, however, failed to stem the rising tide of U.S. exclusionism, and in 1924 the Immigration Act was passed, tightly restricting migration from Japan. Although economic fears and populist racism were the main factors behind the act, its introduction was also publicly justified by eugenicists like Harry H. Laughlin, who worked with Charles Davenport in Chicago and presented evidence before the House of Representatives Committee on Immigration and Naturalization emphasizing the undesirability of "racial mixing." Needless to say, this view did not go unchallenged: liberal geneticists like H. S. Jennings attacked Laughlin's arguments against migration, and the debate surrounding

[14] See **Torii** Ryūzō, "Nihon jinruigaku no hattatsu" (The development of anthropology in Japan) (1927), rpt. in *Torii Ryūzō zenshū* (the complete works of Torii Ryūzō), Vol. 1 (Tokyo: Asahi Shimbunsha, 1975), pp. 459–470, on pp. 460–461.

[15] See Suzuki, *Nihon no yūseigaku* (cit. n. 13), p. 60; and Torii, "Nihon jinruigaku," pp. 467–468.

the Immigration Act thus further inflamed passions over racial issues in the United States.[16]

In Japan, too, the 1924 Immigration Act encouraged debate on eugenic theories. In an article published in 1925, for example, the geneticist **Tanaka** Yoshimaro took up the claims of U.S. eugenicists and addressed the question of the racial inferiority of Japanese migrants. His conclusion was that, although Japanese were inferior to white Americans in terms of physical stature, they were in no way inferior in terms of the spiritual and intellectual qualities that really mattered. However, he went on to observe that there were internal differences between "superior" and "inferior" types of Japanese. In considering migration policy, these differences were important: encouraging "inferior types" to emigrate might inflame international tensions, but allowing the emigration of the "superior" would lower the quality of the national population. All of this provided ammunition for a demand that Japanese scholars should embark on a major program of eugenic research, including "comparative eugenic studies of foreigners and Japanese," "studies on the results of exogamy," and "research into means of promoting the multiplication of superior genes."[17]

These comments encapsulate some of the main elements in the emerging ideology of "racial hygiene": on the one hand, it was deeply concerned with establishing hierarchies of racial superiority and inferiority and debating issues of "racial mixing"; but on the other, it also embodied a profound social elitism directed toward the "lower orders" of the national population itself. Tanaka's comments on emigration reflect a common interwar belief that the poverty of the poor was a result of undesirable inherited characteristics and the widespread view among the intellectual elite that Japanese migrants, many of whom came from relatively poor rural areas, were "of a poor type."[18]

As Japanese migration to the colonial empire and (after 1931) to the puppet state of Manchukuo expanded, questions regarding the effects of emigration and of interracial marriage were debated with increasing passion in national and international forums. At the 1931 conference of the Institute of Pacific Relations, for example, Japanese delegates (who included the prominent Christian intellectual **Nitobe** Inazō, the trade union leader **Suzuki** Bunji, and the political scientist **Tsurumi** Yūsuke) took part in a lengthy discussion of "race mixture" in colonial territories. The conference, with delegates from the United States, Britain, Australia, New Zealand, Canada, the Philippines, China, and Japan, embraced a wide range of varying views on the subject but reached an equivocal conclusion fairly characteristic of the early 1930s "liberal" approach to intermarriage: "the indications are such that for the present it seems fair to assume that the mixed blood in most areas is on the whole under no necessary biological handicap; whether this, however, could hold true of such extreme crosses as the White and Aboriginal in Australia or the Asiatic and Negrito is hard to say."[19]

[16] On the 1907 "gentlemen's agreement" see **Idei** Seishi, "Japan's Migration Problem," *International Labour Review,* 1930, *22:*773–789; on the 1924 Immigration Act see Barkan, *Retreat of Scientific Racism* (cit. n. 2), pp. 195–196.

[17] See Suzuki, *Nihon no yūseigaku* (cit. n. 13), pp. 102–103.

[18] See, e.g., the comment by an unnamed Japanese delegate to the fourth conference of the Institute of Pacific Relations, quoted in Bruno Lasker, ed., *Problems of the Pacific, 1931* (Chicago: Univ. Chicago Press, 1932), p. 393.

[19] *Ibid.,* p. 382.

Interest in the science of race, in other words, did not necessarily imply a thoroughgoing belief in the importance of "racial purity," though it did suggest a belief in the importance of genetics in determining the fate of peoples. During the 1920s, for example, biometry—the detailed measurement and statistical comparison of skulls, facial features, blood types, and the like—came to play an increasingly important part in Japanese anthropology. Biometric research, however, was directed not only at minorities and colonized peoples but also at the majority Japanese population itself. A number of large-scale studies of regional differences in physique were undertaken between the wars and during the Pacific War, and these were used by some scholars to support theories on the mixed racial origins of the Japanese. So, for example, the physical anthropologist **Taniguchi** Konen argued that the inhabitants of southwestern Honshu were taller than average and brachycephalic, reflecting the genetic influence of migrants from the Korean peninsula, while the people of the northeast were shorter and dolichocephalic, reflecting the influence of Ainu heritage.[20]

By the 1930s, though, a more extreme approach to the issue of "racial purity" was also starting to be propagated, particularly by the newly created Japan Association of Racial Hygiene (Nippon Minzoku Eisei Kyōkai), established in 1930 under the leadership of the medical scientist **Nagai** Hisomu. Nagai had studied in Germany, where he was influenced by the theorizing of the eugenicist and (later) Nazi Party member Alfred Ploetz, and he had also visited and admired the Swedish eugenics laboratory in Uppsala: the first state-sponsored research center dedicated to the subject of "racial hygiene." By 1939 the Japanese association, whose stated aim was the "elevation of the quality of the Nippon Race from the standpoint of racial hygiene, thereby to contribute toward the prosperity of the State and the welfare of society," had a membership of thirteen hundred. The list of its founding sponsors includes the names of two future postwar prime ministers: **Yoshida** Shigeru and **Hatoyama** Ichirō. Though not all members of the association shared his views on the subject, Nagai used the organization as a forum for expounding his own outspoken opposition to "racial mixing."[21]

RACE AND WAR

From the late 1930s to the time of World War II, European ideas of race followed divergent trajectories. In Germany, Nazi ideas of racial purity gained increasing influence, and a number of prominent German scholars, including the anthropologist Eugen Fischer and the geneticist Alfred Ploetz—who had previously stressed the importance of race but had not embraced anti-Semitic ideas—became converts to National Socialism and members of the Nazi Party.[22] In Britain and America, on the other hand, the reaction against Nazi theorizing, together with growing doubts about

[20] **Taniguchi** Konen, *Tōyō minzoku no taishitsu* (The physical quality of Oriental peoples) (Tokyo: Sangabō, 1942), pp. 32–36.

[21] National Committee of Japan on Intellectual Cooperation, *Academic and Cultural Organisations in Japan* (Tokyo: Kokusai Bunka Shinkōkai, 1939), pp. 503–504. See also Oguma, *Tanitsu minzoku* (cit. n. 6), pp. 249 (prime ministers), 261 (Nagai); and Suzuki, *Nihon no yūseigaku* (cit. n. 13), pp. 150–154 (Nagai), 148 (prime ministers).

[22] George L. Mosse, *Toward the Final Solution: A History of European Racism* (London: Dent, 1978), p. 82.

the scientific validity of ideas of "race," led to the emergence of influential critiques of scientific racism. Although earlier eugenic ideas were not abandoned overnight, their authority declined, and a number of prominent scientists (like Julian Huxley) shifted their position from a belief in forms of racial segregation to a public questioning of the concept of race itself.

Scholars in Japan found themselves exposed to the influence of both these divergent trends. With increasing political tensions between Japan (on the one hand) and the United States, Britain, and France (on the other), Japan's intellectual links shifted increasingly toward Germany. A growing number of Japanese scholars received part of their academic training at German universities, where some were influenced by the racial ideas of National Socialism. But not all who studied in Germany were receptive to these ideas. Understandably, many Japanese intellectuals were repelled by notions of Aryan supremacy and by the ideology of Adolf Hitler, who had once dismissed the Japanese people as a "culture bearing" but never a "culture creating" race.[23] As a result, some returned from Germany with an entrenched antipathy to the Nazi view of race, an antipathy that made them receptive to the ideas put forward in works like Huxley and Haddon's *We Europeans.* By the time of the outbreak of war in Europe, the friction of these contrasting influences had produced an upsurge of debate within Japan over notions of scientific racism.

In the pages that follow I shall explore the nature and implications of these divergent views by looking at the writings of four of the most prolific and influential contributors to the wartime debate on race: the anthropologist **Nishimura** Shinji, the sociologists **Koyama** Eizō and **Shinmei** Masamichi, and the economic historian **Kada** Tetsuji. The works of these four scholars provide a useful framework for understanding wartime debates on race because, while all explicitly supported Japan's imperial expansion and all focused particularly on the issue of race, they occupied distinctly different positions within the spectrum of racial theorizing. They are also of interest because, although their work was influential at the time, it has since been relatively neglected by historians of ideas (while the work of other contemporary theorists, like the anthropologists **Kiyono** Kenji and **Hasebe** Kotondo, has been more widely discussed).

KOYAMA EIZŌ AND THE GLOBAL ROLE OF THE JAPANESE RACE

During the 1930s, the leading academic center for Japanese anthropological research, Tokyo Imperial University's anthropology department, had undergone an important shift in intellectual perspective. The eclectic mixture of physical and cultural anthropology that had characterized the earlier work of **Tsuboi** Shōgorō and **Torii** Ryūzō was replaced by a growing focus on biometry. The central figure in the shift was the medically trained anthropologist **Hasebe** Kotondo (1882–1969), who was appointed to the chair of anthropology in 1936, but a similar interest in the physical science of race was also developed by a number of Hasebe's younger contemporaries, including the sociologist **Koyama** Eizō (1899–1983).

Koyama had traveled widely in Southeast Asia and the South Pacific and wrote

[23] Adolf Hitler, *Mein Kampf,* trans. John Chamberlain *et al.* (New York: Reynal & Hitchcock, 1939), p. 399.

extensively on the anthropology of Java and Australia. On his return, he joined the faculty of Tokyo Imperial University, where he pursued two distinct, though not unrelated, lines of research. As one of the founding members of the university's Newspaper Research Institute (Shinbun Kenkyūjo), Koyama studied the emerging role of the press, advertising, and propaganda in influencing public opinion. At the same time, he also developed a deep interest in anthropology and racial science and became an active member of the so-called Ape Association—a study group of younger Tokyo Imperial University academics that was founded in 1937 and derived its name from the initials of the English words "anthropology," "prehistory," and "ethnology."[24] Koyama's approach to anthropological problems was shaped by his strong interest in issues of social policy, particularly in promoting the physical and spiritual "health" of the national population, but his understanding of the nature of race and ethnicity was largely derived from the theories of German physical anthropologists and geneticists such as Eugen Fischer and Walter Scheidt.

For Koyama, race was the key to Japan's role as leader of a new Asian order, and therefore (as he wrote in his 1939 *Outline of a Theory of Racial Science* [*Jinshugaku gairon*]), "the study of race [*jinshugaku*] is necessary, not just for reasons of intellectual curiosity, but also because it is the scientific basis that can provide direction for Japan's policy on the [Asian] continent." Interestingly, Koyama accepted Huxley and Haddon's view that there was no such thing as a "pure" race in the modern world: all contemporary races were the result of prehistoric or more recent mixings of peoples, and indeed "advanced" races like the Europeans were particularly likely to have undergone waves of racial intermixture. However, he also argued that, through the processes of genetic selection and survival of the fittest, certain of these racial mixtures had acquired a high degree of homogeneity and long-term historical stability, for which he used the German term *Dauertypus*. One example of this was the Japanese race: many waves of early migrants from Siberia, Korea, China, and Southeast Asia had been thoroughly absorbed by a single "main" bloodline—the "Yamato race"—whose origins were obscure but who had inhabited the archipelago since ancient times.[25]

At this point it is worth drawing a connection between Koyama's ideas and the notion of the Japanese "family state." As we saw at the beginning of this essay, many writers argue that prewar and wartime perceptions of the Japanese nation as a "family" encouraged a belief in ideas of Japanese racial purity and homogeneity. But it is important to remember that the model of the "family" (*ie*) enshrined in state ideology since the Meiji period was not one that necessarily depended on blood relationship. The ideal Japanese *ie* was seen as a lineage whose continuity was ensured by the maintenance of the family name and family traditions. Within this structure, it was entirely acceptable for outsiders to be adopted into the family, on condition that they renounced their past name and identified themselves wholeheartedly with the "main branch" into which they had been adopted. Although some wartime

[24] For an example of Koyama's work see **Koyama** Eizō, *Senden gijutsuron* (Theory of the technology of propaganda) (Tokyo: Kōyō Shoin, 1937). On the "Ape Association" see **Terada** Kazuo, *Nihon no jinruigaku* (Anthropology in Japan) (Tokyo: Shisakusha, 1975), p. 205.

[25] **Koyama** Eizō, *Jinshugaku gairon* (Outline of a theory of racial science) (Tokyo: Nikkō Shoin, 1939), pp. 2 (quotation), 162, 321. (Here and throughout, translations are my own unless otherwise indicated.) See also Koyama, *Minzoku* (cit. n. 12), p. 234.

anthropologists such as **Hasebe** Kotondo came increasingly to emphasize the racial purity of the Japanese, the alternative model proposed by writers like Koyama, where many different races were seen as having been absorbed and assimilated by a single "main branch," was just as compatible with the ideology of the "family state."

Koyama's views on the nature of race led him to propose a comprehensive program of biometric research to determine the makeup and homogeneity of various modern racial groups. To this end, his *Outline of a Theory of Racial Science* provided detailed description of the techniques of skull and skeleton measurement, blood-typing, the assessment of skin pigmentations and hair types, and so forth (complete with photographs of a wide array of specialized implements developed for these intrusive and dehumanizing investigations). The human form, indeed, is dissected—partitioned into separate chapters that classify, define, and label the shapes of heads, eyes, ears, and noses. Koyama recognized, however, that the issue at stake was not merely a biological one. He distinguished between physical "races" (*jinshu*) and cultural "ethnic groups" (*minzoku*) and argued that, although the two categories were closely related, their boundaries did not always coincide. Detailed "scientific" studies of race therefore needed to be supplemented by equally detailed scientific studies of "ethnic spirit" (*minzoku seishin*), which would take the form of large-scale sociological surveys of public opinion and popular values: indeed, during the course of the Pacific War Koyama himself embarked on just such a survey of the nature of Japanese family relationships.

All of this racial and ethnic research, however, had a wider political purpose: it was designed to promote Japan's expansion in Asia. After the outbreak of the Pacific War, Koyama argued that the key to Japan's victory lay in its ability to exploit the vast natural resources of East and Southeast Asia. This would inevitably involve the deployment of Japanese people to key administrative and managerial positions in all parts of the "Greater East Asia Co-Prosperity Sphere." Racial studies were therefore necessary, according to Koyama, both to establish whether Japanese people were suited to life in unfamiliar (particularly tropical) climates and to determine whether Japanese colonists in other parts of Asia and the Pacific should intermarry with "the natives."

Although he recognized that the Japanese people themselves were the product of earlier phases of racial intermixture, Koyama took a negative view of colonial intermarriage. In strictly biological terms, he believed that intermarriage could produce either benign or malign results, depending both on social circumstances and on the nature of the "blood lines" that were mixed in the process. In cultural terms, however, Koyama expressed an intense fear that the unity and strength of the "Japanese spirit" would be diluted by intermarriage. Citing the example of Portuguese colonial emigrants who had been absorbed into the "inferior" populations of their African colonies, he proposed a racial version of Gresham's law: particularly where their numbers were small in relation to the native population, there was always a danger that "superior" colonists, with their low birthrates, would be "bred out" by intermarriage with the fecund but "inferior" colonized. Koyama therefore contrasted the fate of the Portuguese in Africa with the "success" of exclusionist and segregationist policies in the Australian state of Queensland. The implications for Japan were plain: at home, Japan should focus on eugenic policies that would create a large reservoir of "superior" potential migrants to other parts of the Co-Prosperity

Sphere; abroad, racial mixing with the colonized peoples of Japan's expanding empire was to be strictly circumscribed.[26]

Today, **Koyama** Eizō's wartime racial theories, as well his later postwar sociological works, are seldom read, and his name appears only fleetingly in histories of Japanese social thought. His ideas, however, have a historical importance because of their considerable influence on one of the more remarkable wartime writings on race: the six-volume government report entitled *An Investigation of Global Policy with the Yamato Race as Nucleus* (*Yamato minzoku o chūkaku to suru sekai seisaku no kentō*). Compiled in 1943 by the Population Research Bureau of the Ministry of Health and Welfare, this document remained secret during the war and was published in facsimile only in 1981. Although there is no evidence that the report had any great influence outside the ministry, it remains the most comprehensive statement of official wartime views on questions of race and, as such, has attracted considerable attention from historians. Both John Dower and **Oguma** Eiji devote sections of their important studies to the report, and both present it as having been compiled by a committee of anonymous bureaucrats. The sheer size and scope of the document certainly point to multiple authorship: it runs to 3,127 pages and includes chillingly dispassionate discussions of topics such as "the Jewish problem in Japan" (which begins by stressing Japan's long tradition of assimilating, rather than discriminating against, alien races but ends by concluding that "special wartime circumstances" and "the peculiarities of the Jews" make it inevitable that legal restrictions be imposed on Jewish inhabitants of the Japanese empire).[27]

One of the main contributors to the report, though, was undoubtedly **Koyama** Eizō. Koyama had been appointed to a chair at Rikkyō University in 1937, but in 1939 he resigned to enter the bureaucracy as a researcher on issues of population and race. During the early 1940s he served both as chief researcher in the Ministry of Health and Welfare's Population Research Bureau and as a section chief in the Ministry of Education's Ethnic Research Institute (Minzoku Kenkyūjo). His central role in the composition of the *Yamato Race* report is evident from the fact that key sections dealing with definitions of race, issues of intermarriage, and the origins and colonial expansion of the Japanese people are reproduced almost verbatim from his other published works. For example, the definitions of "race" and "ethnicity" in the first volumes paraphrase his definitions in *Outline of a Theory of Racial Science;* the chapter on "racial mixture" is virtually identical with a chapter on the subject in the same volume; and the description of the racial origins of the Japanese repeats word-for-word Koyama's description in his 1942 publication *Problems of Ethnicity and Culture* (*Minzoku to bunka no shomondai*).[28] Koyama's wartime research activi-

[26] Koyama, *Minzoku,* pp. 74, 128–129; and **Koyama** Eizō, *Nampō kensetsu to minzoku jinkō seisaku* (The construction of the southern region and ethnic and population policy) (Tokyo: Dai-Nippon Shuppan, 1944), pp. 641–642.

[27] Kōseishō Kenkyūbu Jinkō Minzokubu, *Minzoku jinkō seisaku kenkyū shiryō 6: Yamato minzoku o chūkaku to suru sekai seisaku no kentō* (Materials for research on ethnic and population policy, 6: An investigation of global policy with the Yamato race as nucleus), Vol. 4 (Tokyo: Bunsei Shoin, 1981), p. 1823. For discussions of the report see Dower, *War without Mercy* (cit. n. 7), pp. 262–290; and Oguma, *Tanitsu minzoku* (cit. n. 6), pp. 253–258.

[28] Cf. Kōseishō Kenkyūbu Jinkō Minzokubu, *Minzoku jinkō seisaku kenkyū shiryō 3: Yamato minzoku o chūkaku to suru sekai seisaku no kentō,* Vol. 1 (Tokyo: Bunsei Shoin, 1981), pp. 28–32, 313–319, 311–312, with Koyama, *Jinshugaku* (cit. n. 25), pp. 8–12, 160–164, and Koyama, *Minzoku* (cit. n. 12), p. 235.

ties, indeed, illustrate an increasingly close interaction between academic research and government policy from the late 1930s onward, an interaction that influenced not only plans for industrial and technological development but also wartime social policy.

SHINMEI MASAMICHI, KADA TETSUJI, AND THE CRITIQUE OF SCIENTIFIC RACISM

However, not all advocates of Japanese imperial expansion shared Koyama's enthusiasm for racial science. Another interesting trend of the late 1930s and early 1940s was the emergence of a current of thought that, while acclaiming Japan's "leading role" in the Greater East Asia Co-Prosperity Sphere, sought to base this expansionist ideology on a more critical approach to questions of race. The two main exponents of this approach were **Shinmei** Masamichi (1898–1984) and **Kada** Tetsuji (1895–1964), both of whom used arguments that, in places, closely resemble Huxley and Haddon's critique of scientific racism.

Shinmei Masamichi had encountered the emerging Nazi ideology of race during a stay in Europe from 1929 to 1931; on his return to Japan, he became involved in heated debates on the issues of race, ethnicity, and nationalism. His major study of the subject, *Race and Society* (*Jinshu to shakai*), was published in 1940 and presented an extended and detailed critique of ideas of Aryan supremacy. Although Shinmei accepted the existence of an ancient Aryan people and the fact that the languages of the ancient Aryans and modern Western Europeans were connected, he firmly rejected the Nazi image of a racially pure and culturally superior "Aryan race." The linguistic connections between ancient Aryans and modern Europeans, he emphasized, did not imply a single biological lineage, and indeed there was every reason to believe that modern Europeans were the product of millennia of racial mixing. Ideas of Aryan supremacy also totally failed to make sense of the emergence of powerful civilizations in countries like China, which could not—by the wildest stretch of the imagination—be described as having "Aryan origins." Throughout *Race and Society,* Shinmei was careful to confine his arguments to debates about the origins and character of Europeans, avoiding direct statements on the "racial" character of the Japanese, but his approach to the question of race clearly had implications that went beyond the matter of Aryanism. For example, he pointed out that "scientific studies of race have shown that the division between races is relative, and that race is not the decisive determinant of society."[29]

In other writings, Shinmei directly addressed problems closer to home. His *Ideal of the East Asia Community* (*Tōa Kyōdōtai no risō*), published in 1939, called for the creation of an East Asian "new order" that would transcend narrow-minded "European" ideologies of racism: "racism is a phenomenon that we must reject. The objective of constructing an East Asian Community requires that we should reject it decisively." Racial differences, he argued, were real but were merely a matter of physiology. Culture, being nonmaterial, could readily pass across the boundaries of race, and therefore there was no reason to assume that racial dividing lines would coincide with cultural dividing lines and no reason to believe that any race was

[29] **Shinmei** Masamichi, *Jinshu to shakai* (Race and society) (1940), rpt. in *Shinmei Masamichi chosakushū* (The collected works of Shinmei Masamichi), Vol. 8 (Tokyo: Seishin Shobō, 1980), p. 142. I have discussed the work of Shinmei and Kada in further detail in Tessa Morris-Suzuki, *Re-Inventing Japan: Time, Space, Nation* (New York: Sharpe, 1998), Ch. 5.

innately superior to others. Citing theories of Japan's own mixed racial origins, Shinmei suggested that Japan itself had ancient traditions of racial tolerance and receptivity to difference that should be revitalized in the newly emerging "East Asian Community."[30]

In the midst of these calls for tolerance and racial equality, however, it is important to notice a clear distinction being drawn between "race" and "ethnic group" [*minzoku*] and between genetics and culture. While Shinmei rejected all theories of inherent racial superiority and inferiority, he unquestioningly accepted the notion of a single line of human progress—and thus also the proposition that some nations or ethnic groups were more culturally and economically advanced than others.[31] So the Japanese, although they had no innate racial right to rule in Asia, did have the right (and indeed the duty) to "lead" other Asian societies because of their higher level of cultural development. Shinmei's work, therefore, demonstrates how a rejection of scientific racism could still be combined with an unmistakably hierarchical vision of the relationship between cultural groups: a universalist human philosophy of social evolution could thus become a justification for Japan's imperial expansion.

A rather similar approach was developed by **Kada** Tetsuji in his 1940 bestseller *Jinshu, minzoku, sensō* (best translated, for reasons I will discuss, as *Race, Nation, and War*). Like Shinmei, Kada had visited Germany and was familiar with Nazi racial theorizing. The provocative title of his book, and his penchant for quoting from Nazi ideologues like Alfred Rosenberg, has led some commentators to suggest that his views were close to those of Nazi scientific racism.[32] In fact, however, it is clear that Kada's aim was to develop a more subtle philosophy of social evolution and national expansionism; one that would appeal to those Japanese intellectuals who were uncomfortable with doctrines of genetic determinism and "racial hygiene."

Kada believed that human beings had an instinctive tendency to react to visible physical differences between "self" and "other," and his study provides a historical outline of the way in which this response had been used by various societies to draw communal boundaries. However, efforts to find a scientific basis for separating one race from another, or for establishing hierarchies of racial superiority and inferiority, had so far all ended in failure. Attempts to define races in terms of skull and skeleton measurements depended on the arbitrary averaging of data from diverse individuals; the relatively new study of blood types had failed to produce decisive evidence of racial boundaries; and approaches that sought to combine physical measurement with notions of cultural difference served only to muddy the waters. Modern racism, Kada argued, was essentially an ideology that had emerged to explain differences in economic development between different countries and to justify the process of European imperial expansion. Thus, while accepting the reality and importance of the concept of race as a social phenomenon, Kada insisted that "contemporary racial theory is not derived from rigorous science. At most, it is nothing more than a propagandist ideology that has dressed itself in scientific clothes."[33]

[30] **Shinmei** Masamichi, *Tōa Kyōdōtai no risō* (The ideal of the East Asia Community) (Tokyo: Nihon Seinen Gaikō Kyōkai Shuppankai, 1939), pp. 94 (quotation), 81–85, 101–106, 108.

[31] *Ibid.*, p. 83.

[32] See Michael Weiner, "Discourses of Race, Nation, and Empire in Pre-1945 Japan," *Ethnic and Racial Studies,* 1995, *18:*433–456, on pp. 440–441.

[33] **Kada** Tetsuji, *Jinshu, minzoku, sensō* (Race, nation, and war) (Tokyo: Keiō Shobō, 1940), pp. 50–54, 45–49, 86–89, 54–58, 65 (quotation).

However, Kada's rejection of scientific racism, like Shinmei's, was accompanied by a clear belief in stages of social progress and in the progressive political role of *minzoku,* a term he uses to mean something closer to the concept of "nation" than that of "ethnic group." Kada, in other words, defines the *minzoku* as a modern phenomenon whose basis was primarily economic: *minzoku* emerged from the evolution of autonomous productive communities, which he calls "basic societies" (*kihon shakai*), and acquired their mature form as a result of the rise of capitalism. As such, *minzoku* might or might not be made up of groups that were linked by blood bonds; whether or not such bonds existed, the need for the modern nation to overcome internal class divisions and create a sense of social cohesion inevitably encouraged national elites to create ideologies—"imagined communities," as it were—of racial homogeneity and superiority.[34]

In Kada's writings, *minzoku* are presented as naturally expansionary communities that are the chief bearers of human progress. Borrowing from Karl Marx, Karl Kautsky, and Ferdinand Tönnies, Kada depicts basic societies as evolving through a fixed series of stages from primitive communism via tribal, classical, and feudal stages to capitalism and the modern nation-state. Once again, therefore, Japan's expansion into Asia—which Kada describes in highly euphemistic terms as the negotiated creation of a mutually beneficial East Asian Community—is justified, not in terms of racial superiority, but in terms of economic dynamism: the fact that the Japanese *minzoku* had reached a higher stage than other Asian peoples in the universal march of progress.[35]

It is important to emphasize that, like **Koyama** Eizō's version of scientific racism, **Shinmei** Masamichi's and **Kada** Tetsuji's critiques of racial science resonated with aspects of official policy. For just as Japanese intellectuals disagreed about racial theories, so different sections of the Japanese bureaucracy also took sharply divergent positions on policies of race. While the Ministry of Health and Welfare was strongly influenced by theories of eugenics, racial hygiene, and segregationism, many officials of the Korean colonial government favored assimilationist policies that included actively encouraging intermarriage between Japanese and Koreans. In 1939, for example, the colonial authorities embarked on a campaign to promote intermarriage as a means of achieving the "unification of Japan and Korea [*Naisen ittai*]" in terms of "form and spirit, blood and flesh." One aspect of this campaign involved the presentation of certificates and awards to model newlywed Japanese-Korean couples. In Manchuria, meanwhile, the Japanese-dominated puppet government proclaimed an official ideology of "racial harmony" between Chinese, Manchus, Japanese, Koreans, and other ethnic groups, an ideology that was maintained in the face of the undeniable reality of massive discrimination against and expropriation of non-Japanese communities (not to mention, of course, the massive exploitation of colonial citizens within metropolitan Japan itself). In fact, **Shinmei** Masamichi was among those recruited by the all-powerful South Manchurian Railway Company to tour Manchuria giving lectures on the evils of racism.[36]

[34] *Ibid.,* pp. 92–94, 116–118.

[35] See, e.g., **Kada** Tetsuji, *Seiji, keizai, minzoku* (Politics, economy, nation) (Tokyo: Keiō Shobō, 1940), pp. 141–143; and Kada, *Gendai no shokumin seisaku* (Contemporary colonial policy) (Tokyo: Keiō Shobō, 1939), pp. 490–491.

[36] See **Shinmei** Masamichi, introduction to *Shinmei Masamichi chosakushū* (cit. n. 29), p. ii. On the certificates and awards see **Suzuki** Yūko, *Jūgun ianfu, naisen kekkon* (The comfort women and

NISHIMURA SHINJI AND THE RISE OF SCIENTIFIC ETHNOCENTRISM

Koyama Eizō's scientific racism and the critiques of race proposed by Shinmei and Kada represent two opposing poles within 1940s Japanese expansionist ideology. In between these poles, however, there was a further trend, increasingly evident from the mid 1930s onward, that is worth discussing because of its continuing influence on postwar Japanese thought. This trend might be described as "the rise of scientific ethnocentrism," and its implications are well illustrated by the work of the prominent cultural anthropologist **Nishimura** Shinji (1879–1943).

Nishimura was of an older generation than Koyama, Shinmei, and Kada, and by the late 1930s he had already made a name for himself as an enthusiastic popularizer of modern racial theories in Japan. His introductory text *A General Theory of Anthropology* (*Jinruigaku hanron* [1929]), for example, introduced readers to the works of Francis Galton, Edward Tylor, Franz Boas, and A. C. Haddon, among others. Nishimura's own approach was eclectic: he explained the nature of "race" by referring particularly to the ideas of three foreign theorists: Joseph Deniker, the nineteenth-century French ethnographer who had first proposed the long-accepted division of skull shapes into the categories dolichocephalic, mesocephalic, and brachycephalic; the Harvard anthropologist Roland Dixon, who had emphasized that "race" was a purely biological phenomenon, to be ascertained by anatomical measurement alone; and the Irish-born linguist, ethnographer, and travel writer Augustus Henry Keane (1833–1912).[37]

Although Keane's writings are now almost totally forgotten, his ideas exerted the greatest influence on Nishimura's concept of race. Keane had rejected the popular nineteenth-century view that physical robustness and brain size were directly correlated with levels of "civilization" or intelligence. However, he saw "race" as involving an inseparable interrelationship between biological and cultural elements. For example, he accepted the widely held belief that European vocal chords were physically incapable of reproducing the sounds found in certain African, Native American, and other languages. In other words, language, the central element in culture and consciousness, was eternally bound up with the genetic heritage of race.[38]

Part of the appeal of Keane's ideas undoubtedly lay in his views on "the Japanese race" itself. Keane had visited Japan twice, with evident enjoyment, and presented the Japanese as the chief exhibits in support of his view that there was no necessary correlation between physical strength and intelligence. The Japanese, he said, were physically "a feeble folk" even by comparison with their Asian neighbors; "on the other hand the Japanese stand intellectually at the head of all Mongolic peoples without exception. In this respect they rank with the more advanced European nations, being highly intelligent, versatile, progressive, quick-witted and brave to a degree of heroism unsurpassed by any race." Encouraged by this analysis, Nishimura enthusiastically adopted Keane's image of a racial "ideal type," embracing both

marriage between Japanese and Koreans) (Tokyo: Miraisha, 1992), pp. 84–85; and Oguma, *Tanitsu minzoku* (cit. n. 6), pp. 250–256.

[37] **Nishimura** Shinji, *Jinruigaku hanron* (A general theory of anthropology) (Tokyo: Tōkyōdō, 1929), pp. 68–76.

[38] A. H. Keane, *Ethnology* (Cambridge: Cambridge Univ. Press, 1901), p. 193.

physical and cultural elements, that could somehow be extracted from the bewildering diversity of real individual human beings.[39]

In his later writings, Nishimura developed this notion in an attempt to define the core characteristics of the Japanese "ideal type." His *Ideal of the Japanese Ethnic Group* (*Nihon minzoku no risō*), first published in 1939, depicted the Japanese (as so many interwar anthropologists had done) as the product of successive waves of migration from Siberia, China, and Southeast Asia: the ancestors of the modern Japanese race, he argued, included a Negrito strain from the South Pacific, proto-Ainu, Tungusic groups, Indo-Chinese, Indonesians, and Han Chinese.[40] These origins were of particular importance, not so much because they determined the genetic makeup of the Japanese people but, rather, because they had helped to mold the essence of Japanese culture. For Nishimura's real interest lay less in analyzing the physical genealogy of the Japanese people than in applying the latest techniques of mass opinion surveys and statistical analysis to the task of measuring their ethnic or national "character."

Following Keane's theories, Nishimura argued that cultural heritage, although not identical with genetic heritage, was closely tied to race. Thus, just as the Japanese had been formed by the mixing of peoples from three main culture zones—the Siberic zone, the Sinic zone (China), and the Indic zone (Southeast Asia)—so the ancient Japanese could be seen as having been characterized by three main cultural traits: adaptability (*tekiōsei*), mobility (*kadōsei*), and receptivity (*juyōsei*). Nishimura related these characteristics, which he claimed to have identified by studying the oldest surviving Japanese myths and legends, to the diverse origins of the Japanese people. For example, adaptability was inherited from the nomadic Tungusic people, who had been able to adjust their lifestyles even to the harshest Siberian conditions; receptivity was evidenced by the speed with which the ancient Japanese adopted rice culture from new waves of Indo-Chinese migrants around the beginning of the common era. Ancient legends also revealed that these core Japanese values were accompanied by other national characteristics, of which the most important were peacefulness, cooperativeness, and morality.[41]

In Nishimura's scheme of things, however, such historical research was only one side of the scientific analysis of "national character": the other side related to contemporary society and involved the testing of value systems by large-scale sample surveys. So, at a time when **Koyama** Eizō was embarking on his statistical survey of Japanese family values, **Nishimura** Shinji was compiling the results of an enquiry designed to discover whether the modern Japanese population retained the core characteristics that he had identified in their ancient ancestors. His target group was one to which he had easy access: the first- and second-year students at his own academic institution, Waseda University in Tokyo, to whom Nishimura distributed 1,511 questionnaires (receiving 1,356 usable responses).

Every student was presented with a series of sixty questions: ten relating to each of the six core characteristics of the Japanese "national character" identified by

[39] *Ibid.,* pp. 316–317; and Nishimura, *Jinruigaku* (cit. n. 37), pp. 74–75.
[40] **Nishimura** Shinji, *Nihon minzoku no risō* (The ideal of the Japanese ethnic group) (Tokyo: Tōkyōdō, 1939), p. 94.
[41] *Ibid.,* pp. 97–105, 108–113.

Nishimura's historical research. Adaptability, for example, was tested by asking such questions as, "Do you think you could live in a cold place like Manchuria or a hot place like the South Seas?" Receptivity was assessed with the help of questions like, "Do you have someone whom you admire and imitate?" Morality was tested by queries such as, "Do you always keep your promises?" The student was to answer each question by choosing a point on a scale from −10 to +10, and the latest statistical techniques were then used to standardize the responses, minimizing the influence of extreme opinions on the final results.[42]

The findings proved gratifying. Nishimura was able to report that "amongst us Japanese, the most outstanding common characteristic is morality, followed by peacefulness and adaptability, and we also demonstrate cooperativeness, receptiveness, and mobility, in descending order of significance. However, since positive values were exhibited for all these characteristics, we can happily conclude that the Japanese are a positive, peaceful, and creative people." Nishimura also used his survey to investigate internal differences within his sample population. First he broke his larger group down by faculty and discovered that commerce students showed the strongest signs of "national character," humanities students the weakest.[43]

Next he separated out his responses according to place of origin, dividing up "Japan proper" (*naichi*) into eight regions and "external territories" (*gaichi*) into five: Hokkaido plus Karafuto, Taiwan, Korea, Manchuria, and China. By now, the sample numbers were becoming disturbingly unbalanced: there were, for example, 490 respondents from the Kanto region of Japan (which includes Tokyo), but only 29 from the southern Japanese island of Shikoku, 27 from Korea, 19 from Manchuria, and 5 from China. Besides, there was no means of determining whether the students' places of origin were related to their "ethnicity"—for example, how many of the students from Taiwan were of Chinese origin and how many of Japanese origin.

Undeterred by these problems, Nishimura pressed on with his numerical calculations, assigning Japan and its colonies a numerical ranking from 1 (highest) to 6 (lowest) for each element of the "national spirit" (see Table 1). This allowed him to conclude that people from

> Japan proper possess the highest indicators of national spirit [*kokuminsei*]; people from Manchuria and Taiwan show levels equal to one another; next come people from Hokkaido, Karafuto, China, and Korea, exhibiting declining levels of [national] characteristics in that order. This indicates that Japanese citizens (Japanese and Koreans) who live in Korea possess a weaker national spirit than those who live in other overseas territories. This is an important enigma that we need to solve.[44]

An even greater enigma, perhaps, was the fact that, according to Nishimura's own statistical tables, people from the island of Shikoku demonstrated considerably less "national spirit" than people from any of the "overseas territories," a finding about which Nishimura's text remains entirely silent.

Nishimura's *Ideal of the Japanese Ethnic Group* was just one example of the growing use of statistical surveys to assess the cultural peculiarities of the Japanese. Although the nature of Japan's "national character" had been debated ever since the

[42] *Ibid.,* pp. 238–262.
[43] *Ibid.,* pp. 301, 265.
[44] *Ibid.,* p. 283.

Table 1. Strength of National Spirit, Ranked by Region

	Adaptability	Mobility	Receptivity	Cooperation	Peacefulness	Morality	Average	Rank
Japan proper	2	3	2	1	2	4	2.33	1
Hokkaido*	4	2	4	3	3	6	3.67	4
Taiwan	3	5	3	4	1	2	3.00	3
Korea	6	4	6	5	5	3	4.83	6
Manchuria	5	1	1	2	4	5	3.00	2
China	1	6	5	6	6	1	4.17	5

SOURCE.—**Nishimura** Shinji, *Nihon minzoku no risō* (The ideal of the Japanese ethnic group) (Tokyo: Tōkyōdō, 1939), p. 282.
*Includes Karafuto.

nineteenth century, it was only in the 1930s that the statistical study of mass society began to give these speculations a scientific veneer. In many ways, this statistical analysis of "ethnic culture," with its arbitrary averaging of preselected groups around predefined criteria of difference, resembled the biometric techniques of scientific racism, which were now beginning to be criticized by writers both in Western Europe and in Japan. While eugenic theories and biometry were abandoned by many theorists after the Pacific War, however, efforts to mobilize psychology, sociology, and statistical survey techniques for the "scientific" analysis of the peculiarities of Japanese culture continued to flourish—and indeed gained new momentum with the rise of so-called *Nihonjinron* ("theories of the Japanese") in the 1960s and 1970s.

POSTWAR LEGACIES

A glance at some of the wartime debates surrounding racial science suggests that the history of racial thought in Japan has been more complex and, in some ways, more troubling than is often assumed. Support for Japan's imperial expansion in Asia was sometimes associated with a literal belief in the racial purity of the Japanese but could also be sustained by quite different perspectives on race. Racial theorists like **Koyama** Eizō accepted the diverse origins of the Japanese people while still believing in the genetic superiority of the modern Japanese and in their global mission to colonize and rule East Asia. Cultural analysts like **Nishimura** Shinji were more cautious about the significance of biometric studies but nonetheless managed to develop their own statistical pseudo-science to support notions of Japan's ethnic superiority. Meanwhile, **Shinmei** Masamichi and **Kada** Tetsuji, while rejecting scientific racism, turned universalist notions of the march of human progress into an elaborate apologia for Japanese imperial expansion in Asia.

In each case, the theorists of the late 1930s and early 1940s drew on a range of European and American sources and yet reworked these sources into theories that were distinctively adapted to the circumstances of wartime Japan. Indeed, rather than simply adopting prevailing European theories of race, they often seem to have deliberately sought out relatively obscure Western hypotheses that were readily malleable to their own perspectives on Japan's political destiny.

It would be more comforting, I think, to believe that Japan's wartime expansion was driven by simple and unquestioned myths of a racially pure "family state" with the emperor as father figure. Such an image allows us to draw reassuring lines between past and present and between Japan and the rest of the world. Even if wartime myths of racial purity are seen as having survived into postwar Japanese society, this image still enables us to envisage such myths simply as bizarre relics of a receding past, whose influence might be expected to weaken over time. A closer look at wartime debates about race, however, forces us to confront the complexities of prejudice, ethnocentrism, and nationalist ideology and the multiplicity of the legacies of midcentury ideology for the late twentieth-century world.

The biometric and eugenicist enthusiasms of writers like **Koyama** Eizō did not disappear after the Pacific War, and their influence has continued to cast a shadow over popular discourse in Japan. During the 1960s and 1970s a variety of books relating the "uniqueness" of the Japanese people to racial peculiarities such as the structure of the brain or the digestive system became best-sellers, and a fascination for relating blood type to personality still survives in the pages of the pocket-book paperbacks and weekly magazines.[45] In academia, however, the association of scientific racism with wartime ideology made many postwar writers wary of biometric approaches to race. Indeed, although the origins of the Japanese continued to be a topic of heated controversy in archaeology and prehistory, one of the most striking features of sociological and political debate during the immediate postwar decades was the almost total avoidance of the subject of race. At the same time, though, the basic intellectual structure of Koyama's approach—in which an originally diverse population is seen as being molded into homogeneity by Darwinian selection and the assimilatory powers of a single central bloodline—seems to me to have remained implicitly influential: many postwar writings similarly see the Japanese as springing from diverse ancient origins but as having achieved a high degree of homogeneity relatively early in their history.

Meanwhile, sociological surveys of national character, of the sort developed both by Koyama and by **Nishimura** Shinji, flourished and multiplied after the war, drawing further inspiration from the postwar U.S. boom in comparative statistical surveys of public opinion and cultural values. In this context, the continuities between wartime and postwar thought are most vividly illustrated by the career of **Koyama** Eizō. In September 1945, two years after the completion of *An Investigation of Global Policy with the Yamato Race as Nucleus,* and just a month after Japan's surrender, Koyama was summoned by the newly established Allied Occupation Headquarters to help create a research group to study Japanese public opinion.[46] As his wartime work had shown, the statistical techniques of racial measurement and classification could be readily adapted to the measurement and classification of the mental characteristics of national or ethnic groups, and the Occupation Administration's public opinion surveys would give him a chance to develop this side of his research interests. Koyama subsequently became the leading pioneer of opinion research techniques in Japan, and in 1949 he was appointed head of the newly created National

[45] E.g., **Tsunoda** Tadanobu, *Nihonjin no nō* (The Japanese brain) (Tokyo: Taishūkan, 1978); and **Suzuki** Yoshimasa, *O-gata ningen* (The O-type human being) (Tokyo: Sanshinsha, 1973).

[46] See **Koyama** Eizō, *Yoron, Shōgyō chōsa no hōhō* (The methodology of opinion and commercial society) (Tokyo: Yūsankaku, 1956), pp. 1–2.

Public Opinion Investigation Office (Kokuritsu Yoron Chōsakyoku). Thus the techniques refined by wartime scientific racism came to be applied to the creation of an intricate and highly influential government system of opinion research, which was used by successive postwar administrations to create, define, and popularize their image of Japanese "national attitudes" and "national character."

It is worth noting that a number of the specific features singled out by wartime writers in their analyses of "national character" continue to be emphasized by writers on Japan's national culture to the present day. Thus, for example, **Nishimura** Shinji's stress on Japan's "adaptability" and "receptivity" is echoed in the recent works of **Ueyama** Shumpei, who describes Japan as a "concave" (that is, receptive and adaptive) society, as opposed to the West's convex (self-assertive and expansionist) society.[47]

Wartime debates, however, also remind us of the dangers of assuming that expansionism and ethnocentrism can simply be equated with a narrowly defined scientific racism. Military expansionism, discriminatory practices, and a sense of national superiority could be justified by ideologies that avoided, and even sometimes explicitly rejected, scientific racism. Thus **Shinmei** Masamichi and **Kada** Tetsuji relied on a unilinear vision of human progress, rather than myths of racial purity, as a basis for their support of Japanese imperialism. This is significant, because the same vision of progress continued to be immensely important to Japanese postwar social and political thought. From this perspective, it becomes easier to understand how an implicitly ethnocentric approach, particularly to Japan's Asian neighbors, remained deeply embedded in the postwar writing even of some liberal intellectuals who would never have dreamed of espousing pseudo-scientific doctrines of scientific racism.

The implications of this history go beyond the bounds of Japan itself. The particular debates that I have discussed here were molded by local conditions: by Japan's intellectual environment, political system, and imperial expansion. But the underlying point has an international relevance. The critique of scientific racism proposed by writers like Julian Huxley and A. C. Haddon in the 1930s was only one small and tentative step toward dismantling the many complex ideological structures through which individuals exalt the superiority of their own group over others, propagate stereotypes, and justify discrimination and national expansionism. Sixty years later, when scientific racism has been banished to the fringes of academia, the reappearing worldwide specter of a popular racism adorned in new rhetorical garb reminds us how far we still have to go in getting to the root of those ideologies.

[47] **Ueyama** Shumpei, *Nihon bunmeishi no kōsō: juyō to sōzō no kiseki* (An outline concept of the history of Japanese civilization: A trajectory of receptivity and creativity) (Tokyo: Kadokawa Shoten, 1990).

Index